Selected Papers from 1st International Electronic Conference on Biological Diversity, Ecology, and Evolution (BDEE 2021)

Selected Papers from 1st International Electronic Conference on Biological Diversity, Ecology, and Evolution (BDEE 2021)

Editor

Michael Wink

MDPI • Basel • Beijing • Wuhan • Barcelona • Belgrade • Manchester • Tokyo • Cluj • Tianjin

Editor
Michael Wink
Heidelberg University
Heidelberg
Germany

Editorial Office
MDPI
St. Alban-Anlage 66
4052 Basel, Switzerland

This is a reprint of articles from the Special Issue published online in the open access journal *Diversity* (ISSN 1424-2818) (available at: https://www.mdpi.com/journal/diversity/special_issues/BDEE_2021).

For citation purposes, cite each article independently as indicated on the article page online and as indicated below:

LastName, A.A.; LastName, B.B.; LastName, C.C. Article Title. *Journal Name* **Year**, *Volume Number*, Page Range.

ISBN 978-3-0365-7362-5 (Hbk)
ISBN 978-3-0365-7363-2 (PDF)

Cover image courtesy of Michael Wink

Contents

About the Editor vii

Luca Luiselli, Gift Simon Demaya, John Sebit Benansio, Fabio Petrozzi, Godfrey C. Akani, Edem A. Eniang, Stephanie N. Ajong, et al.
A Comparative Analysis of the Diets of a Genus of Freshwater Turtles across Africa
Reprinted from: *Diversity* **2021**, *13*, 165, doi:10.3390/d13040165 1

Nicholas Koutsikos, Leonidas Vardakas, Olga-Ioanna Kalantzi and Stamatis Zogaris
Patterns of Spatial Overlap between Non-Indigenous and Critically Endangered Freshwater Fishes from a Mediterranean Biodiversity Hotspot
Reprinted from: *Diversity* **2021**, *13*, 233, doi:10.3390/d13060233 13

Petro M. Tsarenko, Olena P. Bilous, Olha M. Kryvosheia-Zakharova, Halyna H. Lilitska and Sophia Barinova
Diversity of Algae and Cyanobacteria and Bioindication Characteristics of the Alpine Lake Nesamovyte (Eastern Carpathians, Ukraine) from 100 Years Ago to the Present
Reprinted from: *Diversity* **2021**, *13*, 256, doi:10.3390/d13060256 27

Alexander Cotrina Sánchez, Nilton B. Rojas Briceño, Subhajit Bandopadhyay, Subhasis Ghosh, Cristóbal Torres Guzmán, Manuel Oliva, Betty K. Guzman, et al.
Biogeographic Distribution of *Cedrela* spp. Genus in Peru Using MaxEnt Modeling: A Conservation and Restoration Approach
Reprinted from: *Diversity* **2021**, *13*, 261, doi:10.3390/d13060261 49

Mariana A. Tsianou, Maria Lazarina, Danai-Eleni Michailidou, Aristi Andrikou-Charitidou, Stefanos P. Sgardelis and Athanasios S. Kallimanis
The Effect of Climate and Human Pressures on Functional Diversity and Species Richness Patterns of Amphibians, Reptiles and Mammals in Europe
Reprinted from: *Diversity* **2021**, *13*, 275, doi:10.3390/d13060275 65

Sarah E. Fischer, Andrew C. Edwards, Patrice Weber, Stephen T. Garnett and Timothy G. Whiteside
The Bird Assemblage of the Darwin Region (Australia): What Is the Effect of Twenty Years of Increasing Urbanisation?
Reprinted from: *Diversity* **2021**, *13*, 294, doi:10.3390/d13070294 77

Arsen V. Dotsev, Andrey N. Rodionov, Veronika R. Kharzinova, Sergey N. Petrov, Dmitry G. Medvedev, Vugar A. Bagirov, Gottfried Brem, et al.
An Assessment of Applicability of SNP Chip Developed for Domestic Goats in Genetic Studies of Caucasian Tur (*Capra caucasica*)
Reprinted from: *Diversity* **2021**, *13*, 312, doi:10.3390/d13070312 95

Alexandra S. Abdelmanova, Veronika R. Kharzinova, Valeria V. Volkova, Arsen V. Dotsev, Alexander A. Sermyagin, Oksana I. Boronetskaya, Roman Yu. Chinarov, et al.
Comparative Study of the Genetic Diversity of Local Steppe Cattle Breeds from Russia, Kazakhstan and Kyrgyzstan by Microsatellite Analysis of Museum and Modern Samples
Reprinted from: *Diversity* **2021**, *13*, 351, doi:10.3390/d13080351 107

Tatiana Deniskova, Arsen Dotsev, Marina Selionova, Gottfried Brem and Natalia Zinovieva
Biodiversity of Russian Local Sheep Breeds Based on Pattern of Runs of Homozygosity
Reprinted from: *Diversity* **2021**, *13*, 360, doi:10.3390/d13080360 123

Sônia Guimarães Alves and Maria Cristina Gaglianone
Bee Guilds' Responses to Urbanization in Neotropics: A Case Study
Reprinted from: *Diversity* **2021**, *13*, 365, doi:10.3390/d13080365 . . . . **139**

Oksana Nekrasova, Volodymyr Tytar, Mihails Pupins, Andris Čeirāns, Oleksii Marushchak and Arturs Skute
A GIS Modeling Study of the Distribution of Viviparous Invasive Alien Fish Species in Eastern Europe in Terms of Global Climate Change, as Exemplified by *Poecilia reticulata* Peters, 1859 and *Gambusia holbrooki* Girarg, 1859
Reprinted from: *Diversity* **2021**, *13*, 385, doi:10.3390/d13080385 . . . . **153**

Beanélri B. Janecke
Mammal Species Richness at a Catena and Nearby Waterholes during a Drought, Kruger National Park, South Africa
Reprinted from: *Diversity* **2021**, *13*, 387, doi:10.3390/d13080387 . . . . **163**

Pushpalata Kayastha, Milena Roszkowska, Monika Mioduchowska, Magdalena Gawlak and Łukasz Kaczmarek
Integrative Descriptions of Two New Tardigrade Species along with the New Record of *Mesobiotus skorackii* Kaczmarek et al., 2018 from Canada
Reprinted from: *Diversity* **2021**, *13*, 394, doi:10.3390/d13080394 . . . . **179**

Sarahi Jaime, Adrián Cervantes-Martínez, Martha A. Gutiérrez-Aguirre, Eduardo Suárez-Morales, Julio R. Juárez-Pernillo, Elena M. Reyes-Solares and Victor H. Delgado-Blas
Historical Zooplankton Composition Indicates Eutrophication Stages in a Neotropical Aquatic System: The Case of Lake Amatitlán, Central America
Reprinted from: *Diversity* **2021**, *13*, 432, doi:10.3390/d13090432 . . . . **205**

Sergio Albertazzi, Elisa Monterastelli, Manuela Giovanetti, Emanuele Luigi Zenga, Simone Flaminio, Marta Galloni, Marino Quaranta and Laura Bortolotti
Biodiversity Evaluation: From Endorsed Indexes to Inclusion of a Pollinator Indicator
Reprinted from: *Diversity* **2021**, *13*, 477, doi:10.3390/d13100477 . . . . **215**

Valeria Itskovich, Oxana Kaluzhnaya, Olga Glyzina, Ragothaman Prathiviraj, George Seghal Kiran and Joseph Selvin
Microbiome Changes of Endemic Lake Baikal Sponges during Bleaching Syndrome Development
Reprinted from: *Diversity* **2021**, *13*, 653, doi:10.3390/d13120653 . . . . **229**

Paweł K. Bereś, Patrycja Ziętara, Mirosław Nakonieczny, Łukasz Kontowski, Michał Grzbiela and Maria Augustyniak
Cydalima perspectalis in Poland—8 Years of Invasion against the Background of Three Other Invasive Species
Reprinted from: *Diversity* **2022**, *14*, 22, doi:10.3390/d14010022 . . . . **241**

About the Editor

Michael Wink

Michael Wink is a professor of Pharmaceutical Biology at Heidelberg University, Germany, where he has headed the Biology Department at the Institute of Pharmacy and Molecular Biotechnology since 1989. Since October of 2019, he has been working as a senior professor. After studying biology and chemistry at the University of Bonn, he conducted research in Brunswick, Cologne, Munich and Mainz. His fields of work range from phytochemistry, medicinal and poisonous plants, ornithology and conservation to systematics, phylogeny and evolutionary research. He is author/co-author of more than 30 books and over 1000 original papers. He is a visiting professor at universities in China, Thailand, Argentina and Mexico, as well as a member of various scientific advisory boards, editor of several journals (EiC of Diversity) and the recipient of several awards.

diversity

MDPI

Article

A Comparative Analysis of the Diets of a Genus of Freshwater Turtles across Africa

Luca Luiselli [1,2,3,*], Gift Simon Demaya [4], John Sebit Benansio [4,5], Fabio Petrozzi [6], Godfrey C. Akani [1,2], Edem A. Eniang [1,7], Stephanie N. Ajong [1,8], Massimiliano Di Vittorio [9], NioKing Amadi [2] and Daniele Dendi [1,2,3]

1 Institute for Development, Ecology, Conservation & Cooperation, via G. Tomasi di Lampedusa 33, I-00144 Rome, Italy; gakanina2000@yahoo.com (G.C.A.); edemeniang@yahoo.com (E.A.E.); ajong.stephanie@gmail.com (S.N.A.); d.dendi@ideccngo.org (D.D.)
2 Department of Animal and Environmental Biology, Rivers State University of Science and Technology, P.M.B. Port Harcourt 5080, Nigeria; king.amadi@ust.edu.ng
3 Department of Zoology and Animal Biology, University of Lomé, Lomé 01 BP 1515, Togo
4 Department of Wildlife, CNRES, University of Juba, Juba P.O. Box 82, South Sudan; gftsimon@yahoo.co.uk (G.S.D.); sebitbenansio@gmail.com (J.S.B.)
5 AERD—Alliance for Environment and Rural Development, El Hikma Medical Centre Street, Gudele West, Juba Block II. P.O. Box 445, South Sudan
6 Ecolobby, via Edoardo Jenner 70, I-00151 Rome, Italy; fapetrozzi@gmail.com
7 Department of Forestry and Wildlife, University of Uyo, Uyo 520241, Nigeria
8 Department of Fisheries, Lagos State University, Ojo 10201, Nigeria
9 Ecologia Applicata Italia S.R.L., via Jevolella, 2, 90018 Termini Imerese (Palermo), Italy; divittoriomassimiliano@gmail.com
* Correspondence: l.luiselli@ideccngo.org

Abstract: *Pelusios* (Testudines: Pleurodira) is an Afrotropical endemic genus of freshwater turtles that have adapted to a variety of habitats, with savannahs and forests being their two main habitat types. Although considered generally carnivorous, these turtles have rarely been subjected to detailed field surveys for determining their quantitative diet. In this paper, by using both the literature and original data, we analyze the diet of several *Pelusios* populations: three *P. adansonii* populations from South Sudan, one *P. nanus* from Zambia, seven *P. castaneus* from Nigeria, Benin and Togo, and four *P. niger* from Nigeria. All species were omnivorous but with a clear preponderance of the prey items being of animal origin (e.g., amphibians, fish, arthropods and annelids). Saturation curves revealed that the diet composition of all the surveyed populations was adequately assessed, and the diversity profiles indicated that all the populations were relatively similar in terms of overall dietary diversity. General Linear Models (GLM) showed a negative effect of vegetation cover on Anura adult consumption by turtles, and showed that the frequencies of Anura tadpoles, fish, reptiles and birds on *Pelusios* diets increased with the increase in vegetation cover. The GLM model also showed positive effects of individual body size on algae, Bivalvia, reptiles, birds and small mammal consumption by turtles, and underlined that the predation on Arachnida decreased with the increase in turtle body size. In all species, there were no significant intersexual dietary differences, whereas there were substantial ontogenetic dietary changes in three out of four species. Small-sized individuals of *P. castaneus*, *P. niger* and *P. adansonii* tended to feed mainly upon insects, with the adults also taking many fish and adult frogs, and in the case of P. niger, also birds and small mammals. Conversely, in *P. nanus*, the diet composition did not vary substantially from the juvenile to the adult age. All species appeared substantially generalist in terms of their diet composition, although the effects of season (wet versus dry) were not adequately assessed by our study.

Keywords: chelonians; pelomedusidae; foraging ecology

Citation: Luiselli, L.; Demaya, G.S.; Benansio, J.S.; Petrozzi, F.; Akani, G.C.; Eniang, E.A.; Ajong, S.N.; Di Vittorio, M.; Amadi, N.; Dendi, D. A Comparative Analysis of the Diets of a Genus of Freshwater Turtles across Africa. *Diversity* **2021**, *13*, 165. https://doi.org/10.3390/d13040165

Academic Editor: Michael Wink

Received: 10 March 2021
Accepted: 8 April 2021
Published: 12 April 2021

1. Introduction

Many recent studies have focused on the diet of freshwater turtles worldwide, including studies from North America [1,2], South America [3], Europe [4,5], Asia [6], and

Africa [7], but most species remain little known and there is virtually no study summarizing from a quantitative view the dietary characteristics of any turtle family. However, these kinds of reviews/meta-analyses may uncover life-history aspects that remain hidden in individual studies at the local scale, thus considerably enhancing the knowledge on the diversity of ecological strategies of freshwater turtles worldwide.

Pelusios is an Afrotropical endemic genus of freshwater turtles (Testudines: Pleurodira) that have adapted to a variety of habitats, with savannahs and forests being their two main habitat types [8]. Although considered generally carnivorous [8], these turtles have rarely been subjected to detailed field surveys for determining their quantitative diet [9]. Because of their wide distribution and their occurrence in divergent habitat types, the Pelomedusidae species may constitute an ideal model of study for assessing the extent of variation in the foraging ecology of tropical freshwater turtles.

In this paper, by using both the literature and original data, we present a preliminary analysis of the diet of several *Pelusios* populations belonging to four distinct species (Figure 1): three *P. adansonii* populations from South Sudan, one *P. nanus* from Zambia, seven *P. castaneus* from Nigeria, Benin and Togo, and four *P. niger* from Nigeria. Our aims with this paper are also to provide a database that can be used for further, deeper analysis of the feeding habits of *Pelusios* populations across Africa. More explicitly, we ask the following key questions:

(1) What is the taxonomic diet composition of the various *Pelusios* populations studied to date? More specifically, are these turtles essentially carnivorous as previously reported on the basis of general anecdotal literature [8,10]?

(2) Is there any effect of species, sex, or habitat (vegetation cover) on the diet composition of the various populations? This question is relevant because many studies have showed that there are considerable dietary variations associated with these variables in freshwater turtles [11].

(3) Are there any ontogenetic dietary variations in the various species? More specifically, given that there are species that exhibit large sizes as adults but very small size at hatching (*P. niger*) and others that are very small even at the adult stage (*P. nanus*), is the ontogenetic dietary variation more pronounced in the larger than in the smaller species? This question is relevant because previous studies revealed that there are considerable dietary variations associated with body size and ontogenesis in turtles [12,13], whereas in modern reptiles, important ontogenetic dietary changes are generally linked to conspicuous variations in body size between juveniles and adults [14].

Figure 1. The four study species: (**a**) *Pelusios niger* from Nigeria; (**b**) *Pelusios nanus* from Zambia; (**c**) *Pelusios castaneus* from Nigeria; (**d**) *Pelusios adansonii* from South Sudan.

2. Materials and Methods

2.1. Data Sources and Field Protocol

The literature data included three populations of *P. castaneus* and two populations of *P. niger* studied in the Niger Delta, Nigeria [7,9]. The original data came from additional four populations of *P. castaneus*, two of *P. niger*, and one of *P. nanus*. The geographic positions of the various study areas are presented in Figure 2.

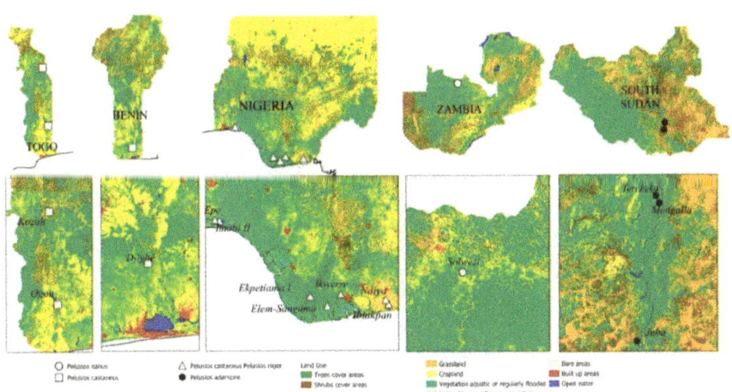

Figure 2. Map of the countries involved in the present study, showing the location of the sites where the diets of *Pelusios* spp. were studied. In Togo and Benin, only *Pelusios castaneus* were analyzed for this study; in Nigeria both *P. castaneus* and *P. niger*, in Zambia only *P. nanus* and in South Sudan only *P. adansonii*. Land use categories are also shown on the maps. Localities for both the literature and original data are pooled in this map.

Overall, original field studies were conducted between 1996 and 2020, in some savanna sites as well as in rainforest sites, in both perennial waterbodies (rivers, streams, lakes) and in temporary ponds (Figure 3). Concerning free-ranging turtles, the methodology used for obtaining the food items were carefully described in [9,15]. All captured turtles were sexed by examining their plastron and caudal shape, measured for curved carapace length, curved carapace width, plastron length and plastron width, and permanently individually marked by unique sequences of notches filed into the marginal scutes. Since the various morphometric measurements were significantly autocorrelated in all populations ($p < 0.0001$), we retained only the curved carapace length for our analyses involving turtle body size.

Figure 3. Typical habitats of *Pelusios* spp. in tropical Africa. (**a**,**c**) habitat types of *Pelusios adansonii* along the White Nile river course in South Sudan; (**b**) habitat type of *Pelusios niger* and *Pelusios castaneus* in Southern Nigeria; (**d**) habitat of *Pelusios nanus* in Zambia.

The dietary study is based on both the stomach analyses of a few dead specimens (generally offered in bush-meat markets or as roadkills), and stomach-flushing (as described in [16]) and fecal pellet analyses of living specimens (specimens were singly kept in plastic boxes until defecation occurred) [16]. Specimens captured into baited traps [15] were not included in the analyses, so we included in this study only those turtles that did not eat "artificially attractive" food during our studies. No specimen was killed or injured by the researchers. Each turtle was sexed by examining tail and plastron morphologies, measured for carapace length, and individually marked by scute notching and with a painted number on the carapace for identification and for excluding risks of data pseudoreplication. Food items or feces of each individual were deposited separately in test tubes (under alcohol) for laboratory dissection, and the reference number of each tube corresponded to the painted number on the turtle carapace. An example of the painted numbering on the turtle carapace is given in Figure 1c.

We included algae in our diet data analyses, although these may have been ingested secondarily by turtles, at least in some instances. Feces were examined under binocular microscope for the identification of any food items.

The diet composition of each population was described as the percentage of stomachs containing a given food item and not on the basis of the total number of items of each food category in the stomachs. This was necessary because it is often impossible to count the number of items from a feces analysis, and because the easiness of identification varied considerably by the various types of food item. Obviously, this methodology may have some shortcomings when comparing the data across studies. In fact, the stomach content corresponds to less processed material than in the feces, thus it could show different % when comparing the resources consumed. Therefore, we cannot exclude that the data obtained for the various dietary studies could be, in part, different also because of the different methodologies applied for gathering the food data, with the evidence on some consumed taxa that could have degraded and not be detected in the feces.

2.2. Statistical Analyses

We evaluated whether our sampling effort captured the true food items' richness and diversity within each study population by building a rarefaction curve for food type discoveries at each site, using the software PAST 4.0.

Generalized Linear Models (GLM, see [17]) were used to test the relationship between body size and vegetation cover in the diet of four species of turtles (*P. niger*, *P. castaneus*, *P. adansonii*, *P. nanus*). In the models, three different vegetation classes (savannah, derived forest and forest) and three body size classes as dependent variables and the frequencies of different prey species as predictors were used. For assigning the vegetation classes to each study site, we considered the dominant vegetation type along the banks of the concerned waterbody where turtles were captured and studied. In the models, computing by an all-effects procedure, the identity link function and a normal distribution of error were used [18].

Parametric tests were used only after having verified the normality and homoscedasticity of the used variables by the Kolmogorov–Smirnov test. Intersexual differences in the frequency of consumption of food items were assessed by an observed versus expected χ^2 test, with the *p*-value generated after 9999 Monte Carlo permutations. For analyses on the ontogenetic changes in diet composition, we divided the turtles into three body size categories: (i) <8 cm carapace length, (ii) 8.1–14 cm, and (iii) >14 cm. Since one of the study species (*P. nanus*) rarely exceeds 10 cm in carapace length and is one of the smallest turtle species in the world [1], for this latter species we had only size categories (i) and (ii) in our samples. In the text, the means are indicated as $\pm$ 1 Standard Deviation (S.D.), and alpha was set at 5%.

3. Results and Discussion

3.1. Diet Composition by Species

Overall, the diet data on 1260 *Pelusios* individuals were collected: 668 were *P. castaneus*, 310 were *P. niger*, 213 were *P. adansonii*, and 69 were *P. nanus*. A total of 705 turtle individuals were captured in Nigeria (395 *P. castaneus*, 310 *P. niger*), 56 in Benin (all *P. castaneus*), 217 in Togo (all *P. castaneus*), 213 in South Sudan (all *P. adansonii*) and 69 in Zambia (all *P. nanus*). The synopsis of the diet composition (% of stomachs containing a given food item) by species and by country/study area is given in Table 1. Saturation curves revealed that the diet composition of all the surveyed populations was adequately assessed (Figure 4), and the diversity profiles indicated that all the populations were relatively similar in terms of overall dietary diversity (Figure 5).

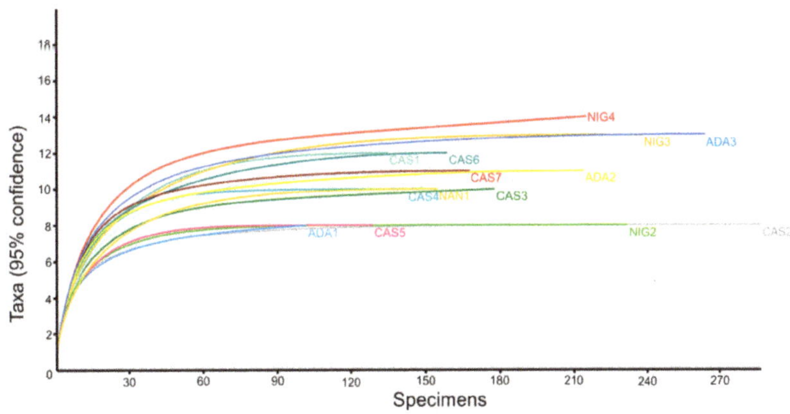

Figure 4. Saturation curves for the diet of the various populations of *Pelusios* spp. analyzed in this paper. NIG = *P. niger*; CAS = *P. castaneus*; ADA = *P. adansonii*; NAN = *P. nanus*. The numbers represent distinct populations within each species. All NIG came from Nigeria; CAS 1–4 from Nigeria, CAS5 from Benin, CAS 6–7 from Togo; ADA 1–3 from South Sudan; and NAN1 from Zambia.

Table 1. Diet composition (% of stomachs containing a given food item) of *Pelusios* species across countries. N = number of individuals examined; for-derived = forest-derived savannah/plantation mosaic.

	Nigeria	Nigeria	Nigeria	Nigeria	Benin	Togo	Togo	Nigeria	Nigeria	Nigeria	Nigeria	South Sudan	South Sudan	South Sudan	Zambia
	Forest	For-Derived	Forest	For-Derived	Savannah	Savannah	Savannah	Forest	For-Derived	Forest	For-Derived	Savannah	Savannah	Forest	Savannah
	castaneus	*castaneus*	*castaneus*	*castaneus*	*castaneus*	*castaneus*	*castaneus*	*niger*	*niger*	*niger*	*niger*	*adansonii*	*adansonii*	*adansonii*	*nanus*
N	217	21	65	92	56	135	82	113	39	77	81	133	41	39	69
Fruits	5.1	19	5	7	0	2.22	0	7.1	7.7	0	0	0	2.44	5.13	0
Seeds	3.7	76.2	10.3	15.9	5.37	15.56	13.4	3.3	30.8	5.2	0	0	7.32	10.26	7.25
Aquatic plants	7.4	66.7	52.8	38.5	55.3	25.19	20.7	7.9	33.3	27.3	16.0	30.8	43.9	38.5	8.7
Algae	0	0	1.4	6.6	0	2.96	0.0	0	0	3.9	13.6	0	0.00	0.00	0.00
Annelida	3.7	42.9	8.3	4.9	12.5	12.59	19.5	10.6	61.5	6.5	7.4	12.8	19.5	46.2	17.4
Gastropoda	5.1	57.1	29.2	36.1	0	4.44	15.9	3.5	69.2	6.5	25.9	1.5	19.5	10.3	4.3
Bivalvia	0.9	0	0	0	0	0.00	0.0	0	0	0.0	1.2	0.0	0.0	0.0	0.0
Arachnida	2.8	0	0	0	0	0.74	0.0	2.6	0	0.0	0.0	0.0	0.0	2.6	30.4
Insecta	6	0	12.5	10.6	12.5	31.85	43.9	0.9	0	27.3	30.9	30.8	51.2	43.6	68.1
Crustacea	18.9	0	11.1	9	10.71	8.89	2.4	23.9	0	42.9	13.6	4.5	14.6	53.8	2.9
Fish	52.5	14.3	45.8	7.4	25	30.37	14.6	69.9	12.8	80.5	58.0	15.8	31.7	20.5	0.0
Anura adults	3.7	4.8	5.6	18	5.37	24.44	25.6	12.4	10.2	20.8	23.5	7.5	14.6	15.4	0.0
Anura eggs	9.2	0	0	0	0	4.44	8.5	15	0	0.0	7.4	0.0	2.4	0.0	4.3
Anura tadpoles	22.6	9.5	0	0	0	0.00	6.1	29.2	10.2	0.0	0.0	0.0	0.0	10.3	10.1
Reptiles	0	0	0	0	0	0	0.0	0	0	3.9	1.2	0.0	0.0	2.6	0.0
Birds	0	0	0	0	0	0	0.0	0	0	9.1	6.2	0.0	0.0	0.0	0.0
Small mammals	0	0	0	0	0	0	0.0	0	0	7.8	9.9	0.0	0.0	0.0	0.0
Indeterminate	0.4	0	0	0	5.36	2.96	3.7	0.9	0	3.9	6.2	3.0	12.2	10.3	5.8

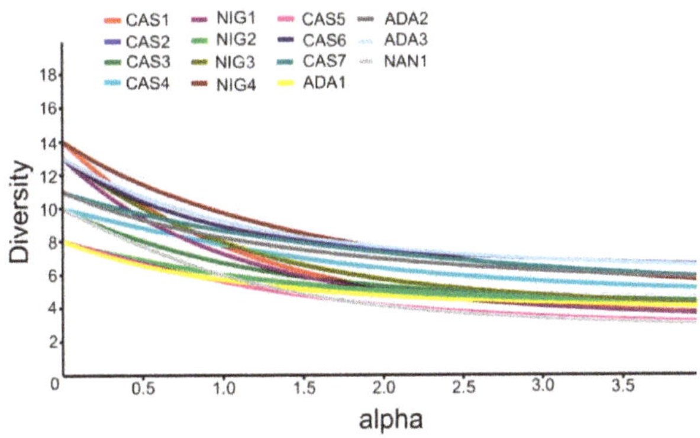

Figure 5. Diversity profiles for the diet of the various populations of *Pelusios* spp. analyzed in this paper. NIG = *P. niger*; CAS = *P. castaneus*; ADA = *P. adansonii*; NAN = *P. nanus*. The numbers represent distinct populations within each species. All NIG came from Nigeria; CAS 1–4 from Nigeria, CAS5 from Benin, CAS 6–7 from Togo; ADA 1–3 from South Sudan; and NAN1 from Zambia.

Pooling data from the various species, it resulted that the main bulk of the *Pelusios* spp. diet consisted of invertebrates (present in 75.2% of the examined specimens, $n = 1260$), followed by plant materials (found in 46.1% of the turtles) and by small vertebrates (22.8%) (Table 1). However, there were remarkable differences between species: *Pelusios niger* fed on larger sized prey types (including terrestrial vertebrates) than the other species, but this was an effect of its much larger body size. Indeed, terrestrial vertebrates were found in three out of four *P. niger* populations, and in up to 9.9% of the examined individuals within each population, whereas they were never observed in other *Pelusios* species, apart from one population of *P. adansonii* (2.6% of the examined individuals) (Table 1). On the other hand, *P. nanus* (the smallest species in the group) fed mainly upon invertebrates and was the only species that had no fish remains in stomachs or feces (Table 1). Fish remains were found in all the other 14 *Pelusios* populations, with frequencies of occurrence ranging from 7.4% (in a *P. castaneus* population from a forest-derived area) to 80.5% (in a *P. niger* population from a rainforest area in Nigeria) (Table 1).

Although not statistically significant at the species level (at least $p > 0.05$ at χ^2 test), the various methods applied to gathering the food data also influenced the taxonomic dietary composition: for instance, tadpoles were detected by stomach flushing, whereas fish and arthropod remains were detected easily also with feces analyses. This pattern was consistent across species and populations.

If we consider, as a metric of dietary preference by *Pelusios* spp., the % frequency of occurrence of a given prey type across populations (calculated based on the number of populations in which at least one individual ate a certain type of food compared to the total number of populations examined ($n = 15$)), it resulted that aquatic plants, Gastropoda, fish and frogs represented the main food categories for these turtles (Figure 6).

The various turtle populations did not show any clear species-specific pattern, but most *P. castaneus* populations clustered together, and two of out of three *P. adansonii* populations clustered together with *P. castaneus*, in a UPGMA tree-diagram with Euclidean distances (Figure 7). A UPGMA tree-diagram with Euclidean distances also showed that forest and forest-derived populations clustered together in terms of taxonomic diet composition, whereas savannah populations formed another well-defined group (Figure 8).

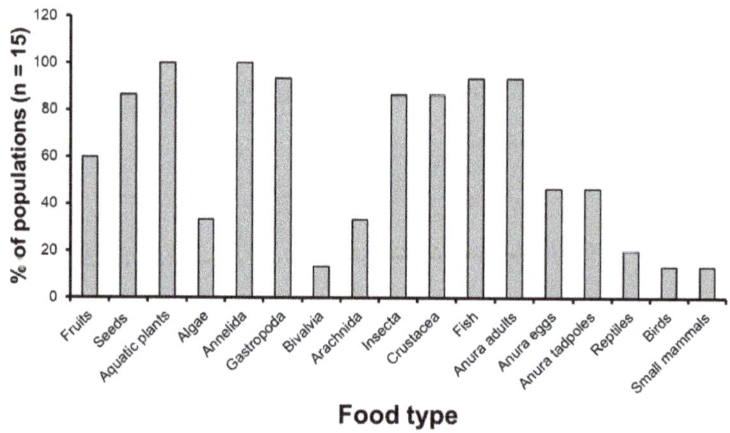

Figure 6. Percentage of *Pelusios* spp. populations that included a given food type in the diet. In this graphic, the percentages are calculated based on the number of populations in which at least one individual in a given population ate a certain type of food. Total number of populations examined: $n = 15$.

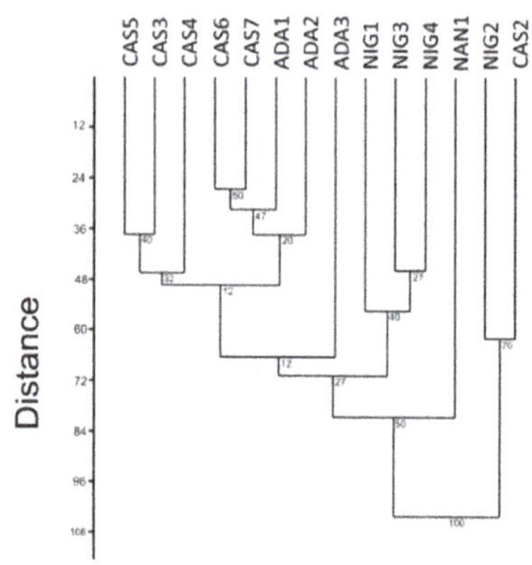

Figure 7. UPGMA, with Euclidean distances and 40 bootstraps as branching measurement, showing the dissimilarities among the various *Pelusios* populations as for their taxonomic composition of the diet is concerned. NIG = *P. niger*; CAS = *P. castaneus*; ADA = *P. adansonii*; NAN = *P. nanus*. The numbers represent distinct populations within each species. All NIG came from Nigeria; CAS 1–4 from Nigeria, CAS5 from Benin, CAS 6–7 from Togo; ADA 1–3 from South Sudan; and NAN1 from Zambia.

3.2. Effects of Vegetation Cover and Turtle Body Size

Our GLM results (Table 2) showed a negative effect of vegetation cover on Anura adults' consumption by turtles, and showed that the frequencies of Anura tadpoles, fish, reptiles and birds on *Pelusios* diets increased with the increase in vegetation cover. The GLM model also showed positive effects of individual body size on algae, Bivalvia, reptiles, birds and small mammals' consumption by turtles, and underlined that the predation on Arachnida decreased with the increase in turtle body size (Table 3).

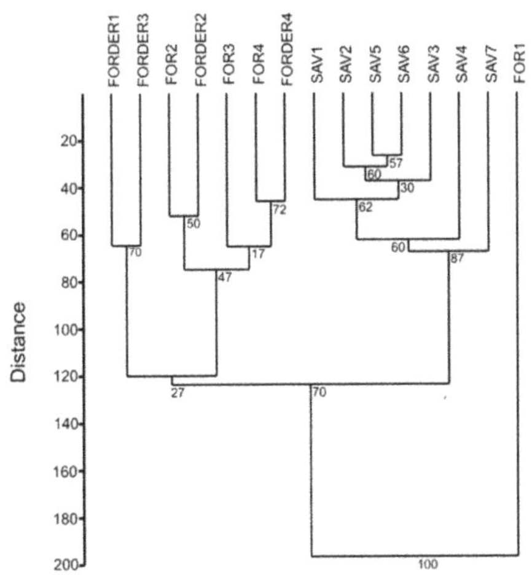

Figure 8. UPGMA, with Euclidean distances and 40 bootstraps as branching measurement, showing the dissimilarities among the main habitats of the various study areas where *Pelusios* populations were studied for determining their taxonomic composition of the diet. FOR = forest; FORDER = forest-derived; SAV = savannah. FOR 1–4 and FORDER 1–4 were from Nigeria; SAV1 from Benin, SAV 2–3 from Togo, SAV 4–6 from South Sudan and SAV7 from Zambia.

Table 2. Output of the General Linear Model (GLM) on the relationship between vegetation cover and diet in four species of *Pelusios* from tropical Africa. Only significant variables are presented in this table.

	Estimate	St. Error	Wald	*p*
Anura tadpoles	0.035197	0.009961	12.48426	0.000410
Anura adults	−0.13081	0.019494	45.02788	0.000000
Fish	0.009662	0.002308	17.52001	0.000028
Reptiles	0.274646	0.056558	23.58069	0.000001
Birds	0.210985	0.025179	70.21284	0.000000

Table 3. Output of the GLM on the relationship between turtle body size and diet in four species of *Pelusios* from tropical Africa. Only significant variables are presented in this table.

	Estimate	St. Error	Wald	*p*
Algae	0.112020	0.009507	138.8375	0.000000
Bivalvia	1.514130	0.231157	42.90536	0.000000
Arachnida	−0.060554	0.015102	16.0784	0.000061
Reptiles	0.389812	0.000062	38,937,600	0.000000
Birds	0.098272	0.032757	9.00000	0.002700
Small mammals	0.084896	0.028299	9.00000	0.002700

3.3. *Effects of Sex and Ontogenetic Changes in the Diet Composition*

The summary of the food data collected by country and by sex, for the various *Pelusios* species, are presented in the Online Supplemental Tables S1–S5. We did not uncover any significant intersexual difference in the frequency of occurrence of the various prey items within the various *Pelusios* populations and by country (Table 4).

Table 4. Summary of the statistical results on the intersexual differences in taxonomic dietary composition within *Pelusios* populations from the various countries of Africa that were analyzed for this review. χ^2 test is calculated on the frequencies of occurrence of each food type category by sex, within each population studied in a given country.

Species	Country	No. Males	No. Females	χ^2	df	p
P. castaneus	Nigeria	211	178	19.3	13	0.113
P. castaneus	Benin	27	29	5.6	7	0.590
P. castaneus	Togo	103	114	6.4	13	0.929
P. niger	Nigeria	162	148	22.2	16	0.106
P. adansonii	South Sudan	122	91	15.8	13	0.262
P. nanus	Zambia	30	39	8.4	9	0.494

In terms of ontogenetic diet composition, the data for the various species are presented in the Online Supplementary Material Tables S6–S9). For these analyses, we pooled data from males and females because of no significant intersexual differences in prey composition (see Table 4 and text at above). Concerning *P. castaneus*, the contingency table analysis revealed that there were significant differences in the frequency of consumption of the various food types by turtle size category (Table 5), with small individuals consuming significantly more insects than the two other size categories, intermediate sized and large individuals more fish and adult anurans (Table S6). These two latter categories were similar in terms of taxonomic composition of the diet, but large individuals fed upon adult anurans more frequently than intermediate sized individuals (Table S6). In *P. niger*, there was also a statistically significant ontogenetic dietary change (Table 5), with adult anurans, birds and small mammals being preyed upon particularly by large sized individuals, fish by both large and intermediate sized individuals, and insects by small sized individuals (Table S7). Insects also dominated in the diet of small-sized *P. adansonii*, whereas insects and fish were the main prey category for the large sized individuals, and aquatic plants, insects and annelids in the diet of average sized individuals (Table S8). Overall, also in this species, the ontogenetic dietary divergence was statistically significant (Table 5). Conversely, there was no significant ontogenetic divergence in *P. nanus* (Table 5), with insects being the main food type for the two size categories considered in this study (Table S9).

Table 5. Summary of the statistical results on the ontogenetic body-size-related differences in the taxonomical dietary composition within *Pelusios* populations from the various countries of Africa that were analyzed for this review. χ^2 test is calculated on the frequencies of occurrence of each food type category by turtle size category, within each population studied.

Species	<8 cm (N)	8–14 cm (N)	>14 cm (N)	χ^2	df	p
P. castaneus	158	389	121	426.3	28	<0.0001
P. niger	99	101	110	162.5	34	<0.0001
P. adansonii	58	122	33	94.3	26	<0.0001
P. nanus	31	38	not available	11.25	9	0.259

Our comparative data showed therefore that, whereas intersexual dietary divergence is virtually non-existent among *Pelusios* populations, there were instead significant ontogenetic diet variations in three out of four species. Both patterns are likely linked to the fact that intersexual dietary divergence should be expected in species with remarkable sexual size dimorphism [19] or in species with "telescopic growth" from newborn to adult age [14,20,21], whereas in our studied *Pelusios* species (apart from *P. niger* with males being much larger than females) the two sexes were relatively similar in size. In this regard, it is also interesting to mention that, consistent with theory [19–21], the species with the smallest absolute body size (*P. nanus*) did not show any ontogenetic dietary shifts.

We can therefore hypothesize that, in Pleurodira species, the diet composition (1) should be similar between males and females in those species exhibiting a minor sexual size dimorphism but not in those where one sex may reach much larger size than the other, and

(2) should be more ontogenetically variable in those species reaching larger adult sizes but still with small sized hatchlings. For instance, we expect a significant sexual size dimorphism in species such as *Pelusios sinuatus*, where the males may reach 35 cm and the females 55 cm carapace length and the size divergence between hatchlings and adults is remarkable [8,22,23].

4. Conclusions

In conclusion, our study revealed that all species of *Pelusios* analyzed here were substantially generalist in terms of their diet composition, although the effects of season (wet versus dry) and of other co-variates that are usually informative for field studies of turtles, including microhabitat characteristics and proximate meteorological conditions, were not assessed by our study. In addition, we showed that all species were omnivorous but with a clear preponderance of the prey items being of animal origin (amphibians, fish, arthropods and annelids), thus confirming earlier anecdotal accounts [8,10] and accounts from captivity (e.g., see <https://www.encyclo-fish.com/EN/paludarium/animals/pelusios-castaneus.php> (accessed on 10 April 2021). The relative head size and shape probably influenced the ingestion performance of the various species [24–27] and may possibly produce mechanisms of intraspecific niche partitioning [24,25]: indeed, when considering only the prey items that were found almost intact in the flushed stomachs, the species with the most massive head (*P. niger*) at a given body size was particularly able to ingest very large prey items compared to other species, and the larger-sized category of *P. niger* individuals also fed frequently upon larger size vertebrates, including even birds and small mammals. The ecological consequences (minimization of interspecific competition strength) of these differences in ingestion performance should be further analyzed by ad hoc studies.

Supplementary Materials: The following are available online at https://www.mdpi.com/article/10.3390/d13040165/s1, Table S1: Diet composition (% of stomachs containing a given food item) of male and female *Pelusios* species in Nigeria, Table S2: Diet composition (% of stomachs containing a given food item) of male and female *Pelusios castaneus* in Benin, Table S3: Diet composition (% of stomachs containing a given food item) of male and female *Pelusios castaneus* in Togo, Table S4: Diet composition (% of stomachs containing a given food item) of male and female *Pelusios adansonii* in South Sudan, Table S5: Diet composition (% of stomachs containing a given food item) of male and female *Pelusios nanus* in Zambia, Table S6: Diet composition (% of stomachs containing a given food item) of the three body size categories of *Pelusios castaneus* (all countries being pooled), Table S7: Diet composition (% of stomachs containing a given food item) of the three body size categories of *Pelusios niger* (Nigeria), Table S8: Diet composition (% of stomachs containing a given food item) of the three body size categories of *Pelusios adansonii* (South Sudan), Table S9: Diet composition (% of stomachs containing a given food item) of the three body size categories of *Pelusios nanus* (Zambia).

Author Contributions: Conceptualization, L.L.; methodology, L.L.; formal analysis, L.L., M.D.V.; investigation, G.S.D., J.S.B., S.N.A., N.A.; resources, L.L.; data curation, F.P., D.D.; writing—original draft preparation, L.L.; writing—review and editing, all authors; supervision, L.L., G.C.A., E.A.E.; funding acquisition, L.L., F.P.; All authors have read and agreed to the published version of the manuscript.

Funding: Chelonian Research Foundation, Mohamed Bin Zayed Species Conservation Fund, Conservation International and Turtle Conservation Fund (to L.L. and F.P.).

Institutional Review Board Statement: Not applicable.

Informed Consent Statement: Not applicable.

Data Availability Statement: Data are presented in the paper and in the Online Supplemental Materials; additional data are available from authors on request.

Acknowledgments: We thank J. E. Fa, G.H. Segniagbeto, E, M. Hema, and S. Gonedele Bi, for helpful collaboration during the various research phases of this study. Two anonymous reviewers are thanked for their helpful comments on the submitted draft.

Conflicts of Interest: The authors declare no conflict of interest.

References

1. Aresco, M.J.; Travis, J.; MacRae, P.S. Trophic interactions of turtles in a north Florida lake food web: Prevalence of omnivory. *Copeia* **2015**, *103*, 343–356. [CrossRef]
2. Mccoy, C.J.; Flores-Villela, O.A.; Vogt, R.C.; Pappas, M.; McCoy, J.K. Ecology of riverine turtle communities in the southern United States: Food resource use and trophic niche dimensions. *Chel. Cons. Biol.* **2020**, *19*, 197–208.
3. Lara, N.R.F.; Marques, T.S.; Montelo, K.M.; de Ataides, A.G.; Verdade, L.M.; Malvásio, A.; de Camargo, P.B. A trophic study of the sympatric Amazonian freshwater turtles *Podocnemis unifilis* and *Podocnemis expansa* (Testudines, Podocnemidae) using carbon and nitrogen stable isotope analyses. *Can. J. Zool.* **2012**, *90*, 1394–1401. [CrossRef]
4. Balzani, P.; Vizzini, S.; Santini, G.; Masoni, A.; Ciofi, C.; Ricevuto, E.; Chelazzi, G. Stable isotope analysis of trophic niche in two co-occurring native and invasive terrapins, *Emys orbicularis* and *Trachemys scripta elegans*. *Biol. Inv.* **2016**, *18*, 3611–3621. [CrossRef]
5. Ducotterd, C.; Crovadore, J.; Lefort, F.; Guisan, A.; Ursenbacher, S.; Rubin, J.F. The feeding behaviour of the European pond turtle (*Emys orbicularis*, L. 1758) is not a threat for other endangered species. *Glob. Ecol. Conserv.* **2020**, *23*, e01133. [CrossRef]
6. Sung, Y.H.; Hau, B.C.; Karraker, N.E. Diet of the endangered big-headed turtle *Platysternon megacephalum*. *Peer J.* **2016**, *4*, e2784. [CrossRef]
7. Luiselli, L. Food habits of the pelomedusid turtle *Pelusios castaneus castaneus* in southeastern Nigeria. *Chel. Cons. Biol.* **1998**, *3*, 106–107.
8. Branch, B. *Tortoises, Terrapins and Turtles of Africa*; New Holland Publishing: Cape Town, South Africa, 2008.
9. Luiselli, L.; Akani, G.C.; Politano, E.; Odegbune, E.; Bello, O. Dietary shifts of sympatric freshwater turtles in pristine and oil-polluted habitats of the Niger Delta, southern Nigeria. *Herpetol. J.* **2004**, *14*, 57–64.
10. Ernst, C.H.; Altenburg, R.G.; Barbour, R.W. *Turtles of the World*; Smithsonian Press: Washington, DC, USA, 1998.
11. McCauley, S.J.; Bjorndal, K.A. Response to dietary dilution in an omnivorous freshwater turtle: Implications for ontogenetic dietary shifts. *Physiol. Biochem. Zool.* **1999**, *72*, 101–108. [CrossRef]
12. Bouchard, S.S.; Bjorndal, K.A. Ontogenetic diet shifts and digestive constraints in the omnivorous freshwater turtle *Trachemys scripta*. *Physiol. Biochem. Zool.* **2006**, *79*, 150–158. [CrossRef]
13. Claude, J.; Pritchard, P.H.; Tong, H.; Paradis, E.; Auffray, J.-C. Ecological Correlates and Evolutionary Divergence in the Skull of Turtles: A Geometric Morphometric Assessment. *Syst. Biol.* **2004**, *53*, 933–948. [CrossRef]
14. Burke, R.L.; Luiselli, L. Ontogenetic niche shifts: Modern reptiles change diets with age too. *Science* **2021**, *371*, 941–944.
15. Luiselli, L.; Akani, G.C.; Ajong, S.N.; George, A.; Di Vittorio, M.; Eniang, E.A.; Dendi, D.; Hema, E.M.; Petrozzi, F.; Fa, J.E. Predicting the structure of turtle assemblages along a megatransect in West Africa. *Biol. J. Linn. Soc.* **2020**, *130*, 296–309. [CrossRef]
16. Luiselli, L.; Amori, G. Diet. In *Reptile Ecology and Conservation: A Handbook of Techniques*; Dodd, C.K., Jr., Ed.; Oxford University Press: Oxford, UK, 2016; pp. 97–109.
17. Hosmer, D.W.; Lemeshow, S. *Applied Logistic Regression Analysis*, 2nd ed.; John Wiley and Sons: New York, NY, USA, 2000.
18. McCullagh, P.; Nelder, J.A. *Generalized Linear Models*; Chapman and Hall/CRC: London, UK, 1989.
19. Shine, R. Intersexual dietary divergence and the evolution of sexual dimorphism in snakes. *Am. Nat.* **1991**, *138*, 103–122. [CrossRef]
20. Houston, D.; Shine, R. Sexual dimorphism and niche divergence: Feeding habits of the *Arafura filesnake*. *J. Anim. Ecol.* **1993**, *62*, 737–748. [CrossRef]
21. Shine, R.; Harlow, P.S.; Keogh, J.S. The influence of sex and body size on food habits of a giant tropical snake, *Python reticulatus*. *Funct. Ecol.* **1998**, *12*, 248–258. [CrossRef]
22. Gerlach, J.; Rhodin, A.G.J.; Pritchard, P.C.H.; van Dijk, P.P.; Saumure, R.A.; Buhlmann, K.A.; Iverson, J.B. Pelusios subniger parietalis Bour 1983–Seychelles black mud turtle. In *Conservation Biology of Freshwater Turtles and Tortoises: A Compilation Project of the IUCN/SSC Tortoise and Freshwater Turtle Specialist Group*; Chelonian Research Foundation: Lunenburg, MA, USA, 2008.
23. Ceballos, C.P.; Adams, D.C.; Iverson, J.B.; Valenzuela, N. Phylogenetic patterns of sexual size dimorphism in turtles and their implications for Rensch's Rule. *Evol. Biol.* **2013**, *40*, 194–208. [CrossRef]
24. Tucker, A.D.; Fitzsimmons, N.N.; Gibbons, J.W. Resource partitioning by the estuarine turtle *Malaclemys terrapin*: Trophic, spatial, and temporal foraging constraints. *Herpetologica* **1995**, *51*, 167–181.
25. Bulté, G.; Gravel, M.A.; Blouin-Demers, G. Intersexual niche divergence in northern map turtles (*Graptemys geographica*): The roles of diet and habitat. *Can. J. Zool.* **2008**, *86*, 1235–1243. [CrossRef]
26. Herrel, A.; O'Reilly, J.C. Ontogenetic scaling of bite force in lizards and turtles. *Physiol. Bioch. Zool.* **2006**, *79*, 31–42. [CrossRef]
27. Bulté, G.; Irschick, D.J.; Blouin-Demers, G. The reproductive role hypothesis explains trophic morphology dimorphism in the northern map turtle. *Funct. Ecol.* **2008**, *22*, 824–830. [CrossRef]

diversity

MDPI

Article

Patterns of Spatial Overlap between Non-Indigenous and Critically Endangered Freshwater Fishes from a Mediterranean Biodiversity Hotspot †

Nicholas Koutsikos [1,2,*], Leonidas Vardakas [1,3], Olga-Ioanna Kalantzi [2] and Stamatis Zogaris [1]

1 Hellenic Centre for Marine Research, Institute of Marine Biological Resources and Inland Waters, Anavissos, 19013 Attica, Greece; lvard@hcmr.gr (L.V.); zogaris@hcmr.gr (S.Z.)
2 Department of Environment, University of the Aegean, 81100 Mytilene, Greece; kalantzi@aegean.gr
3 Department of Forestry and Management of the Environment and Natural Resources, Democritus University of Thrace, 193 Pantazidou St., 68200 Orestiada, Greece
* Correspondence: nkoutsik@hcmr.gr; Tel.: +30-22910-76393

† Parts of this article have been presented at the 1st International Electronic Conference on Biological Diversity, Ecology and Evolution, 15–31 March 2021, as "*Overlapping Areas of Non-Indigenous and Critically Endangered Freshwater Fishes: Setting Conservation Priorities in Greece*". Available online: https://bdee2021.sciforum.net/.

Abstract: Non-indigenous fish species (NIFS) can cause severe ecological impacts on the invaded ecosystems and are considered as one of the leading factors of freshwater biodiversity loss. Unraveling the spatial overlap between NIFS and critically endangered (CR) fish species can contribute to targeted conservation actions to minimize the potential negative effects. In this study, we applied geostatistical analyses to investigate the spatial overlap of NIFS against fish species that are designated under the CR status according to the IUCN and the Hellenic Zoological Society (HZS) Red Lists. Distributional data (presence–absence) from 800 records of 52 NIFS were compiled for both lentic and lotic ecosystems of Greece. Our results indicate that freshwater ecosystems under high NIFS richness were located mainly in lowland areas and often near large cities and ecosystems with well-developed commercial and recreational fisheries. On the contrary, low NIFS richness was observed in mountainous regions and in relatively small river basins. Overlapping areas of CR species with moderate to high NIFS richness (1.5–4.3 NIFS per 1 km^2) were relatively high (~50%). A quarter of the overlapping areas (24.8%) fall within NATURA 2000 network, where legal management bodies could implement specialized programs to minimize the negative impacts. However, the majority of CR fish species' distribution remains in unprotected areas indicating that protected areas should be re-designed to include areas containing freshwater species under the highest threatened category. Our findings demonstrate that whole assemblages of fishes are rapidly changing as NIFS spread into Greece and many freshwater ecosystems of outstanding biodiversity conservation value are under significant invasion pressure.

Keywords: invasive species; alien; translocated; critically endangered; freshwater fishes; freshwater ecosystems; conservation; biodiversity; red list

Citation: Koutsikos, N.; Vardakas, L.; Kalantzi, O.-I.; Zogaris, S. Patterns of Spatial Overlap between Non-Indigenous and Critically Endangered Freshwater Fishes from a Mediterranean Biodiversity Hotspot †. *Diversity* **2021**, *13*, 233. https://doi.org/10.3390/d13060233

Academic Editor: Michael Wink

Received: 21 April 2021
Accepted: 23 May 2021
Published: 26 May 2021

1. Introduction

Recent scientific studies have indicated that biodiversity is declining at an extremely rapid rate, suggesting that the sixth mass extinction of species is already under way [1–3]. Indeed, according to the International Union for the Conservation of Nature (IUCN), more than 35,500 species (or 28% of all assessed species) are threatened with extinction worldwide, while at least 1677 species out of 15,060 assessed are threatened with extinction in Europe. Freshwater fishes are considered among the most threatened species worldwide [4,5]. The most recent IUCN Red List includes 20,109 species belonging to the class of Actinopterygii, 51.8% of which inhabit freshwater ecosystems (10,434 species), where 2234 species (i.e., 22%) of the freshwater species are at risk of global extinction as they are under a threatened category (CR: Critically Endangered (576 species; 5.5%), EN: Endangered (857 species; 8.2%), VU: Vulnerable (961 species; 9.2%)). In Europe, freshwater fishes display the third highest percentage of taxa which are at risk; that is, in order of extinction risk: freshwater mollusks (59%), endemic trees (58%), and freshwater fish (40%). Pollution is one of the major threats that significantly affects freshwater fish species in the European region [6]; however, currently it is not considered the leading factor for species extinctions. Other pressures such as over-abstraction, combined with the increasing frequency of drought events due to climate change [7], the disruption of river connectivity due to the construction of multiple barriers [8], and the introduction of invasive alien species severely affect the viability of the native freshwater fish fauna of Europe [9].

Freshwater fish species inhabiting Mediterranean inland water ecosystems are considerably more vulnerable compared to the species located in the northern temperate regions of Europe since they are forced to survive under diverse and unstable hydrological regimes [10]. In addition, a large number of the fish species occurring in Mediterranean countries display a range-restricted geographical distribution, as they may occur in a single river or lake [11], thus making them even more vulnerable to additive threats. Greece, located in the eastern Mediterranean region, holds a unique ichthyofaunal diversity within Europe and displays one of the highest levels of fish species endemism in the Mediterranean [10,12]. The complex geological processes of the wider area of the Balkans and the eastern Mediterranean has allowed multiple fish species colonizations, long-term survival of ancient taxa in aquatic refugia, and enhanced speciation due to long-term biogeographical barriers that enhance hydrographic isolation among very different biogeographic areas [12,13]. These factors are mainly responsible for the increased diversity and high degree of endemicity in Greece's freshwater fish fauna.

As elsewhere in the Mediterranean basin, anthropogenic alterations such as habitat degradation and fragmentation are the main threats for freshwater fishes in Greece [13]. These anthropogenic stresses are also augmented by the introduction of alien and intra-country translocated fishes (i.e., species transferred out of their natural distributional range but within the country limits) [14]. Non-indigenous species can have severe ecological impacts on the recipient ecosystems they invade. Due to the possible absence of natural predators they are able to increase in numbers and disperse to new areas. In addition, they can hybridize with related native species, spread diseases, compete and displace native species, and alter the structure and function of ecosystems, even leading to local extinctions of native species [15,16].

The aim of this study was to identify the potential overlapping areas of non-indigenous fish species (hereafter NIFS) against fish species that are designated under critically endangered-CR status according to the IUCN and the Hellenic Zoological Society (HZS) Red Lists. This applied geographical analysis aims to support conservation planning and actions to minimize the potential negative effects that NIFS may pose to freshwater biodiversity.

2. Materials and Methods

2.1. Study Area and Species Pool

According to the most recent checklist to date [17], 160 fish species have been recorded in freshwater ecosystems of Greece, where 137 are native. The country presents a substantial proportion of country-specific endemics, 47 in total (or 34% of the native fish fauna). Moreover, a further 10% of the freshwater fish species are "near endemic"; that is, occurring also in shared transboundary freshwater lake basins. By excluding aliens and species of marine origin, the percentage of endemic and near-endemic species rises to 57%. In Greece, 51 fish species are considered threatened at a global scale (i.e., CR, Critically Endangered (20); EN, Endangered (15); and VU, Vulnerable (16)) corresponding to 31.8% of all native inland water fish species in Greece based on the most recent IUCN Red List and distinctions made in Barbieri et al. [17].

2.2. Data Acquisition

We obtained geographic range data for fish species designated as critically endangered (CR) based on the IUCN Red List integrated with the critically endangered fish species listed by the Hellenic Zoological Society (HZS) Red List (Table 1). The distributional range of CR species have been developed during the conservation status assessment of EU species of conservation concern [18]. Furthermore, we included three additional fish species not yet evaluated since they display extremely restricted distributional ranges in Greece.

Table 1. Freshwater fish species of Greece listed as Critically Endangered (CR) under IUCN and/or the HZS Red Lists. Categories: CR, Critically Endangered; EN, Endangered; VU, Vulnerable; LC, Least Concern; and DD, Data Deficient. Transl. in brackets denotes species that are introduced as translocated populations, not the original wild stock. Species with an asterisk are currently presumed extirpated or extinct in Greece.

Species	Authority	IUCN Red List	HZS Red List
Acipenser naccarii *	Bonaparte, 1836	CR	[Transl.]
Acipenser stellatus *	Pallas, 1771	CR	DD
Acipenser sturio *	Linnaeus, 1758	CR	DD
Alburnus macedonicus	Karaman, 1928	CR	CR
Alburnus vistonicus	Freyhof and Kottelat, 2007	CR	CR
Alosa vistonica *	Economidis and Sinis, 1986	CR	CR
Aphanius almiriensis	Kottelat, Barbieri, and Stoumboudi, 2007	CR	CR
Barbus euboicus	Stephanidis, 1950	CR	CR
Barbus pergamonensis	Karaman, 1971	LC	CR
Caspiomyzon graecus	Renaud and Economidis, 2010	-	-
Caspiomyzon hellenicus	Vladykov, Renaud, Kott, and Economidis, 1982	CR	CR
Cobitis stephanidisi	Economidis, 1992	CR	CR
Eudontomyzon sp. Almopaios	Provisionally in Barbieri et al. 2015	-	-
Huso huso *	(Linnaeus, 1758)	CR	[Transl.]
Knipowitschia goerneri	Ahnelt, 1991	DD	CR
Knipowitschia milleri	(Ahnelt and Bianco, 1990)	CR	VU
Oxynoemacheilus theophilii	Stoumboudi, Kottelat, and Barbieri, 2006	LC	CR
Pelasgus epiroticus	(Steindachner, 1896)	CR	CR
Pelasgus laconicus	(Kottelat and Barbieri, 2004)	CR	CR
Pungitius hellenicus	Stephanidis, 1971	CR	CR
Salaria economidisi	Kottelat, 2004	CR	LC
Scardinius graecus	Stephanidis, 1937	CR	VU
Squalius sp. Evia	Provisionally in Kottelat and Freyhof 2007	CR	-
Valencia letourneuxi	(Sauvage, 1880)	CR	CR
Valencia robertae	Freyhof, Kärst, and Geiger, 2014	-	-

In total, 25 freshwater fish species were included in our dataset: 19 by IUCN, three species under the HZS Red List (*Barbus pergamonensis, Knipowitschia goerneri, Oxynoemacheilus theophilii*), and three additional species which are not yet formally evaluated (*Caspiomyzon graecus, Eudontomyzon* sp. Almopaios, *Valencia robertae*), displaying however restricted distributional ranges (Table 1). The European eel, *Anguilla anguilla* (L.), a widespread euryhaline fish of marine origin, was excluded from our dataset due to its broad distributional range in the country, considering that this species could confound our analyses.

The geographical distribution of NIFS in Greece was compiled based on data from two different sources: a) a bibliographical survey for lentic ecosystems and b) survey data from standardized field surveys in lotic ecosystems (rivers, streams, canals, and springs) in the framework of various national and local projects. The derived matrix summarized records from 169 lakes, 154 with at least one NIFS and 15 with none, within the Greek territory (Figure 1). Data of fish species in lentic waters—lakes, reservoirs and ponds—were obtained from both field surveys and bibliographical data between the years 2001 and 2020 (107 artificial and 62 natural). Fish sampling data in lotic waters were acquired from research surveys conducted between the years 2001–2017 by the Hellenic Centre for Marine Research (HCMR); these cover the entire mainland as well as the major islands of Greece.

Most of the sampled data have been recently published [14,19]. Field samplings were conducted primarily through a standardized electrofishing procedure following the FAME research project guidelines [20] with some modifications; for a detailed description of the sampling procedure see [21]; in some cases seine nets and other methods were also used to ascertain fish presence. In total, 265 lotic sites with at least one NIFS and 366 sites without NIFS in Greece were compiled (Figure 1). Barbieri et al. [17] was used for species taxonomy and nomenclature. Species that are introduced by humans beyond their native freshwater ecoregion, but are native to a part of the country, are designated as translocated species [22]. NIFS include both alien and translocated species in all analyses.

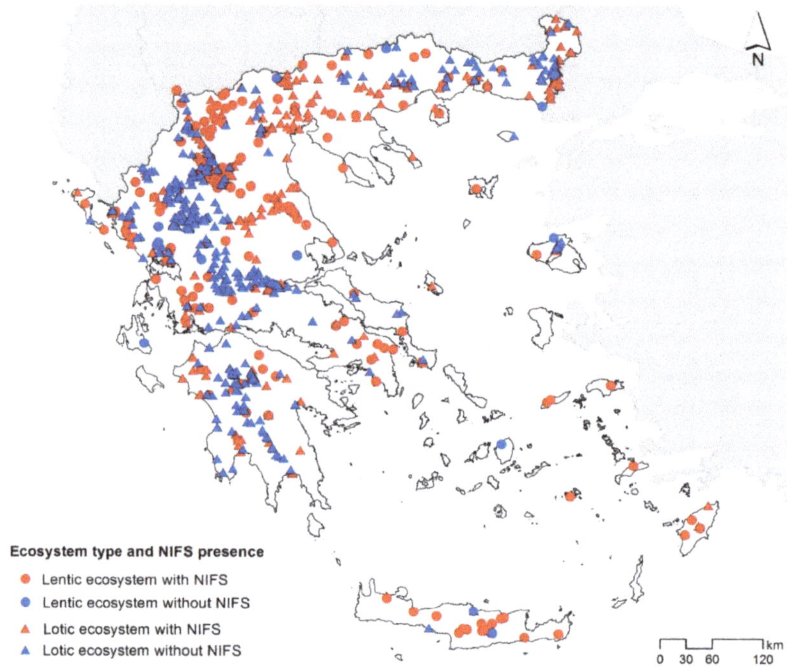

Figure 1. Map of Greece indicating the location of lentic (points) and lotic (triangles) ecosystems used in this study. Red symbols indicate locations occupied with NIFS and blue NIFS-free locations.

2.3. *Spatial Analysis*

Spatial analyses of species distributional data were performed in order to investigate the potential overlapping areas of NIFS against fish species that are designated under a CR status. The geographical range of each CR fish species was projected within grid cells of 1×1 km in a geographical information system (ESRI—ArcGIS v. 10.4; Figure 2), and the final matrix generated by the total number of CR species in each cell. The predictive distributional patterns of NIFS richness was generated by using presence/absence NIFS data through the Kriging interpolation analysis in ArcGIS, initially separately for lotic and lentic ecosystems to identify possible differences or limitations attributed to the ecosystems investigated. The overlapping "conflict areas" were spatially delimited by the produced NIFS distributional patterns within the geographical range of CR fish species. Finally, we compared the distribution of CR fish species richness of the overlapping areas for protected areas (i.e., Greece's Natura 2000 network) versus areas outside protected regions.

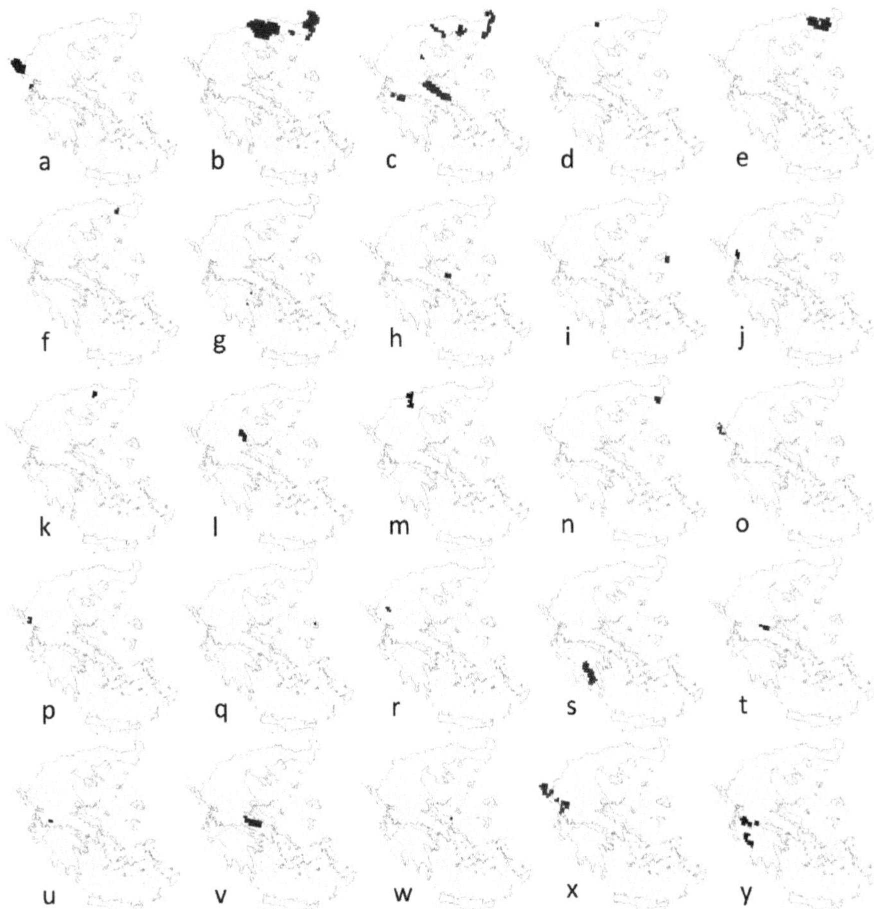

Figure 2. Geographical ranges of CR fish species in Greece projected in 1 x 1 km grid cells. Background colored areas show terrestrial (green) and marine (blue) EU Natura 2000 protected areas. Codes: (**a**) *A. naccarii;* (**b**) *A. stellatus;* (**c**) *A. sturio;* (**d**) *A. macedonicus;* (**e**) *A. vistonicus;* (**f**) *A.vistonica;* (**g**) *A.almiriensis;* (**h**) *B. euboicus;* (**i**) *B. pergamonensis;* (**j**) *C. graecus;* (**k**) *C. hellenicus;* (**l**) *C. stephanidisi;* (**m**) *Eudontomyzon* sp. *Almopaios;* (**n**) *H. huso;* (**o**) *K. goerneri;* (**p**) *K. milleri;* (**q**) *O. theophilii;* (**r**) *P. epiroticus;* (**s**) *P. laconicus;* (**t**) *P. hellenicus;* (**u**) *S. economidisi;* (**v**) *S. graecus;* (**w**) *Squalius.* sp. *Evia;* (**x**) *V. letourneuxi;* (**y**) *V. robertae.*

3. Results

In total, 800 records were utilized from 169 lakes and 631 lotic sites from 51 river basins. By using solely lotic data from field surveys, our results indicated low NIFS richness (0–1.40 NIFS per 1 km^2) throughout the country, with the exception of specific hotspots in central and northern Greece (Figure 3a). On the contrary, by utilizing bibliographic data from lentic ecosystems the patterns of predictive NIFS richness was predominately high (2.7–4.3 NIFS per 1 km^2) in most areas of the country, followed by moderate NIFS predicted richness (1.5–2.6 NIFS per 1 km^2) in the remaining parts (Figure 3b).

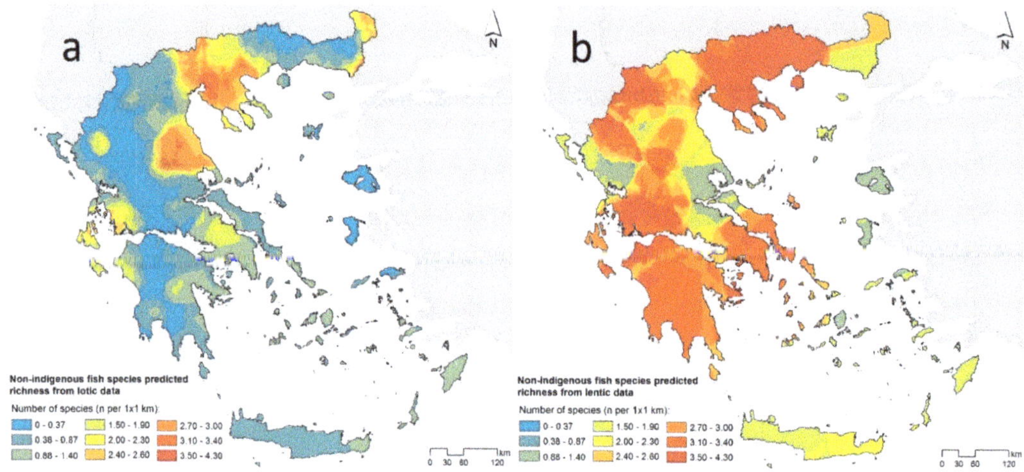

Figure 3. Patterns of the predicted richness of non-indigenous fish species derived from (**a**) lotic data and (**b**) lentic data.

Overall, we recorded 52 NIFS in 154 lentic and 51 lotic ecosystems (265 sites) of which 17 were categorized as alien and 35 as translocated. The five most widespread NIFS within Greece included four aliens, namely *Gambusia holbrooki* occurring in 223 locations (53.1%), *Carassius gibelio* occurring in 187 locations (44.52%), *Lepomis gibbosus* occurring in 113 locations (26.9%), *Pseudorasbora parva* occurring in 93 locations (22.14%), and one translocated species *Cyprinus carpio* occurring in 109 locations (25.95%). NIFS richness ranged from single species (176 sites) up to 12 fish species (one site—Lake Pamvotis). Our results indicate that freshwater ecosystems under high NIFS richness are located mainly in lowland areas of western, central, and northern Greece (Figure 3a), usually near large cities and the presence of lentic ecosystems with well-developed commercial and recreational fisheries. On the contrary, areas with low NIFS richness were observed in mountainous regions and within small river basin areas in southern Greece and the Aegean islands (Figure 4a).

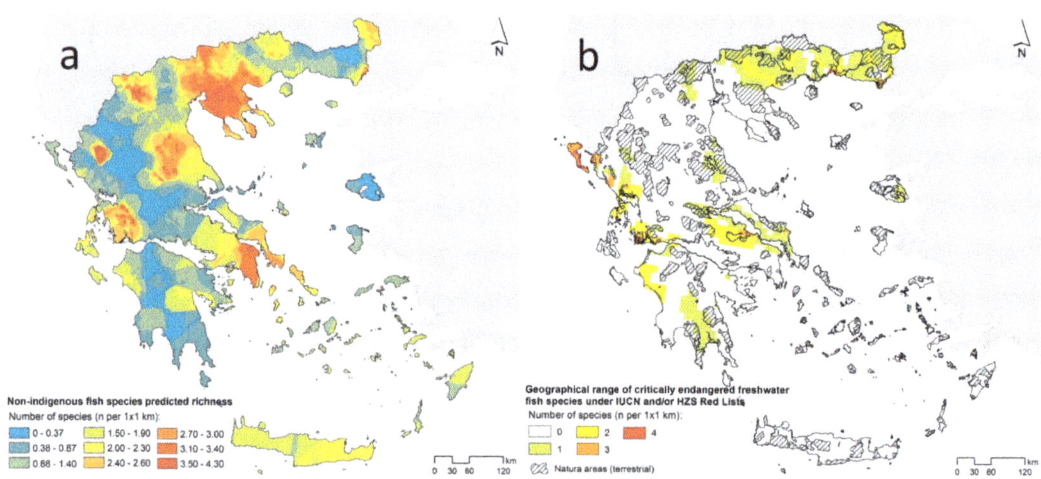

Figure 4. (**a**) Patterns of predicted richness of non-indigenous fish species (indicating high invasion pressure) and (**b**) geographical distribution range of critically endangered freshwater fish species and protected areas (Natura 2000 network).

The distribution of the CR freshwater fish species in Greece was scattered throughout the aquatic ecosystems of the country (Figure 2), covering approximately 28% (36,708 km^2) of the entire area of Greece (Table 2). In most areas, only one CR fish species occurs or is known to have occurred (69.9%); however, in some ecosystems two or more species co-occur (Figure 4b; Table 2). The highest species richness of CR fish species was observed in the north and northeastern, western, and central parts of Greece, while no CR fish species were recorded in the mountainous areas of northern and central Greece, in the majority of the Aegean or Ionian Islands, or the island of Crete (Figure 4b).

Table 2. Distributional coverage of critically endangered (CR) freshwater fish species in Greece.

CR Category (n of Species)	Area (km^2)	Percent (%)
1	25,658	69.9
2	8925	24.3
3	1796	4.9
4	329	0.9
Total	**36,708**	**100**

Overlapping areas of CR species with moderate to high NIFS richness (1.5–4.3 NIFS per 1 km^2) were relatively high (~50%) (Table 3). These areas were located in the western, central-east, and northern part of Greece (Figure 4). Shared absences indicating both low CR species and NIFS and thus low overlapping areas were observed in the mountainous regions of central Greece, the Aegean Islands, and the Island of Crete (Table 3; Figure 5). The above is to be expected since most Aegean Islands (with the exception of the Island of Lesbos) as well as the Island of Crete lack any CR species.

Table 3. The proposed class boundaries (low, moderate, high) of the overlapping areas of non-indigenous and critically endangered freshwater fish species, the actual areas, and the total percentage for each category.

Category			Area (km^2)	Percent (%)	Total Area (km^2)	Total Percent (%)
Low		0–0.37	4270	11.6	18.408	50.1
		0.38–0.87	5906	16.1		
		0.88–1.40	8232	22.4		
Moderate		1.50–1.90	7154	19.5	14.997	40.9
		2.00–2.30	3926	10.7		
		2.40–2.60	3917	10.7		
High		2.70–3.00	1637	4.5	3.303	9.0
		3.10–3.40	1371	3.7		
		3.50–4.30	295	0.8		
Total			**36,708**	**100**		

Only, five CR species were recorded exclusively in areas with low NIFS richness (*Barbus pergamonensis, Knipowitschia goerneri, Knipowitschia milleri, Oxynoemacheilus theophilii* and *Pungitius hellenicus*). The CR fish species co-occurring in areas with moderate to high NIFS richness were: *Pelasgus epiroticus* (Lake Pamvotis), *Valencia robartae, Salaria economidisi* (Acheloos basin), *Scardinius graecus* (Lakes Yliki and Paralimni), *Cobitis stephanidisi* (Lake Karla basin), *Alburnus macedonicus* (Lake Doirani), *Barbus euboicus, Squalius* sp. *Evia* (streams of Euboea Island) and *Caspiomyzon hellenicus* (Strymon basin) (Figure 5; Table 4).

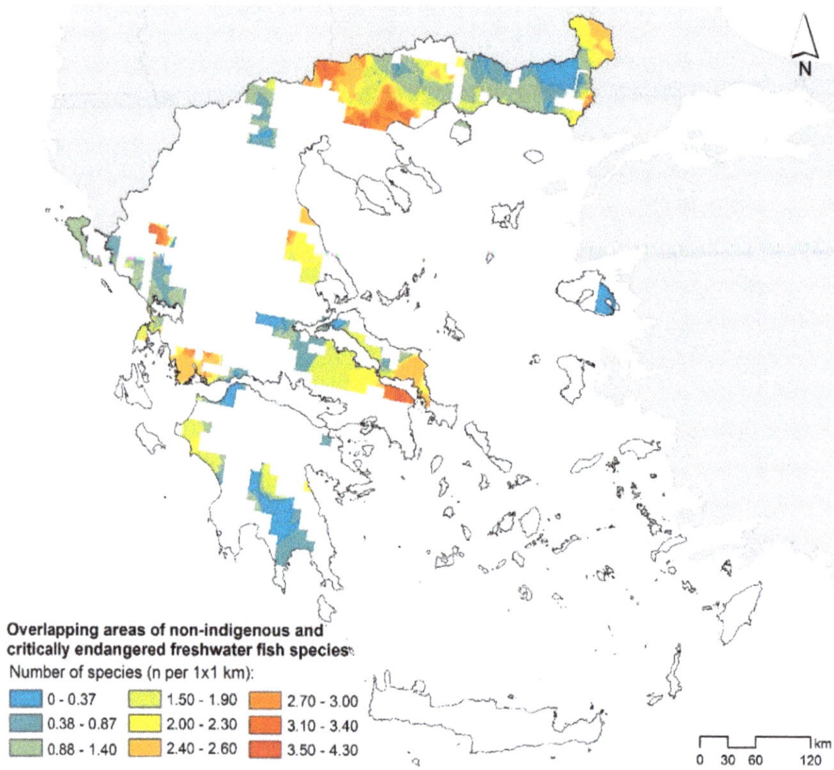

Figure 5. Overlapping areas between the distributions of NIFS and critically endangered freshwater fish species.

A quarter of the overlapping areas (24.8%) fall well within the NATURA 2000 network (Figure 6a), where management bodies could implement control or mitigation programs to minimize the negative impacts from NIFS to native biodiversity. However, the majority of the distribution areas of CR fish species in Greece are located outside of protected areas (75.2%; Figure 6a) and any additional anthropogenic stressors have the potential to increase pressure to the populations of these species. Moreover, the vast majority of the overlapped areas within the protected zones encompassed only one or two CR fish species per 1 km^2: 78.4% and 17.2%, respectively (Figure 6b), while three or four CR fish species per 1 km^2 covered considerably smaller areas (4.4% in total; Figure 6b). In addition, the area coverage of moderate to high NIFS richness was similar within the unprotected (49.3%) and protected areas (44.5%) (Figure 6c,d), potentially indicating that no effective preventive measures are applied to prevent NIFS spread in protected areas.

Table 4. The distributional coverage and the overlapping areas per critically endangered freshwater fish species for each NIFS predicted richness category (low, moderate, high).

Species	Area (km^2)	Area per Category (%)		
		Low (0–1.40)	Moderate (1.5–2.60)	High (2.7–4.3)
Acipenser naccarii	1026	85.96	14.04	-
Acipenser stellatus	11,614	31.90	48.90	19.20
Acipenser sturio	7369	29.00	58.30	12.70
Alburnus macedonicus	277	-	53.43	46.57
Alburnus vistonicus	4187	97.04	2.96	-
Alosa vistonica	433	94.46	5.54	-
Aphanius almiriensis	153	63.40	36.60	-
Barbus euboicus	562	20.28	65.30	14.41
Barbus pergamonensis	461	100	-	-
Caspiomyzon graecus	602	96.51	3.32	0.17
Caspiomyzon hellenicus	499	60.32	39.68	-
Cobitis stephanidisi	1296	-	94.21	5.79
Eudontomyzon sp. Almopaios	1473	91.99	8.01	-
Huso huso	292	20.21	79.79	-
Knipowitschia goerneri	301	100	-	-
Knipowitschia milleri	403	100	-	-
Oxynoemacheilus theophilii	100	100	-	-
Pelasgus epiroticus	400	9.50	18.50	72.00
Pelasgus laconicus	3120	87.66	12.34	-
Pungitius hellenicus	737	100	-	-
Salaria economidisi	202	-	83.17	16.83
Scardinius graecus	2343	30.99	69.01	-
Squalius sp. Evia	53	-	100	-
Valencia letourneuxi	2152	90.20	9.80	-
Valencia robertae	2428	28.54	66.85	4.61

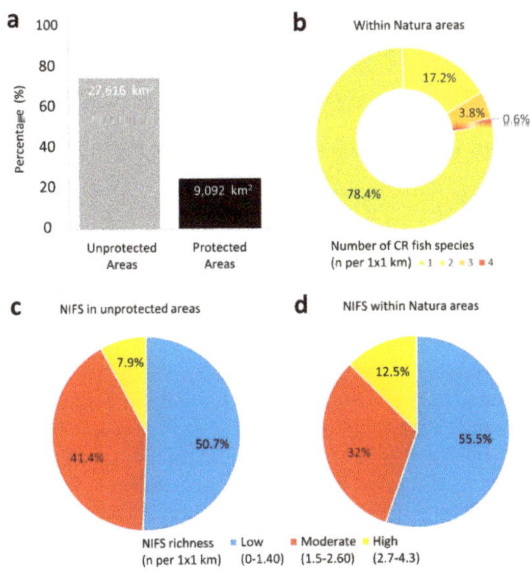

Figure 6. (**a**) The overlapping areas of NIFS and CR freshwater fish species in unprotected areas (grey) and protected areas (black); (**b**) the area coverage (%) of CR freshwater fish species richness within the protected areas; (**c**) the area coverage (%) of NIFS richness in unprotected areas; and (**d**) within protected areas.

4. Discussion

Until fairly recently, Greece was a country that had relatively few alien fish species compared to other European states [23]. However, nowadays, whole riverine fish assemblages are rapidly changing as non-indigenous fish species spread into several lotic and lentic ecosystems [24]. This study aimed to identify the potential overlapping areas of NIFS against fish species that are under the highest extinction threat category (CR) in Greece. The findings of this study demonstrate that many freshwater ecosystems in the country are under significant invasion pressure and concurrently exhibit high biodiversity conservation value; this problem is now pressing throughout the European Mediterranean basins [25,26].

Predictive patterns of NIFS richness varied largely by utilizing different data sources (i.e., bibliographic data for lentic and field surveys for lotic). These immense differences in NIFS predicted richness between lotic and lentic ecosystems can be attributed to several factors. For instance, bibliographic data usually overestimate the total species richness of an ecosystem, since species that were once reported could now have been extirpated. This is actually true for aliens in several Greek lentic systems, which were once stocked with fishes that were unable to develop viable populations (i.e., *Acipenser baeri*, *Acipenser gueldenstaedtii*, *Ctenopharyngodon idella*, *Hypophthalmichthys nobilis*, *Oncorhynchus kisutch*, *Oncorhynchus mykiss*) [17]. Thus, these species will be included in checklists and eventually in future analyses despite the fact that they could be currently extirpated. In this study, our combined analysis is indicative of the situation at the country-wide level, as far as possible, in all inland freshwater ecosystems. Correspondingly, results of how bibliographical data may depict overestimated NIFS richness in Greece were reported by a recent study [22], indicating that alien fish species were relatively restricted in Greek rivers, based on sampling data in contrast with bibliographic data.

Similar to other recent studies, the most common NIFS within the overlapping areas were Greece's four most widespread and abundant aliens *G. holbrooki*, *C. gibelio*, *L. gibbosus*, and *P. parva* as well as the translocated *C. carpio*, *Salmo farioides*, and *Economidichthys pygmaeus* [14,22]. Evidently, the most recorded NIFS display limnophylic (lacustrine) life history strategies. Indeed, apart from the introductions of cold-water species such as salmonids which are primarily conducted in the upper catchments of rivers and streams, most species introductions in Greece concern warm-water lacustrine species in lowland rivers and lentic ecosystems [27]. In most sites, NIFS richness comprised single species; however, in a single case NIFS richness raised up to 12 fish species (Lake Pamvotis), indicating high invasive pressure. Historically, Lake Pamvotis included only four native species; however, during the 1930s until the late 1990s several NIFS were introduced for purposes of eutrophication control or fisheries enhancement [24]. Overall, lowland riverine areas and lentic ecosystems with well-developed commercial and recreational fisheries indicated higher NIFS richness in comparison with small streams in higher altitudes or in arid regions (e.g., southern Greece and the Aegean islands).

According to the latest Red List assessment at the European level, more than 37% of the freshwater fish species are considered as threatened; 15% as Vulnerable, 10% as Endangered, and 12% as CR status [11]. The IUCN analysis shows that Greece hosts the most species under a threatened status and the most critically endangered freshwater fish species in Europe [28]. Despite this fact, few conservation actions have been applied to protect the CR freshwater fish species of Greece, while how NIFS affect their viability is almost totally neglected. According to current distributional records, two CR species (*P. epiroticus* and *C. stephanidisi*) co-occurring in areas with moderate to high NIFS richness are considered on the brink of extinction [17], while an additional species (*Alosa vistonica*) which indicated low to moderate overlapping areas with the distributions of NIFS has recently been assessed as extinct [29].

Despite the fact that our analysis utilizes broad-scale data of NIFS and CR fish species inhabiting both lentic and lotic ecosystems, we acknowledge its limitations by not incorporating the connectivity of aquatic ecosystems into the approach [30,31]. Generally, geostatistical techniques are suitable for modeling features that are inherently continuous [32], which contrasts with the fragmentation observed in freshwater ecosystems and the geographic isolation of the islands. However, in an attempt to mitigate the latter issues we used an extensive sampling/site network, thus narrowing the number of maximum neighbors and minimizing the distance points used to perform the interpolation. Furthermore, we acknowledge that some formerly widespread species, with marginal distributions in the country, are now probably totally extirpated as wild native populations from Greece. For instance, regions which include wild populations of sturgeons (*Huso huso*, *Acipenser sturio*, *Acipener naccarri*, *Acipenser stellatus*) [17] may have an overestimated distributional coverage bias of CR fish species.

Even though there has been a rapid growth in developing protected areas worldwide [33], such efforts are not usually optimally designed for freshwater biodiversity. Many conservationists accept that freshwater conservation must relate to a separate ecological realm, beyond the terrestrial or marine, whose specific recognition may have important consequences for both biodiversity conservation and the wider water management issues [34]. A recent study that aimed to overlap the ranges of threatened species, based on the IUCN Red List and Natura 2000 delineations in Greece [35], reported that species belonging to the class of Actinopterygii (including exclusively freshwater species) were covered fairly well within protected areas. However, the aforementioned study also included freshwater fish species within the threatened categories of Endangered and Vulnerable, thus this increase in spatial overlap with protected areas of the Natura 2000 sites is expected, since more species occupy wider areas. In our study, by focusing on the highest extinction threat category (CR), we revealed that the largest part of the distributional ranges of CR freshwater fish species fall outside protected areas. Finally, it is evident, based on the absence of relevant scientific literature, that conservation actions targeting CR freshwater fish species in Greece have rarely been applied. In fact, out of the 25 CR freshwater fish species, concrete conservation actions have been applied only for four species mainly by conducting conservation translocations (*A. naccarii*, [36]; *Pungitius hellenicus*, [37], *V. letourneuxi*, and *V. robertae*; [38]).

IUCN and national red listing assessment schemes are scheduling to modify a number of listed species categories as new reviews of vulnerability status are revised (i.e., some species that are Critical may be downgraded to Endangered and so forth). Although the new red listing revisions may alter the status of some species, in our opinion, most of the species utilized in our study should remain as important guiding species for conservation planning. The Hellenic Zoological Society's Greek Red List procedure was last reviewed in 2009, lagging far behind many other EU states; efforts to revise it have recently been initiated through funding scheduled from the Hellenic Ministry of Environment and Energy. Recently, a new review of Mediterranean freshwater fish species has also begun by IUCN experts and is expected to be completed in 2021. In the current work, we solely utilized the highest extinction threat level category as a proxy indicator for identifying areas of outstanding conservation interest in a rapid screening process that may be easily repeated and transferable to other states and for future monitoring.

Conservation planning in its entirety is a highly complex process encompassing several steps in order to develop and implement the protection, conservation, and enhancement of natural resources. As freshwater protected area effectiveness is usually challenging [39], especially regarding NIFS threats [40], the need for well-designed protected areas and management plans for freshwater biodiversity is a necessity for critically endangered fishes and the ecosystems that host them in Greece. Unraveling the spatial overlap between NIFS and critically endangered fish species can support the first stages of targeted conservation planning and contribute to preliminary actions to minimize the potential negative effects. Future studies should aim to a) identify and assess broader important fish areas for conservation and b) assess how the combined effects of various stressors (i.e., NIFS, water abstraction, pollution, river fragmentation, etc.) can affect the populations of the threatened freshwater fish species of Greece under a climate change context. Moreover, there is a need for high-quality comprehensive reviews regarding the ichthyofaunal compositions of Greek lentic ecosystems, which will exclude all species with questionable occurrences and clarify the actual and current status of NIFS introductions. Finally, future work should also address the analysis of species traits of the mixed fish assemblages (native and NIFS) to better understand the functional organization of these novel ecosystems.

Author Contributions: Conceptualization, N.K., L.V., and S.Z.; methodology, N.K., L.V., and S.Z.; formal analysis, N.K. and L.V.; investigation and data curation, N.K., S.Z., and L.V.; writing—original draft preparation, N.K., S.Z., and L.V.; writing—review and editing, N.K., S.Z., L.V., and O.-I.K.; visualization, N.K. and L.V.; supervision, O.-I.K. and S.Z.; project administration, O.-I.K., N.K., and S.Z. All authors have read and agreed to the published version of the manuscript.

Funding: This research is co-financed by Greece and the European Union (European Social Fund-ESF) through the Operational Programme «Human Resources Development, Education and Lifelong Learning 2014–2020» in the context of the project "Interdisciplinary assessment of non-indigenous fish fauna in the freshwater ecosystems of Greece" (MIS 5047785).

Institutional Review Board Statement: Not applicable.

Data Availability Statement: Not applicable.

Conflicts of Interest: The authors declare no conflict of interest. The founding sponsors had no role in the design of the study; in the collection, analyses, or interpretation of data; in the writing of the manuscript; and in the decision to publish the results.

References

1. Ceballos, G.; Ehrlich, P.R.; Barnosky, A.D.; García, A.; Pringle, R.M.; Palmer, T.M. Accelerated modern human–induced species losses: Entering the sixth mass extinction. *Sci. Adv.* **2015**, *1*, e1400253. [CrossRef] [PubMed]
2. Johnson, C.N.; Balmford, A.; Brook, B.W.; Buettel, J.C.; Galetti, M.; Guangchun, L.; Wilmshurst, J.M. Biodiversity losses and conservation responses in the Anthropocene. *Science* **2017**, *356*, 270–275. [CrossRef]
3. Turvey, S.T.; Crees, J.J. Extinction in the Anthropocene. *Curr. Biol.* **2019**, *29*, R982–R986. [CrossRef]
4. Duncan, J.R.; Lockwood, J.L. Extinction in a field of bullets: A search for causes in the decline of the world's freshwater fishes. *Biol. Conserv.* **2001**, *102*, 97–105. [CrossRef]
5. Darwall, W.; Smith, K.; Allen, D.; Seddon, M.; Mc Gregor Reid, G.; Clausnitzer, V.; Kalkman, V. Freshwater biodiversity—A hidden resource under threat. In *The 2008 Review of the IUCN Red List of Threatened Species*; Vié, J.C., Hilton-Taylor, C., Stuart, S.N., Eds.; IUCN: Gland, Switzerland, 2009; pp. 43–53.
6. Schinegger, R.; Palt, M.; Segurado, P.; Schmutz, S. Untangling the effects of multiple human stressors and their impacts on fish assemblages in European running waters. *Sci. Total Environ.* **2016**, *573*, 1079–1088. [CrossRef]
7. Jarić, I.; Heger, T.; Monzon, F.C.; Jeschke, J.M.; Kowarik, I.; McConkey, K.R.; Pyšek, P.; Sagouis, A.; Essl, F. Crypticity in biological invasions. *Trends Ecol. Evol.* **2019**, *34*, 291–302. [CrossRef] [PubMed]
8. Deinet, S.; Scott-Gatty, K.; Rotton, H.; Twardek, W.M.; Marconi, V.; McRae, L.; Baumgartner, L.J.; Brink, K.; Claussen, J.E.; Cooke, S.J.; et al. *The Living Planet Index (LPI) for Migratory Freshwater Fish*; Technical Report; World Fish Migration Foundation: Groningen, The Netherlands, 2020.
9. Piria, M.; Copp, G.H.; Dick, J.T.; Duplić, A.; Groom, Q.; Jelić, D.; Caffrey, J.M. Tackling invasive alien species in Europe II: Threats and opportunities until 2020. *Manag. Biol. Invasions* **2017**, *8*, 273–286. [CrossRef]
10. Darwall, W.; Carrizo, S.; Numa, C.; Barrios, V.; Freyhof, J.; Smith, K. *Freshwater Key Biodiversity Areas in the Mediterranean Basin Hotspot: Informing Species Conservation and Development Planning in Freshwater Ecosystems*; IUCN: Cambridge, UK, 2014.
11. Freyhof, J.; Brooks, E. *European Red List of Freshwater Fishes*; Publications Office of the European Union: Luxembourg, 2011.
12. Economou, A.N.; Giakoumi, S.; Vardakas, L.; Barbieri, R.; Stoumboudi, M.T.; Zogaris, S. The freshwater ichthyofauna of Greece-an update based on a hydrographic basin survey. *Mediterr. Mar. Sci.* **2007**, *8*, 91–166. [CrossRef]
13. Economidis, P.S.; Banarescu, P.M. The distribution and origins of freshwater fishes in the Balkan Peninsula, especially in Greece. *Int. Rev. Ges. Hydrobiol. Hydrogr.* **1991**, *76*, 257–284. [CrossRef]
14. Koutsikos, N.; Vardakas, L.; Vavalidis, T.; Kalogianni, E.; Dimitriou, E.; Kalantzi, O.I.; Zogaris, S. Defining non-indigenous fish assemblage types in Mediterranean rivers: Network analysis and management implications. *J. Environ. Manag.* **2021**, *278*, 111551. [CrossRef]
15. Moyle, P. Novel aquatic ecosystems: The new reality for streams in California and other Mediterranean climate regions river research and applications. *River Res. Appl.* **2013**, *30*, 1335–1344. [CrossRef]
16. Buoro, M.; Olden, J.D.; Cucherousset, J. Global Salmonidae introductions reveal stronger ecological effects of changing intraspecific compared to interspecific diversity. *Ecol. Lett.* **2016**, *19*, 1363–1371. [CrossRef] [PubMed]

17. Barbieri, R.; Zogaris, S.; Kalogianni, E.; Stoumboudi, M.T.; Chatzinikolaou, Y.; Giakoumi, S.; Kapakos, Y.; Kommatas, D.; Koutsikos, N.; Tachos, V.; et al. *Freshwater Fishes and Lampreys of Greece: An Annotated Checklist. Monographs on Marine Sciences No. 8*; HCMR: Athens, Greece, 2015.
18. Chatzinikolaou, Y.; Zogaris, S.; Leonardos, I.; Bobori, D.; Koutrakis, E.; Tachos, V.; Anagnopoulos, N.; Loukatos, A.; Konstantinidis, E.; Bourdaniotis, N.; et al. Geographical mapping in GIS of potential distribution of each species and documented technical report. In *The Framework of the Project "Monitoring and Evaluation of Conservation Status of EU's Interest Icthyofanuna Species in Greece"*; GMEECC, Joint Venture Anagnopoulos Nikolaos–EPEM S.A.–Konstantinidis Ilias: Athens, Greece, 2014. (In Greek)
19. Zogaris, S.; Tachos, V.; Economou, A.N.; Chatzinikolaou, Y.; Koutsikos, N.; Schmutz, S. A model-based fish bioassessment index for Eastern Mediterranean rivers: Application in a biogeographically diverse area. *Sci. Total Environ.* **2018**, *622*, 676–689. [CrossRef] [PubMed]
20. Schmutz, S.; Cowx, I.G.; Haidvogl, G.; Pont, D. Fish-based methods for assessing European running waters: A synthesis. *Fish. Manag. Ecol.* **2007**, *14*, 369–380. [CrossRef]
21. Economou, A.N.; Zogaris, S.; Vardakas, L.; Koutsikos, N.; Chatzinikolaou, Y.; Kommatas, D.; Tachos, V. Developing policy-relevant river fish monitoring in Greece: Insights from a nation-wide survey. *Mediterr. Mar. Sci.* **2016**, *17*, 302–322. [CrossRef]
22. Koutsikos, N.; Zogaris, S.; Vardakas, L.; Kalantzi, O.I.; Dimitriou, E.; Economou, A.N. Tracking non-indigenous fishes in lotic ecosystems: Invasive patterns at different spatial scales in Greece. *Sci. Total Environ.* **2019**, *659*, 384–400. [CrossRef] [PubMed]
23. Bianco, P.G. Potential role of the palaeohistory of the Mediterranean and Paratethys basins on the early dispersal of Euro-Mediterranean freshwater fishes. *Ichthyol. Explor. Freshw.* **1990**, *1*, 167–184.
24. Leonardos, I.D.; Kagalou, I.; Tsoumani, M.; Economidis, P.S. Fish fauna in a Protected Greek lake: Biodiversity, introduced fish species over a 80-year period and their impacts on the ecosystem. *Ecol. Freshw. Fish.* **2008**, *17*, 165–173. [CrossRef]
25. Sommerwerk, N.; Wolter, C.; Freyhof, J.; Tockner, K. Components and drivers of change in European freshwater fish faunas. *J. Biogeogr.* **2017**, *44*, 1781–1790. [CrossRef]
26. Grapci-Kotori, L.; Vavalidis, T.; Zogaris, D.; Šanda, R.; Vukić, J.; Geci, D.; Ibrahimi, H.; Bilalli, A.; Zogaris, S. Fish distribution patterns in the White Drin (Drini i Bardhë) river, Kosovo. *Knowl. Manag. Aquat. Ecosyst.* **2020**, *421*, 29. [CrossRef]
27. Piria, M.; Simonović, P.; Kalogianni, E.; Vardakas, L.; Koutsikos, N.; Zanella, D.; Joy, M.K. Alien freshwater fish species in the Balkans—Vectors and pathways of introduction. *Fish. Fish.* **2018**, *19*, 138–169. [CrossRef]
28. International Union for Conservation of Nature's Red List of Threatened Species. Available online: https://www.iucnredlist.org/ (accessed on 17 January 2021).
29. Bobori, D. *Research and Proposals for the Protection of Near-Extinct Endemic Species of the Ichthyofaunal in Lakes Vistonis and Mitrikou (Alosa vistonica & Alburnus vistonicus)*; Management Body of Delta Nestos-Vistonis-Ismarida: Thessaliniki, Greece, 2015; Unpublished Report. (In Greek)
30. Moilanen, A.; Leathwick, J.; Elith, J. A method for spatial freshwater conservation prioritization. *Freshw. Biol.* **2008**, *53*, 577–592. [CrossRef]
31. Beger, M.; Grantham, H.; Pressey, R.L. Conservation planning for connectivity across marine, freshwater, and terrestrial realms. *Biol. Conserv.* **2010**, *143*, 565–575. [CrossRef]
32. Hengl, T.; Sierdsema, H.; Radović, A.; Dilo, A. Spatial prediction of species' distributions from occurrence-only records: Combining point pattern analysis, ENFA and regression-kriging. *Ecol. Modell.* **2009**, *220*, 3499–3511. [CrossRef]
33. Pimm, S.L.; Jenkins, C.N.; Abell, R.; Brooks, T.M.; Gittleman, J.L.; Joppa, L.N.; Sexton, J.O. The biodiversity of species and their rates of extinction, distribution, and protection. *Science* **2014**, *344*, 6187. [CrossRef] [PubMed]
34. van Rees, C.B.; Waylen, K.A.; Schmidt-Kloiber, A.; Thackeray, S.J.; Kalinkat, G.; Martens, K.; Domisch, S.; Lillebø, A.I.; Hermoso, V.; Grossart, H.-P.; et al. Safeguarding freshwater life beyond 2020: Recommendations for the new global biodiversity framework from the European experience. *Conserv. Lett.* **2020**, *14*, e12771. [CrossRef]
35. Spiliopoulou, K.; Dimitrakopoulos, P.G.; Brooks, T.M.; Kelaidi, G.; Paragamian, K.; Kati, V.; Oikonomou, A.; Vavylis, D.; Trigas, P.; Lymberakis, P.; et al. The Natura 2000 network and the ranges of threatened species in Greece. *Biodivers. Conserv.* **2021**, *30*, 1–17. [CrossRef]
36. Paschos, I.; Nathanailides, C.; Kagalou, I.; Leka, E.; Tsoumani, M.; Perdikaris, C. The prospects for restoring the nearly extinct population of the Adriatic sturgeon *Acipenser naccarii* Bonaparte 1836 (Acipenseridae) in Greece. *Aqua. J. Ichthyol. Aquat. Biol.* **2003**, *7*, 123–132.
37. Zogaris, S.; Skoulikidis, N.; Dimitriou, E. River and wetland restoration in Greece: Lessons from biodiversity conservation ini-tiatives. In *The Rivers of Greece*; Skoulikidis, N.T., Dimitriou, E., Karaouzas, I., Eds.; Springer: Berlin/Heidelberg, Germany, 2017.
38. Kalogianni, E.; Koutsikos, N.; Vardakas, L.; Giakoumi, S.; Chatzinikolaou, Y.; Oikonomou, A. Impacts of the alien mosquitofish on the abundance and condition of two Mediterranean native fish. *Med. Mar. Sci.* **2019**, *20*, 727–735. [CrossRef]
39. Acreman, M.; Hughes, K.A.; Arthington, A.H.; Tickner, D.; Dueñas, M.A. Protected areas and freshwater biodiversity: A novel systematic review distils eight lessons for effective conservation. *Conserv. Lett.* **2020**, *13*, e12684. [CrossRef]
40. Ruchin, A.B.; Osipov, V.V.; Fayzulin, A.I.; Bakin, O.V.; Tselishcheva, L.G.; Bayanov, N.G. Chinese sleeper (*Perccottus glenii* Dybowski, 1877)(Pisces, Odontobutidae) in the reserves and National Parks of the middle and lower Volga (Russia): Mini-review. *Aquac. Aquar. Conserv. Legis.* **2019**, *12*, 1114–1124.

diversity

MDPI

Article

Diversity of Algae and Cyanobacteria and Bioindication Characteristics of the Alpine Lake Nesamovyte (Eastern Carpathians, Ukraine) from 100 Years Ago to the Present

Petro M. Tsarenko [1], Olena P. Bilous [2,*], Olha M. Kryvosheia-Zakharova [1], Halyna H. Lilitska [1] and Sophia Barinova [3]

1 M.G. Kholodny Institute of Botany, National Academy of Sciences of Ukraine, Tereschenkivska Str. 2, 01004 Kyiv, Ukraine; ptsar@ukr.net (P.M.T.); olha_krivosheia@ukr.net (O.M.K.-Z.); dunaliella@ukr.net (H.H.L.)
2 Institute of Hydrobiology, National Academy of Sciences of Ukraine, Geroiv Stalingrada 12, 04210 Kyiv, Ukraine
3 Institute of Evolution, University of Haifa, Abba Khoushi Ave, 199, Mount Carmel, Haifa 3498838, Israel; sophia@evo.haifa.ac.il
* Correspondence: bilous_olena@ukr.net; Tel.: +380-977810018

Abstract: The species diversity and changes in the structural dynamics of the algal flora from the alpine lake Nesamovyte has been studied for 100 years. During the period of investigations, 234 species (245 infraspecific taxa) were revealed to cover more than 70% of the modern species composition of the studied lake. The modern biodiversity of algae is characterized by an increase in the number of widespread forms, a change from the baseline "montane" complex in comparison to the beginning of the 20th century. Nevertheless, the Nesamovyte Lake still has a unique algae composition that is typical for high-mountainous European lakes. The presence of a different complex of conventionally arctic species of algae, in particular, diatoms is discussed. Structural changes in the taxonomic composition of the algal flora of the lake as well as in the complex of the leading genera, species and their diversity are revealed. An ecological analysis of the algal species composition of the lake showed vulnerability and degradation to the ecosystem of the lake. On this basis, the issue regarding the question of protection and preservation of the algae significance and uniqueness of the flora of algae in the Nesamovyte Lake are discussed.

Keywords: diversity; algae; alpine lake; bioindication; ecological characteristic; ecosystem; Nesamovyte Lake; Eastern Carpathians

Citation: Tsarenko, P.M.; Bilous, O.P.; Kryvosheia-Zakharova, O.M.; Lilitska, H.H.; Barinova, S. Diversity of Algae and Cyanobacteria and Bioindication Characteristics of the Alpine Lake Nesamovyte (Eastern Carpathians, Ukraine) from 100 Years Ago to the Present. *Diversity* **2021**, *13*, 256. https://doi.org/10.3390/d13060256

Academic Editor: Michael Wink

Received: 29 April 2021
Accepted: 5 June 2021
Published: 8 June 2021

1. Introduction

The lakes in the mountainous regions of the Eastern Carpathians (as well as in similar areas in the Alpine-Carpathian mountain system) play a leading role in the formation of the algal flora of the region and its conservation of rare species, and are indicators of the local ecosystems. These lakes contain specific regional complexes of algae species that were formed under specific conditions: the origin of the definite waterbodies, their trophic state, water chemistry and the level of anthropogenic load [1–15]. Other comprehensive studies of lakes in the mountain systems of Europe (the Alps, Pyrenees, Balkans, Tatras, Western Carpathians) are considered to be relevant and the results of the studies are used in international projects such as European Mountain lake Ecosystems: Regionalisation, diaGnostics & Socio-economic Evaluation (EMERGY), Alpine Lakes: Paleolimnology and Ecology (AL:PE2), the European Mountain Lake Research (MOLAR) project and many others [6,16–28]. These results have also been implemented in the EU Water Framework Directive 2000/60 [29]. The basis of the paradigm in studying mountain lake ecosystems is to gain knowledge regarding the genesis of waterbodies of this type, stratigraphy and paleolimnology, chemistry and general patterns of their functioning, the presence and conservation of biotic component and the type of waterbody itself and ecosystems in general, climate impact and its further transformation, as well as socio-economical aspects, etc. However, the basic question still remains regarding the study and analysis of aquatic organisms—their diversity, productivity, relationships, indicators, species specificity and uniqueness. Species diversity of these organisms in general and algae in particular in the Central and Southern European ecosystems of mountain lakes were carefully studied during the end of the XIX–XXth century [30–43]. The species composition of algae found in the mountain lakes in the Carpathian-Tatra region (western part of the Carpathian mountain system) was studied in detail [9,10,35,37,44–60]. However, the eastern part of the region (Eastern Carpathians and especially the Ukrainian Carpathians, which cover an area of 24,000 km^2 and have a length of 280 km, from the Polish to the Romanian border) is characterized by incomplete data regarding the algal composition and the lack of information according to individual taxonomic groups and types of waterbodies so far [61]. The situation has improved since the signing in 2014 of the Association Agreement between Ukraine and the European Union and the implementation of the provisions and standards of the Water Framework Directive of the European Union [29], in particular, those related to hydro-biological and hydro-morphological assessment of the water bodies and water management practices of Ukraine. In the last decade, several works had been published reporting the species diversity of the lakes in the forest zone of the Ukrainian Carpathians (Synevyr, Gropa, Maricheika, Hirske Oko and some others) and some information on the high-altitude lakes in the foothills of this mountain region [11,12,59,62–64]. Current intensified recreational pressure on the lakes and general hydrological changes in the region allow us to form a scientific working hypothesis assuming that the ecosystem of the lakes in the region is changing. The additional scientific question is whether the new results of ecological analysis based on a full taxonomic list of algae show any changes in comparison to previous results, focused on some groups of algae [61,65].

The importance of this study is supported by a threat to the existence of oligo-mesotrophic highland waterbodies and their peculiar species composition of aquatic organisms [11,66]. To confirm the hypothesis, the available data on algae species of the Nesamovyte Lake (Chornohora mountain group) were summarized from scientific papers covering over 100 years of research [2,3,63,67].

This study aims to document the algal and cyanobacterial composition of the Nesamovyte Lake using three datasets collected in the last hundred years and outlining the changes in the ecological state of the lake due to the bio-indication characteristics of the discovered species composition.

2. Materials and Methods

The Nesamovyte Lake is one of the highest mountain waterbodies of its type in Ukraine, located in the central, highest mountainous part of the Eastern (Ukrainian) Carpathians– Polonyno-Chornogora geomorphological area [68], belonging to Carpathian-Danube algofloristic sub-provinces according to algofloristic zoning of Ukraine [69]. The lake is located in the Carpathian National Nature Park (48°07′36.6″ N, 24°32′26.4″ E) and is situated in the glacial mountain trench on the eastern slope of Mount Turkul (Chornohora Ridge, Eastern Carpathians, and is approximately 5 km from the highest mountain in Ukraine—Hoverla) at an altitude of 1 748 m above sea level (a.s.l.) belongs to the water basin of the Prut River (Figure 1). According to the origin of the lake, it can be called a polymictic tarn that refers the following explanation 'a proglacial mountain lake, pond or pool, formed in a cirque excavated by a glacier' [70].

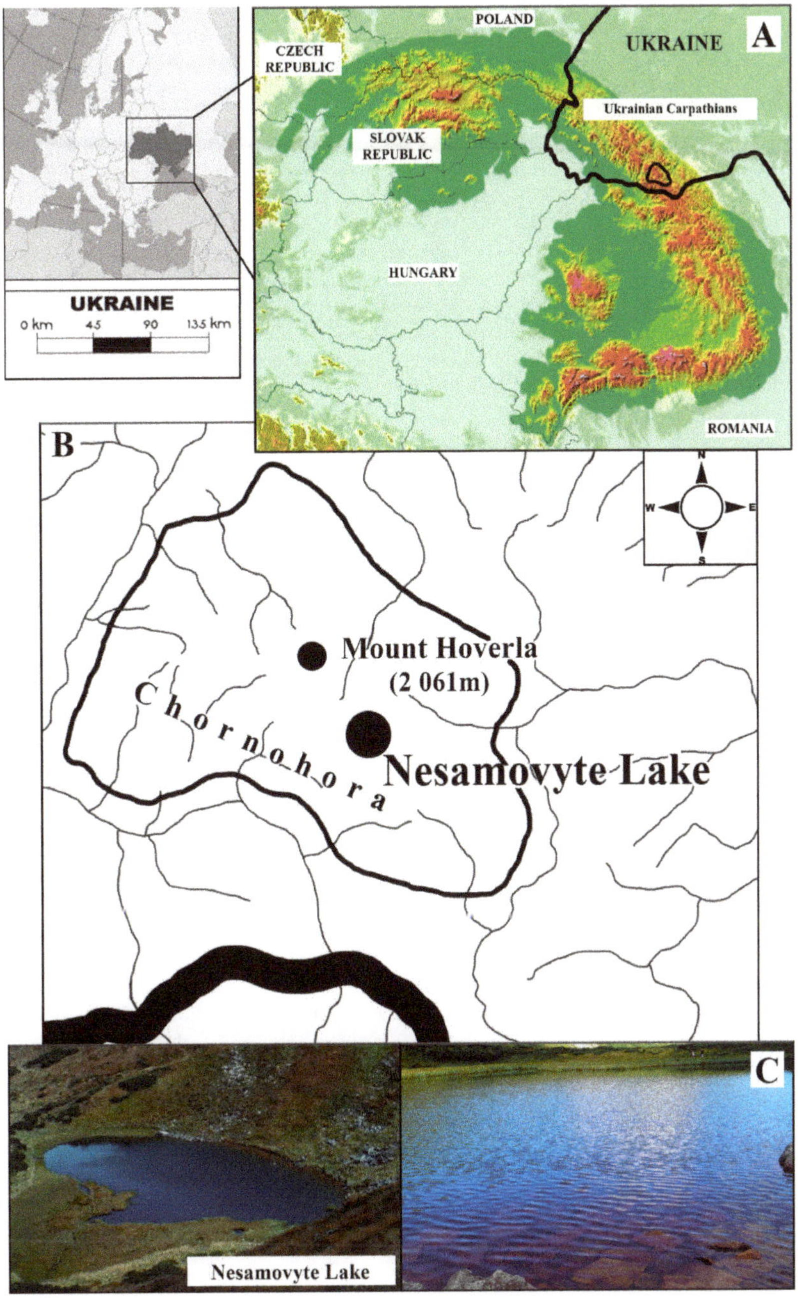

Figure 1. Map of the Carpathian region and the location of the Nesamovyte Lake. (**A**)—the Carpathian Mountains with its Eastern part the Ukrainian Carpathians. (**B**)—map-scheme of Chornohora Massif with the location of the lake. (**C**)—littoral zone of the lake.

The lake covers an area of about 0.4 hectares (about 88 × 45 m) and has a depth of about 2 m and, according to the EC Water Framework Directive [29], belongs to the category of small and very small lakes. In general, the lake water is generated by rain and melting snow with a sandy muddy bottom (rocky from geological rocks of sandy flysch near the southern and western coast and sandy-muddy—in the northern part), it has a long period of freezing (October-May up to a depth of 1.5 m). It belongs to the cold climatic zone of the Carpathians. The summer water temperature normally is 18.3–(12.4) ±3.2 °C. The lake has a slow underground runoff in the north-western direction, through a narrow channel on the alloy (swamp intrusion). This part of the lake is intensively overgrown by *Carex rostrata* Stokes and *Sphagnum* spp. Other moss-like organisms and sedge-sphagnum alloy cover the surface of the lake up to 35 cm from the shore and occupies about 40% of the lake bed area and reaches 1 m in thickness. The water in the lake corresponds now to the hydrocarbonate-sulphate-sodium class (Table 1) and belongs to zones of saprobity from oligo- to β-mesosaprobic [71–75]. The hydrochemical formula of water of the lake [76] is as follows: M0.018 SO4 HCO3 42/Na 79 Ca 21. According to EUNIS classification, [77], the Nesamovyte lake belongs to biotopes C1.1 or B1.1.1 (permanent oligotrophic lakes, ponds, puddles).

The material for this study was built on the analysis of the first record (1910–1920) for the Nesamovyte Lake published by Wołoszyńska, work with samples from 1967–1978 (collectors Prof. Z.I. Asaul and Prof. G.M. Palamar-Mordvintseva) and our previous investigations (2013–2018).

The species list provided by Wołoszyńska (collected in 1910) [67] was carefully analysed and describes the period between 1910–1920.

From 1967 to 1978, samples from Algoteca funds of M.G. Kholodny Institute of Botany of NAS of Ukraine (AKW)—NN 16855–16898 (samples from the year 1967–1978, collectors Prof. Z.I. Asaul and Prof. G.M. Palamar-Mordvintseva) were studied (44 samples). In addition, the data based on these samples published from the period of 1967–1978 [2–5] were included in the current analysis.

To characterize the modern period (2013–2021), 18 samples of net plankton (50–100 L of water filtered out), periphyton from vascular plants (dead and living parts of herbaceous plants (*Carex rostrata*) and pine branches (*Pinus mugo* Turra) and squeezes from the moss (*Sphagnum* spp.) were collected along the perimeter of the lake in August 2013–2018. To study the species composition of *Bacillariophyta*, the method of forming a combined sample from different studied substrates was used. This study is based on the living algal material from plankton (algal cultures, with the addition of BBM-medium) [78] and samples that were fixed with 4% formaldehyde solution.

The obtained results are comparable because almost identical methods for collecting and fixation of the algal material were used in all the studied periods (1910–1920, 1967–1978, 2013–2021).

The algae were studied and identified using light (LM) and scanning electron (SEM) microscopy, the permanent slides of diatoms were made according to the standard procedure using 35% H_2O_2 [79]. For LM investigations, the diatoms were fixed in the synthetic mounting medium Naphrax (refractive index 1.74) and investigated under a BX-53 microscope (Olympus, Tokyo, Japan). The slides are stored in the Algotheca of M.G. Kholodny Institute of Botany, NAS of Ukraine (AKW). For SEM analysis, the samples were put to specimen stubs, dried, covered with gold (10–20 nm) JFC-1600 sputter coater, and examined in a scanning electron microscope JSM-6060LA (JEOL, Tokyo, Japan) at the Institute of Botany in the Center of Electron Microscopy. The resulting micrographs were processed using the software packages Axiovision 4.3.7. (Carl Zeiss MicroImaging GmbH, Jena, Germany) and GIMP 2.8.10.m (Free Software Foundation, Inc., Boston, MA, USA).

Table 1. Hydrochemical characteristics of the Nesamovyte Lake [1].

Variable	pH	O_2, mg L^{-1}	Total Suspended Solids (TSS), mg L^{-1}	Total Dissolved Solids (TDS), ppm	Ca^{2+}, mg L^{-1}	Na^{+}, mg L^{-1}	HCO_3^{-}, mg L^{-1}	SO_4^{2-} mg L^{-1}	Cl^{-}, mg L^{-1}	NO_3, mg L^{-1}	Cd, μg L^{-1}	Cr, μg L^{-1}	Pb^{2+}, μg L^{-1}
Value	6.2–6.4 ± 0.4	7–10.7	1.6 ± 0.4	8.2–9.8 (12–98.6)	1.30 ± 0.3	1.93 ± 0.24	3.66 ± 0.145	3.98–15.0	0.71–1.42	2.0 ± 0.4	0.26	2.37	0.39

[1] according to [66,70,72] and our research.

Identification of the species diversity was carried out according the Süßwasserflora von Mitteleuropa [80–85] with some newer updates from Diatoms of Europe [86–89], Diatomeen im Süßwasser-Benthos von Mitteleuropa [90] and Flora of algae of Ukraine [91–97] as well as with updates from [98–102], and electronic resources [103]. The identified taxa, as well as all algal species lists from previous years of studies of the territory were validated using the AlgaeBase system [104] and "Algae of Ukraine ... " [105] monographic series.

Ecological bioindicator species analysis was based both on the historical data (1910–1920 and 1967–1978 collections) and the results of our studies (2013–2021) [2–4,61,63,65,67,106,107]. The following ecological characteristics were used: Habitat preference, streaming and oxygenation, pH [108], salinity, trophic state and class of organic pollution [109–112]. Identification of organic pollution by the values of saprobity indices and indicator groups, equated to water quality classes [113,114]. Considering the groups of indicators that we identified, the intervals of the quality classes were distributed as follows: I—0–0.5 (x, x-o); II—0.6–1.5 (o-x, x-b, o, o-b), III—(b-o, o-a, b, b-a) and IV—(a) [112,115–117]. Ecological features of the species were presented according to [112,117,118].

The Cluster analysis of algal composition was carried out using the Paleontological Statistics Software (PAST, Palaeontological Association, Hammer & Harper, Oslo, Denmark, Galway, Norway, Denmark, Ireland) to measure the degree to which species composition was similar among the studied periods (1910–1920, 1967–1978, 2013–2021) [119]. For this purpose, the presence/absence of data were used in the meaning of the Sørencen coefficient, calculated in the program as Bray-Curtis Similarity index [120].

3. Results

3.1. *Species Composition of Algae during 100-Year Investigated Period*

The taxonomic structure of the algal composition of the Nesamovyte Lake is a summary of the available data on the species composition of algae for 100 years of investigations during the following study periods: 1910–1920, 1967–1978 to 2013–2021 [2,4,61,63,65,67,106,121]. It is represented by 234 species (245 infraspecific taxa (inft) composing eight divisions, 15 classes, 33 orders, 55 families, 100 genera of Cyanobacteria and algae (Table 2, Figure 2).

Table 2. Systematical composition of algae of the Nesamovyte Lake (1910–2021).

Division	1910–1920			1967–1978			2013–2021			1910–2021		
	Genera	Sp (inft)	%	Genera	Sp (inft)	%	Genera	Sp (inft)	%	Genera	Sp (inft)	%
Cyanobacteria	2	2	3.0	–	–	–	3	3	1.8	5	5	2.1
Euglenozoa	–	–	–	6	10	14.6	5	8	4.8	8	12	5.1
Ochrophyta	1	1	1.5	2	3	4.9	1	1	0.6	3	4	1.7
Cryptophyta	–	–	–	1	1	2.4	2	2	1.2	2	2	0.9
Bacillariophyta	7	15 (16)	23.4	–	–	–	44	115 (117)	69.7	46	122 (125)	52.1
Miozoa	2	2	3.0	–	–	–	–	–	–	2	2	0.9
Chlorophyta	4	4	6.0	–	–	–	11	12	7.3	13	14	6.0
Charophyta	16	43 (45)	64.1	8	27 (31)	58.5	14	24 (25)	14.6	21	73 (81)	31.2
Total	32	67 (70)	100	17	41 (45)	100	80	165 (168)	100	100	234 (245)	100

The basis of this taxonomic and species diversity for the lake is formed by Bacillariophyta and Charophyta, uniting together giving about 57% generic diversity and over 83% species and infraspecific composition of algae in different habitats of the Nesamovyte Lake. Less diverse, but indicative for presenting the dynamics of changes in taxonomic structure over time, presented the divisions of Chlorophyta (6.0%) and Euglenozoa (5.1%). A slight variety of species is inherent for the divisions' Cyanobacteria and Ochrophyta (Chrysophyceae) (2.1–1.7%), as for Cryptophyta and Miozoa (Dinophyta)—they are noted as having a few representatives (>1.0%).

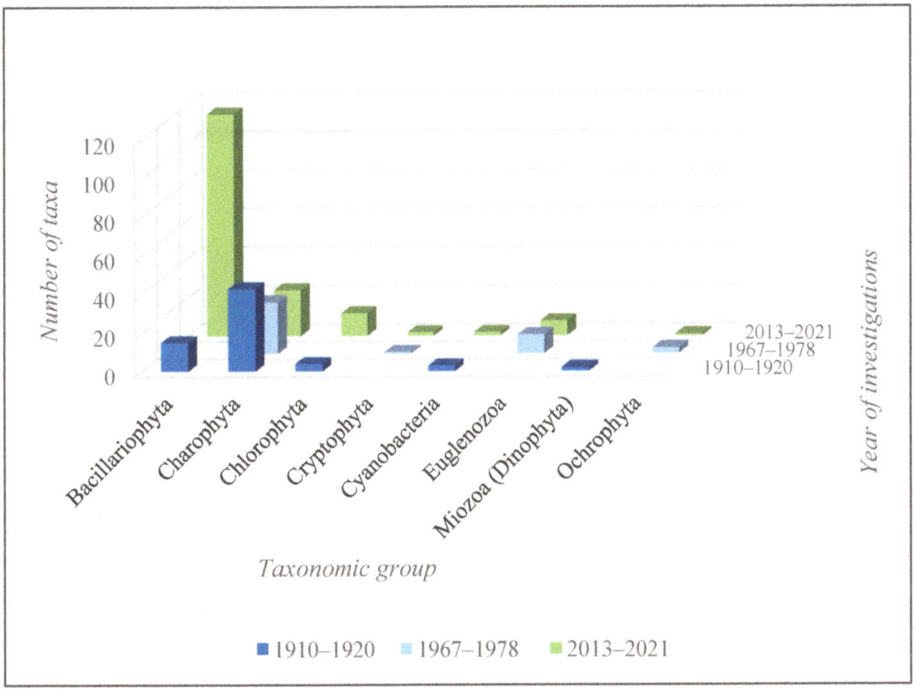

Figure 2. Species composition of algae and cyanobacteria in the Nesamovyte Lake (split by taxonomic groups and three stages of research in 1910–2021).

At the same time, the leading set of orders of these taxonomic groups is formed by representatives of Desmidiales (15 genera—66 species—75 inft), Naviculales (10 genera—50 species), Cymbellales (5–25), Achnanthales (5–11), Eunotiales (1–11), Sphaeropleales (6–7), Fragillariales (4–9), Zygnematales (6–6) and Euglenidida (2–5). These orders account for around 56% of the generic and species composition of algae in the Nesamovyte Lake. Leading genera of the algal composition are *Staurastrum* Meyen ex Ralfs (21–22 inft), *Pinnularia* Ehrenberg (14–16), *Euastrum* Ehrenberg ex Ralfs (12–15), *Navicula* Bory (12), *Cosmarium* Corda ex Ralfs (12), *Eunotia* Ehrenberg (11), *Gomphonema* Ehrenberg (10), *Encyonema* Kützing (5), *Planothidium* Round & L. Bukhtiyarova (5) and *Closterium* Nitzsch ex Ralfs (4). The named 10 genera cover about half (45.5%) of the whole diversity of species composition of this waterbody. However, almost 60% of genera are represented only by one species, which could serve as a characteristic feature of the studied lake flora.

3.2. Comparison between Studied Periods of Species Composition of Algae during 100-Year Investigated Period

Comparative floristic analyses for three periods of investigations showing the similarity of algae composition is presented in Figure 3. For building the graph, the species lists for the periods 1910–1920, 1967–1978 and 2013–2021 were compared. The graph demonstrates a low similarity between the studied periods, however, some higher similarity for 2013–2021 and 1910–1920 can be noted in contrast to the middle period of the investigations.

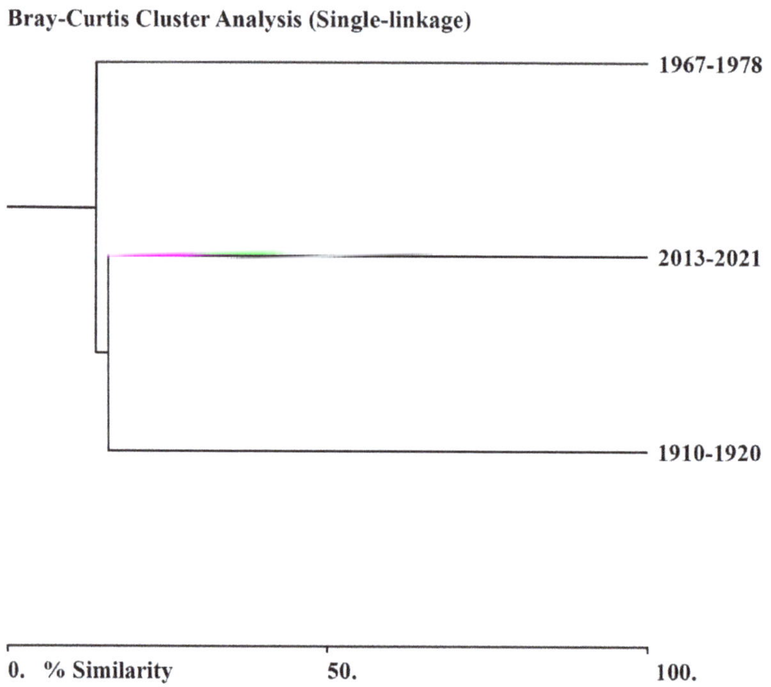

Figure 3. Bray-Curtis tree of species composition comparison in three periods of the Nesamovyte Lake study (1910–1920; 1967–1978; 2013–2021).

The first studies of the species composition of algae in this lake were conducted in the early twentieth century by Professor J. Wołoszyńska (based on the 1910 samples of Prof. M. Raciborski) [67]. With the help of this study, the presence of only 67 species (70 inft), in particular Bacillariophyta (15–16 inft), Charophyta (43–45 inft), Chlorophyta (4), Miozoa (Dinophyta) (2), Ochrophyta (Chrysophyceae) (1) and Cyanobacteria (2) were noted.

Leading taxonomic groups by species richness were charophytes (desmids—64.1%) and diatoms (23.4%), combining more than 87% of the whole composition of algae of this waterbody. At the same time, the author noted that diatoms do not make a "*significant contribution to the diversity of species composition of the lake*" ([67], p. 144) (probably meaning a much lower percentage of diatoms). Some distinguishing features for the uniqueness of the diatom composition at the genera level—*Eunotia*, *Pinnularia* and *Neidium* Pfitzer—were underlined.

Half a century later (1967–1978 studied period) the targeted studies of species diversity of euglenoid and desmids algae [2–4] from different ecotopes of the lake revealed 102 species of algae. The core divisions of Charophyta (66) and Euglenozoa (9), were added by different divisions presented by Cyanobacteria (3), Cryptophyta (1), Ochrophyta (2), Miozoa (2), Chlorophyta (4) and Bacillariophyta (15) (Table 2).

Our investigations of the modern composition of algae in the Nesamovyte Lake (2013–2021) confirm its species diversity (165 species—168 inft, belonging to 80 genera, 47 families, 28 orders, 11 classes and seven divisions—Table 2) and provide insights into the taxonomic composition and leading complexes of the species. According to the results of this study, the species composition was presented by Bacillariophyta (115 species—117 inft, namely ~70% of total species composition), Charophyta (24–25, ~15%) and Chlorophyta (12, ~7.3%), sparse—Euglenozoa (5–8, ~5.0%) and the lower—Cyanobacteria (3, ~1.8%) and Cryptophyta (2 > 1%).

The basis of the species composition of Bacillariophyta is formed by the following orders: *Naviculales* (38.8%), *Cymbellales* (21.5%), *Achnanthales* (10.3%), *Eunotiales* (8.6%), *Fragilariales* (5.2%), and among families—*Naviculaceae* (12.1%), *Cymbellaceae* (12.1%), *Pinnulariaceae* (11.3%), *Gomphonemataceae* (9.5%), *Eunotiaceae* (8.6%), *Achnanthaceae* (7.7%) [61,65,121]. These families comprise more than 61% of the species composition of diatoms in the Nesamovyte Lake. The genera characterized by high species diversity are the following: *Pinnularia* (12 species (14 inft)—12.1%), *Navicula* (11 species—9.5%), *Eunotia* (10 species—8.6%), *Gomphonema* (10 species—8.6%), *Encyonema* (5 species—4.3%). The complex of leading species of diatoms from different ecotopes (mainly plankton) of the lake with high quantitative indicators of development was formed by *Tabellaria flocculosa* (Roth) Kützing, *Eunotia minor* (Kützing) Grunow and *Frustulia crassinervia* (Brébisson) Lange-Bertalot et Krammer. Their abundance varied from 4 to 5 according to Starmach scale [122]. In addition, the regionally rare species of the flora of Ukraine, which are known to be from this lake were revealed—*Cymbella lange-bertalotii* Krammer, *Encyonema neogracile* Krammer, *Eunotia tetraodon* Ehrenberg, *Pinnularia macilenta* Ehrenberg, *P. subanglica* Krammer (Figure 4), *Skabitchewskia peragalloi* (Brun & Héribaud) Kuliskovskiy & Lange-Bertalot, and *Pinnularia falaiseana* Krammer. Some of them have a pronounced disjunctive distribution in the world and are considered as rare species [65].

A high variety of modern species composition found for Bacillariophyta [65] contrasts sharply with the data at the beginning of the XXth century [67], when only 16 species were reported (17 inft). Moreover, for the modern period of studies, the presence of arctic diatom species that also are regionally rare for Ukraine [105], and in particular, *Cavinula pseudoscutiformis* (Hustedt) D.G. Mann & Stickle, *Pinnularia rhombarea* Krammer, *P. rupestris* Hantzsch, *P. subanglica* Krammer were found (Figure 4).

According to the results of the "green" phyla algae flora (Charophyta and Chlorophyta), the comparison modern and studied periods of 50 (1967–1978) and 100 years [3,4,67] were conducted. A decrease in the species diversity of Charophyta (from 43 species (45 inft) in 1910–1920 to 27 (31)—in 1967–1978 and up to 24 (25) nowadays) and increase of Chlorophyta (four species—13 species, correspondingly for 1910–2021) were noticed [61,121]. Representatives of the classes Zygnemaphyceae (14.6%) and Chlorophyceae (7.3%) comprise about one-fifth of all the algal species in the Nesamovyte Lake nowadays. A significant role in forming this diversity belongs to the representatives of the order Desmidiales (11.6%), and species diversity of *Zygnematales* and *Sphaeropleales*—low and is between 3.7% and 3.0%, correspondingly. Among genera by species diversity differ *Euastrum* Ehrenberg ex Ralfs and *Staurodesmus* Teiling ex Compere (five species—3.0%, each of them). Eighteen (18) genera are represented only by one species each. In turn, compared to the beginning of the XXth century, the presence of species of genera has not been confirmed for *Actinotaenium* (Nägeli) Teiling, *Cylindrocystis* Meneghini ex De Bary, *Micrasterias* Meneghini ex De Bary, *Netrium* (Nägeli) Itzigsohn & Rothe, *Teilingia* Bourrelly and *Coelastrum* Nägeli, and for half a century (1967–1978)—species of genera *Penium* Brébisson ex Ralfs, *Sphaerozosma* Corda ex Ralfs, *Spirotaenia* Brébisson and *Tortitaenia* Brook.

At the same time, rare species of genera have now been identified: *Hyalotheca* Ehrenberg ex Ralfs, *Euastrum* Ehrenberg ex Ralfs and *Tetmemorus* Ralfs ex Ralfs. These species are inherent to the flora of water bodies in mountainous regions in general [34]. During the current research period (2013–2021), the presence of rare coccoid green algae (in particular, *Pediastrum braunii* Wartmann = *P. tricornutum* Borge var. *alpinum* Schmidle) had also been unconfirmed. In addition to rare forms of these three genera, the following algae species were found now for Nesamovyte lake the first time: representatives of filamentous charophyte algae and mucilage-forming green coccoid, flagellar algae and cyanobacteria (*Mougeotia* C. Agardh, *Spirogyra* Link, *Zygnema* C. Agardh, *Oedogonium* Link ex Hirn, *Botryococcus* Kützing, *Chlamydomonas* Ehrenberg, *Mucidosphaerium* C. Bock, Pröschold & Krienitz, *Chlorella* Beijerinck, *Mychonastes* P.D. Simpson & S.D. Van Valkenburg, *Westella* De Wildeman, *Anabaena* Bory ex Bornet & Flahault).

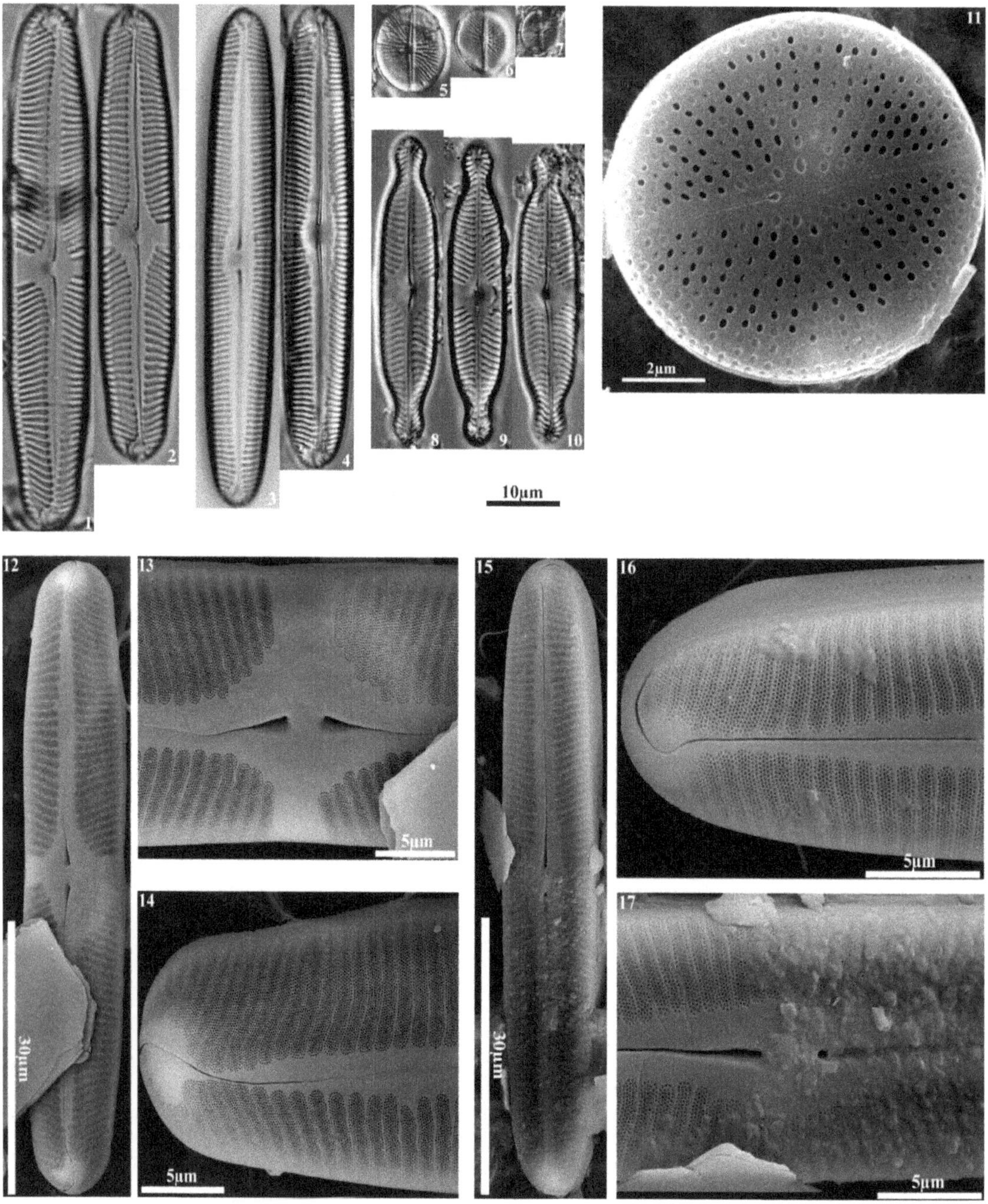

Figure 4. Regionally rare diatom species from the Nesamovyte Lake: 1–2, 12–14—*Pinnularia rhombarea*, 3–4, 15–17—*P. rupestris*, 5–7, 11—*Cavinula pseudoscutiformis*, 8–10—*P. subanglica*.

According to the results of comparative analysis of species composition of Charophyta division in the Nesamovyte Lake for 100 years [67], a change in the complex of leading groups of Desmidiales/Zygnematales according to quantitative characteristics (frequency of occurrence) was observed. The communities *Cylindrocystis brebissonii* (Ralfs) De Bary, *Actinotaenium cucurbita* (Brébisson ex Ralfs) Teiling ex Růžička, *Cosmarium staurastriforme* Gutwinski, *C. venustum* (Brébisson) W. Archer var. *excavatum* (Eichler et Gutwinski) West et G.S. West, *Euastrum insigne* Hassall ex Ralfs, *E. humerosum* Ralfs var. *humerosum* and var. *subintermedium* Schröder, *E. didelta* (Turpin) Ralfs, *Staurastrum muricatiforme* Schmidle in 1910–1920, have changed to the similar *Staurastrum senarium* Ralfs f. *senarium* and f. *tatricum* Raciborski, *Euastrum pinnatum* Ralfs, *E. humerosum* var. *humerosum* and var. *affine* (Ralfs) Raciborski, *E. didelta* Ralfs, *E. amoenum* F. Gay in 60–70-es of XX[3]. The modern grouping of these algae has happened in the beginning of the XXIst century and the community was formed by *Hyalotheca dissiliens* Brébisson ex Ralfs, *Netrium digitus* (Ehrenberg ex Ralfs) Itzigsohn et Rothe emend. Ohtani, *Euastrum humerosum* var. affine, *E. ansatum* Ehrenberg ex Ralfs and *Staurastrum polytrichum* (Perty) Rabenhorst.

The comparison between the modern period and total list of algae showed that the species diversity of algae nowadays makes up over 70% of the total number of the found species and is characterized by an increase in the number of widespread forms. At present, the basis of the taxonomic and species diversity of the lake is being formed by Bacillariophyta and Charophyta, which in total exceeds about 57% of the genera amount and over 84% of the species and infraspecific composition of algae. Less diversely represented are Chlorophyta and Euglenozoa (>6.0%), and Cyanobacteria and Ochrophyta—are quite low (>2.1%). The dynamics of change in the composition of the algae over the period of 100-years shows an increase of Bacillariophyta (from ~23% to above 70%) and the change in the leading taxonomic group—Charophyta.

The appearance and floristic significance of Euglenozoa (~5.0%) over the last half-century—as one of the indicators to the increased degree of the trophic state of the waterbody—was also recorded. In turn, the preservation of secondary importance in terms of species composition groups as Cyanobacteria, Ochrophyta and Cryptophyta were noted. Representatives of Miozoa (Dinophyta) according to comparison with the latest investigations were not identified. In addition, the presence of widespread species of filamentous charophytes and mucilage forming green coccoid and filamentous algae has been noted (*Mougeotia, Spirogyra, Zygnema, Oedogonium, Botryococcus, Chlamydomonas, Mucidosphaerium, Chlorella, Mychonastes*). Also, some members of Euglenozoa group were found [61].

The "blooming" in the water of the Nesamovyte lake was also tested. It is assumed, that the blooming is caused by the mass development of green colonial coccoid algae *Botryococcus terribilis* Komarek et Marvan (Trebouxiales, Trebouxiophyceae), and which was noted for the first time during the summer of 2015 [106,123].

Our previous investigations [61,63,65] for the comparison of the floral community similar to the Chornogora mountain group revealed a low level of similarity and spatial separation among the lakes of the group (Figure 5). However, the Jaccard coefficient with 43% distinguished a group of lakes—Nesamovyte, Bolotne Oko and Tsyclop with similar algal composition.

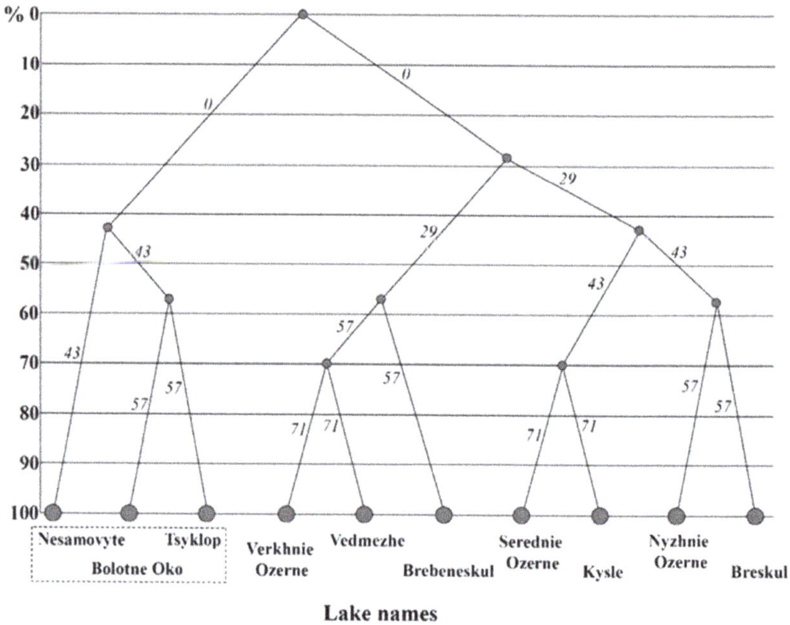

Figure 5. Dendrogram of floristic similarity of species composition of algae in the Nesamovyte Lake with regionally close lakes of the Chornogora mountain group (Jaccard coefficient).

3.3. Ecological Characteristics of the Nesamovyte Lake Due to Algal Preferences

Environmental analysis is based on the types of indicators, which are grouped according to the following characteristics: habitat preferences, streaming and oxygenation, pH, salinity, trophic state, organic pollution (water quality class) (Figure 6). With the help of ecological preferences of species grouped by the previously mentioned time intervals (I—1910–1920; II—1967–1978; III—2013–2021), the ecological characteristics of these periods are also presented.

Characteristics of habitat preferences are determined on a base of indicator species (according to [112]). From 1910 to 1920, species preferring the attached to the substrate way of existence, were formed by benthic species (B) that amounted to 19 taxa or 43.2% (from the total number of indicators of habitat preference) and plankto-benthic (P–B)—15 taxa or 34.1%. For 1967–1978, 17 benthic (B) taxa dominated, composing 53.1% of the total amount of indicators of habitat preferences. However, the amount of plankto-benthic (P–B) taxa were less—eight taxa or 25%. During the modern period (2013–2021), the benthic algae (B) prevailed composing 61 taxa or 50% from total indicator number of habitat preferences, also plankto-benthic (P–B)—47 taxa or 38.5%. Over the studied periods, the restructuring of dominant groups of habitat preference indicators were reported.

Streaming and oxygenation analysis for 1910–1920 revealed the indicators of medium-mobile waters, medium enriched with oxygen or standing-streaming (st-str) composing 10 taxa or 45.5% from the total number of indicators of this ecological preference. Somewhat less amount (seven taxa or 31.8%) was formed by aerophytic forms (ae) that are also called pseudoaerial species. For 1967–1978, indicators of standing-streaming (st-str) prevailed and their amount was 53.8% or seven taxa. Analysing the modern period (2013–2021), indicators of standing-streaming (st-str) waters, also prevailed and their amount equalled 58 taxa-indicators (conjugates, diatoms and green algae), composing 58% from the total amount of taxa of indicators of this group.

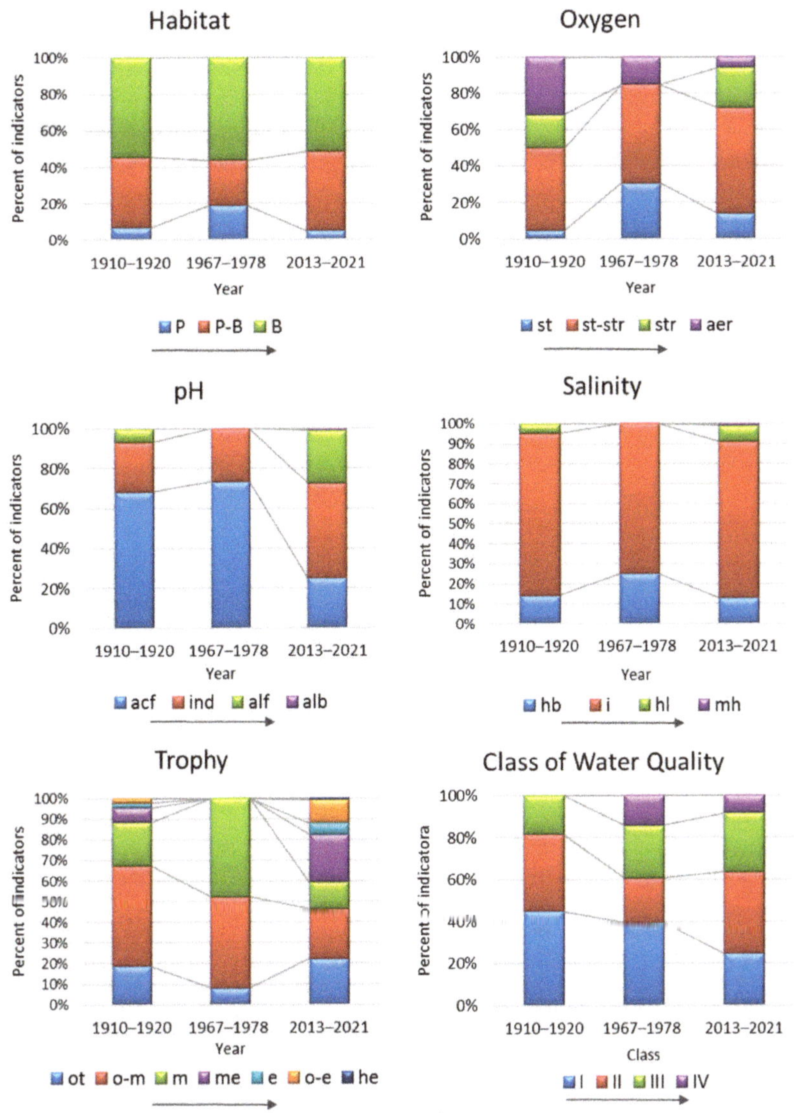

Figure 6. Bioindication plots for ecological analysis of *Habitat preference* (Habitat): P—planktonic, P-B—plankto-benthic, B—benthic; *Streaming and oxygenation* (Oxygen): st—organisms that favour standing water with low oxygenation, st-str—organisms that favour low-flow, moderately oxygenated water; str—organisms that favour streaming water with a high level of oxygenation, aer—aerophyles; *pH*: acf—acidophiles, ind—pH-indifferents; alf—alkaliphiles; alb—alkalibiontes; *Salinity*: hb—halophobes, i—indifferents, hl—halophiles, mh—mesohalobes, *Trophic state* (Trophy): ot—oligotraphentes, o-m—oligo-mesotraphentes, m—mesotraphentes; me—meso-eutraphentes; e-eutraphentes; o-e—from oligo- to eutraphentes; he—hypereutraphentes and *Class of water quality* (according to organic pollution) in the Nesamovyte Lake (for the periods: I—1910–1920, II—1967–1978, III—2013–2021.

Indication of pH (according to [108,112]) for 1910–1920 revealed the prevalence of acidophils, which included 30 taxa, composing 68.2% of the total amount of indicators of this group. For 1967–1978, acidophils also prevailed containing 22 taxa, that was equal to 73.3% from the total amount of the taxa. For modern period (after 2013), the indifferents (ind) prevailed and their amount equalled 50 taxa or 43.1% with the relatively large number of alkaliphiles (alf), composing 31 taxa or 26.7%. Noteworthy is the number of acidophilic (acf) species that is formed by 29 taxa or 25% of the total number of indicators of this group for this period of investigations.

The salinity indicators (according to [109,112]) covered from 8.9% to 60.5% of the total amount of species for each studied period. During all the years of investigations, the significant prevailing of indifferents (i) was recorded.

The trophic state of the Nesamovyte Lake in 1910–1920 indicated prevailing of oligo-mesotraphentic (o-m) indicators that covered 21 taxa or 48.8% from the total amount of taxa for this year. In 1967–1978, the mesotraphentic (m) indicator taxa, which numbered 12 taxa or 48% along with oligo-mesotraphentic (o-m) composing 11 taxa or 44% prevailed. During the modern period (2013–2021), the following groups formed majority of indicators: oligo-mesotraphentic (o-m): 26 taxa or 24.1%, meso-eutraphentic (me): 25 taxa or 23.1% and oligotraphentic (ot): 24 taxa or 22.2%. It is also worth noting the presence of hypereutraphetic (he) species as well as an increase in the number of eutraphetic (e) species for this period that was not recorded from the earliest periods.

Characterization of organic pollution (saprobity) for 1910–1920, indicated class I water quality formed by 17 indicator taxa composing 44.7% from all groups of indicators for this period. In 1967–1978, indicators of class I predominated (11 taxa or 39.3% of indicators of characterizing parameter of this year). However, the presence of indicators of class IV water quality was also noticed. During the modern period, indicators of class II of water quality (46 taxa or 39% of indicators characterizing this parameter for this year) prevailed with the presence of indicators of the class IV quality.

4. Discussion

4.1. The Comparison of the Nesamovyte Lake Diversity with Similar Lakes in Adjacent Areas

A comprehensive study of the entire species composition of algae revealed the high diversity and richness in the algae of Lake Nesamovyte, its specificity and uniqueness as well as the need for conservation and protection. The data on similar lakes of the Carpathian-Tatra region (including the Western Carpathians) covering different taxonomic groups of algae and the dynamic changes in their species composition over a long period of studies or even under modern conditions are limited. Tatra National Park, Slovakia [47–49] and Tatrzański Park Narodowy, Poland [35,37,46] are regionally close regions with available taxonomical characteristics of diversity of all divisions of algae. However, these studies differ in the affiliation of the studied lakes to other climatic zones of mountain groups, excellent altitude gradient location and hydrology of lakes and the period of the study, which makes it impossible to compare correctly the available results. Besides, a variety of studies of mountain lakes in this region focus on the study of general hydrobiological features of these reservoirs, and in terms of diversity—aimed at the careful analysis of the species composition of only one taxonomic group of algae of modern and fossil algal composition of adjacent areas or one ecotope of mountain reservoirs [8,14,52–55,57,124,125]. This purposeful nature of the taxonomic study also complicates the comparative aspect of studying the diversity and specific features of the species composition of the algae flora of the Nesamovite Lake.

4.2. Comparison between Studied Periods of Species Composition of Algae during the 100-Year Investigated Period

The total dataset regarding the composition of algae covering a 100-year period of investigations presents unique information for a comparative study between the different decades of studies and reveals some defining features in each. The low similarity between years of studies could be a result of the changes that happened with the Nesamovyte lake. Ecological analysis allows us to determine possible reasons for such changes. Some similarity between 2013–2021 and 1910–1920, is in contrast to the 1967–1978 years of investigation (Figure 3) and is attributable to the nature of the study of species diversity during the middle period. The coverage of only two taxonomic groups of algae—Euglenozoa [2] and Desmidiales [3] were conducted during 1967–1978, while the studies of the species diversity at the beginning of XXth century and XXIst centuries covered the diversity of all taxonomic groups of algae.

The species composition of the lake during 1910–1920 revealed the presence of specific 27 alpine Holarctic species of desmids and coccoid green algae. It also confirmed uniqueness as well as some affinity to the "Montane-Alpine" component of the species composition of the studied "flora" of the Eastern Carpathians with similar lakes of the Tatra, Sudetes and Alps. At the same time, the author provides the common species lists of algae with the lakes of Tatra, Sudetes and Alps (*Cylindrocystis brebissonii* (Ralfs) De Bary, *Penium cylindrus* Brébisson ex Ralfs (=*Penium cylindrus* var. *subtruncatum* Schmidle), *Tetmemorus laevis* Ralfs ex Ralfs (=*T. laevis* var. *ornatus* Schmidle), *Tetmemorus brebissonii* Ralfs var. *minor* De Bary, *Actinotaenium cuccurbita* (Brébisson ex Ralfs) Teiling ex Růžička (=*Cosmarium cucurbita* Brébisson var. *cucurbita* et var. *attenuatum* G.S. West), *Euastrum montanum* West et G.S. West (=*Cosmarium subreinschii* Schmidle var. *boldtianum* Schmidle), *Euastrum insigne* Hassall ex Ralfs, *Staurastrum scabrum* Brébisson, *S. muricatiforme* Schmidle and other) and alpine holarctic species of charophytes (desmids) and green (coccoid) algae—*Cosmarium nasutum* Nordstedt f. *tatrica* Gutwinski, *C. staurastriforme* Gutwinski, *C. polonicum* Raciborski (=C. vogesiacum Lemaire), *Euastrum aboense* Elfving, *E. binale* (Turpin) Ralfs var. *papilliferum* Gutwinski, *Pediastrum tricornutum* Borge var. *alpinum* Schmidle, *Teilingia granulata* (J.Roy & Bisset) Bourrelly (=*Sphaerozosma granulatum* Roy et Bisset var. *trigranulatum* West), *Staurastrum subavicula* (West) West et G.S. West f. *tyrolense* (Schmidle) G.W.Prescott, C.E.M.Bicudo & W.C.Vinyard (=*Staurastrum vastum* Schmidle var. *tyrolense* Schmidle), *Tortitaenia alpina* (Schmidle) Brook [67] (p. 147).

The studies during 1967–1978 [2–4] showed certain changes in the taxonomic structure of the "flora" and the importance of the following groups: euglenoid algae in different ecotopes of the lake as well as the distinctive high diversity of desmids [3,61].

The modern species composition of algae in the Nesamovyte Lake (2013–2021) has extremely high rates of species diversity noted for Bacillariophyta. This fact makes the modern period significantly different from the two previous periods of studying the floristic composition of algae during 1910–1920 [67] and 1967–1978 [2–4], i.e., 100 and 50 years ago respectively [11]. However, this fact can probably be explained by the current purposeful study of the species composition of diatoms and apparently the influence of a set of ecological and geographical reasons for such transformation, in particular the possible settlement of various ecotopes of the lake by the widespread forms as well as mesotrophic and eutrophic species of the plains area. The reason for the detected diversity of diatoms can possibly be explained by the change in the hydrochemical parameters of the lake. Comparisons by our study with the previous studies as well as inter-comparison of the previous studies revealed more than an 8-fold increase in TDS level over the past 50 years (12–98.6 mg L^{-1}) and a 4-fold increase in the number of sulfates (up to 15.0 mg L^{-1}). Species of this taxonomic group were found to be resistant to heavy metal ions Cd—0.26 mg L^{-1}, Cr—2.37 mg L^{-1}, Pb—0.39 mg L^{-1} [70,72,74]. Furthermore, high values of the diversity of Bacillariophyta as the leading taxonomic group is distinctive of the highlands of Europe [14,26,52,54,55,60] and correspond to our obtained results. Besides, oligotrophic lakes of the alpine zone are refugium for the conservation of rare and conditionally endemic species [15,39,126], and therefore the presence of European and regionally rare species of Bacillariophyta in the ecotopes of the Nesamovyte Lake is an additional argument for the protection and preservation of the ecosystem of this lake and its diversity as a habitat for their existence.

Materials used in our study for the diversity of the "green" phyla algal flora (Charophyta and Chlorophyta) of this lake during the last decade of the XXI confirm the richness of their species composition and floristic importance in the ecosystem of the lake (Table 2) compared to the period of 50 (1967–1978) and 100 years [3,4,67]. The species composition of Chlorophyta in alpine lakes is characterized by a low level of diversity in general, but often reveals features of uniqueness and ecological specificity [34,49,58,127], which was reported at the beginning of the twentieth century at the first stage of investigations of algae in oligotrophic lake Nesamovyte [61,67]. The findings of noted algae from "green" phyla indicate the probable degradation of the alpine oligotrophic ecosystem of the lake with the earliest possible data (early twentieth century), its transformation to oligo-mesotrophic status nowadays and the settlement of its ecotopes by common species with a wide ecological amplitude [11,61,66,121].

In addition, the comparison of the species diversity of algae in the beginning of XXwith a modern period revealed changes in the "Montane" complex. This resulted in the disappearance of the majority (over 80%) of the taxa of the "Montane-Alpine" complex that is common for high-altitude lakes of the Alpine-Carpathian region and suggested the mentioned above degradation of the ecosystem. Besides, this fact once again proved the loss of the alpine oligotrophic indicator species [15,126]. However, the presence of a unique algal composition typical for alpine lakes in Europe—a complex of conditionally arctic algae species, including diatoms was noted [26,39,67,121].

The change in ecological conditions became first evident in the 1960s during focused investigations of different ecotopes of the lake for Euglenozoa and Cyanobacteria [2,61,121]. These groups were considered as the typical representatives of the flora in alpine lakes of Europe and indicated a change of the ecosystem of lakes with high levels of organic pollution. This fact confirms a probable increase of trophic state in the Nesamovyte Lake [124]. Such evidence may indicate a degradation of the ecosystem of the high-altitude oligotrophic lake, its transformation to the mesotrophic type and its settlement by representatives with a wide ecological amplitude that started somewhere in between the two studied periods from 1910–1920 to 1967–1978 [61]. And nowadays, the confirmation of such changes in the ecosystem, its violation and increase in trophic state is in our opinion supported by the fact of "blooming" of the lake with green colonial coccoid algae *Botryococcus terribilis*, which was previously noted only in mesotrophic lakes in Australia [128].

An analysis of geographically close and hydrologically related lakes of the Chornogora mountain group (Figure 5) testified to its high level of floristic originality, the pronounced difference between algal communities and uniqueness. The group of lakes (Nesamovyte, Bolotne Oko, Tsyclop) is characterized by the distinct species composition (in particular the leading taxonomic groups—green, charophyte algae and diatoms). In turn, they are close because of the presence of a significant number of regionally rare species. That could also serve as an additional argument for the protection and preservation of the ecosystem of this waterbody [61,63,65].

4.3. Ecological Characteristics of the Nesamovyte Lake

Our ecological investigations published in previous works [61,65] were based on the analysis of selected groups of algae, and the complex species composition is characterized for the first time.

Indication of habitat preferences (with reference to the type of substrate) and changes in the amount of P-B group gives us an assumption that some hydrological changes in the ecosystem of the Nesamovyte Lake has occurred between 1910–1920 and 1967–1978. In turn, the same changes also appeared in the modern period. Such conclusion can be connected also with the focus of investigation into some groups of algae and thus it should be checked by way of further analysis.

Streaming and oxygenation analysis, in the meaning of oxygen concentration and preference of species to mixing of water in lake, also could confirm the changes in oxygen concentration are due to hydrological or hydromorphological changes [112]. However, the small amount of indicator taxa suggests caution before coming to any conclusions. Even though the amount of pseudoaerial forms seems quite strange for the result of the aquatic investigation and may be connected with the sampling method. The water was sampled in a littoral zone, which is characterized by wave activity or in the ebb and flow zone where high humidity creates certain conditions for the development of a high number of pseudoaerial forms. Analysis of this group of indicators in 1967–1978 showed absence of the organisms that favour streaming water with a high level of oxygenation. This fact can be connected with a focused study of conjugates, however further analysis of indicator taxa in the modern period (2013–2021) testified the stable presence of organisms that favour standing water with low oxygenation. Thus, it also provides a thesis that some changes with the ecosystem of the lake happened.

The characteristics of pH indicator taxa between different years of investigations revealed some changes in water pH. The prevailing of acidophiles that could characterize the acidic conditions in 1910–1920 indicates a marsh type in the studied waterbody (one of the banks had marsh waters) or input of such type of waters into it [112], because this group of algae is characterized by the ability to survive under pH of 5–6. During the 1967–1978 period, the water pH was on the same level, however, the absence of alkaliphiles reveals some changes that have happened at that time. And in the modern period, both hydrological and ecological conditions of the lake have significantly changed. The evidence of it is the high amount of alkaliphiles, testifying to the change of pHto the low-alkaline side under pH7–8. However, the low number of acidophil taxa is still explained by the presence of marsh type on one bank in the Nesamovyte Lake confirmed by a swampy area with overgrowth of sedge-sphagnum flotation (personal observation). The conducted analysis indicates that the area with marsh conditions has decreased.

Indicator characteristics of salinity displays chloride concentration in the Nesamovyte Lake which allows us to characterize their concentration in the waterbody [112]. As noticeable for all studied periods, the salinity level in the Nesamovyte lake was stable and did not change significantly.

Trophic state analysis [111] indicated changes from an almost minimal level of trophism to some increase in 1967–1978 and a far worst position in 2013–2021. The emergence of hypereutraphetic (he) and an increase in the number of eutraphetic (e) species characterizes the general trend towards the deterioration of the trophic state of the investigated lake. The possible reason for this could be connected to the increasing tourism near the lake.

Characterization of water saprobity indicators as well as saprobity indices made it possible to characterize organic pollution and correlate it with the classification adopted in Ukraine, highlighting water quality classes for the Nesamovyte Lake [113,114]. It revealed the changes in water quality from class I in 1910–1920 to a degradation of the ecological condition of the lake confirmed by the changes in the proportion between classes and appearance of indicators of class IV in 1967–1978 with the same tendency for the deterioration in the modern period.

Thus, the conducted investigation testified a gradual increase in organic pollution from 1910–1920 to the modern period which was also underlined earlier during ecological analyses of the selected taxonomic groups [61,65].

5. Conclusions

The comparison between three periods of investigations of the alpine Nesamovyte lake: 1910–1920, 1967–1978 and 2013–2021 revealed significant structural changes in its taxonomic composition of algae, as well as a complex of leading genera and species.

Ecological analysis of the species composition of algae was based on its ecological characteristics and confirmed specific changes that took place with the ecosystem of the lake during different historical periods. The changes in hydrological characteristics, species diversity of algae, systematic structure, communities of dominant and indicator species, the trophic state and organic pollution over a 100-year period revealed the anthropogenic influence in the ecosystem of the lake. The obtained results confirmed the opinion [42,129–131] on the sensitivity and vulnerability of this type of water bodies and its role as an early response-indicator to a change in climate, environmental conditions and an increased level of anthropogenic impact, as well as the need for efforts to preserve this ecosystem of the Eastern Carpathians.

Supplementary Materials: The following are available online at https://www.mdpi.com/article/10.3390/d13060256/s1, Table S1: Species composition and ecological preferences of algae and cyanobacteria of the Nesamovyte Lake in three studied periods (1910–1920; 1967–1978; 2013–2021).

Author Contributions: Conceptualization, P.M.T. and O.P.B.; validation, P.M.T., O.P.B., S.B.; investigation, O.M.K.-Z. and H.H.L.; writing—original draft preparation, P.M.T. and O.P.B.: data curation, P.M.T.; writing—review and editing, P.M.T.; visualization, S.B., O.M.K.-Z.; supervision, P.M.T. All authors have read and agreed to the published version of the manuscript.

Funding: This research received no external funding.

Data Availability Statement: Species composition and ecological preferences of algae and cyanobacteria of the Nesamovyte Lake in three studied periods are available in Supplementary Table S1.

Acknowledgments: The authors acknowledge financial and organizational support given by the administration and the employees of the National Academy of Science of Ukraine and Academy of Science of Poland for financial support of the Polish-Ukrainian project of scientific cooperation. We are grateful to I.I. Chorney, V.V. Budjak and O.I. Khudyi (Yuri Fedkovych Chernivtsi University), as well as T. Mykitchak (Institute of Ecology of the Carpathians of the NAS of Ukraine, Lviv) for organizing and conducting field trips to the Ukrainian Carpathians and for assistance in the work. Our gratitude extends to J. Tunovsky (Institute of Freshwater Fish Research, Poland) for the provided hydroecological indicators. Also, authors are very grateful to Taras Kazantsev and Trevor Williams for the proofreading of the English text.

Conflicts of Interest: The authors declare no conflict of interest.

References

1. Nauwerk, A. Beobachtungen über das Phytoplankton klaren Hochgebirgseen. *Schweiz. Zeinftchrift Hydrol.* **1966**, *28*, 4–28.
2. Asaul, Z.I. Euglenophyta of high mountain lakes of the Ukrainian Carpathians. *Ukr. Bot. J.* **1969**, *26*, 8–13. (In Ukrainian)
3. Palamar-Mordvintseva, G.M. Analysis of Desmidiales flora of the Ukrainian Carpathians. *Ukr. Bot. J.* **1978**, *35*, 29–38. (In Ukrainian)
4. Palamar-Mordvintseva, G.M. *Desmidial Algae of Ukrainian SSR (Morphology, Systematic, Paths of Evolution, Flora and Geographical Distribution)*; Naukova Dumka: Kiev, Ukraine, 1982; pp. 1–240. (In Russian)
5. Palamar-Mordvintseva, G.M.; Tsarenko, P.M.; Nikiforov, V.V.; Prikhodko, E.M.; Nikiforova, V.G. Algae of lake Gropa (National park "Synevir", Ukrainian Carpathians). *Algologia* **1992**, *2*, 73–86. (In Russian)
6. Fott, J. (Ed.) *Limnology of Mountain Lakes (Developments in Hydrobiology)*; Springer: Dordrecht, The Netherlands, 1994; p. 190. [CrossRef]
7. Kamenik, C.; Schmidt, R.; Kum, G.; Psenner, R. The Influence of Catchment Characteristics on the Water Chemistry of Mountain Lakes. *Arct. Antarct. Alp. Res.* **2001**, *33*, 404–409. [CrossRef]
8. Buczkó, K.; Wojtal, A. A new Kobayasiella species (Bacillariophyceae) from Lake Saint Anna's sub-recent deposinft in the Eastern Carpathian Mountains, Europe. *Nova Hedwig.* **2007**, *84*, 155–166. [CrossRef]
9. Robinson, C.T.; Kawecka, B.; Füreder, L.; Peter, A. Biodiversity of flora and fauna in Alpine waters. In *Alpine Waters*; Bundi, U., Ed.; Handbook Environmental Chemistry 6; Springer: Berlin/Heidelberg, Germany, 2010; pp. 193–223.
10. Cărăus, I. Algae of Romania. A distributional checklist of actual algae. Version 2.4. *Studii si Cercet. Biol.* **2017**, *7*, 1–1002.
11. Mykitchak, T. (Ed.) *Ecosystems of Lenthic Water Bodies of Chornohora Massif (Ukrainian Carpathians)*; ZUKS: Lviv, Ukraine, 2014; pp. 1–288. (In Ukrainian)
12. Tsarenko, P.M.; Lilitska, H.H. Algofloristical date of the lake Maricheyka (Chornohora massif, Ukrainian Carpathians). In Proceedings of the 3 International Scientific and Practical Conference Regional Aspects of Floristic and Faunistic Researches, Materials of the First International Scientific and Practical Conference, Putyla-Chernivtsi, Ukraine, 13–14 May 2016; Druk Art: Chernivtsi, Ukraine, 2016; pp. 33–35. (In Ukrainian).
13. Malakhov, Y.; Kryvosheia, O.; Tsarenko, P. Microalgae of protected lakes of northwestern Ukraine. *Pol. Bot. J.* **2017**, *62*, 61–76. [CrossRef]
14. Ognjanova-Rumenova, N.; Wojtal, A.Z.; Sienkiewicz, E.; Botev, I.; Trichkova, T. Biodiversity of high mountain lakes in Europe with special regards to Rila Mountains (Bulgaria) and Tatra Mountains (Poland). *Diatoms Fundam. Appl.* **2019**, 335–354. [CrossRef]
15. Pastorino, P.; Prearo, M. High-Mountain Lakes, Indicators of Global Change: Ecological Characterization and Environmental Pressures. *Diversity* **2020**, *12*, 260. [CrossRef]
16. Mosello, R.; Wathne, B.M.; Lien, L.; Birks, H.J.B. AL:PE projects: Water chemistry and critical loads. *Water Air Soil Pollut.* **1995**, *85*, 493–498. [CrossRef]

17. Mosello, R. The MOLAR Project: Atmospheric deposition and lake water chemistry. *J. Limnol.* **1999**, *58*, 88–106.
18. Mosello, R.; Lami, A.; Marchetto, A.; Rogora, M.; Wathne, B.; Lien, L.; Catalan, J.; Camarero, L.; Ventura, M.; Psenner, R.; et al. Trends in the chemical composition of high altitude lakes selected for the MOLAR project. *Wat. Air Soil Pol. Focus* **2002**, *2*, 75–89. [CrossRef]
19. Cantonati, M.; Poikane, S.; Pringle, C.M.; Stevens, L.E.; Turak, E.; Heino, J.; Richardson, J.S.; Bolpagni, R.; Borrini, A.; Cid, N.; et al. Characteristics, Main Impacts, and Stewardship of Natural and Artificial Freshwater Environments: Consequences for Biodiversity Conservation. *Water* **2020**, *12*, 260. [CrossRef]
20. Wathne, B.M.; Patrick, S.; Cameron, N. (Eds.) *AL:PE—Acidification of Montain Lakes: Paleolimnology and Ecology. Part 2. Remote Mountain Lakes as Indicators of Air Pollution and Climate Change*; NIVA Report 3528-97; NIVA: Oslo, Norway, 1997; p. 525.
21. Wathne, B.M.; Rosseland, B.O. *Measuring and Modeling the Dynamic Response of Remote Mountain Lake Ecosystems to Environmental Changes: A Programme of Mountain Lake Research-MOLAR, Final Report 4/1999, March 1996-March 1999*; NIVA Report 96061-1; NIVA: Oslo, Norway, 1999; p. 201.
22. Straškrabová, V.; Callieri, C.; Carrillo, P.; Cruz-Pizarro, L.; Fott, J.; Hartman, P.; Macek, M.; Medina-Sánchez, J.M.; Nedoma, J.; Šimek, K. Investigations on pelagic food webs in mountain lakes—Aims and methods. *J. Limnol.* **1999**, *58*, 77–87. [CrossRef]
23. Patrick, S.; Battarbee, R.W.; Wathne, B.; Psenner, R. Measuring and modelling the dynamic response of remote mountain lake ecosystems to environmental change: An introduction to the MOLAR project. In *Hydrology, Water Resources and Ecology in Headwaters, Proceedings of the HeadWater'98 Conference, Merano, Italy, 20–23 April 1998*; Kovar, K., Tappeiner, U., Peters, N.E., Craig, R.G., Eds.; No. 248; International Association of Hydrological Sciences Publication, UKCEH Wallingford: Oxfordshire, UK, 1998; pp. 403–410.
24. Battarbee, R.W.; Kernan, M.; Rose, N. Threatened and stressed mountain lakes of Europe: Assessment and progress. *Aquat. Ecosyst. Health* **2009**, *12*, 118–128. [CrossRef]
25. Battarbee, R.W.; Thompson, R.; Catalan, J.; Grytnes, J.A.; Birks, H.J.B. Climate variability and ecosystem dynamics of remote alpine and arctic lakes: The MOLAR project. *J. Paleolimnol.* **2002**, *28*, 1–6. [CrossRef]
26. Kawecka, B.; Galas, J. Diversity of epilithic diatoms in high mountain lakes under the stress of acidification (Tatra Mts, Poland). *Ann. Limnol. Int. J. Limnol.* **2003**, *39*, 239–253. [CrossRef]
27. Cantonati, M.; Zorza, R.; Bertoli, M.; Pastorino, P.; Salvi, G.; Platania, G.; Prearo, M.; Pizzul, E. Recent and subfossil diatom assemblages as indicators of environmental change (including fish introduction) in a high-mountain lake. *Ecol. Indic.* **2021**, *125*, 107603. [CrossRef]
28. Catalan, J.; Barbieri, M.G.; Bartumeus, F.; Bitušík, P.; Botev, I.; Brancelj, A.; Cogălniceanu, D.; Manca, M.; Marchetto, A.; Ognjanova-Rumenova, N.; et al. Ecological thresholds in European alpine lakes. *Freshw. Biol.* **2009**, *54*, 2494–2517. [CrossRef]
29. Directive 2000/60/EC of the European Parliament and of the Council of 23 October 2000 Establishing a Framework for Community Action in the Field of Water Policy. Available online: https://eur-lex.europa.eu/legal-content/en/ALL/?uri=CELEX%3A32000L0060 (accessed on 7 June 2021).
30. Schmidle, W. Beiträge zur alpinen Algenflora. Separat-Abdruck aus der Österr. *Bot. Z.* **1896**, *7*, 1–40.
31. Schröder, B. Neue Beiträge zur Kenntnis der Algen des Riesengebirges. Forschungsber. *Biol. Stat. Plön.* **1898**, *6*, 9–47.
32. Lemmermann, E. Zur Algenflora des Riesengebirges. *Forsch. aus der Biol. St. Plön.* **1896**, *Teil IV*, 88–133.
33. Gutwiński, R. Flora algarum montium Tatrensium. *Bull. Intern. Acad. Sci. Cracovie Cl. Sci. Math. Nat.* **1909**, 415–560.
34. Messikommer, E. Beitrag zur Kenntnis der Algenflora und Algenvegetation des Hochgebirges um Davos. *Beiträge zur Geobot. Landesaufn. Schweiz* **1942**, *24*, 1–452.
35. Szklarczyk-Gazdowa, C. Phytoplankton of some Tatra lakes. *Acta Soc. Bot. Pol.* **1960**, *29*, 597–624. (In Polish) [CrossRef]
36. Lhotsky, O. Bibloigrafia algologiezna Tatr. *Acta hydrobiol.* **1971**, *13*, 477–490. (In Polish)
37. Starmach, K. Benthic algae of the Great Lake in the Valley of the Five Polish Lakes in the Tatra Mountains. *Fragm. Flor.Geobot.* **1973**, *19*, 481–511. (In Polish)
38. Starmach, K. Stan badań algologicznych w Tatrach. *Kosmos* **1957**, *25*, 142–151.
39. Lange-Bertalot, H.; Metzeltin, D. *Indicators of oligotrophy. Iconographia Diatomologica*; Lange-Bertalot, H., Ed.; Koeltz Science Books: Königstein, Germany, 1996; pp. 1–390.
40. Tolotti, M. Phytoplankton and littoral epilithic diatoms in high mountain lakes of the Adamello-Brenta Regional Park (Trentino, Italy) and their relation to trophic status and acidification risk. *J. Limnol.* **2001**, *60*, 171–188. [CrossRef]
41. Tolotti, M. Littoral diatom communities in high mountain lakes of the Adamello-Brenta Regional Park (Trentino, Italy) and their relation to acidification. In *Lange-Bertalot-Festschrift*; Jahn, R., Kociolek, J.P., Witkowski, A., Comperé, P., Eds.; Gantner: Ruggell, Liechtenstein, 2001; pp. 327–352.
42. Tolotti, M.; Cantonati, M. Diatomee litorali. In *I laghi del Parco Naturale Adamello-Brenta*; Cantonati, M., Tolotti, M., Lazzara, M., Eds.; Museo Tridentino di Scienze Naturali: Trento, Italy, 2002; pp. 201–224.
43. Šiško, M.; Kosi, G. Algae. In *High Mountain Lakes in the Eastern Part of the Julian Alps*; Brancelj, A., Ed.; ZRC Publishing: Ljubljana, Slovenia, 2002; pp. 111–128.
44. Wasylik, K. Remnants of algae in Tatra Montains. Kim. Zagospod. *Diem Gorskich PAN* **1965**, *11*, 39–58.
45. Kawecka, B. Aufwuchsalgen Aug Potamogeton sp. *Im See Morskie Ooo. Acta Hydrobiol.* **1966**, *8*, 321–328. (In Polish)
46. Siemińska, J. Algae. In *Tatrzański Park Narodowy*, 2nd ed.; Szafer, W., Ed.; Zakład Ochrony Przyrody: Kraków, Poland, 1962; pp. 305–316. (In Polish)

47. Juriš, Š.; Kováčik, Ľ. Prispevok k poznaniu fytoplanktonu jazier vo Vysokych Tatrach. *Zbor. Slov. Nar. Muz.* **1987**, *33*, 23–40.
48. Hindák, F.; Kováčik, L. Súpis siníc a rias Tatranského Národného parku. *Zbornik Prác o Tatranskom Národnom Parku* **1993**, *33*, 235–279.
49. Lukavsky, J. Algal flora of lakes in the High Tatra Mountains (Slovakia). *Hydrobiologia* **1994**, *274*, 65–74. [CrossRef]
50. Fott, J.; Blažo, M.; Stuchlik, E.A.; Struneský, O. Phytoplankton in three Tatra Mountain lakes of different acidification status. *J. Limnol.* **1999**, *58*, 107–116. [CrossRef]
51. Kopaček, J.; Hardekopf, D.; Majer, M.; Pšenáková, P.; Stuchlík, E.; Veselý, J. Response of alpine lakes and soils to changes in acid deposition: The MAGIC model applied to the Tatra Mountain region, Slovakia-Poland. *J. Limnol.* **2004**, *63*, 143–156. [CrossRef]
52. Štefková, E. Epilithic diatoms of mountain lakes of the Tatra Mountains (Slovakia). *Biol. Bratisl.* **2006**, *61* (Suppl. 18), S101–S108. [CrossRef]
53. Kawecka, B.; Robinson, C.T. Diatom communities of lake/stream networks in the Tatra Mountains, Poland, and the Swiss Alps. *Oceanol. Hyrdrobiological Stud.* **2008**, *37*, 21–35. [CrossRef]
54. Buczkó, K.; Magyari, E.K.; Soróczki-Pintér, É.; Bálint, M. Diatom-based evidence for abrupt climate changes during the Late Glacial in the Southern Carpathian Mountains. *Cent. Eur. Geol.* **2009**, *52*, 249–268. [CrossRef]
55. Fránková, M.; Bojková, J.; Poulíčková, A.; Hájek, M. The structure and species richness of the diatom of the Western Carpathian spring fens along the gradient of mineral richness. *Fottea* **2009**, *9*, 355–368. [CrossRef]
56. Kot, M. *Życie Tatrzańskich Wód*; Wydawnictwa Tatrzańskiego Parku Narodowego: Zakopane, Poland, 2009.
57. Lenarczyk, J. Taxonomic diversity of green algae (Chlorophyta) in six high altitude lakes of the Polish Tatra Mountains. *Fragm. Florist. Geobot. Pol.* **2012**, *19*, 503–523.
58. Lenarczyk, J.; Tsarenko, P. Some rare and interesting green algae (Chlorophyta) from subalpine Tatra lakes (High Tatra Mountains, Poland). *Oceanol. Hydrobiol. Stud.* **2013**, *42*, 225–232. [CrossRef]
59. Lenarczyk, J.; Łukaszek, M.; Tsarenko, P.; Lenzenweger, R. Cosmarium species (Desmidiaceae) in the lakes of the Western and Eastern Carpathians—The Tatra Mountains (Poland) and the Chornohora Mountains (Ukraine). *Studia Univ. Babeş-Bolyai Biol.* **2017**, *62*, 150–151.
60. Wojtal, A.; Ognjanova-Rumenova, N.; Wetzel, C.A.; Hintz, F.; Piatek, J.; Kapetanovic, T.; Ector, L.; Buczko, K. Diversity of the genus Genkalia (Bacillariophyta) in boreal and mountain lakes—Taxonomy, distribution and ecology. *Fottea* **2014**, *14*, 225–239. [CrossRef]
61. Tsarenko, P.; Wolowski, K.; Lenarczyk, J.; Bilous, O.; Lilitska, O. Green and charophytic algae of the high-mountain Nesamovyte and Brebeneskul lakes (Eastern Carpathians, Ukraine). *Plant Fungal Syst.* **2019**, *64*, 25–32. [CrossRef]
62. Tsarenko, P.M.; Palamar-Mordvyntseva, G.M. Algal diversity of the lakes of NNP "Synevyr". In *National Natural Park "Synevyr"*; Solomakha, V.A., Ed.; Plant world. Iss. 10; Fitosotsiotsentr: Kyiv, Ukraine, 2016; pp. 257–266. (In Ukrainian)
63. Tsarenko, P.; Lilitska, H.; Kapustin, D.; Honcharenko, V. Algoflora. In *Ekosystems of Lentic Water Bodies of Chornohora Massif (Ukrainian Carpathians)*; Mykitchak, T., Ed.; ZUKS: Lviv, Ukraine, 2014; pp. 47–60.
64. Tsarenko, P.M.; Kryvosheia, O.M.; Lilitska, H.H. Algae of lake "Hirske oko" (Chevchyn Mts, Ukrainian Carpatian). In Proceedings of the IV International Conference Regional Aspects of Floristical and Faunistical Research, Putyla-Chernivtsi, Ukraine, 28–29 April 2017; pp. 36–39. (In Ukrainian).
65. Kryvosheia, O.M.; Tsarenko, P.M. Bacillariophyta in the High-Mountain Lakes of Chornogora Range in Ukrainian Carpathians. *Int. J. Algae* **2018**, *20*, 239–264. [CrossRef]
66. Mykitchak, T.I. Transformation of ecosystems glacial lakes in Ukrainian Carpathians. *Ecol. Noospherol.* **2017**, *28*, 28–36. [CrossRef]
67. Wołoszyńska, J. Jeziorka czarnohorskie. *Rozprawy wydz. matem.-Przyrodn. PAN. Ser. B* **1920**, *60*, 141–153. (In Polish)
68. Tsys, P.N. Vodorozdilno-Verkhovynsk region. In *Phisiko-Geographical Zoning of Ukrainian SSR*; Popov, V.P., Marinich, A.M., Lanko, A.I., Eds.; Vysshaia Shkola: Kiev, Ukraine, 1968; pp. 608–613.
69. Palamar-Mordvintseva, G.M.; Tsarenko, P.M. Algofloristic zoning of Ukraine. *Int. J. Algae* **2015**, *25*, 303–328. [CrossRef]
70. Bennett, M.R.; Glasser, N.F. (Eds.) *Glacial Geology: Ice Sheets and Landforms*, 2nd ed.; John Wiley: London, UK, 2009; 400p.
71. Miller, H.P. Glacial lakes of Chornohora massif. *Visnyk Lviv. Universytetu. Geogr.* **1964**, *54*, 44–52. (In Ukrainian)
72. Terek, J. Zooplankton of mountain lakes near Hoverla. In *Fauna of the Eastern Carpatihians: Modern status and Protection*; Uzhorod University Press: Uzhorod, Ukraine, 1993; pp. 294–296.
73. Korchemliuk, M.V. *Quality Indicators of Waters. Carpathian National Natural Park*; Foliant: Jaremche, Ukraine, 2009; pp. 247–256. (In Ukrainian)
74. Mykitchak, T.; Kokish, A. Physico-geographycal characteristics of lentic water bodies of Chornohora. In *Ekosystems of Lentic Water Bodies of Chornohora Massif (Ukrainian Carpathians)*; Mykitchak, T., Ed.; ZUKS: Lviv, Ukraine, 2014; pp. 23–46.
75. Mykitchak, T.; Reshetylo, O.; Popelnytska, O.; Kostyuk, A. Antropogenic transformation of Chornohora lentic ecosystems. In *Ekosystems of lentic water bodies of Chornohora massif (Ukrainian Carpathians)*; Mykitchak, T., Ed.; ZUKS: Lviv, Ukraine, 2014; pp. 235–254. (In Ukrainian)
76. Kolodiy, V.V.; Demedyuk, M.S. Hydrochemy of lakes of Ukrainian Carpathians. *Rep. UkrSSR Acad. Sciences. Ser. B Geolog. Chemic. Biolog. Sci.* **1990**, *10*, 11–17. (In Ukrainian)
77. Davies, C.E.; Moss, D.; Hill, M.O. *EUNIS Habitat Classification Revised 2004*; European Environment Agency European Topic Centre on Nature Protection and Biodiversity: Paris, France, 2004; pp. 1–307.
78. Andersen, R.A. (Ed.) *Algal Culturing Techniques*; Elsevier Academic Press: London, UK, 2005.

79. Prygiel, J.; Coste, M. *Guide Methodologique Pour la mise en Oeuvre de l'indice Biologique Diatomées;* Agences de l'Eau Cemagref: Bordeaux, France, 2000; pp. 1–134. (In French)
80. Krammer, K. Bacillariophyceae 1. Naviculaceae. In *Süßwasserflora von Mitteleuropa;* Gustav Fischer Verlag: Stuttgart, Germany, 1986; pp. 1–876.
81. Krammer, K. Bacillariophyceae. 4. Achnanthaceae. Kritische Erganzungen zu Navicula (Lineolatae) und Gomphonema. In *Süßwasserflora von Mitteleuropa;* Gustav Fischer Verlag: Stuttgart, Germany, 1991; pp. 1–437.
82. Krammer, K.; Lange-Bertalot, H. Bacillariophyceae. 3. Centrales, Fragilariaceae, Eunotiaceae. In *Süßwasserflora von Mitteleuropa;* Gustav Fischer Verlag: Stuttgart, Germany, 1991; pp. 1–600.
83. Komarek, J.; Anagnostidis, K. Cyanoprokaryota. 1. Chroococcales. In *Susswasserflora von Mirreleuropa;* Bd. 19/1; Gustav Vischer Verlag: Jena, Germany, 1998; pp. 1–548.
84. Komárek, J.; Anagnostidis, K. Cyanoprokaryota. Teil 2. Oscillatoriales. In *Süßwasserflora von Mitteleuropa;* Bd 19/2; Elsevier: München, Germany, 2005; pp. 1–759.
85. Komárek, J. Cyanoprokaryota 3. Teil/3rd Part: Heterocystous genera. In *Süsswasserflora von Mitteleuropa;* Budel, B., Gärtner, G., Krienitz, L., Schagerl, M., Eds.; Springer Spektrum: Heidelberg, Germany, 2013; pp. 1–1130.
86. Lange-Bertalot, H.; Hofmann, G.; Werum, M.; Cantonati, M. *Freshwater Benthic Diatoms of Central Europe: Over 800 Common Species Used in Ecological Assessment;* Koeltz Botanical Books: Schmitten-Oberreifenberg, Germany, 2017; pp. 1–942.
87. Krammer, K. The Genus Pinnularia. In *Diatoms of Europe;* A.R.G. Gantner Verlag: Königstein, Germany, 2000; Volume 1, pp. 1–703.
88. Krammer, K. Cymbella. In *Diatoms of Europe;* A.R.G. Gantner Verlag K.-G.: Ruggell, Liechtenstein, 2002; Volume 3, pp. 1–584.
89. Krammer, K. Cymbopleura, Delicata, Navicymbula, Gomphocymbellopsis, Afrocymbella. In *Diatoms of Europe;* A.R.G. Gantner Verlag K.-G.: Ruggell, Liechtenstein, 2003; Volume 4, pp. 1–530.
90. Lange-Bertalot, H.; Hofmann, G.; Werum, M. *Diatomeen im Süßwasser-Benthos von Mitteleuropa;* Gantner Verlag: K.-G., Ruggell, Liechtenstein, 2011; pp. 1–908.
91. Vetrova, Z.I. Euglenophyta. In *Flora of Algae of the Continental Water Bodies of the Ukrainian SSR. 1;* Naukova Dumka: Kiev, Ukraine, 1986; pp. 1–348. (In Russian)
92. Vetrova, Z.I. Euglenophyta. Order 1. Euglenales, Family 2. Euglenaceae. Genus 1. Trachelomonas. Group II. Genus 2. Strombomonas. Genus 3. Euglena). In *Flora of Algae of the Continental Water Bodies of Ukraine. 1;* Naukova Dumka: Kiev, Ukraine, 1993; pp. 1–260. (In Russian)
93. Vetrova, Z.I. Euglenofitovye vodorosli. In *Flora of Algae of the Continental Water Bodies of Ukraine. 2;* Lileia Press: Kiev, Ukraine, 2004; pp. 1–272. (In Russian)
94. Palamar-Mordvintseva, G.M. Desmidiales. Gonatozygaceae, Peniaceae, Closteriaceae, Desmidiaceae. In *Flora of Algae of the Continental Water Bodies of Ukraine;* Fasc. 1, Pars 1; Akademperiodica Press: Kiev, Ukraine, 2003; pp. 1–355. (In Ukrainian)
95. Palamar-Mordvintseva, G.M. Desmidiales. In *Flora of Algae of the Continental Water Bodies of Ukraine. Fasc. 1, Pars 2;* Akademperiodyca Press: Kyiv, Ukraine, 2005; pp. 1–573. (In Ukrainian)
96. Kovalenko, O.V. Cyanoprokaryota. Ordo Chroococcales. In *Flora of Algae of Ukraine. 1. Fasc. 1;* VELES: Kyiv, Ukraine, 2009; pp. 1–397. (In Ukrainian)
97. Palamar-Mordvintseva, G.M.; Petlovany, O.A. Streptophyta. Familia Mesotaeniaceae. In *Flora of Algae of Ukraine. 12. Fasc. 1;* VELES: Kyiv, Ukraine, 2009; pp. 1–158. (In Ukrainian)
98. Tsarenko, P.M. *Short Identification Book of Chlorococcales of the Ukraine SSR;* Naukova Dumka: Kiev, Ukraine, 1990; pp. 1–208. (In Russian)
99. Krammer, K. Die cymbelloiden Diatomeen. Eine Monographie der weltweit bekannten Taxa. 1. Allegmeins und Encyonema Part. In *Bibliotheca Diatomologica;* J. Cramer: Berlin/Stuttgart, Germany, 1997; pp. 1–382.
100. Reichardt, E. Zur Revision der Gattung Gomphonema. In *Iconographia Diatomologia;* Gantner Verlag K.-G.: Ruggell, Liechtenstein, 1999; Volume 8, pp. 1–916.
101. Mann, D.G.; Thomas, S.J.; Evans, K.M. Revision of the diatom genus Sellaphora: A first account of the larger species in the British Isles. *Fottea* **2008**, *8*, 15–78. [CrossRef]
102. Kulikovskiy, M.S.; Glushchenko, A.M.; Genkal, S.I.; Kuznetsova, I.V. *Identification Book of Diatoms from Russia;* Filigran: Yaroalvl, Russia, 2016; pp. 1–804. (In Russian)
103. Diatoms of the United States. Diatom Identification Guide and Ecological Resource. 2021. Available online: https://westerndiatoms.colorado.edu (accessed on 15 April 2021).
104. Guiry, M.D.; Guiry, G.M. AlgaeBase. World-Wide Electronic Publication, National University of Ireland, Galway. 2021. Available online: http://www.algaebase.org (accessed on 15 April 2021).
105. Tsarenko, P.M.; Wasser, S.; Nevo, E. (Eds.) *Algae of Ukraine: Diversity, Nomenclature, Taxonomy, Ecology and Geography. Vol. 1. Cyanoprocaryota, Euglenophyta, Chrysophyta, Xanthophyta, Raphidophyta, Phaeophyta, Dinophyta, Cryptophyta, Glaucocystophyta and Rhodophyta;* Gantner Verlag: Ruggell, Liechtenstein, 2006; pp. 1–713.
106. Tsarenko, P.M.; Khudyi, A.M.; Tunovsky, J. On the structure of phyto- and zooplankton communities of Nesamovite Lake in the Ukrainian Carpathians. In Proceedings of the V International Conference "Lake Ecosystems: Biological Processes, Anthropogenic Transformation, Water Quality", Minsk-Naroch, Belarus, 12–17 September 2016; Mikheeyeva, T.M., Ed.; BSU: Minsk, Belarus, 2016; pp. 187–189. (In Russian).

107. Wołoszynska, J. Asterionella formosa Hass. var. tatrica n. var. w jeziorach tatrza'nskich. *Acta Soc. Bot. Pol.* **1934**, *11*, 137–151. (In Polish) [CrossRef]
108. Hustedt, F. Systematische und ökologische Untersuchungen über die Diatomeen-Flora von Java, Bali und Sumatra nach dem Material der Deutschen Limnologischen Sunda-Expedition. Allgemeiner Teil. I. Systematischer Teil, Schluss. *Arch. für Hydrobiol. Suppl.* **1938**, *15*, 393–506.
109. Hustedt, F. Die Diatomeenflora des Flüßsystems der Weser im Gebiet der Hansestadt Bremen. *Abh. Hrsg. vom Nat. Verein zu Brem.* **1957**, *34*, 181–440.
110. Sládeček, V. System of water quality from the biological point of view. Archiv für Hydrobiologie. *Beih. Ergeb. Limnol.* **1973**, *7*, 1–218.
111. Van Dam, H.; Mertens, A.; Sinkeldam, J. A coded checklist and ecological indicator values of freshwater diatoms from the Netherlands. *Neth. J. Aquat. Ecol.* **1994**, *28*, 117–133.
112. Barinova, S.S.; Belous, Y.P.; Tsarenko, P.M. *Algal Indication of Water Bodies in Ukraine*; Haifa University Press: Haifa, Israel; Haifa University Press: Kiev, Ukraine, 2019; pp. 1–367. (In Russian)
113. Barinova, S.S.; Medvedeva, L.A.; Anisimova, O.V. *Diversity of Algal Indicators in the Environmental Assessment*; Pilies Studio: Tel Aviv, Israel, 2006; pp. 1–498. (In Russian)
114. Romanenko, V.D.; Zhukynsky, V.M.; Oksiyuk, O.P.; Yatsyk, A.V.; Chernyavska, A.D.; Vasenko, O.G.; Vernichenko, G.A. *Procedure of the Determination and Use of Ecological Norms of the Quality of the Surface Waters of Land and Estuaries of Ukraine*; Minekoresursiv Ukrayiny Press: Kyiv, Ukraine, 2001; pp. 1–48. (In Ukrainian)
115. Belous, Y.P.; Barinova, S.S.; Klochenko, P.D. Phytoplankton of the middle section of the Southern Bug River as the index of its ecological state. *Hydrobiol. Journ.* **2013**, *49*, 29–42. [CrossRef]
116. Bilous, O.; Barinova, S.; Klochenko, P. The role of phytoplankton in the ecological assessment of the Southern Bug River middle reaches (Ukraine). *Fundam. Appl. Limnol.* **2014**, *184*, 277–295. [CrossRef]
117. Bilous, O.P.; Barinova, S.S.; Ivanova, N.O.; Huliaieva, O.A. The use of phytoplankton as an indicator of internal hydrodynamics of a large seaside reservoir–case of the Sasyk Reservoir, Ukraine. *Ecohydrol. Hydrobiol.* **2016**, *6*, 160–174. [CrossRef]
118. Bilous, O.P.; Wojtal, A.Z.; Ivanova, N.O.; Tsarenko, P.M.; Burova, O.V.; Barinova, S.S. Benthic Diatom Composition in Coastal Zone of Black Sea, Sasyk Reservoir (Ukraine). *Diversity* **2020**, *12*, 458. [CrossRef]
119. Hammer, O.; Harper, D.A.T.; Ryan, P.D. Past: Paleontological Statistics Software Package for Education and Data Analysis. *Palaeontol. Electron.* **2001**, *4*, 1–9.
120. Bray, R.J.; Curtis, J.T. An ordination of the upland forest communities of southern Wisconsin. *Ecol. Monogr.* **1957**, *27*, 325–349. [CrossRef]
121. Tsarenko, P.M.; Bilous, O.P.; Kryvosheia-Zakharova, O.M.; Lilitska, H.H.; Malakhov, Y.; Soininen, J. Changes in algal species composition of alpine lake Nesamovyte (Eastern Carpathians, Ukraine) from 100 years ago to present. *Proccedings* **2021**, in press.
122. Wasser, S.P.; Kondrat'eva, N.V.; Masyuk, N.P. (Eds.) *Vodorosli: Spravochnik (Algae: A Reference Book)*; Naukova Dumka: Kiev, Ukraine, 1989; pp. 1–608. (In Russian)
123. Tsarenko, P.M.; Lilitska, H.H.; Khudyi, O.I.; Tunovsky, J. Unusual "bloom" of water in the Nesamovyte Lake (Chornohora, Ukrainian Carpathians). In Proceedings of the 2 International Scientific and Practical Conference Regional Aspects of Floristic and Faunistic Researches, Materials of the First International Scientific and Practical Conference, Putyla-Chernivtsi, Ukraine, 24–25 April 2015; Druk ART: Chernivtsi, Ukraine, 2015; pp. 452–454.
124. Poniewozik, M.; Wołowski, K. Euglenoids (Euglenophyta) in water-peatbog habitats of the Tatra Mts. *Fragm. Florist. Geobot. Polon.* **2017**, *24*, 283–294.
125. Sochuliaková, L.; Sienkiewicz, E.; Hamerlík, L.; Svitok, M.; Fidlerová, D.; Bitušík, P. Reconstructing the Trophic History of an Alpine Lake (High Tatra Mts.) Using Subfossil Diatoms: Disentangling the Effects of Climate and Human Influence. *Water Air Soil Pollut.* **2018**, *229*, 289. [CrossRef] [PubMed]
126. Cantonati, M.; Lange-Bertalot, H. Diatom biodiversity of springs in the Berchtesgaden National Park (North-Eastern Alps, Germany), with the ecological and morphological characterization of two species new to science. *Diatom Res.* **2010**, *25*, 251–280. [CrossRef]
127. Beck-Mannaghetta, G. Neue Grünalgae aus Kärnten. *Archiv Protistenkd.* **1926**, *55*, 173–183.
128. Komárek, J.; Marvan, P. Morphological differences in natural populations of the genus Botryococcus (Chlorophyceae). *Archiv für Protistenkd.* **1992**, *141*, 65–100. [CrossRef]
129. Rogora, M.; Arese, C.; Balestrini, R.; Marchetto, A. Climate control on sulphate and nitrate concentrations in alpine streams of Northern Italy along a nitrogen saturation gradient. *Hydrol. Earth Syst. Sci.* **2008**, *12*, 371–381. [CrossRef]
130. Bundi, U. Synthesis: Features of Alpine Waters and Management Concerns. In *Alpine Waters 2010*; Springer: Berlin/Heidelberg. Germany, 2010; pp. 1–14.
131. Tiberti, R.; Tartar, G.A.; Marchetto, A. Geomorphology and hydrochemistry of 12 Alpine lakes in the Gran Paradiso National Park, Italy. *J. Limnol.* **2010**, *69*, 242–256. [CrossRef]

diversity

MDPI

Article

Biogeographic Distribution of *Cedrela* spp. Genus in Peru Using MaxEnt Modeling: A Conservation and Restoration Approach

Alexander Cotrina Sánchez [1,*], Nilton B. Rojas Briceño [1,*], Subhajit Bandopadhyay [2], Subhasis Ghosh [3], Cristóbal Torres Guzmán [1], Manuel Oliva [1], Betty K. Guzman [1] and Rolando Salas López [1]

1 Instituto de Investigación para el Desarrollo Sustentable de Ceja de Selva (INDES-CES), Universidad Nacional Toribio Rodríguez de Mendoza de Amazonas, Chachapoyas 01001, Peru; cristobal.torres@untrm.edu.pe (C.T.G.); soliva@indes-ces.edu.pe (M.O.); betty.guzman@untrm.edu.pe (B.K.G.); rsalas@indes-ces.edu.pe (R.S.L.)

2 Department of Ecology and Environmental Protection, Poznan University of Life Sciences, Piatkowska 94, 60-649 Poznan, Poland; subhajit.bandopadhyay@up.poznan.pl

3 Geoinformatics and Remote Sensing Cell, West Bengal State Council of Science & Technology, Government of West Bengal, Kolkata 700091, India; mail.subhasis@yahoo.com

* Correspondence: alexander.cotrina@untrm.edu.pe (A.C.S.); nrojas@indes-ces.edu.pe (N.B.R.B.)

Abstract: The increasing demand for tropical timber from natural forests has reduced the population sizes of native species such as *Cedrela* spp. because of their high economic value. To prevent the decline of population sizes of the species, all *Cedrela* species have been incorporated into Appendix II of the Convention on International Trade in Endangered Species of Wild Fauna and Flora (CITES). The study presents information about the modeled distribution of the genus *Cedrela* in Peru that aims to identify potential habitat distribution of the genus, its availability in areas protected by national service of protected areas, and highlighted some areas because of their conservation relevance and the potential need for restoration. We modeled the distribution of the genus *Cedrela* in Peru using 947 occurrence records that included 10 species (*C. odorata*, *C. montana*, *C. fissilis*, *C. longipetiolulata*, *C. angustifolia*, *C. nebulosa*, *C. kuelapensis*, *C. saltensis*, *C. weberbaueri*, and *C. molinensis*). We aim to identify areas environmentally suitable for the occurrence of *Cedrela* that are legally protected by the National Service of Protected Areas (PAs) and those that are ideal for research and restoration projects. We used various environmental variables (19 bioclimatic variables, 3 topographic factors, 9 edaphic factors, solar radiation, and relative humidity) and the maximum entropy model (MaxEnt) to predict the probability of occurrence. We observed that 6.7% (86,916.2 km^2) of Peru presents a high distribution probability of occurrence of *Cedrela*, distributed in 17 departments, with 4.4% (10,171.03 km^2) of the area protected by PAs mainly under the category of protection forests. Another 11.65% (21,345.16 km^2) of distribution covers areas highly prone to degradation, distributed mainly in the departments Ucayali, Loreto, and Madre de Dios, and needs immediate attention for its protection and restoration. We believe that the study will contribute significantly to conserve *Cedrela* and other endangered species, as well as to promote the sustainable use and management of timber species as a whole.

Keywords: CITES; endangered species; SDM; degraded amazon; machine learning

Citation: Cotrina Sánchez, A.; Rojas Briceño, N.B.; Bandopadhyay, S.; Ghosh, S.; Torres Guzmán, C.; Oliva, M.; Guzman, B.K.; Salas López, R. Biogeographic Distribution of *Cedrela* spp. Genus in Peru Using MaxEnt Modeling: A Conservation and Restoration Approach. *Diversity* **2021**, *13*, 261. https://doi.org/10.3390/d13060261

Academic Editors: Maurizio Rossetto and Michael Wink

Received: 1 May 2021
Accepted: 7 June 2021
Published: 10 June 2021

Publisher's Note: MDPI stays neutral with regard to jurisdictional claims in published maps and institutional affiliations.

1. Introduction

Forest covers have been reduced drastically in the Peruvian Amazon region over the last few decades as a result of agricultural expansion and livestock activities, deforestation, mining, and urban expansion [1,2]. In Peru, 2,433,314 ha of Amazonian forests have been lost during 2001–2019 [3]. Although the tropical Amazon forest covers about 60% of Peru [4], it has now been highly fragmented because of the forest harvesting activities. The need for more agricultural land also promoted heavy migratory agricultural practices, [5] eliminating approximately 0.5 ha of forest cover for crop production [6,7]. As a result of such growing land-use changes induced by migratory agriculture and cattle ranching, many native species, including genus *Cedrela*, are now experiencing massive destruction of their habitats [8]. In addition, the selective falling of trees, mainly of species having high economic values, has also caused the near extinction of many vegetation species such as mahogany (*Swietenia macrophylla*) and cedar (*Cedrela odorata*) [9].

Cedrela is a genus of tropical trees that includes species such as *C. odorata* L. and *C. fissilis* Vell., which had been collected for wood for more than 500 years in Central and South America, with *C. odorata* being the second most demanded tropical wood [10–13]. Worldwide, this genus has 17 recognized species [13,14], out of which Peru alone has 10. Hence, Peru can be considered as a center of diversity for *Cedrela* species [15], which currently includes three endemic species with restricted distribution, i.e., *C. molinensis*, *C. longipetiolulata*, and *C. weberbaueri* [16]. However, because of the high economic value of the genus *Cedrela* species, their usage had started increasing since the end of the 1980s, mainly in Mexico, Brazil, Peru, and Bolivia [17,18]. Such overexploitation eventually resulted in the near-extinction of the *Cedrela* population and made the international conservation community call for its greater protection under the Convention on International Trade in Endangered Species of Wild Fauna and Flora (CITES). In Peru, the National Forest and Wildlife Service (SERFOR) has also recently incorporated the populations of genus *Cedrela* (*C. odorata, C. montana, C. fissilis, C. longipetiolulata, C. angustifolia, C. nebulosa, C. kuelapensis, C. Saltensis, C. weberbaueri*, and *C. molinensis*) in Appendix II of CITES on 28 August 2020.

This alarming situation indicates a strong need for further research studies that may effectively contribute to decision making related to the sustainability and conservation of biodiversity of the *Cedrela* and its habitat. Species distribution models (SDMs) are tools that combine species presence data with factors such as bioclimatic, edaphic, topographic, etc. and allow more effective and generous support for species conservation, biogeography, and climate change actions [19–23]. SDMs have made it possible to identify the distribution of timber forest species [24,25], other endemic species [26], wildlife [27,28], etc. on a regional scale facilitating proper identification, protection, and conservation of the endangered ones [29,30]. Among all the available SDMs, the maximum entropy algorithm (MaxEnt) [31] is one of the most widely used algorithms to find out the distribution of species under current and future conditions [32,33]. This way, MaxEnt allows habitat mapping and produces credible, defensible, and repeatable information, which contributes to a structured and transparent process of sustainable natural resources management by predicting the possible degradation of potential forest areas with species under risk in the future climate change scenarios [34].

After identifying the potential distribution areas of a species, the areas having the best aptitude to carry out reforestation or recovery initiatives of degraded areas are needed to be quantified and monitored properly. Such restoration is of great interest since 13.78% (177,592.82 km^2) of the Peruvian territory has been identified as degraded areas as a consequence of deforestation, livestock activities, agriculture, mining, forest fires, etc. [35]. The strategies to be implemented must be oriented to the restoration and/or conservation of threatened species that are widely distributed over the geographic spaces integrated into the territorial order using environmental services, ecotourism, management of renewable resources, and productive practices promoted through Protected Natural Areas (PNAs) initiative [36].

The study has two main objectives—firstly, to model the biogeographic distribution of 10 available species of genus *Cedrela* (i.e., *C. odorata, C. montana, C. fissilis, C. longipetiolulata, C. angustifolia, C. nebulosa, C. kuelapensis, C. Saltensis, C. weberbaueri* and *C. molinensis*) over the Peruvian territory using the MaxEnt model in a current scenario, and secondly, to identify the locations of *Cedrela* within the designated conservation areas (to evaluate its effectiveness in conserving the species' habitat) and degraded areas (to implement forest restoration practices using these species). The study considered sample location information of the *Cedrela* species (947 geographical records) and 33 different variables (19 bioclimatic variables, 3 topographic, 9 edaphic, solar radiation, and relative humidity).

2. Materials and Methods

2.1. Study Area

This study covers the entire territory of Peru (1,300,000 km^2 approx.) located between the parallels of 0°03′00″ and 18°30′0″ south and the meridians of 68°30′00″ and 81°30′00″ west, sharing borders with Ecuador and Colombia to the north, Brazil to the east, Bolivia to the southeast, Chile to the south, and the Pacific Ocean to the west. The altitudinal gradient of this region starts from 0 m above sea level (a.s.l.) in the north and goes up to 6800 m above sea level (Mataraju Mountain). Almost 60% of the study area is covered by the Amazon rainforest, which is characterized by heavy rainfall and high temperatures, except for its southernmost part, which has cold winters and seasonal rainfall. The Protected Natural Areas (PNAs) belong to the National System of Natural Areas Protected by the State (SINANPE) [36]. These broadly include regional conservation areas, private conservation areas, national sanctuary, historic sanctuary, wildlife refuge, national reserve, communal reserve, national park, and hunting and protection forest scattered all over the study region (Figure 1).

2.2. Dataset and Methodological Design

The methodological framework used in the present study has been described graphically in Figure 2. From the cartographic standardization through the rescaling in the raster calculator of Qgis 3.16, 33 variables at a spatial resolution of 250 m were derived as input for use in modeling with MaxEnt. The bioclimatic information under current conditions (average 1970–2000) with a spatial resolution of 30 s (~1 km) was obtained from Woldclim version 2.1 (https://www.worldclim.org/data/worldclim21.html; accessed on 5 January 2021) [37]. Topographic factors such as elevation, slope, slope, and ground orientation were obtained from the 90 m spatial resolution DEM generated by the Shuttle Radar Topography Mission (SRTM) [38], United States Geological Survey (USGS) (http://srtm.usgs.gov; accessed on 28 December 2020). The edaphic variables were collected from SoilGrids 0.5.3 (http://soilgrids.org; accessed on 15 January 2021) with a spatial resolution of 250 m.

Likewise, the geographic occurrence data of 10 target species of the genus *Cedrela* to be used in the MaxEnt model were obtained from GBIF's Global Biodiversity Information Service (https://www.gbif.org/; accessed on 1 February 2021) through "Species Explorer" plug-in of QGIS software. The registration information of CITES species was obtained from the Ministry of the Environment of Peru (https://geoservidor.minam.gob.pe/recursos/intercambio-de-datos/; accessed on 18 February 2021). Finally, to identify the locations of *Cedrela* habitats within the protected areas, and the areas prone to degradation but having high suitability for genus *Cedrela* habitat, the modeled potential distribution result was overlaid with the degraded areas identified by the Ministry of Environment (MINAM) and the spaces conserved by the National Service of Natural Areas Protected by the Peruvian State (SERNANP). These degraded areas were identified by the ministry mainly based on deforestation, soil erosion, forest fires, mining, illegal logging, etc.

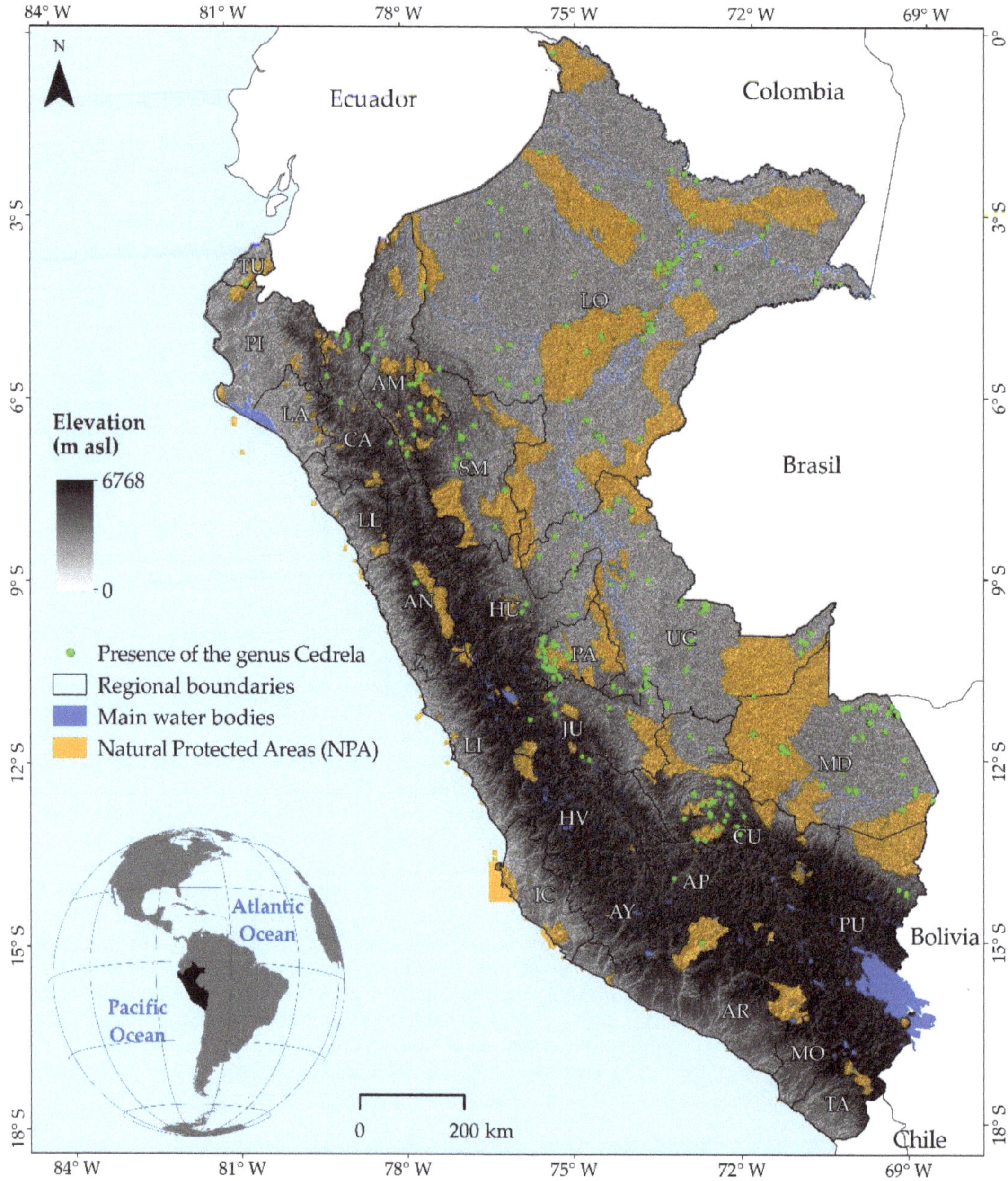

Figure 1. Study area and presence of *Cedrela* species.

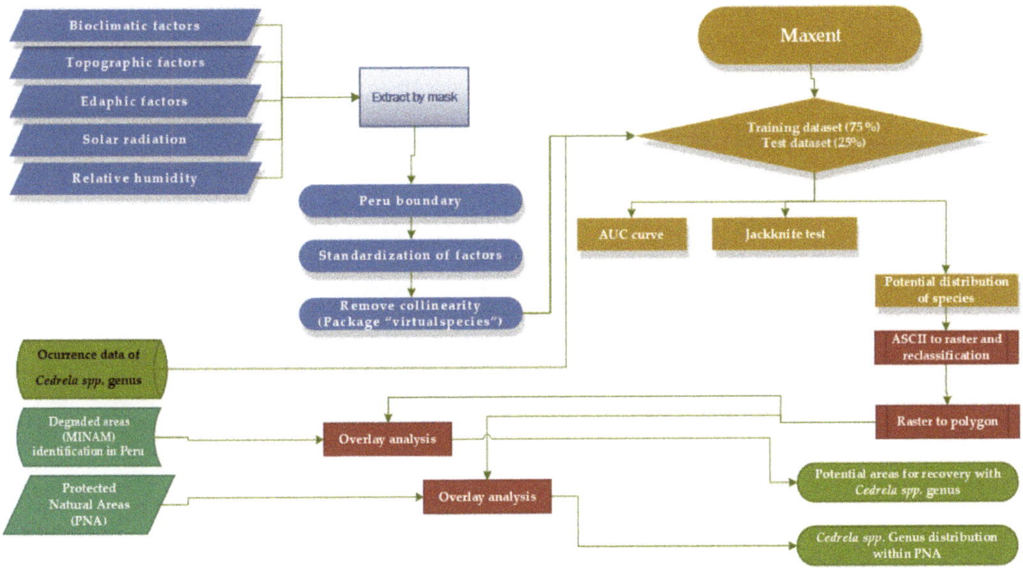

Figure 2. Methodological process for the biogeographic modeling of the genus *Cedrela* in Peru.

2.3. Geographical Records of Forest Species

The geographic coordinates of the 10 species of the genus *Cedrela* (Table 1) were obtained using the GBIF and Species Explorer plug-ins in QGIS 3.16 software. It was also complemented with the records of the presence of *Cedrela*, obtained from the Ministry of Environment of Peru. The CITES species information was collected from the systematization of forest inventories, review of national herbaria which is available in its geoservidor (https://geoservidor.minam.gob.pe/recursos/intercambio-de-datos/; accessed on 18 February 2021), and information related to the species of the genus *Cedrela* was separated. The data were then re-sampled at a spatial resolution of 250 m [39] by visually excluding those samples that were falling within lagoons, rivers, and roads, or urban areas. Finally, 947 resulting data were exported into CSV to be used for modeling in MaxEnt (https://biodiversityinformatics.amnh.org/open_source/maxent/; accessed on 10 November 2020).

Table 1. The number of records of 10 species of the genus *Cedrela* used in biogeographic modeling.

N°	Family	Genere	Species	Records Number
1	Meliaceae	*Cedrela*	*fissilis*	42
2			*kuelapensis*	16
3			*molinensis*	1
4			*montana*	30
5			*nebulosa*	32
6			*odorata*	787
7			*saltensis*	6
8			*angustifolia*	18
9			*longipetiolulata*	8
10			*weberbaueri*	7
		Total		947

2.4. Bioclimatic, Physiographic, and Soil Variables

The spatial distribution of species within a geographic area dependson the interaction with several environmental factors that contribute to their development and coexistence [40]. Considering this, 33 variables were selected (Table 2) to carry out the modeling. These variables include 19 bioclimatic and solar radiation obtained from WorldClim 2.1 (https://www.worldclim.org/; accessed on 5 January 2021) [37]; 3 topographic derived from digital elevation model (DEM) obtained from the United States Geological Survey (USGS) web portal (http://srtm.usgs.gov; accessed on 28 December 2020); the relative humidity obtained from the Climate Research Unit (CRU) [41] (www.cru.uea.ac.uk; accessed on 1 May 2021); and 9 soil properties collected from SoilGrids 0.5.3 (http://soilgrids.org; accessed on 15 January 2021) [42]. All variables were rescaled into a spatial resolution of 250 m to overcome the issues such as collinearity between variables, which causes overfitting problems, increases uncertainty, and decreases the statistical power of the model [43]. Therefore, using the function "remove collinearity" from the package "virtual species" [44] in R 3.6, the variables were grouped (clustering) according to the Pearson correlation coefficient, and only variables having Pearson's r $\geq$ 0.7 were considered. This threshold is an acceptable measure to minimize the multicollinearity of fitted models [43].

Table 2. Variables for MaxEnt modeling of *Cedrela* in Peru.

Variable	Units	Symbol	Δ Earnings in Jackknife [1]	Clúster
		Bioclimatic		
Annual Mean Temperature	°C	bio01	0.7379	1
Mean Diurnal Range	°C	bio02	0.7627	7
Isothermality		bio03	0.9150	4
Temperature Seasonality	**°C**	**bio04**	**0.7097**	**9**
Max Temperature of Warmest Month	°C	bio05	0.6811	1
Min Temperature of Coldest Month	°C	bio06	0.7068	1
Annual Temperature Range	°C	bio07	0.7655	9
Mean Temperature of Wettest Quarter	°C	bio08	0.7608	1
Mean Temperature of Driest Quarter	°C	bio09	0.7107	1
Mean Temperature of Warmest Quarter	°C	bio10	0.7606	1
Mean Temperature of Coldest Quarter	°C	bio11	0.7067	1
Annual Precipitation	**mm**	**bio12**	**0.6231**	**3**
Precipitation of Wettest Month	mm	bio13	0.7674	2
Precipitation of Driest Month	mm	bio14	0.5525	3
Precipitation Seasonality	mm	bio15	0.6692	9
Precipitation of Wettest Quarter	mm	bio16	0.7524	2
Precipitation of Driest Quarter	mm	bio17	0.5481	3
Precipitation of Warmest Quarter	mm	bio18	0.7915	2
Precipitation of Coldest Quarter	**mm**	**bio19**	**0.5147**	**3**
		Topographic		
Elevation above mean sea level	**msnm**	**dem**	**0.6709**	**7**
Slope of the terrain	**°**	**slope**	**0.9104**	**7**
Cardinal orientation of the slope	**°**	**aspect**	**1.0117**	**5**
		Edaphic at 0.30 m		
pH in H_2O	**pH × 10**	**ph**	**0.6543**	**7**
Cation exchange capacity	**cmol kg^{-1}**	**cec**	**0.7898**	**6**
Organic carbon	**g kg^{-1}**	**soc**	**0.8094**	**4**
Bulk density of the fine earth fraction	**cg/cm^3**	**bdod**	**0.8881**	**8**
Volumetric fraction of coarse fragments	cm^3/dm^3 (vol %)	cfvo	0.7051	7
Total nitrogen	**cg/kg**	**nitrogen**	**0.8375**	**6**
Clay content	%	clay	0.8743	4
Sand content	**%**	**sand**	**0.8155**	**7**
Slime content	**%**	**silt**	**0.7970**	**2**
Solar radiation	MJ m^{-2} day^{-1}	srad	0.6801	9
Relative humidity	%	rhm	0.7777	3

[1] In bold, the variables with less variation between the regularized training data and single variable for each cluster used for MaxEnt model are shown.

To select an important variable for each cluster, a preliminary MaxEnt model was run (the configuration is explained in Section 3.2.) using all the variables. The variable with the best performance in the Jackknife test [25] was selected (i.e., the smallest difference in regularized training gains obtained from a model generated with all criteria except that of interest and a model generated only with the criterion of interest [21] (Table 2).

2.5. Execution of the Model

The biogeographic distribution model for the 10 species of the genus *Cedrela* was performed using a maximum entropy algorithm [31] which estimates the probability of potential distribution of each species from the presence data (location) using the open-source software MaxEnt ver. 3.4.1 (https://biodiversityinformatics.amnh.org/open_source/maxent/; access on 10 November 2020). For the validation of this model, 75% of the randomly selected presence data were used for training purposes, and 25% were used for validation [31]. The algorithm was run using 100 repetitions in 5000 iterations with different random partitions (Bootstrap method), and other configurations (i.e., extrapolation, graph drawing, etc.) were kept as default [45].

The resulting model was validated based on the area under the curve (AUC) calculated from the operating characteristic of the receptor (ROC) [31,46,47]. According to the AUC values, five performance levels were differentiated: excellent (>0.9), good (0.8–0.9), accepted (0.7–0.8), poor (0.6–0.7) and invalid (<0.6) [46,48]. We used the logistic output format to obtain the model of the 10 evaluated species by generating a raster of continuous values in a range from 0 to 1. The raster obtained was further reclassified into four ranges: (1) "high potential" habitat (>0.6), (2) "moderate" habitat (0.4–0.6), (3) "low potential" habitat (0.2–0.4), and (4) "no potential" habitat (<0.2) [24,25,28,48].

2.6. Identification of Potential Areas for Restoration and Conservation

Subsequently, the areas of high distribution potential were overlapped with the Protected Natural Areas (PNA) information obtained from GeoServer (https://geo.sernPNA.gob.pe/visorsernPNA/; access on 18 February 2021) of the National Service of Natural Areas, which is protected by the State (SERNPNA) to promote conservation of the genus *Cedrela*, currently considered as endangered and overexploited in Peru.

Similarly, the raster layer (30 m resolution) of degraded areas as identified by the Ministry of the Environment of Peru (MINAM) in 2019 was also obtained from its geoservidor, (https://geoservidor.minam.gob.pe/recursos/intercambio-de-datos/; access on 18 February 2021) and overlapped with the potential distribution of *Cedrela*. Finally, the distribution surfaces of the 10 species within the PAs and degraded areas were quantified. This way, the analysis had made it possible to identify the protected areas that conserve the genus *Cedrela* and those degraded spaces that could be restored with one or more of the species under study.

3. Results

3.1. Model Performance and the Importance of Environmental Variables

Model performance evaluation aims to estimate the accuracy of machine learning-based prediction models and ensures confidence in the results obtained. The performance of this model obtained an area under the curve (AUC) value of 0.866 (Figure 3a), which is considered good (0.8 < AUC < 0.9). The response curves (Figure 3b–n) reflect the dependence of predicted suitability, both on the selected variable and on dependencies induced by correlations between the selected variable and other variables. Overall, 83% of the potential distribution of *Cedrela* was found to be driven mainly by four environmental variables, i.e., bio19 (precipitation of coldest quarter), soc (organic carbon), dem (elevation above mean sea level), and cec (cation exchange capacity) (Table 3). On the other hand, silt (slime content), bdod (bulk density of the fine earth fraction), and nitrogen were the three environmental variables that contributed the least. Figure 3o shows the results of jackknife test of variable importance. The environmental variable that reported the highest gain when used in isolation was bio19, which therefore appeared to have the most useful information by itself. The environmental variable that decreased the gain the most on its omission was dem, which therefore appeared to have the most information that was not present in the other variables. Likewise, the Jackknife test (Figure 3o) identified that the variables bio 19 (coldest quarter precipitation), bio 12 (annual precipitation), soil pH, and elevation (DEM) contributed highly to the biogeographic distribution model of the *Cedrela* species.

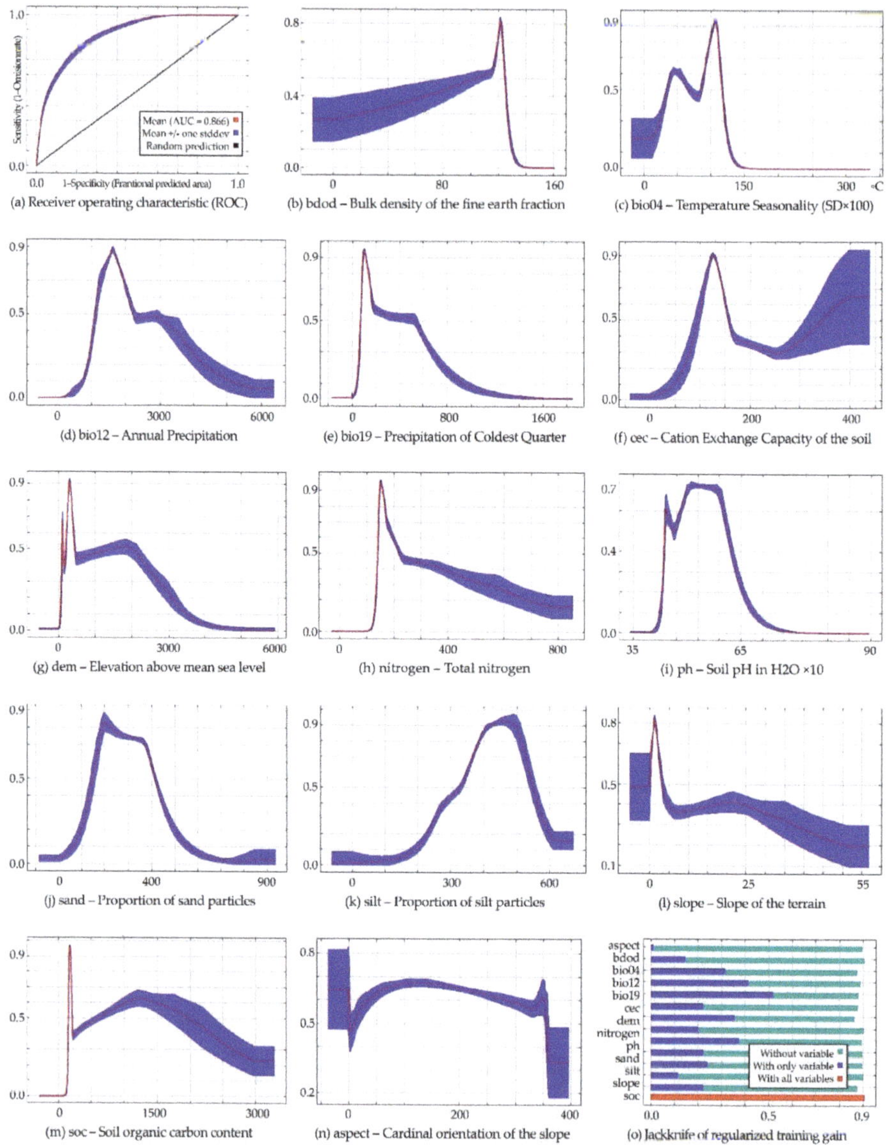

Figure 3. Model performance based on the area under the curve (AUC) (**a**), mean response curves of the 100 replicated MaxEnt runs (red) and standard deviation (blue), showing the relationships between environmental variables and the probability of the presence of the *Cedrela* (**b**–**n**), and Jackknife test of environmental variables importance to MaxEnt model of the *Cedrela* (**o**).

Table 3. Relative contributions (%) of environmental variables to the MaxEnt model of the genus *Cedrela* in Peru.

Variable	Percent Contribution	Permutation Importance
bio19	51.3	19.7
soc	18.3	6
dem	6.9	25.8
cec	6.5	2.3
bio12	4.6	19.2
bio04	3.7	9.1
ph	2.6	4.6
aspect	1.3	1.1
slope	1.3	1.2
sand	1.3	4.7
silt	0.9	4
bdod	0.7	1.1
nitrogen	0.7	1.1

3.2. *Potential Distribution of the Genus Cedrela*

The areas of a high probability of occurrence of genus *Cedrela* under present climatic and environmental conditions were identified mainly across the lowland Amazonia, covering 86,916.2 km^2 (6.7%) area of the study region. This potential habitat distribution covers about 17 regions of the Peruvian territory (Figure 4) with a high concentration in the departments of Ucayali (23,322.04 km^2), Loreto (22,842.3 km^2), and Madre de Dios (20,755.7 km^2) (Table 4).

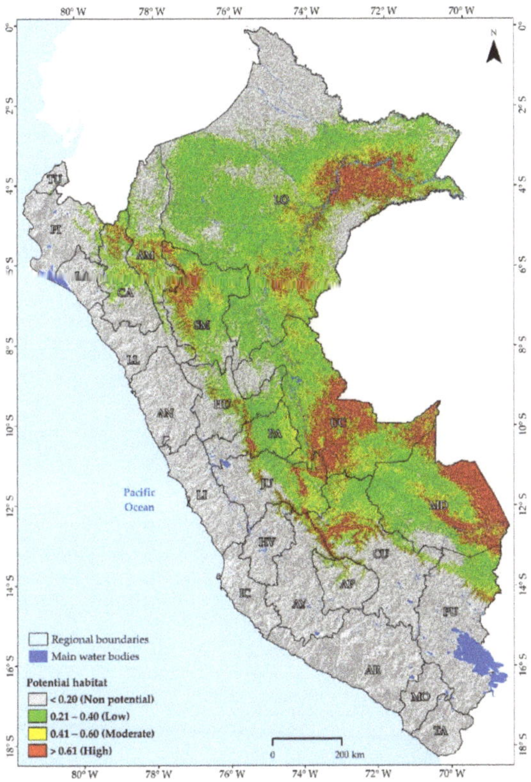

Figure 4. Distribution of the biographical model of the genus *Cedrela* in Peru.

Table 4. Area of distribution of the biographical model of the genus *Cedrela* in Peru.

Region/Country	Geographic Area km^2	Low		Moderate		High	
		km^2	%	km^2	%	km^2	%
Amazonas	39,306.46	12,692.78	32.3	6339.41	16.1	1925.42	4.9
Ancash	35,962.25	7.18	0	0.16	0	0.00	0
Apurimac	21,114.18	1719.01	8.1	476.60	2.3	148.55	0.7
Ayacucho	43,503.84	1150.37	2.6	1306.67	3	985.08	2.3
Cajamarca	33,044.68	8512.56	25.8	4444.75	13.5	1140.41	3.5
Cusco	72,076.2	15,217.89	21.1	11,450.66	15.9	5145.99	7.1
Huancavelica	22,065.07	444.75	2	260.12	1.2	194.12	0.9
Huánuco	37,200.53	12,279.44	33	2274.37	6.1	867.62	2.3
Junín	43,997.3	11,298.72	25.7	8212.08	18.7	3340.96	7.6
La Libertad	25,295.94	1025.67	4.1	270.80	1.1	39.08	0.2
Lambayeque	14,342.31	294.65	2.1	20.76	0.1	0.19	0
Loreto	375,115.94	161,464.72	43	53,151.91	14.2	22,842.30	6.1
Madre De Dios	85,045.87	34,710.95	40.8	20,326.56	23.9	20,755.70	24.4
Pasco	24,113.95	10,797.92	44.8	4771.88	19.8	979.54	4.1
Piura	36,065.1	1953.37	5.4	225.54	0.6	14.43	0
Puno	67,962.79	8270.85	12.2	3048.19	4.5	947.78	1.4
San Martin	50,961.26	23,558.38	46.2	13,728.27	26.9	4267.01	8.4
Tumbes	4690.28	122.44	2.6	0.00	0	0.00	0
Ucayali	105,341.77	38,186.29	36.2	32,564.53	30.9	23,322.04	22.1
Peru	1,288,564.27	343,708	26.7	16,2873	12.6	86,916.2	6.7

3.3. *High-Priority Areas for Research, Conservation, and Restoration*

The study identified that 4.4% (10,171.03 km^2) of the areas of the high-occurrence probability of genus *Cedrela* was distributed in the designated Peruvian conservation areas (Figure 5a), out of which the PNA cover of 35.5% (8995.64 km^2) was distributed among the reserved zones (85.18 km^2), national sanctuary (130.76 km^2), historic sanctuary (20.18 km^2), wildlife refuge (0.13 km^2), national reserve (2323.46 km^2), communal reserve (1023.63 km^2), national park (5000.93) and protection forest (411.37 km^2). The distribution also included conservation areas administered by regional governments (1020.71 km^2), and by individuals or institutions at a private level through private conservation areas, whose high-occurrence potential covered a total of 154.68 km^2 (Table 5) area of the study region.

Table 5. Total potential distribution area predicted that is protected by the modalities of Protected Natural Area in Peru.

PNA Modalities	Geographic Area (km^2)	Low		Moderate		High	
		km^2	%	km^2	%	km^2	%
Reserved Zone	6257.55	1799.22	28.8	358.28	5.7	85.18	1.4
Regional Conservation Areas	33,253.79	16,206.54	48.7	4309.78	13.0	1020.71	3.1
Private Conservation Areas	3963.05	764.34	19.3	536.13	13.5	154.68	3.9
National sanctuary	3173.66	1169.43	36.8	1226.79	38.7	130.76	4.1
Historic Sanctuary	412.79	50.46	12.2	31.67	7.7	20.18	4.9
Wildlife Refuge	207.75	44.25	21.3	20.11	9.7	0.13	0.1
National Reserve	46,528.52	15,987.96	34.4	9005.68	19.4	2323.46	5.0
Communal Reserve	21,665.88	7247.05	33.4	5636.40	26.0	1023.63	4.7
National Park	103,943.67	53,959.42	51.9	21,151.83	20.3	5000.93	4.8
Hunting	1247.35	7.51	0.6	0.00	0.0	0.00	0.0
Protection Forest	3899.87	1563.68	40.1	1175.82	30.2	411.37	10.5
PNA Peru	231,672.07	98,799.85	42.6	43,452.49	18.8	10,171.03	4.4

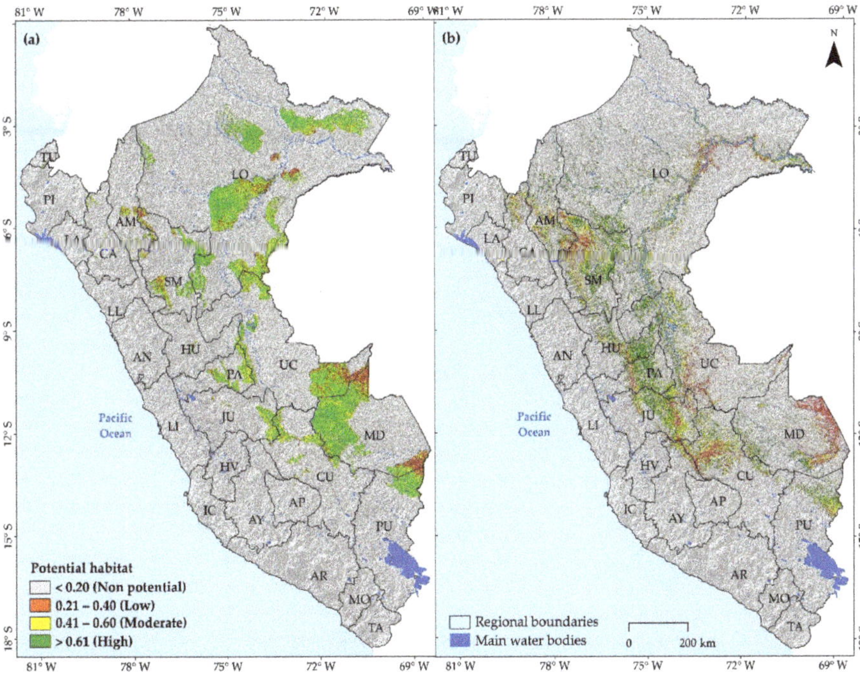

Figure 5. Priority areas (**a**) for research and conservation practices, and (**b**) for the restoration of degraded Peruvian areas with genus *Cedrela*.

After compiling the potential habitat distribution results with the information on degraded areas, a high-distribution potentiality of genus *Cedrela* was observed over 20,857.0 km^2 areas of the central and western parts of Peru (accounts for 11.4% of the study area) that are highly prone to degradation (Table 6). In other words, with proper conservation and management practices in these areas, 11.4% of degraded Peruvian Amazon can be potentially restored.

Table 6. Departments with potential area of recovery of degraded areas with *Cedrela* in Peru.

Region	Degraded Area (km^2)	Low		Moderate		High		Total	
		Km^2	%	Km^2	%	Km^2	%	Km^2	%
Amazonas	11,210	3758.4	33.5	2186.62	19.5	881.56	7.9	6826.58	60.9
Apurimac	146.36	10.93	7.5	4.58	3.1	2.05	1.4	17.56	12.0
Ayacucho	1983.93	444.99	22.4	601.95	30.3	450.5	22.7	1497.44	75.5
Cajamarca	3254.27	1013.86	31.2	927.43	28.5	359.49	11.0	2300.78	70.7
Cusco	14,955.42	5001.71	33.4	4241.09	28.4	2622.01	17.5	11,864.81	79.3
Huancavelica	255.84	62.53	24.4	34.26	13.4	34.1	13.3	130.89	51.2
Huanuco	12,492.15	6630.71	53.1	1057.65	8.5	487.87	3.9	8176.23	65.5
Junin	12,312.95	5537.25	45.0	3807.1	30.9	1449.93	11.8	10,794.28	87.7
La Libertad	1430	173.07	12.1	115.79	8.1	19.68	1.4	308.54	21.6
Lambayeque	1169.75	5.29	0.5	0.04	0.0	0	0.0	5.33	0.5
Loreto	45,320.82	19,959.46	44.0	8105.48	17.9	3635.89	8.0	31,700.83	69.9
Madre de Dios	16,224	4671.55	28.8	3127.58	19.3	5604.83	34.5	13,403.96	82.6
Pasco	7317.36	4689.39	64.1	1394.6	19.1	494.75	6.8	6578.74	89.9
Piura	2890.04	215.5	7.5	40.47	1.4	6.45	0.2	262.42	9.1
Puno	6561.03	7.64	0.1	17.85	0.3	15.47	0.2	40.96	0.6
San Martin	20,356.17	3530.22	17.3	3328.42	16.4	1621.72	8.0	8480.36	41.7
Tumbes	256.35	12.59	4.9	0	0.0	0	0.0	12.59	4.9
Ucayali	21,792.53	8418.18	38.6	5072.2	23.3	3170.73	14.5	16,661.11	76.5
Peru	183,288.15	64,143.3	35.0	34,063.1	18.6	20,857	11.4	11,9063.4	65.0

4. Discussion

4.1. Potential Distribution of Genus Cedrela

Our study is the first attempt that makes use of SDMs as a probabilistic decision-making tool [49], which allows the prediction and identification of geographic spaces of the genus *Cedrela* [50] through maximum entropy modeling technique [51]. The proposed model can be applied at regional [24,25] to national scale [52–54] that will significantly contribute to the decision-making system for the Peruvian Amazon authorities. Our model is evident with higher accuracies represented by the strong AUC values of 0.866. Among different topographic and bioclimatic variables, altitude emerges as the most significant variable, which proved to be a determining factor in distribution ranges [24,25]. Species such as *C. montana* and *C. lilloi*, are mostly located and distributed at the higher altitudes. However, the distribution of species depends on the biogeographic conditions and also has a strong influence on historical or evolutionary constraints along with biogeographical, physiological, and ecological factors [55]. In this study, we observed that the 10 species of the genus *Cedrela* covered 17 departments related to the National Forestry and Wildlife Service, as of 2021 [16], and evaluated the location of botanical collections and inventories of the species [13,53,54]. Overall, the modeled distribution of genus *Cedrela* will also help to understand the historic evaluation of genus *Cedrela* species under a spatiotemporal framework. Therefore, we believe that our modeling framework will help in the future in order to establish forest management strategies.

4.2. Conservation and Restoration of Genus Cedrela

Peru is one of the most megadiverse countries in the world and is enriched with the biogeographic distribution of various species that requires the implementation of adequate strategies for species conservation [56,57]. Among the 10 species of *genus Cedrela, C. odorata* is currently one of the important timber species, threatened by deforestation and unsustainable logging [58]. However, the PAs that harbor *C. odorata*, together with the other species of the *Cedrela* genus (10,171.03 km^2), will allow the implementation of mechanisms to maintain its population and genetic diversity, given that the PAs constitute territorial protection reserves [59,60]. Similarly, the degraded areas are the result of anthropogenic pressure and forest fires in 2019 in Peru, occupying an area of 183,288.15 km^2 [35]. Among the degraded region, 11.4% of the area is currently having a high probability of recovery through the plantation of species of high economic value such as the *Cedrela* genus. The *Cedrela* genus needs to be protected from selective logging and overexploitation over time [9,12,18,61]. This is possible through the implementation of sustainable forest management strategies [62], strengthening forest monitoring and surveillance actions [48,63], and forming strategic alliances for conservation to protect these vulnerable species [56].

This study modeled the potential distribution of the genus *Cedrela* in Peru under current climatic conditions and identified which part of this potential distribution is protected by conservation areas or coincides with degraded areas. However, future studies could evaluate the distribution in future conditions of climate change, similar to Rojas et al. [25], who studied five timber forest species in Amazonas (northeastern Peru). However, it should be noted that species distribution models in climate change scenarios should be interpreted with caution since they may overestimate the decline or increase, by not considering the qualities of the species to adapt in situ to new conditions or persist outside the conditions in which they have been observed [64,65]. Despite the above, the relatively stable distribution sites (same current and future potential) of species would be of interest and essential to ensure the success of any conservation or restoration initiative.

5. Conclusions

The current biogeographic distribution of the 10 genus *Cedrela* (*C. odorata*, *C. montana*, *C. fissilis*, *C. longipetiolulata*, *C. angustifolia*, *C. nebulosa*, *C. kuelapensis*, *C. saltensis*, *C. weberbaueri* and *C. molinensis*) using MaxEnt, covers around 6.7% of Peru, found in 17 departments under Ucayali, Loreto, and Madre de Dios that are more likely to be located in the Amazon region. Likewise, the Natural Protected Areas categorized by the Peruvian State play a fundamental role in allowing the conservation of 4.4% of the potentially high-distribution regions of its territory. Such regions have high potentiality forthe genus *Cedrela* plantations, and such plantations could possibly be protected through appropriate conservation strategies.

Our research has also allowed us to quantify that 11.4% of the degraded areas identified in Peru as of 2019, can possibly be recovered through the plantation of one or more species of *Cedrela* genus. Therefore, our study has a strong potentiality to serve as a tool for identifying geographic spaces of genus *Cedrela* under a spatiotemporal framework in order to conserve or recover its local populations in areas degraded by anthropic or natural factors.

Author Contributions: Conceptualization, A.C.S., N.B.R.B., B.K.G. and R.S.L.; formal analysis, M.O. and B.K.G.; funding acquisition, C.T.G. and M.O.; investigation, A.C.S., N.B.R.B., S.B., S.G., M.O., B.K.G. and R.S.L.; methodology, A.C.S., N.B.R.B., S.B., S.G., C.T.G., B.K.G. and R.S.L.; project administration, C.T.G., M.O. and R.S.L.; resources, C.T.G., M.O., R.S.L.; software, A.C.S. and N.B.R.B.; supervision, A.C.S., S.B., S.G. and R.S.L.; validation, A.C.S., N.B.R.B., C.T.G., M.O., B.K.G. and R.S.L.; writing—original draft preparation, A.C.S., N.B.R.B. and S.B.; writing—review and editing, A.C.S., N.B.R.B., S.B. and S.G. All authors have read and agreed to the published version of the manuscript.

Funding: This work was supported by the SNIP project N° 316114 "Service Creation Project of the Biodiversity and Conservation of Wild Species Genetic Resources Laboratory at the Toribio Rodríguez de Mendoza National University—Amazonas Region".

Institutional Review Board Statement: Not applicable.

Informed Consent Statement: Not applicable.

Data Availability Statement: Not applicable.

Acknowledgments: The authors appreciate and acknowledge the support of the Research Institute for the Sustainable Development of the Eyebrow of the Jungle (INDES-CES) of the National University Toribio Rodriguez de Amazonas (UNTRM).

Conflicts of Interest: The authors declare no conflict of interest.

References

1. Bruun, T.B.; Elberling, B.; de Neergaard, A.; Magid, J. Organic carbon dynamics in different soil types after conversion of forest to agriculture. *L. Degrad. Dev.* **2013**, *26*, 272–283. [CrossRef]
2. Sánchez-Cuervo, A.M.; de Lima, L.S.; Dallmeier, F.; Garate, P.; Bravo, A.; Vanthomme, H. Twenty years of land cover change in the southeastern Peruvian Amazon: Implications for biodiversity conservation. *Reg. Environ. Chang.* **2020**, *20*, 1–14. [CrossRef]
3. GeoBosques Bosque y Pérdida de Bosque. Available online: http://geobosques.minam.gob.pe/geobosque/view/perdida.php (accessed on 25 February 2020).
4. Flores, Y. *Crecimiento y Productividad de Plantaciones Forestales en la Amazonía Peruana*; Lima, Perú, 2010.
5. Dourojeanni, R.M. Aprovechamiento del barbecho forestal en áreas de agricultura migratoria en la Amazonía peruana. *Rev. For. Perú* **2016**, *14*, 1–33. [CrossRef]
6. Suárez de Freitas, G. *Reducción de la Deforestación (Principalmente en la Amazonía) en el Contexto del Cambio Climático y de un Enfoque de Crecimiento Verde*; MINAM: Lima, Perú, 2017.
7. Marquardt, K.; Pain, A.; Bartholdson, Ö.; Rengifo, L.R. Forest dynamics in the Peruvian Amazon: Understanding processes of change. *Small-Scale For.* **2019**, *18*, 81–104. [CrossRef]
8. Lombardi, I. *Las Poblaciones del Género Cedrela en el Perú*; UNALM: Lima, Perú, 2014; ISBN 9786124147357.
9. Escobal, J.; Aldana, U. Are nontimber forest products the antidote to rainforest degradation? Brazil nut extraction in Madre De Dios, Peru. *World Dev.* **2003**, *31*, 1873–1887. [CrossRef]
10. Lamb, A.F.A. *Fast Growing Timber Trees of the Lowland Tropics. Cedrela Odorata*; Commonwealth Forestry Institute: Oxford, UK, 1968; Volume 2, p. 46.
11. Cintrón, B.B. Cedrela odorata L. Cedro hembra, Spanish cedar. In *Silvics of North America*; United States Department of Agriculture (USDA): Washington, DC, USA, 1990; Volume 2, pp. 250–257.
12. Navarro, C.; Montagnini, F.; Hernández, G. Genetic variability of Cedrela odorata Linnaeus: Results of early performance of provenances and families from Mesoamerica grown in association with coffee. *For. Ecol. Manag.* **2004**, *192*, 217–227. [CrossRef]
13. Pennington, T.D.; Muellner, A.N. *A Monograph of CEDRELA (Meliaceae)*, 1st ed.; DH Books: England, UK, 2010; ISBN 9780953813476.

14. Palacios, W.A.; Santiana, J.; Iglesias, J. A new species of cedrela (Meliaceae) from the eastern flanks of ecuador. *Phytotaxa* **2019**, *393*, 84–88. [CrossRef]
15. MINAM. *Evaluación Dendrológica y Anatómica del las Especies del Género Cedrela*; MINAM: Lima, Perú, 2017.
16. SERFOR. *Estado Situacional del Género Cedrela en el Perú*; SERFOR: Lima, Perú, 2021.
17. Rosser, A.; Haywood, M. *Guidance for CITES Scientific Authorities*; IUCN: Gland, Switzerland; Cambridge, UK, 2002.
18. Cerrillo, R.M.N.; Agote, N.; Pizarro, F.; Ceacero, C.J.; Palacios, G. Elements for a non-detriment finding of *Cedrela* spp. in Bolivia-A CITES implementation case study. *J. Nat. Conserv.* **2013**, *21*, 241–252. [CrossRef]
19. Guisan, A.; Tingley, R.; Baumgartner, J.B.; Naujokaitis-Lewis, I.; Sutcliffe, P.R.; Tulloch, A.I.T.; Regan, T.J.; Brotons, L.; Mcdonald-Madden, E.; Mantyka-Pringle, C.; et al. Predicting species distributions for conservation decisions. *Ecol. Lett.* **2013**, *16*, 1424–1435. [CrossRef] [PubMed]
20. Fagundes, C.K.; Vogt, R.C.; De Marco Júnior, P. Testing the efficiency of protected areas in the Amazon for conserving freshwater turtles. *Divers. Distrib.* **2016**, *22*, 123–135. [CrossRef]
21. Guisan, A.; Zimmermann, N.E. Predictive Habitat Distribution Models in Ecology. *Ecol. Model.* **2000**, *135*, 147–186. [CrossRef]
22. Guisan, A.; Thuiller, W. Predicting species distribution: Offering more than simple habitat models. *Ecol. Lett.* **2005**, *8*, 993–1009. [CrossRef]
23. Peterson, A.T.; Soberón, J.; Pearson, R.G.; Anderson, R.P.; Martínez-Meyer, E.; Nakamura, M.; Araújo, M.B. *Ecological Niches and Geographic Distributions (MPB-49)*; Princeton University Press: Princeton, NJ, USA, 2011; ISBN 9780691136868.
24. Cotrina, D.A.; Castillo, E.B.; Rojas, N.B.; Oliva, M.; Guzman, C.T.; Amasifuen, C.A.; Bandopadhyay, S. Distribution models of timber species for forest conservation and restoration in the Andean-Amazonian landscape, North of Peru. *Sustainability* **2020**, *12*, 7945. [CrossRef]
25. Rojas, N.B.; Cotrina, D.A.; Castillo, E.B.; Oliva, M.; Salas, R. Current and future distribution of five timber forest species in Amazonas, Northeast Peru: Contributions towards a restoration strategy. *Diversity* **2020**, *12*, 305. [CrossRef]
26. Abdelaal, M.; Fois, M.; Fenu, G.; Bacchetta, G. Using MaxEnt modeling to predict the potential distribution of the endemic plant Rosa arabica Crép. in Egypt. *Ecol. Inform.* **2019**, *50*, 68–75. [CrossRef]
27. Naveda-Rodríguez, A.; Vargas, F.H.; Kohn, S.; Zapata-Ríos, G. Andean Condor (Vultur gryphus) in Ecuador: Geographic distribution, population size and extinction risk. *PLoS ONE* **2016**, *11*, e0151827. [CrossRef]
28. Meza Mori, G.; Barboza Castillo, E.; Torres Guzmán, C.; Cotrina Sánchez, D.A.; Guzman Valqui, B.K.; Oliva, M.; Bandopadhyay, S.; Salas López, R.; Rojas Briceño, N.B. Predictive modelling of current and future potential distribution of the spectacled bear (*Tremarctos ornatus*) in Amazonas, Northeast Peru. *Animals* **2020**, *10*, 1816. [CrossRef]
29. Gilani, H.; Arif Goheer, M.; Ahmad, H.; Hussain, K. Under predicted climate change: Distribution and ecological niche modelling of six native tree species in Gilgit-Baltistan, Pakistan. *Ecol. Indic.* **2020**, *111*, 106049. [CrossRef]
30. Qin, A.; Liu, B.; Guo, Q.; Bussmann, R.W.; Ma, F.; Jian, Z.; Xu, G.; Pei, S. Maxent modeling for predicting impacts of climate change on the potential distribution of Thuja sutchuenensis Franch., an extremely endangered conifer from southwestern China. *Glob. Ecol. Conserv.* **2017**, *10*, 139–146. [CrossRef]
31. Phillips, S.J.; Anderson, R.P.; Schapire, R.E. Maximum entropy modeling of species geographic distributions. *Ecol. Modell.* **2006**, *190*, 231–259. [CrossRef]
32. Hernandez, P.A.; Graham, C.H.; Master, L.L.; Albert, D.L. The effect of sample size and species characteristics on performance of different species distribution modeling methods. *Ecography* **2006**, *29*, 773–785. [CrossRef]
33. Aguirre-Gutiérrez, J.; Carvalheiro, L.G.; Polce, C.; van Loon, E.E.; Raes, N.; Reemer, M.; Biesmeijer, J.C. Fit-for-purpose: Species distribution model performance depends on evaluation criteria—Dutch Hoverflies as a case study. *PLoS ONE* **2013**, *8*, e63708. [CrossRef] [PubMed]
34. Sofaer, H.R.; Jarnevich, C.S.; Pearse, I.S.; Smyth, R.L.; Auer, S.; Cook, G.L.; Edwards, T.C.; Guala, G.F.; Howard, T.G.; Morisette, J.T.; et al. Development and delivery of species distribution models to inform decision-making. *Bioscience* **2019**, *69*, 544–557. [CrossRef]
35. MINAM. *Estudio para la Identificación de Áreas Degradadas y Propuesta de Monitoreo*; MINAM: Lima, Perú, 2017.
36. MINAM-SERNANP. *Áreas Naturales Protegidas Del Perú (2011–2015)—Conservación para el Desarrollo Sostenible*; MINAM-SERNANP: Lima, Perú, 2016; Volume 1.
37. Fick, S.E.; Hijmans, R.J. WorldClim 2: New 1-km spatial resolution climate surfaces for global land areas. *Int. J. Climatol.* **2017**, *37*, 4302–4315. [CrossRef]
38. Hennig, T.A.; Kretsch, J.L.; Pessagno, C.J.; Salamonowicz, P.H.; Stein, W.L. The shuttle radar topography mission. *Lect. Notes Comput. Sci. (Subser. Lect. Notes Artif. Intell. Lect. Notes Bioinform.)* **2001**, *2181*, 65–77. [CrossRef]
39. Boria, R.A.; Olson, L.E.; Goodman, S.M.; Anderson, R.P. Spatial filtering to reduce sampling bias can improve the performance of ecological niche models. *Ecol. Modell.* **2014**, *275*, 73–77. [CrossRef]
40. Stevens, G.C. The latitudinal gradient in geographical range: How so many species coexist in the tropics. *Am. Nat.* **1989**, *133*, 240–256. [CrossRef]
41. New, M.; Lister, D.; Hulme, M.; Makin, I. A high-resolution data set of surface climate over global land areas. *Clim. Res.* **2002**, *21*, 1–25. [CrossRef]

42. Hengl, T.; De Jesus, J.M.; Heuvelink, G.B.M.; Gonzalez, M.R.; Kilibarda, M.; Blagotić, A.; Shangguan, W.; Wright, M.N.; Geng, X.; Bauer-Marschallinger, B.; et al. SoilGrids250m: Global gridded soil information based on machine learning. *PLoS ONE* **2017**, *12*, e0169748. [CrossRef] [PubMed]
43. Dormann, C.F.; Elith, J.; Bacher, S.; Buchmann, C.; Carl, G.; Carré, G.; Marquéz, J.R.G.; Gruber, B.; Lafourcade, B.; Leitão, P.J.; et al. Collinearity: A review of methods to deal with it and a simulation study evaluating their performance. *Ecography* **2013**, *36*, 027–046. [CrossRef]
44. Leroy, B.; Meynard, C.N.; Bellard, C.; Courchamp, F. virtualspecies, an R package to generate virtual species distributions. *Ecography* **2016**, *39*, 599–607. [CrossRef]
45. Otieno, B.A.; Nahrung, H.F.; Steinbauer, M.J. Where did you come from? Where did you go? Investigating the origin of invasive Leptocybe species using distribution modelling. *Forests* **2019**, *10*, 115. [CrossRef]
46. Manel, S.; Williams, C.; Ormerod, S.J. Evaluating presence—Absence models in ecology: The need to account for prevalence. *J. Appl. Ecol.* **2001**, *38*, 921–931. [CrossRef]
47. Hanley, J.A.; McNeil, B.J. The meaning and use of the area under a Receiver Operating Characteristic (ROC) Curve1. *Radiology* **1982**, *143*, 29–36. [CrossRef]
48. Zhang, K.; Zhang, Y.; Tao, J. Predicting the potential distribution of Paeonia veitchii (Paeoniaceae) in China by incorporating climate change into a maxent model. *Forests* **2019**, *10*, 190. [CrossRef]
49. Soberon, J.; Peterson, A.T. Interpretation of models of fundamental ecological niches and species' distributional areas. *Biodivers. Inform.* **2005**, *2*, 3392–3396. [CrossRef]
50. Hernández, J.; Reynoso, R.; Hernández, A.; García, X.; Hernández-Máximo, E.; Cob, J.; Sumano, D. Historical, current and future distribution of Cedrela odorata in Mexico. *Acta Bot. Mex.* **2018**, *2018*, 117–134. [CrossRef]
51. Elith, J.; Graham, C.H.; Anderson, R.P.; Dudík, M.; Ferrier, S.; Guisan, A.; Hijmans, R.J.; Huettmann, F.; Leathwick, J.R.; Lehmann, A.; et al. Novel methods improve prediction of species' distributions from occurrence data. *Ecography* **2006**, *29*, 129–151. [CrossRef]
52. Pecchi, M.; Marchi, M.; Burton, V.; Giannetti, F.; Moriondo, M.; Bernetti, I.; Bindi, M.; Chirici, G. Species distribution modelling to support forest management. A literature review. *Ecol. Modell.* **2019**, *411*, 108817. [CrossRef]
53. Laurente, M. Modeling the effects of climate change on the distribution of *Cedrela odorata* L. "Cedro" in the Peruvian Amazon. *Biologist* **2015**, *13*, 213–224.
54. Reynel, C.; Pennington, T.; Pennington, R.; Flores, C.; Daza, A. *Árboles Útiles de la Amazonía Peruana y Sus Usos*; Herbario de la Facultad de Ciencias Forestales de la Universidad Agraria La Molina, Royal Botanic Gardens Kew, Royal Botanic Gardens Edinburgh e ICRAF: Lima, Peru, 2003; ISBN 9972-9733-1-X.
55. OSINFOR. *Modelamiento Espacial de Nichos Ecológicos para la Evaluación de Presencia de Especies Forestales Maderables en la Amazonía Peruana*; OSINFOR: Lima, Perú, 2013.
56. Fajardo, J.; Lessmann, J.; Bonaccorso, E.; Devenish, C.; Muñoz, J. Combined use of systematic conservation planning, species distribution modelling, and connectivity analysis reveals severe conservation gaps in a megadiverse country (Peru). *PLoS ONE* **2014**, *9*, e0122159. [CrossRef]
57. Rodriguez, L.O.; Young, K.R. Biological diversity of Peru: Determining priority areas for conservation. *Ambio* **2000**, *29*, 329–337. [CrossRef]
58. Bonilla-Bedoya, S.; Estrella-Bastidas, A.; Molina, J.R.; Herrera, M.Á. Socioecological system and potential deforestation in Western Amazon forest landscapes. *Sci. Total Environ.* **2018**, *644*, 1044–1055. [CrossRef] [PubMed]
59. Lombardi, I.; Barrena, V.; Huerta, P.; Carranza, J.; Vallejo, S. *Propuesta Para la Recuperación de las Poblaciones de Cedrela spp. en el Perú*; Universidad Nacional Agraria La Molina: Lima, Perú, (s.f.).
60. Possingham, H.P.; Wilson, K.A. Protected Areas: Goals, Limitations, and Design. In *Principies of Conservation Biology*, 3rd ed.; Groom, M.J., Meffe, G.K., Carroll, C.R., Eds.; Sinauer Associates, lnc.: Sunderland, MA, USA, 2006.
61. Romijn, E.; Lantican, C.B.; Herold, M.; Lindquist, E.; Ochieng, R.; Wijaya, A.; Murdiyarso, D.; Verchot, L. Assessing change in national forest monitoring capacities of 99 tropical countries. *For. Ecol. Manag.* **2015**, *352*, 109–123. [CrossRef]
62. Araujo, M.; Pearson, R.; Thuiller, W.; Erhard, M. Validation of species-climate impact models under climate change. *Glob. Chang. Biol.* **2005**, *11*, 1504–1513. [CrossRef]
63. Godsoe, W.; Franklin, J.; Blanchet, F.G. Effects of biotic interactions on modeled species' distribution can be masked by environmental gradients. *Ecol. Evol.* **2017**, *7*, 654–664. [CrossRef]
64. Lamont, B.B.; Connell, S.W. Biogeography of Banksia in southwestern australia. *J. Biogeogr.* **1996**, *23*, 295–309. [CrossRef]
65. Sarmiento, F.O.; Kooperman, G.J. A Socio-hydrological perspective on recent and future precipitation changes over tropical montane cloud forests in the Andes. *Front. Earth Sci.* **2019**, *7*, 324. [CrossRef]

diversity

MDPI

Article

The Effect of Climate and Human Pressures on Functional Diversity and Species Richness Patterns of Amphibians, Reptiles and Mammals in Europe

Mariana A. Tsianou *, Maria Lazarina, Danai-Eleni Michailidou, Aristi Andrikou-Charitidou, Stefanos P. Sgardelis and Athanasios S. Kallimanis

Department of Ecology, School of Biology, Aristotle University of Thessaloniki, GR-54124 Thessaloniki, Greece; mlazarin@bio.auth.gr (M.L.); dmichail@bio.auth.gr (D.-E.M.); andrikouc@bio.auth.gr (A.A.-C.); sgardeli@bio.auth.gr (S.P.S.); kalliman@bio.auth.gr (A.S.K.)

* Correspondence: mtsianou@bio.auth.gr

Abstract: The ongoing biodiversity crisis reinforces the urgent need to unravel diversity patterns and the underlying processes shaping them. Although taxonomic diversity has been extensively studied and is considered the common currency, simultaneously conserving other facets of diversity (e.g., functional diversity) is critical to ensure ecosystem functioning and the provision of ecosystem services. Here, we explored the effect of key climatic factors (temperature, precipitation, temperature seasonality, and precipitation seasonality) and factors reflecting human pressures (agricultural land, urban land, land-cover diversity, and human population density) on the functional diversity (functional richness and Rao's quadratic entropy) and species richness of amphibians (68 species), reptiles (107 species), and mammals (176 species) in Europe. We explored the relationship between different predictors and diversity metrics using generalized additive mixed model analysis, to capture non-linear relationships and to account for spatial autocorrelation. We found that at this broad continental spatial scale, climatic variables exerted a significant effect on the functional diversity and species richness of all taxa. On the other hand, variables reflecting human pressures contributed significantly in the models even though their explanatory power was lower compared to climatic variables. In most cases, functional richness and Rao's quadratic entropy responded similarly to climate and human pressures. In conclusion, climate is the most influential factor in shaping both the functional diversity and species richness patterns of amphibians, reptiles, and mammals in Europe. However, incorporating factors reflecting human pressures complementary to climate could be conducive to us understanding the drivers of functional diversity and richness patterns

Keywords: taxonomic diversity; functional richness; Rao's quadratic entropy; climatic variables; generalized additive mixed models; macroecological patterns; diversity drivers

Citation: Tsianou, M.A.; Lazarina, M.; Michailidou, D.-E.; Andrikou-Charitidou, A.; Sgardelis, S.P.; Kallimanis, A.S. The Effect of Climate and Human Pressures on Functional Diversity and Species Richness Patterns of Amphibians, Reptiles and Mammals in Europe. *Diversity* **2021**, *13*, 275. https://doi.org/10.3390/d13060275

Academic Editor: Michael Wink

Received: 27 May 2021
Accepted: 16 June 2021
Published: 18 June 2021

Publisher's Note: MDPI stays neutral with regard to jurisdictional claims in published maps and institutional affiliations.

1. Introduction

The multiple processes and factors acting simultaneously and shaping biodiversity patterns at different scales constitute a long-standing puzzle for ecologists and biogeographers [1,2]. Taxonomic diversity has been extensively used to unravel underlying mechanisms that structure communities and drive diversity patterns across scales [3]. However, other aspects of biodiversity such as functional diversity reflecting species' functional traits within communities and ecosystems [4] might provide a more detailed and integrated interpretation of diversity patterns and species composition [5,6]. Functional traits can be defined as the main dimensions of the real ecological niche [7], mediating species responses to environmental conditions and reflecting the way organisms respond to environmental variation (e.g., habitat or climate preference) [8]. Functional diversity is quantified by arranging species in a functional trait space according to their functional trait values [9]. This representation allows us to measure various aspects of functional diversity such as functional richness (i.e., the overall volume of trait space (niche) occupied by species in a community) [10] or Rao's quadratic entropy (i.e., the functional distance between all pairs of species within a community) [9]. Therefore, the analysis of functional diversity patterns along environmental gradients, including land-use changes and human pressures, is a valuable tool to unlock the role of different factors shaping biodiversity patterns and to predict possible shifts in ecosystem functioning under the prism of global change [11].

Climate is a strong driver of species distributions and diversity patterns [10,12–15]. Research has shown that climatic stress gradients can limit functional diversity; for example, only species with certain adaptations can exist in harsh environmental conditions, and as a result, more functionally similar species coexist there [13,14,16,17]. On the other hand, climate seasonality can facilitate species coexistence despite their different ecological niches [18,19] or limit functional diversity when species with narrower niches coexist in areas of higher energy availability (and/or of lower environmental seasonality) [20,21]. Beyond climatic variables, landscape and contemporary human imprint play important roles in shaping species distributions [22] and functional diversity patterns [11,23]. A negative association between human imprint and functional diversity has been reported [24,25]; for example, highly urbanized communities show significantly decreased functional diversity in contrast to natural environments [26,27], while higher vulnerability characterizes functionally dissimilar species of intensified agricultural areas [28]. However, there are counterexamples which show that urban and agricultural land changes influence community structuring and thus functional diversity by favoring specific adaptations of species to cope with the new environments [11]. Yet, our understanding of the contribution of human pressures to large-scale patterns of functional diversity and their relative importance compared to other mechanisms such as climate remains fragmentary [11,23,29]. Furthermore, functional diversity patterns have generally been investigated in single taxonomic groups, despite the fact that different taxa may play similar and/or complementary ecological and functional roles [11,30]. Comparative analyses of the functional roles and functional diversity patterns of different taxonomic groups are scarce and focused primarily on local scales [11,26], while the same question for broad scales is still in its infancy (see [30]).

Functional traits and functional diversity infer a linkage between biodiversity and ecosystem functioning [31,32]. In this context, examining how functional diversity changes with environmental conditions can shed light on the impacts of climate, landscape, and human imprint on ecosystem processes. Here, we examine the taxonomic and functional diversity patterns of three taxonomic groups (amphibians, reptiles, and mammals) across Europe and explore the effect of climate (temperature, precipitation, temperature seasonality, and precipitation seasonality) and human pressures (agricultural and urban land area, land-cover diversity, and human population density) on their diversity patterns. Amphibians, reptiles, and mammals might either have similarities in key functional roles (e.g., amphibians and reptiles) or differ in ecological roles (mammals) while also having a direct link (e.g., one taxonomic group as a feeding resource for another). Given the scale and extent of our study, we expect that climate will have a stronger influence on these patterns [15,16] in contrast to the human pressures which mainly act at local scales ([11,23]; see [33]).

2. Materials and Methods

2.1. Species Distribution Data

We compiled distributional data for 68 amphibian species, 107 reptile species, and 176 mammal species in Europe (2488 grid cells) from two atlases, *The Atlas of European Amphibians and Reptiles* [34] and *The Atlas of European Mammals* [35]. Both atlases provide distributional presence/absence data on equal area grid cells of 50 km × 50 km, based on field surveys, published records, and national atlases projected on the WGS84 coordinate reference system.

2.2. Trait Data

We compiled trait datasets for amphibians, reptiles, and mammals in Europe using several available databases (published papers, books, electronic databases; Supplementary Material, Table S1). The selection of traits was based on the completeness of the availability of species' trait data and on their previous use in quantifying amphibian, reptile, and mammal functional diversity [19,29]. Our trait selection process resulted in the following five functional trait categories (with sub-categories): (a) body length (body mass for mammals), (b) clutch size (litter size for mammals and some viviparous amphibians and reptiles), (c) activity time (nocturnal, diurnal, crepuscular only for mammals), (d) diet type (herbivore, insectivore, molluscivore (only for amphibians and reptiles), carnivore, and omnivore), (e) habit (aquatic, fossorial, ground dwelling, and above-ground dwelling/arboreal). Traits a and b were considered as numerical variables, and traits c–e as binary variables.

2.3. Functional Diversity Indices

The functional diversity of each grid cell was estimated by metrics previously evaluated for their relationship with species richness [36], and amongst them [37] were (a) functional richness, defined by the convex hull volume occupied by the species of each grid cell; and (b) functional Rao's quadratic entropy, measured as the distance between two randomly selected species within the grid cell. We applied a Gower distance matrix to capture both the numeric and binary variables in our trait dataset and then performed principal coordinates analysis (PCoA) to ordinate species along the major axes and arrange them in a multidimensional functional trait space. Metrics were calculated using the "dbFD" function in the R package "FD" [38]. Furthermore, we estimated species richness per grid cell.

2.4. Environmental Data

Four climatic variables retrieved from the WorldClim climate database [39] (mean annual temperature, annual precipitation, temperature seasonality, and precipitation seasonality) and four variables related to human pressures (agricultural land area, urban land area, land-cover diversity, and population density) were investigated to understand their relationship with functional richness and Rao's quadratic entropy. Landscape data (agricultural land area, urban land area, and land-cover diversity) were provided by the land-cover dataset CLC2000 [40], and human population density was obtained from HYDE Gridded Population version 3.1 [41]. The environmental data were reprojected and resampled to the same projection and resolution as the distribution data in QGIS version 2.10.0.

2.5. Statistical Analysis

We applied generalized additive mixed models (GAMMs) with the "gamm" function of the "mgcv" R package [42], predicting the species richness and the functional diversity (functional richness and Rao's quadratic entropy separately) of amphibians, reptiles, and mammals as a function of the climatic predictors and those related to human pressures. Specifically, we built three models for each diversity metric: (a) a climatic model that included only climatic variables, (b) a land–human model including land-use-related variables and human population density, and (c) an overall model including all the predictors. We used Poisson error distribution for species richness and Gaussian error distribution for functional diversity indices. To account for spatial autocorrelation, we included the spatial correlation structure of coordinates (Gaussian distribution). All the predictors were modelled as smooth predictors with penalized thin plate regression splines, using three knots per spline. Prior to modelling, we checked for multicollinearity among variables by applying the variance inflation factor (VIF). Since all VIF values scored <10 [43], we included all variables in the model. Total precipitation and precipitation seasonality were square-root transformed and land-cover diversity, agricultural area extent, and human population density were log10 transformed prior to analysis to improve normality. Grid cells with less than 50% land-cover were excluded from the analysis.

3. Results

Species richness patterns varied across different taxonomic groups. Amphibians and mammals showed similar patterns with higher species richness in Central Europe, while reptiles followed a different pattern showing higher species richness in Southern Europe (Figure 1a–c). Amphibians and reptiles exhibited similar spatial patterns of functional richness in Europe, where they manifest a clear latitudinal pattern with the lowest functional richness values found in northern Europe, that is, regions with low temperatures (Figure 1d,e). Interestingly, arid areas along the coastline of Southeastern Europe were poorer in terms of functional richness for amphibians than for reptiles, with reptiles having the highest values in these regions. Contrastingly, mammal functional richness showed a more uniform pattern across Europe, with moderate to high functional richness values across all of Europe (Figure 1f). Lower values of mammal functional richness were found in the coastline of the Mediterranean region (Figure 1f). Rao's quadratic entropy patterns exhibited similar results, with those of functional richness patterns for all the three examined taxa showing high and significant associations (amphibians: $R^2 = 0.72$, $p < 0.001$; reptiles: $R^2 = 0.88$, $p < 0.001$; mammals: $R^2 = 0.54$, $p < 0.001$). Cross-taxon relationships between amphibians and reptiles indicated significant moderate associations for functional richness ($R^2 = 0.27$, $p < 0.001$), but also for Rao's quadratic entropy ($R^2 = 0.40$, $p < 0.001$). On the other hand, mammal functional diversity was significantly but weakly related with amphibian (functional richness: $R^2 = 0.09$, $p < 0.001$; Rao's quadratic entropy: $R^2 = 0.02$, $p < 0.001$) and reptile (functional richness: $R^2 = 0.13$, $p < 0.001$; Rao's quadratic entropy: $R^2 = 0.14$, $p < 0.001$) functional diversity (Figure 1d–f).

The climatic model performed better than the land–human model in all cases, and the overall model generally outperformed both the climate and land–human models (Table 1). In the case of functional richness, the explanatory power of the climatic model was (in descending order of fit): $R^2 = 0.59$ for reptiles, $R^2 = 0.42$ for amphibians, and $R^2 = 0.25$ for mammals. Meanwhile, the corresponding values of the land–human model were (in descending order of fit): $R^2 = 0.37$ for amphibians, $R^2 = 0.33$ for reptiles, and $R^2 = 0.10$ for mammals. The overall model including all the predictors was the best-fitting model in all cases, scoring slightly better than the climatic model. The highest explanatory power of the overall model was observed for reptiles, followed by amphibians and mammals, in all diversity metrics (Table 1, Figure 2), indicating that the variables related to landscape and human pressures play a subordinate role to climatic variables.

Table 1. Performance (R^2) of the generalized additive mixed models explaining the species richness, functional richness and Rao's quadratic entropy patterns of three taxonomic groups (amphibians, reptiles, and mammals) in Europe (grid cell size 50 km × 50 km). Temperature, precipitation, temperature seasonality, and precipitation seasonality were used to quantify the effects of climate in the climatic model, while land-cover diversity, agricultural land area, urban land area, and human population density were used to quantify the effects of land-cover and human pressures in the land–human model. All variables were used in the overall model.

Diversity Aspect	Taxonomic Group	Climatic Model	Land–Human Model	Overall Model
Species richness				
	Amphibians	0.41	0.35	0.44
	Reptiles	0.60	0.34	0.65
	Mammals	0.30	0.19	0.35
Functional richness				
	Amphibians	0.42	0.37	0.47
	Reptiles	0.59	0.33	0.61
	Mammals	0.25	0.10	0.30
Rao's quadratic entropy				
	Amphibians	0.48	0.43	0.50
	Reptiles	0.61	0.38	0.61
	Mammals	0.23	0.10	0.27

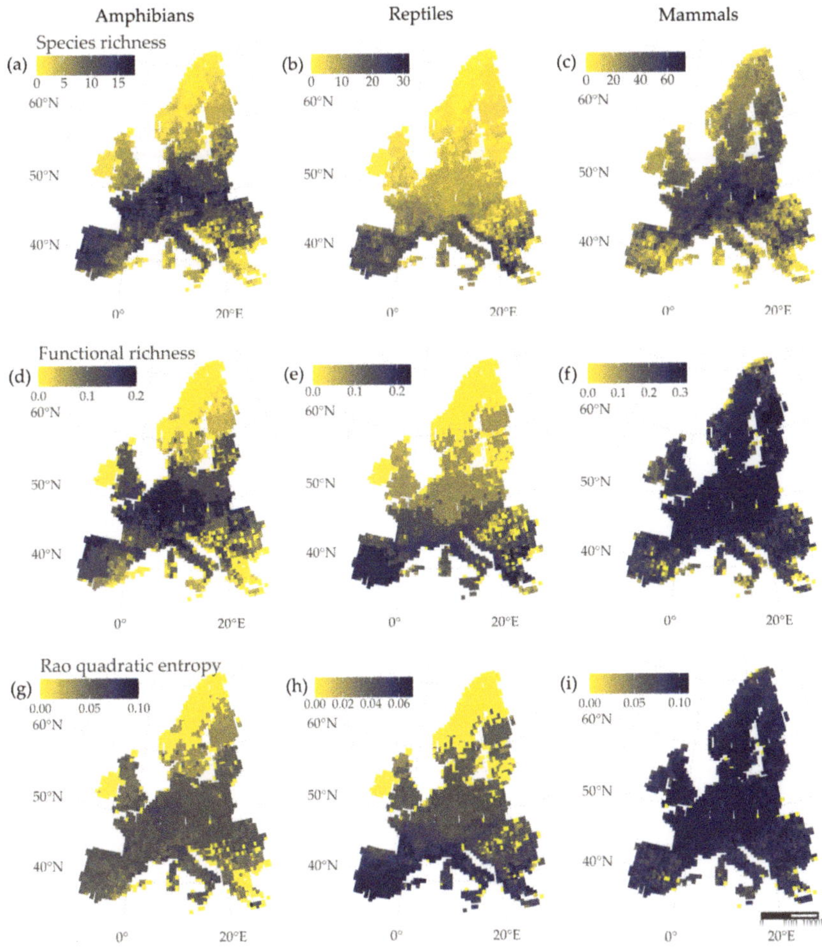

Figure 1. The species richness (**a–c**), functional richness (**d–f**), and Rao's quadratic entropy (**g–i**) distribution patterns of amphibians, reptiles, and mammals in Europe (50 km × 50 km grid cell size).

The performance and shape of the relationship between the predictors (overall model) and each functional diversity metric and species richness for each taxon are summarized in Figure 2. Temperature and precipitation seasonality were significantly related to the species richness and functional diversity of all taxa. The species richness (as in the case of both indices of functional diversity) of amphibians and mammals had a unimodal relationship with temperature, while reptile species richness and functional diversity increased with it. Precipitation seasonality exhibited a convex relationship (suggesting the existence of a bimodal relationship, with the curve showing only a part of the variability) with diversity, independently of taxon or metric, although relatively little variation of their values was observed. Temperature seasonality significantly affected the species richness of all taxa, but only mammal and reptile functional richness was positively related to temperature seasonality. Reptile species and functional richness increased significantly with precipitation, and amphibians and mammals showed a unimodal relationship. A unimodal relationship was also found for Rao's quadratic entropy of mammals. Rao's quadratic entropy of amphibians decreased with increasing percentage of urban land area. All richness and diversity measures tended to have negative relationships with the percentage of urban area. However, only in the case of reptile functional richness and Rao's quadratic entropy of amphibians was this negative relationship significant. Furthermore, the species and functional richness of amphibians and mammals tended to increase with the percentage of agricultural area, while a linear decreasing relationship was observed in the case of reptiles. Species richness for all taxa increased significantly with land-cover diversity. In addition, functional diversity measures tended to have a positive relationship with land-cover diversity, although these relationships were significant for reptiles, as well as for mammal functional richness. Finally, mammal species richness and amphibian Rao's quadratic entropy increased with human population density.

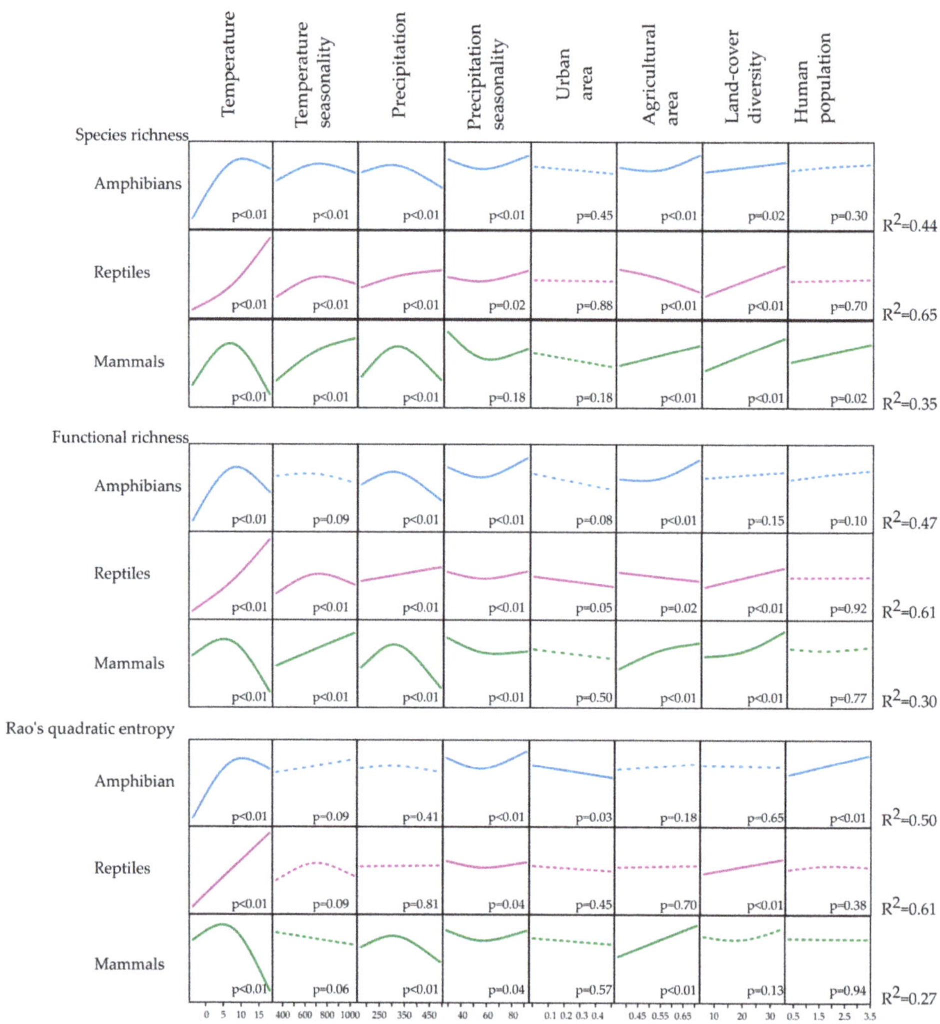

Figure 2. Summary plot showing the results of generalized additive mixed models (performance, direction, and significance of the relationship) predicting species richness, functional richness, and Rao's quadratic entropy of amphibian, reptiles, and mammals in Europe as function of climatic variables (temperature, precipitation, temperature seasonality, and precipitation seasonality) and variables related to land use and human pressures (land-cover diversity, agricultural land area, urban land area, and human population density). Continuous lines show significant associations while dotted lines show non-significant associations. Performance (R^2) is shown for the overall model in which all variables were used.

4. Discussion

4.1. Climatic Gradients

Climate overrides the effects of land use and human population density on shaping the functional diversity and species richness patterns of amphibians, reptiles, and mammals across Europe. The land–human model exhibited low explanatory power, but the land use variables and human population density slightly reinforced the explanatory power of the overall model in almost all cases (see the reptile Rao's quadratic entropy). Therefore, climate is the prevalent driver of functional diversity and species richness patterns [15,16], and although land use and human population density act at local scales [23,26], their imprint can be detected at broader scales [22].

Amphibian and reptile functional and taxonomic diversity varied along the climatic gradients, with different aspects of diversity within each taxon responding similarly to factors. The amphibian and reptile communities were species-poorer and functionally less diverse in Northern Europe, while reptile communities were species-richer and more diverse than amphibians in Southeastern Europe. The critical thermal minima of amphibians are lower than those of other ectothermic vertebrates, such as reptiles [44–46]. Therefore, the cooler temperatures of Northern Europe seem to exceed these thermal ranges and only a few species can survive there, which is reflected in their distribution [47] and functional diversity patterns [29]. In Southeastern Europe (e.g., along the coastline of the Mediterranean region), areas are characterized by high temperatures and moderate levels of precipitation, but higher seasonality. Amphibians and reptiles, as ectotherms, depend on the ambient temperature to thermoregulate, and other aspects of their physiology and behavior (e.g., reproduction) depend on the temperature and precipitation [48]. Amphibian distribution is associated more strongly with precipitation at the fine spatial scale [49]. Their body structure (i.e., the water permeability of their skin) is linked to thermoregulation, and the precipitation-dependent aspects of their ecology (e.g., most amphibians reproduce in the water) renders precipitation a crucial factor for their distribution [49,50]. Therefore, lower precipitation combined with higher temperatures (i.e., higher aridity) seems to restrict the number and range of the amphibian traits, resulting in higher trait similarity, and thus lower functional diversity [29]. In contrast, reptiles depend more strongly on temperature [49] and seem to be equipped with specific traits to cope well with high aridity [33].

Mammalian diversity patterns seem to tell a slightly different story. Their species richness varied along the climatic gradient, but mammalian functional diversity showed little variation and higher values compared to amphibians and reptiles across all of Europe. Some functionally homogeneous communities were detected scattered in Southern Europe but also in Northern Europe (particularly Norwegian coastal areas) and Ireland. The climatic and overall models performed less well for this taxon, implying that mammalian diversity is more weakly associated with climate compared to ectothermic taxa [49]. Furthermore, mammalian species richness is driven by environmental factors related to water–energy dynamics, for example, actual evapotranspiration and primary productivity in Europe [51] and globally [52], which were not included in the present study. On the other hand, functional diversity depends mostly on evolutionary time [52], and this perhaps is reflected in the lower predictive performance of our models. In temperate areas, the available time for niches to evolve, along with competition, energy availability, and adaptations to the environment, have resulted in functional divergence [19].

Seasonality was significantly related to the species richness and functional diversity of amphibians, reptiles, and mammals of Europe, confirming previous broad-scaled research on other taxa [14,16,19,29,53]. The species richness of ectothermic taxa had a unimodal relationship with temperature seasonality, while both species richness and functional diversity exhibited an approximately inverse unimodal relationship with precipitation seasonality. Amphibians are favored by low levels of precipitation seasonality that are observed in Central Europe, but also by high levels of precipitation seasonality that are observed in Southern Europe; however, in the latter region, temperature was a strong constraining factor. Ochoa-Ochoa, Mejía-Domínguez, Velasco, Marske, and Rahbek [53] explored the amphibian functional diversity in America and reported a positive association between amphibian communities with precipitation and low precipitation seasonality, while temperature was not a significant driver. It is possible that precipitation seasonality acts together with temperature, resulting in functionally even amphibian communities in areas with low precipitation seasonality in Europe [29] and low to moderate temperatures. Low to moderate levels of temperature seasonality favored reptile species richness and functional diversity, while although the effect of precipitation seasonality was significant, reptile functional diversity varied little with it. Considering the positive strong effect of temperature on reptile communities and that temperature seasonality is higher at higher latitudes [54], climatic requirements were mirrored in the higher functional diversity in Southern Europe. At higher northern latitudes, reptile assemblages are more functionally constrained by seasonal changes, and the species pool consists of more similar traits [29]. Mammal species and functional richness increased with temperature seasonality and tended to decrease with precipitation seasonality, that is, Central Europe offers more suitable climatic conditions. Interestingly, the role of seasonality in mammals suggests either that species with a variety of trait values could co-occur regionally at higher latitudes [19] or that there is a potential discrimination between coastal and mountainous areas in Southern Europe associated with high temperature seasonality (dry/warm vs. cold/wet).

4.2. Anthropogenic Gradients

Beyond the prominent role of climate, land-use-related variables and human imprint emerged as weaker but significant determinants of the functional diversity and species richness of all examined taxa. Unsurprisingly, all taxa benefited from the land use diversity, as it reflects the habitat heterogeneity, that is, the availability of niches, which allows the coexistence of greater numbers of species with diverse functional traits [48,51]. Regarding human imprint, long-term human occupancy structures communities, with human pressures acting either as drivers or filters of functional diversity [24,25,33]. Amphibian (Rao's quadratic entropy) and reptile (functional richness) functional diversity decreased with the percentage of urban land area. Urbanization has a strong negative effect on amphibian and reptile assemblages, with few species having specific adaptations being able to secure their survival in urban areas [55], as in all fragmented and largely human-modified landscapes [11], and might result in the functional homogenization of communities [27]. Although amphibian functional diversity decreased with urban area, it increased with human population density, as has previously been shown for the species richness of different taxonomic groups [56,57]. Regarding the possible mechanisms for the positive richness, human population density invoking the suitability of climatic conditions, resource availability, and spatial heterogeneity [56] might explain the more functionally diverse amphibian communities in areas of higher human population density. On the other hand, agricultural area enhanced the richness (species and functional) of amphibians and disfavored reptiles. In a recent review [55], a non-significant, albeit negative, effect of agriculture on the species richness of these taxa was reported, but in the present study this was confirmed only for reptile functional diversity. Amphibians might benefit from the availability of water related to agriculture, or in some cases the matrix structure (e.g., possible landscape heterogeneity generated by the combination of agricultural and natural habitats) [55,58,59]. Inconsistent with recent research which reports that the trophic structure (an aspect of functional diversity) of communities highly exposed to human impacts is more simplified in terms of predicted structures with climate [60], here we found that mammalian richness and functional diversity were promoted by both the percentage of agricultural area and human population density. Low- to medium-intensity agriculture might conserve the diversity and functions of reptile communities [33,61], while the reported positive association between mammalian richness and human population density in Europe [56] seems to also apply to their functional diversity.

Although amphibians, reptiles, and mammals differ in their ecological roles, given the cross-taxon differentiations, these taxa shared some similarities in their responses and therefore in key functional roles [30]. Further investigation of mechanisms which drive functional diversity patterns considering data on different taxonomic groups could reveal further insights into how species functional roles can either complement or not respond to environmental variation. Climate, land uses, and other human-related factors influence species assemblages synergistically, making it difficult to decipher their individual effects on distribution patterns [33,62]; however, this might be related to the scale of the analysis. Here, we found that some human–landscape factors significantly affected diversity patterns at broad spatial scales. The consistency of the results was also extended to the different aspects of functional diversity (functional richness and Rao's quadratic entropy). Despite the acknowledged relationship between different functional diversity metrics [36,37], applying such analyses could help us to deepen our knowledge of how different aspects of functional diversity respond to environmental variation.

5. Conclusions

Our study highlighted the roles of climate and variables related to land uses and human pressures on shaping the species richness and functional diversity of amphibians, reptiles, and mammals in Europe. We found a strong effect of climate, with the role of human imprint being significant but of lower impact. The effect of urban land area and human population density on functional diversity patterns reported here might have irreversible negative impacts on taxonomic groups such as amphibians, thus resulting in the impaired provision of ecosystem services. However, our study highlights the importance of some human-related factors (e.g., agricultural area) that could preserve communities' functions under specific circumstances. In the era of global change, neglecting human imprint may lead us to misinterpret the effects of environmental variation on the distribution of species and traits [25]. Functional diversity, a significant dimension of biodiversity, bridges ecosystem functioning and community responses to environmental change [32]. Biodiversity hotspots for terrestrial vertebrates may be extensively influenced by climate change, especially in the Mediterranean bioregion [63]. Traditional conservation practices should also implement new approaches (e.g., including research on functional traits and functional diversity) to optimize conservation planning, and thus preserve ecosystem functioning. Enhancing our understanding of the determinants and processes that govern functional diversity patterns is valuable for maintaining ecosystem resilience and stability under the prism of climate and land use change.

Supplementary Materials: The following are available online at https://www.mdpi.com/article/10.3390/d13060275/s1, Table S1: Sources of information used to compile the trait databases of studied taxa.

Author Contributions: Conceptualization, M.A.T., M.L., A.A.-C. and A.S.K.; methodology, M.A.T., M.L. and D.E.-M.; software, M.A.T., M.L. and D.E.-M.; validation, M.A.T., M.L., D.E.-M., A.A.-C. and A.S.K.; formal analysis, M.A.T., M.L. and D.E.-M.; investigation, M.A.T. and M.L.; resources, A.S.K.; data curation, M.A.T., M.L. and D.E.-M.; writing—original draft preparation, M.A.T. and M.L.; writing—review and editing, M.A.T., M.L., D.E.-M., A.A.-C., S.P.S. and A.S.K.; visualization, M.L. and D.E.-M.; supervision, A.S.K.; project administration, A.S.K.; funding acquisition, A.S.K. All authors have read and agreed to the published version of the manuscript.

Funding: Hellenic Foundation for Research and Innovation (H.F.R.I.) under the "First Call for H.F.R.I. Research Projects to support Faculty members and Researchers and the procurement of high-cost research equipment grant".

Institutional Review Board Statement: Not applicable.

Data Availability Statement: Publicly available datasets were analysed in this study. This data can be found here: http://na2re.ismai.pt/; https://www.european-mammals.org/php/mapmaker.php; https://worldclim.org/; https://www.pbl.nl/en/image/data; https://land.copernicus.eu/pan-european/corine-land-cover/clc-2000.

Acknowledgments: This research work was supported by the Hellenic Foundation for Research and Innovation (H.F.R.I.) under the "First Call for H.F.R.I. Research Projects to support Faculty members and Researchers and the procurement of high-cost research equipment grant" (Project Number: HFRI-FM17-2024 Mapping Functional Diversity Drivers, Impacts and Threats-MAPFUN). The results presented in this work were produced using the Aristotle University of Thessaloniki (AUTh) High Performance Computing Infrastructure and Resources. The authors would like to acknowledge the support provided by the IT Center of the Aristotle University of Thessaloniki (AUTh) throughout the process of this research work. We would like to thank the anonymous reviewers for their helpful comments.

Conflicts of Interest: The authors declare no conflict of interest.

References

1. Gaston, K.J. Global patterns in biodiversity. *Nature* **2000**, *405*, 220–227. [CrossRef]
2. Field, R.; Hawkins, B.A.; Cornell, H.V.; Currie, D.J.; Diniz-Filho, J.A.F.; Guégan, J.F.; Kaufman, D.M.; Kerr, J.T.; Mittelbach, G.G.; Oberdorff, T. Spatial species-richness gradients across scales: A meta-analysis. *J. Biogeogr.* **2009**, *36*, 132–147. [CrossRef]
3. Willig, M.R.; Kaufman, D.M.; Stevens, R.D. Latitudinal gradients of biodiversity: Pattern, process, scale, and synthesis. *Annu. Rev. Ecol. Evol. Syst.* **2003**, *34*, 273–309. [CrossRef]
4. Petchey, O.L.; Gaston, K.J. Functional diversity: Back to basics and looking forward. *Ecol. Lett.* **2006**, *9*, 741–758. [CrossRef] [PubMed]
5. Ricklefs, R.E. Disintegration of the ecological community: American Society of Naturalists Sewall Wright award winner address. *Am. Nat.* **2008**, *172*, 741–750. [CrossRef]
6. Münkemüller, T.; Gallien, L.; Pollock, L.J.; Barros, C.; Carboni, M.; Chalmandrier, L.; Mazel, F.; Mokany, K.; Roquet, C.; Smyčka, J. Dos and don'ts when inferring assembly rules from diversity patterns. *Glob. Ecol. Biogeogr.* **2020**, *29*, 1212–1229. [CrossRef]
7. Díaz, S.; Kattge, J.; Cornelissen, J.H.; Wright, I.J.; Lavorel, S.; Dray, S.; Reu, B.; Kleyer, M.; Wirth, C.; Prentice, I.C. The global spectrum of plant form and function. *Nature* **2016**, *529*, 167–171. [CrossRef]
8. Violle, C.; Navas, M.L.; Vile, D.; Kazakou, E.; Fortunel, C.; Hummel, I.; Garnier, E. Let the concept of trait be functional! *Oikos* **2007**, *116*, 882–892. [CrossRef]
9. Villéger, S.; Mason, N.W.; Mouillot, D. New multidimensional functional diversity indices for a multifaceted framework in functional ecology. *Ecology* **2008**, *89*, 2290–2301. [CrossRef]
10. Mason, N.W.; Mouillot, D.; Lee, W.G.; Wilson, J.B. Functional richness, functional evenness and functional divergence: The primary components of functional diversity. *Oikos* **2005**, *111*, 112–118. [CrossRef]
11. Concepción, E.D.; Götzenberger, L.; Nobis, M.P.; de Bello, F.; Obrist, M.K.; Moretti, M. Contrasting trait assembly patterns in plant and bird communities along environmental and human-induced land-use gradients. *Ecography* **2017**, *40*, 753–763. [CrossRef]
12. Mouillot, D.; Bellwood, D.R.; Baraloto, C.; Chave, J.; Galzin, R.; Harmelin-Vivien, M.; Kulbicki, M.; Lavergne, S.; Lavorel, S.; Mouquet, N. Rare species support vulnerable functions in high-diversity ecosystems. *PLoS Biol.* **2013**, *11*, e1001569. [CrossRef]
13. Schleuter, D.; Daufresne, M.; Veslot, J.; Mason, N.W.; Lanoiselée, C.; Brosse, S.; Beauchard, O.; Argillier, C. Geographic isolation and climate govern the functional diversity of native fish communities in European drainage basins. *Glob. Ecol. Biogeogr.* **2012**, *21*, 1083–1095. [CrossRef]
14. Swenson, N.G.; Enquist, B.J.; Pither, J.; Kerkhoff, A.J.; Boyle, B.; Weiser, M.D.; Elser, J.J.; Fagan, W.F.; Forero-Montaña, J.; Fyllas, N. The biogeography and filtering of woody plant functional diversity in North and South America. *Glob. Ecol. Biogeogr.* **2012**, *21*, 798–808. [CrossRef]
15. Bello, F.d.; Lavorel, S.; Lavergne, S.; Albert, C.H.; Boulangeat, I.; Mazel, F.; Thuiller, W. Hierarchical effects of environmental filters on the functional structure of plant communities: A case study in the French Alps. *Ecography* **2013**, *36*, 393–402. [CrossRef]
16. Shiono, T.; Kusumoto, B.; Maeshiro, R.; Fujii, S.J.; Götzenberger, L.; de Bello, F.; Kubota, Y. Climatic drivers of trait assembly in woody plants in Japan. *J. Biogeogr.* **2015**, *42*, 1176–1186. [CrossRef]
17. Mason, N.W.; Irz, P.; Lanoiselée, C.; Mouillot, D.; Argillier, C. Evidence that niche specialization explains species–energy relationships in lake fish communities. *J. Anim. Ecol.* **2008**, *77*, 285–296. [CrossRef] [PubMed]
18. Canavero, A.; Arim, M.; Brazeiro, A. Geographic variations of seasonality and coexistence in communities: The role of diversity and climate. *Austral Ecol.* **2009**, *34*, 741–750. [CrossRef]
19. Safi, K.; Cianciaruso, M.V.; Loyola, R.D.; Brito, D.; Armour-Marshall, K.; Diniz-Filho, J.A.F. Understanding global patterns of mammalian functional and phylogenetic diversity. *Philos. Trans. R. Soc. B Biol. Sci.* **2011**, *366*, 2536–2544. [CrossRef]
20. Brown, J.S.; Kotler, B.P.; Porter, W.P. How foraging allometris and resource dynamics could explain Bergmann's rule and the body-size diet relationship in mammals. *Oikos* **2017**, *126*, 224–230. [CrossRef]
21. Evans, K.L.; Warren, P.H.; Gaston, K.J. Species-energy relationships at the macroecological scale: A review of the mechanisms. *Biol. Rev.* **2005**, *80*, 1–25. [CrossRef] [PubMed]
22. Xu, C.; Huang, Z.Y.; Chi, T.; Chen, B.J.; Zhang, M.; Liu, M. Can local landscape attributes explain species richness patterns at macroecological scales? *Glob. Ecol. Biogeogr.* **2014**, *23*, 436–445. [CrossRef]

23. Sams, M.; Lai, H.; Bonser, S.; Vesk, P.; Kooyman, R.; Metcalfe, D.; Morgan, J.; Mayfield, M. Landscape context explains changes in the functional diversity of regenerating forests better than climate or species richness. *Glob. Ecol. Biogeogr.* **2017**, *26*, 1165–1176. [CrossRef]
24. Rapacciuolo, G.; Marin, J.; Costa, G.C.; Helmus, M.R.; Behm, J.E.; Brooks, T.M.; Hedges, S.B.; Radeloff, V.C.; Young, B.E.; Graham, C.H. The signature of human pressure history on the biogeography of body mass in tetrapods. *Glob. Ecol. Biogeogr.* **2017**, *26*, 1022–1034. [CrossRef]
25. Santini, L.; González-Suárez, M.; Rondinini, C.; Di Marco, M. Shifting baseline in macroecology? Unravelling the influence of human impact on mammalian body mass. *Divers. Distrib.* **2017**, *23*, 640–649. [CrossRef]
26. Concepción, E.D.; Moretti, M.; Altermatt, F.; Nobis, M.P.; Obrist, M.K. Impacts of urbanisation on biodiversity: The role of species mobility, degree of specialisation and spatial scale. *Oikos* **2015**, *124*, 1571–1582. [CrossRef]
27. Sol, D.; Trisos, C.; Múrria, C.; Jeliazkov, A.; González-Lagos, C.; Pigot, A.L.; Ricotta, C.; Swan, C.M.; Tobias, J.A.; Pavoine, S. The worldwide impact of urbanisation on avian functional diversity. *Ecol. Lett.* **2020**, *23*, 962–972. [CrossRef] [PubMed]
28. Flynn, D.F.; Gogol-Prokurat, M.; Nogeire, T.; Molinari, N.; Richers, B.T.; Lin, B.B.; Simpson, N.; Mayfield, M.M.; DeClerck, F. Loss of functional diversity under land use intensification across multiple taxa. *Ecol. Lett.* **2009**, *12*, 22–33. [CrossRef]
29. Tsianou, M.A.; Kallimanis, A.S. Geographical patterns and environmental drivers of functional diversity and trait space of amphibians of Europe. *Ecol. Res.* **2020**, *35*, 123–138. [CrossRef]
30. Cooke, R.S.; Bates, A.E.; Eigenbrod, F. Global trade-offs of functional redundancy and functional dispersion for birds and mammals. *Glob. Ecol. Biogeogr.* **2019**, *28*, 484–495. [CrossRef]
31. McGill, B.J.; Enquist, B.J.; Weiher, E.; Westoby, M. Rebuilding community ecology from functional traits. *Trends Ecol. Evol.* **2006**, *21*, 178–185. [CrossRef] [PubMed]
32. Mori, A.S.; Furukawa, T.; Sasaki, T. Response diversity determines the resilience of ecosystems to environmental change. *Biol. Rev.* **2013**, *88*, 349–364. [CrossRef] [PubMed]
33. Barnagaud, J.Y.; Geniez, P.; Cheylan, M.; Crochet, P.A. Climate overrides the effects of land use on the functional composition and diversity of Mediterranean reptile assemblages. *Divers. Distrib.* **2021**, *27*, 50–64. [CrossRef]
34. Sillero, N.; Campos, J.; Bonardi, A.; Corti, C.; Creemers, R.; Crochet, P.-A.; Isailović, J.C.; Denoël, M.; Ficetola, G.F.; Gonçalves, J. Updated distribution and biogeography of amphibians and reptiles of Europe. *Amphib. Reptil.* **2014**, *35*, 1–31. [CrossRef]
35. Mitchell-Jones, A.J.; Amori, G.; Bogdanowicz, W.; Krystufek, B.; Reijnders, P.; Spitzenberger, F.; Stubbe, M.; Thissen, J.; Vohralik, V.; Zima, J. *The Atlas of European Mammals*; Academic Press: London, UK, 1999; Volume 3.
36. Mouchet, M.A.; Villéger, S.; Mason, N.W.; Mouillot, D. Functional diversity measures: An overview of their redundancy and their ability to discriminate community assembly rules. *Funct. Ecol.* **2010**, *24*, 867–876. [CrossRef]
37. Tsianou, M.A.; Kallimanis, A.S. Different species traits produce diverse spatial functional diversity patterns of amphibians. *Biodivers. Conserv.* **2016**, *25*, 117–132. [CrossRef]
38. Laliberté, E.; Legendre, P.; Shipley, B.; Laliberté, M.E. Package 'FD'. Measuring Functional Diversity from Multiple Traits, and Other Tools for Functional Ecology; R Package Version. 2014. Available online: https://cran.r-project.org/web/packages/FD/FD.pdf (accessed on 18 June 2021).
39. Hijmans, R.J.; Cameron, S.E.; Parra, J.L.; Jones, P.G.; Jarvis, A. Very high resolution interpolated climate surfaces for global land areas. *Int. J. Climatol. J. R. Meteorol. Soc.* **2005**, *25*, 1965–1978. [CrossRef]
40. Heymann, Y. *Corine Land Cover: Technical Guide*; Commission of the European Communities: Luxembourg, 1994.
41. Klein Goldewijk, K.; Beusen, A.; Van Drecht, G.; De Vos, M. The HYDE 3.1 spatially explicit database of human-induced global land-use change over the past 12,000 years. *Glob. Ecol. Biogeogr.* **2011**, *20*, 73–86. [CrossRef]
42. Wood, S.; Wood, M.S. Package 'mgcv'. *R Package Version* **2015**, *1*, 29.
43. Quinn, G.P.; Keough, M.J. *Experimental Design and Data Analysis for Biologists*; Cambridge University Press: Cambridge, UK, 2002.
44. Buckley, L.B.; Hurlbert, A.H.; Jetz, W. Broad-scale ecological implications of ectothermy and endothermy in changing environments. *Glob. Ecol. Biogeogr.* **2012**, *21*, 873–885. [CrossRef]
45. Sunday, J.M.; Bates, A.E.; Dulvy, N.K. Global analysis of thermal tolerance and latitude in ectotherms. *Proc. R. Soc. B Biol. Sci.* **2011**, *278*, 1823–1830. [CrossRef] [PubMed]
46. Sunday, J.M.; Pecl, G.T.; Frusher, S.; Hobday, A.J.; Hill, N.; Holbrook, N.J.; Edgar, G.J.; Stuart-Smith, R.; Barrett, N.; Wernberg, T. Species traits and climate velocity explain geographic range shifts in an ocean-warming hotspot. *Ecol. Lett.* **2015**, *18*, 944–953. [CrossRef] [PubMed]
47. Tsianou, M.A.; Koutsias, N.; Mazaris, A.D.; Kallimanis, A.S. Climate and landscape explain richness patterns depending on the type of species' distribution data. *Acta Oecol.* **2016**, *74*, 19–27. [CrossRef]
48. Rodríguez, M.Á.; Belmontes, J.A.; Hawkins, B.A. Energy, water and large-scale patterns of reptile and amphibian species richness in Europe. *Acta Oecol.* **2005**, *28*, 65–70. [CrossRef]
49. Aragón, P.; Lobo, J.M.; Olalla-Tárraga, M.Á.; Rodríguez, M.Á. The contribution of contemporary climate to ectothermic and endothermic vertebrate distributions in a glacial refuge. *Glob. Ecol. Biogeogr.* **2010**, *19*, 40–49. [CrossRef]
50. Buckley, L.B.; Jetz, W. Insularity and the determinants of lizard population density. *Ecol. Lett.* **2007**, *10*, 481–489. [CrossRef] [PubMed]

51. Fløjgaard, C.; Normand, S.; Skov, F.; Svenning, J.C. Deconstructing the mammal species richness pattern in Europe–towards an understanding of the relative importance of climate, biogeographic history, habitat heterogeneity and humans. *Glob. Ecol. Biogeogr.* **2011**, *20*, 218–230. [CrossRef]
52. Oliveira, B.F.; Machac, A.; Costa, G.C.; Brooks, T.M.; Davidson, A.D.; Rondinini, C.; Graham, C.H. Species and functional diversity accumulate differently in mammals. *Glob. Ecol. Biogeogr.* **2016**, *25*, 1119–1130. [CrossRef]
53. Ochoa-Ochoa, L.M.; Mejía-Domínguez, N.R.; Velasco, J.A.; Marske, K.A.; Rahbek, C. Amphibian functional diversity is related to high annual precipitation and low precipitation seasonality in the New World. *Glob. Ecol. Biogeogr.* **2019**, *28*, 1219–1229. [CrossRef]
54. Carrara, R.; Vázquez, D.P. The species–energy theory: A role for energy variability. *Ecography* **2010**, *33*, 942–948. [CrossRef]
55. Cordier, J.M.; Aguilar, R.; Lescano, J.N.; Leynaud, G.C.; Bonino, A.; Miloch, D.; Loyola, R.; Nori, J. A global assessment of amphibian and reptile responses to land-use changes. *Biol. Conserv.* **2021**, *253*, 108863. [CrossRef]
56. Araújo, M.B. The coincidence of people and biodiversity in Europe. *Glob. Ecol. Biogeogr.* **2003**, *12*, 5–12. [CrossRef]
57. Luck, G.W. A review of the relationships between human population density and biodiversity. *Biol. Rev.* **2007**, *82*, 607–645. [CrossRef] [PubMed]
58. Guerra, C.; Aráoz, E. Amphibian diversity increases in an heterogeneous agricultural landscape. *Acta Oecol.* **2015**, *69*, 78–86. [CrossRef]
59. Brüning, L.Z.; Krieger, M.; Meneses-Pelayo, E.; Eisenhauer, N.; Pinilla, M.P.R.; Reu, B.; Ernst, R. Land-use heterogeneity by small-scale agriculture promotes amphibian diversity in montane agroforestry systems of northeast Colombia. *Agric. Ecosyst. Environ.* **2018**, *264*, 15–23. [CrossRef]
60. Mendoza, M.; Araujo, M.B. Climate shapes mammal community trophic structures and humans simplify them. *Nat. Commun.* **2019**, *10*, 5197. [CrossRef]
61. De Solan, T.; Renner, I.; Cheylan, M.; Geniez, P.; Barnagaud, J.Y. Opportunistic records reveal Mediterranean reptiles' scale-dependent responses to anthropogenic land use. *Ecography* **2019**, *42*, 608–620. [CrossRef]
62. Ferger, S.W.; Peters, M.K.; Appelhans, T.; Detsch, F.; Hemp, A.; Nauss, T.; Otte, I.; Böhning-Gaese, K.; Schleuning, M. Synergistic effects of climate and land use on avian beta-diversity. *Divers. Distrib.* **2017**, *23*, 1246–1255. [CrossRef]
63. Maiorano, L.; Amori, G.; Capula, M.; Falcucci, A.; Masi, M.; Montemaggiori, A.; Pottier, J.; Psomas, A.; Rondinini, C.; Russo, D. Threats from climate change to terrestrial vertebrate hotspots in Europe. *PLoS ONE* **2013**, *8*, e74989.

diversity

MDPI

Article

The Bird Assemblage of the Darwin Region (Australia): What Is the Effect of Twenty Years of Increasing Urbanisation?

Sarah E. Fischer [1,*], Andrew C. Edwards [1], Patrice Weber [1], Stephen T. Garnett [2] and Timothy G. Whiteside [3]

1 Research Institute for the Environment and Livelihoods, College of Engineering, IT & Environment, Charles Darwin University, Darwin, NT 0909, Australia; andrew.edwards@cdu.edu.au (A.C.E.); patrice.weber@cdu.edu.au (P.W.)
2 College of Engineering, IT & Environment, EIE Environment, Charles Darwin University, Darwin, NT 0909, Australia; stephen.garnett@cdu.edu.au
3 Environmental Research Institute of the Supervising Scientist, Darwin, NT 0820, Australia; tim.whiteside@environment.gov.au
* Correspondence: sarah.fischer@cdu.edu.au

Abstract: There has been considerable urban development in the Darwin region over the last twenty years; as for most fauna in Australia since colonisation, the potential effects to the bird assemblage were expected to be disastrous. To provide a broad overview of changes, bird survey data from 1998 and 2018 were extracted from BirdLife Australia's 'Atlas of Australian Birds' database. A total of 165 species were categorised into primary food source feeding guilds and levels of food specialisation. This was integrated into ArcGIS along with land use change mapping from 1998 and 2018 to investigate its impact on bird assemblages. There was no significant change in overall species numbers when all sites were analysed. However, when sites were separated into those with increased urbanisation or decreased greenspace, several sites showed a significant change in the number of species. For the majority of species, analysis of primary food types found no difference in the proportion of species within the assemblages between 1998 and 2018, regardless of the level of urbanisation or greenspace; the exception being those species that primarily feed on insects, where the difference was just significant. An analysis using bird community data sorted into levels of food specialisation also found no difference between 1998 and 2018 despite habitat changes. These findings suggest that although there has been considerable urban development in the Darwin region, bird communities are remaining relatively stable.

Keywords: urban birds; bird assemblages; urbanisation; landscape ecology; land use change; geographic object-based image analysis (GEOBIA); Australian monsoonal tropics

Citation: Fischer, S.E.; Edwards, A.C.; Weber, P.; Garnett, S.T.; Whiteside, T.G. The Bird Assemblage of the Darwin Region (Australia): What Is the Effect of Twenty Years of Increasing Urbanisation? *Diversity* **2021**, *13*, 294. https://doi.org/10.3390/d13070294

Academic Editor: Juan J. Morrone

Received: 30 April 2021
Accepted: 21 June 2021
Published: 28 June 2021

1. Introduction

Global population trends show that humans are moving from rural areas into cities at a considerable rate, and once sparsely populated regions are being transformed to cope with the influx of people [1,2]. Subsequently, urbanisation is now widely considered a major threat to biodiversity conservation [3–5].

Species' responses to urbanisation can vary significantly [6]; however, prior to the 1990s, very few ecological studies were undertaken in urban areas due to them being considered unviable habitats for faunal populations and, ergo, immaterial from a conservation standpoint [7]. It is now shown that urban areas support an array of species that have been able to tolerate or adapt to the highly fragmented new environments but, due to a lack of appropriate knowledge, the success of urban conservation programmes may be hampered [4]. There is little doubt that avian assemblages are greatly altered by urbanisation; many bird species unable to adapt to the suddenly changed environment will often move out of an area resulting in diminished biodiversity and allowing invasive species to move in [4,8,9]. Other species can benefit from urbanisation; these are often birds with a more generalist diet and life-history traits that are conducive to living in a fragmented habitat [10].

Whilst studies of bird populations in urban areas abound in the temperate zones of both the Northern and Southern hemispheres, research into tropical urban avian communities is scant [11]. This trend is continued in Australian studies with the majority of research being undertaken in the temperate regions, predominantly in the two heavily populated states of New South Wales and Victoria. In contrast, this study is situated in the monsoonal tropics of northern Australia; specifically, in Darwin, the capital of the Northern Territory (NT). There has been targeted research on a range of species and thorough overviews of bird distributions within the region [12–14], with more detailed studies of shorebirds [15–18] and mangrove assemblages [19–21]. To date, however, there have been no studies of trends in terrestrial bird assemblages as the city has grown. Given the increased rate of urbanisation in the global tropics [22], coupled with high levels of biodiversity in the zone [11,23–25], there is an increasing need to gather data to better understand how bird communities are coping with the rising encroachment of human habitation. This is particularly interesting in Darwin as, unlike all other urban centres in Australia, Darwin has no introduced birds, so all the adaptation is being undertaken by native species as their environment is changed.

In the nearly 45 years since Darwin was severely damaged by Cyclone Tracy, where 70–80% of dwellings were destroyed [26,27], the city has grown from a perceived 'frontier town' to be a modern capital city. This development has seen an increase of over 20 new suburbs in the greater Darwin area and surrounds, and land once considered bush or rural properties is now being subdivided into urban blocks. One of the most common effects of urbanisation is the increasing prevalence of exotic species [28–30]; however, this has not occurred in the Darwin region with only four species listed as 'foreign invaders', none of which have established permanent breeding populations [31].

In this paper, we shall test this theory by investigating the response of the terrestrial avian assemblages in the Darwin region to urban expansion and land use change over the past 20 years.

2. Materials and Methods

2.1. Study Area

Darwin (12.4634° S, 130.8456° E) is situated on the north coast of Australia, a landscape dominated by tropical savanna (Figure 1).

The population of approximately 140,000 [32] constitutes an increase of nearly 60,000 people in twenty years [33]. The climate is monsoonal and experiences distinct annual dry (May to September) and wet (November to April) seasons with transitional periods in between. Mean annual rainfall is approximately 1700 mm; the mean minimum and maximum temperatures range from 19.3° to 25.3° and 30.6° to 33.3°, respectively [34]. Compared to other major Australian capital cities, housing density is low; under 20 private dwellings per square kilometre as opposed to between 150 to 200 per square kilometre in Sydney and Melbourne [35]. A combination of urban and periurban environments in the Darwin region provides resources for avian populations that are typically unavailable outside of this area in the dry season.

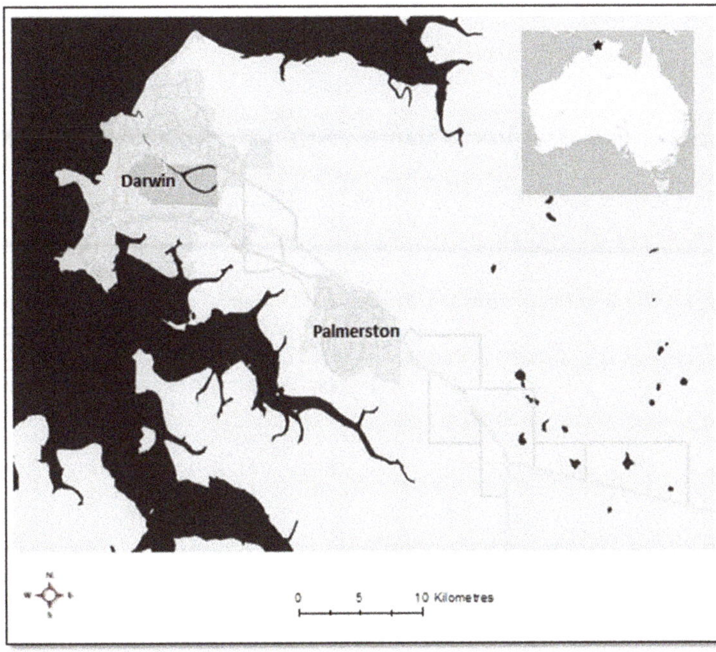

Figure 1. Study area of the Greater Darwin region (map data sources: Esri, DeLorme, HERE, USGS, Intermap, iPC, NRCAN, Esri Japan, METI, Esri China (Hong Kong), Esri (Thailand), MapmyIndia, Tomtom).

2.2. Spatial Data

To investigate broad-scale changes in urbanisation, Landsat satellite imagery of the study area from April 1998 and 2018 was obtained from the United States Geological Survey's Global Visualization Viewer (GloVis), with 1998 imagery obtained from the Landsat 4 and 5 Thematic Mapper (TM) satellite and 2018 imagery taken by the Landsat 8 Operational Land Imager (OLI) and Thermal Infrared Sensor (TIRS) (Tables A1 and A2). April marks the end of the wet season in Darwin, allowing for images with minimal cloud cover and maximum vegetation growth. Images were clipped in ArcGIS version 10.4.1 [36] to a shapefile of Darwin region localities provided by the Northern Territory Government Department of Environment and Natural Resources and then imported into SAGA GIS version 7.3.0 [37]. Classification of land use types was carried out via geographic object-based image analysis (GEOBIA). In traditional pixel-based image classification, classes are assigned per pixel; however, GEOBIA uses segmentation and classification to better replicate what the human eye perceives [38,39]. The segmentation process combines pixels of similar spectral properties into objects in the form of polygons, and these can then be classified using either supervised or unsupervised techniques [40]. After automated segmentation in SAGA GIS, the two clipped images were classified by assigning 'training sites' (essentially selecting a minimum number of polygons and ascribing them a land use type) and then running a supervised classification. The resulting vector layers were then manually edited using the original satellite image to reassign any misclassified polygons. The initial uncorrected GEOBIA and user-corrected images were then re-imported back into ArcGIS, where fifty accuracy assessment points were randomly generated and an error matrix was constructed to assess both the producer (SAGA) and user (human) accuracy when assigning classification. For all map classifications satellite imagery, aerial photographs and Google Earth Pro version 7.3.2.5776 (64 bit) were used to assist in the accuracy assessment; however, due to the retrospective nature of the earlier imagery, only the 2018 images could be further checked, if required, using ground control points.

2.3. Bird Survey Data

Survey data were extracted from the BirdLife Australia 'Atlas of Australian Birds' database (hereafter referred to as the 'Bird Atlas') for the years 1998 and 2018. Several types of surveys compiled these data: systematic bird surveys of 2 ha, 5 km and 500 m; unstandardised bird surveys, either along a fixed route or incidental; bird list surveys and the Shorebird 2020 surveys (a record of shorebird sightings not necessarily in coastal habitats). All records include a location, latitude and longitude, dates and species common names. In most records, a time is recorded and whether there is any breeding activity. Sighting notes of interest are sometimes included.

As the focus of the project was on terrestrial, predominantly diurnal species, Bird Atlas records were excluded if the species was almost exclusively nocturnal, was a waterbird or seabird (except Magpie Geese, *Anseranas semipalmata*), or the species was considered 'vagrant'. Using information from BirdLife Australia [41], the Atlas of Living Australia [42] and Australian Bird Data Version 1.0 [43], the feeding preferences of species were categorised from most preferred to occasional.

To give a general overview of any assemblage changes, the records of 1998 and 2018 were categorised into the following primary food sources: fruit, insects, invertebrates, nectar, omnivore, raptor, scavenger, seed, vegetation or vertebrate. If a species was considered to feed on two types of food source equally, both were considered the primary food source. Species were sorted by their level of specialisation: whether they had one, two, three or more food sources (Table A3).

2.4. Integration of Data

Following Hahs and McDonnell [44] and Conole and Kirkpatrick [30], the final edited spatial images were re-imported into ArcGIS and a 1 $\times$ 1 km grid was overlaid. The modified Bird Atlas data from 1998 and 2018 were then added as point layer files. As with Conole and Kirkpatrick [30], the locations were taken from the coordinates provided. The grid cells that contained records from both 1998 and 2018 were extracted for each year, and the level of land use type in each grid cell was calculated. Land use types were combined and simplified into greenspace (woodland, grass and forest), coastal (mangrove, sea and sand), urbanised (suburban, periurban, building and road), water and bare earth. Percentage change of greenspace and urbanisation between 1998 and 2018 was calculated for the relevant grid cells.

2.5. Statistical Analysis

Data were analysed using R version 3.6.1 [45].

Analysis of variance (ANOVA) was used to investigate whether the site and year had an effect on overall species numbers; paired *t*-tests were then used to examine the significance of any changes between 1998 and 2018 in those sites where urbanisation had increased or greenspace had decreased. The analysis was repeated to assess species' primary food source categories.

Finally, effects of land change on feeding specialisation numbers in bird assemblages were also explored. To do this, assemblage species proportions of one, two, three or more food sources were analysed for grid cells that were determined to have had an increase in urbanisation, a decrease in urbanisation, no change in the amount of urbanisation, an increase in greenspace or a decrease in greenspace.

3. Results

3.1. Spatial Data

Results of the GEOBIA found that urbanised areas increased from 114 to 151 km^2, a percentage increase of almost 3% of the total land mass measured. Greenspace, too, increased by just under 5%; 986 to 1045 km^2.

For the 1998 imagery, overall accuracy was 75% (producer accuracy 72%; user accuracy 84%). The 2018 imagery returned an overall accuracy of 87% (84% producer and 90% user accuracies). This difference in accuracy was expected due to the quality differences in the Landsat 5 versus Landsat 8 imagery. Final corrections were made to the classified imagery before extraction of the relevant land use data for analysis (Figure 2).

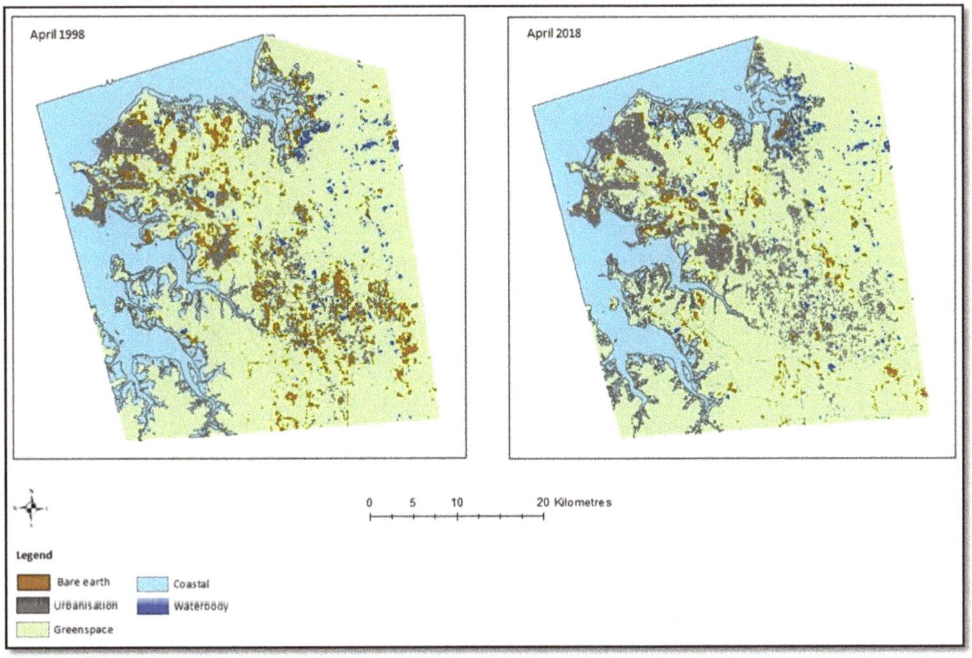

Figure 2. Final classified images of land use change with combined land classes.

3.2. Bird Atlas Data

Overall, there were 23 grid cells that contained Bird Atlas records from both 1998 and 2018. Of these, nine showed an increase in urbanisation ranging from 0.9 to 58.2%; eight indicated a decrease in urbanisation and six were recorded as no change. A decrease in greenspace was found in 15 grid cells, and 8 were determined to have had an increase. No species considered primarily a scavenger was found in the 23 grid cells.

3.3. Overall Change in Species Numbers

In total, 165 individual species were represented in the bird records used. Of these, 23 species were recorded in 1998 only and 19 in 2018 only (Table A4). All sites showed a change in species numbers between 1998 and 2018; however, when all sites were included, there was no significant change in total species numbers (F = 1.1909, p = 0.3999 and F = 1.9395, p = 0.1707, respectively).

Focussing on those sites where urbanisation had increased or greenspace had decreased over the 20-year period, changes in species numbers were more distinctive. Although sites are listed numerically, this is for brevity, and they are not necessarily the same location.

Where urbanisation had increased, of the nine sites analysed, five sites showed an increase in overall species numbers with four of these increases found to be statistically significant ($p < 0.05$). Examination of sites where species numbers decreased found two where the decrease was significant and two where there was no statistical significance (Table 1).

Table 1. Results of paired *t*-test for difference in the overall number of bird species per site between 1998 and 2018 in areas where urbanisation has increased (df = 8).

Site	Percentage Increase in Urbanisation	Percentage Change in Species Numbers	*t*	*p*-Value
1	15.0	537.5	−3.739	0.006
2	0.9	92.0	−2.842	0.022
3	35.1	164.7	−3.563	0.007
4	58.2	−74.5	3.162	0.013
5	7.4	−42.1	2.177	0.061
6	5.8	336.4	−4.331	0.003
7	4.2	−16.2	0.603	0.563
8	8.2	−84.6	2.475	0.038
9	39.1	17.2	−1.552	0.159

In sites where there was a decrease in greenspace, the majority (10 out of 15) displayed an increase in species numbers, but of these, only three were found to be significant. The remaining five sites where a decrease in species was recorded again found three with statistical significance (Table 2).

Table 2. Results of paired *t*-test for difference in the overall number of bird species per site between 1998 and 2018 in areas where greenspace has decreased (df = 14).

Site	Percentage Decrease in Greenspace	Percentage Change in Species Numbers	*t*	*p*-Value
1	2.9	537.5	−3.739	0.006
2	2.6	155.6	−1.974	0.084
3	13	61.8	−1.452	0.185
4	1.7	72.7	−0.828	0.431
5	27.2	164.7	−3.563	0.007
6	13.4	−74.5	3.162	0.013
7	14.2	336.4	−4.331	0.003
8	5.3	−90.5	4.122	0.003
9	31.1	−16.2	0.603	0.563
10	15.6	−84.6	2.475	0.038
11	0.1	−17.2	0.533	0.609
12	15.5	63.6	−0.972	0.359
13	10.7	17.2	−1.552	0.159
14	0.6	56.3	−1.455	0.184
15	1.7	192.3	−1.954	0.087

3.4. Changes to Species Numbers with Regard to Primary Food Sources

Species for each site were categorised by their primary food source, and changes in numbers initially were record as either an increase, decrease or no change for each food source.

In areas where urbanisation had increased, site 6 (increase in urbanisation of 5.8%) showed an increase in every species type, whereas site 4, which had the highest increase in urbanisation of any site (58.2%), displayed a decrease in almost every species type bar one, in which there was no change (Table 3).

Table 3. Change in species numbers across sites where urbanisation has increased; I = increase, D = decrease and NC = no change.

Primary Food Source	Site								
	1	**2**	**3**	**4**	**5**	**6**	**7**	**8**	**9**
Fruit	I	I	I	D	D	I	I	D	I
Insect	I	I	I	D	I	I	I	D	I
Invertebrate	I	D	I	D	D	I	D	D	D
Nectar	I	I	I	D	NC	I	I	NC	I
Omnivore	I	I	NC	D	NC	I	D	D	D
Raptor	I	I	I	D	D	I	D	NC	I
Seed	I	I	I	D	D	I	I	D	I
Vegetation	D	I	NC	NC	NC	I	NC	NC	D
Vertebrate	I	I	I	D	D	I	D	NC	NC

Subsequent analysis found that there was no significant change in species numbers by primary food source overall (Table 4).

Table 4. Change in species numbers across sites where greenspace has decreased; I = increase, D = decrease and NC = no change.

Primary Food Source	Site														
	1	**2**	**3**	**4**	**5**	**6**	**7**	**8**	**9**	**10**	**11**	**12**	**13**	**14**	**15**
Fruit	I	I	D	I	I	D	I	D	I	D	NC	I	I	D	I
Insect	I	I	I	I	I	D	I	D	I	D	D	I	I	I	I
Invertebrate	I	I	I	I	I	D	I	D	D	D	I	D	D	I	I
Nectar	I	NC	I	D	I	D	I	D	I	NC	D	I	I	I	I
Omnivore	I	I	NC	D	I	D	I	D	D	D	I	D	D	NC	I
Raptor	I	I	I	D	I	D	I	D	D	NC	D	I	I	I	I
Seed	I	I	I	D	I	D	I	D	I	D	NC	I	I	I	I
Vegetation	D	NC	D	NC	I	NC	I	NC	NC	NC	D	D	D	NC	NC
Vertebrate	I	D	D	D	I	D	I	D	D	NC	NC	I	NC	NC	I

When investigating those sites where greenspace had decreased, one site (7) showed an increase in every species type despite having a greater reduction in greenspace (14.2%) than either site 6 (13.4%) or site 8 (5.3%), both of which showed a reduction in every species type except those that feed primarily on vegetation (Table 4).

Unlike those sites where urbanisation had increased, analysis of changes in species numbers in the sites where greenspace had decreased found a significant change in those species whose primary food source was insects ($p < 0.05$). Some change was noted for raptors too, but this was not quite statistically significant ($p = 0.06$) (Table 5).

Table 5. Results of paired t-test for difference in the number of bird species by primary food source between 1998 and 2018 in areas where greenspace has decreased (df = 14).

Primary Food Source	t	p-Value
Fruit	−0.556	0.589
Insect	−2.519	0.025
Invertebrate	−1.017	0.326
Nectar	−1.167	0.263
Omnivore	−0.654	0.524
Raptor	−2.049	0.060
Seed	−1.311	0.211
Vegetation	1.740	0.104
Vertebrate	−0.688	0.503

3.5. Changes to Bird Assemblages with Regard to Primary Food Sources

Alongside changes to the number of species within bird communities over time, it was also pertinent to investigate whether the proportion of feeding guild types was altered within assemblages due to changing habitats, specifically in those sites where urbanisation increased or greenspace decreased.

A comparison of the proportions of species type for each site where urbanisation was found to have increased is shown in Figure 3.

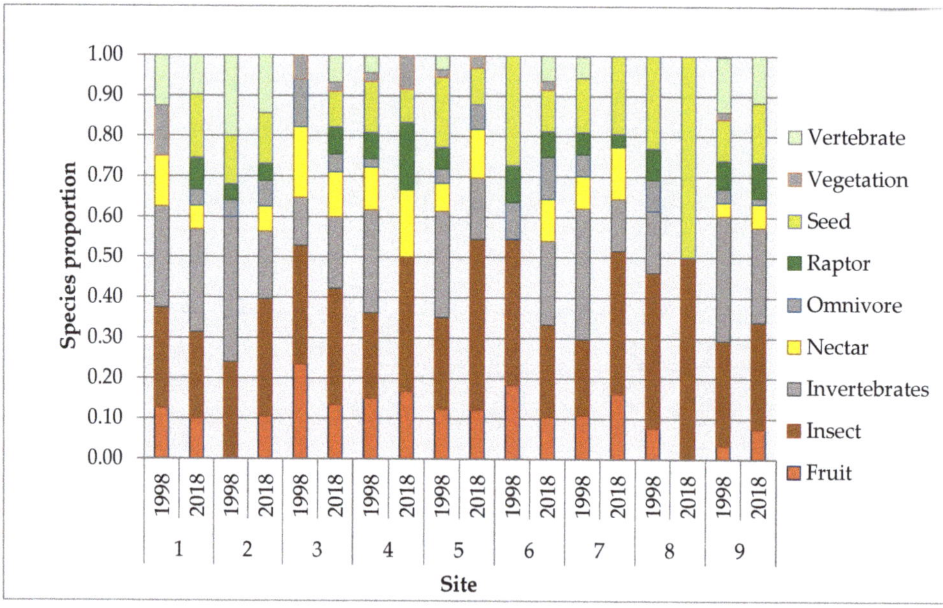

Figure 3. Proportion of species by primary food source at sites where urbanisation has increased.

Subsequent analysis showed an increase in urbanisation resulted in no significant difference in the proportion of species of different primary food sources between 1998 and 2018.

However, in grid cells that had a decreased amount of greenspace, a slight significant difference was shown in those species that chiefly feed on insects, and there is a suggestion of change in those species that primarily feed on fruit, although this was not found to be significant (Table 6).

Table 6. Results of paired t-test for difference in the proportion of bird species by primary food source between 1998 and 2018 in areas where greenspace has decreased (df = 14).

Primary Food Source	t	p-Value
Fruit	1.849	0.086
Insect	−2.156	0.049
Invertebrate	0.072	0.943
Nectar	0.547	0.593
Omnivore	1.520	0.151
Raptor	−1.346	0.200
Seed	0.458	0.654
Vegetation	1.517	0.152
Vertebrate	1.341	0.201

A comparison of species proportion by primary food source for those grid cell sites where greenspace had decreased over the 20-year period is shown in Figure 4.

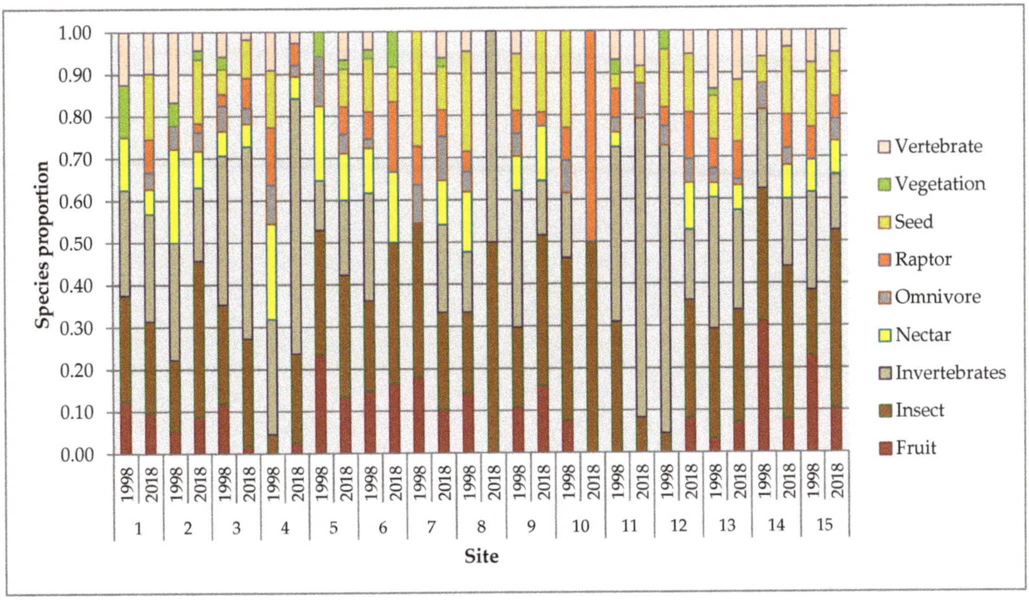

Figure 4. Proportion of species by primary food source at sites where greenspace has decreased.

3.6. Levels of Specialisation

The levels of feeding specialisation within bird communities between 1998 and 2018 showed no significant change despite varying levels of urbanisation and greenspace (Table 7).

Table 7. Results of paired *t* test for difference in the proportion of feeding specialisation in bird assemblages with changing land use.

Habitat Change	df	*t*	*p*-Value
Urbanisation increase	26	0	1
Urbanisation decrease	23	−1.048	1
No change in urbanisation	17	0.018	0.986
Greenspace increase	23	0.002	0.999
Greenspace decrease	44	0.005	0.999

4. Discussion

Of the 258 birds species recorded in the Darwin area (escaped or introduced species and vagrants excluded), approximately 40% are considered resident [46] with the remaining species described as mobile. It has previously been thought that mobile species cope better with urbanisation than sedentary species [47,48], although other studies have suggested that highly mobile species may be more vulnerable to habitat changes due to their dependence on larger habitat patches and generally far-reaching home ranges [49–51]. Furthermore, there is also evidence to suggest that the fragmentation of habitats caused by urbanisation is going some way to changing the status of some birds from 'visitor' to 'resident' [52–54].

That there is a significant difference, albeit slight, in the proportion of those species that primarily feed on insects in areas where greenspace has decreased may possibly be attributed to the establishment of urban gardens. Although the size of the greenspace has decreased overall in these locations, in 73% of the grid cells analysed, the proportion of species that preferred insects increased; in contrast, only 47% of grid cells contained an increase in fructivores. It may be that the plants in these gardens attract a larger array of insects, thus providing more appealing spaces for insectivores. Numerous studies have indicated that the bird species that tend to thrive in urban environments are those that nest in cavities or canopies (sites less likely to be disturbed by human activity) and are more likely to be granivorous or omnivorous, whereas those species that avoid urban areas are those that predominantly nest in shrubs or trees or at ground level and have a more specialised diet, frequently insectivores [29,30,55]. However, these characteristics have been determined from research undertaken in either the Northern Hemisphere or the temperate zones of the Southern Hemisphere, places where garden vegetation is often vastly different to that of the original habitat. Suburban gardens of the Darwin region often contain native plant species alongside exotics, and this mixture may be enough to maintain, or even increase, insect numbers.

That there is no significant difference in the levels of feeding specialisation in bird assemblages is something of a surprise, as it would be fair to assume that more generalised species would be increasing and those species with one preferred food source would be forced out, particularly in areas of urban increase. Homogenisation of habitat is an oft-quoted result of urbanisation [5,56,57], but from this broad overview, it appears not to be affecting the composition of bird communities in the Darwin region. Further research into the characteristics of species making up urban assemblages will go some way to better understanding why this may be.

Invasion by feral species is a frequent result of an increased human presence in an area [29,30,58]. Many bird species that are unable to adapt to the suddenly changed environment will often move out of an area resulting in diminished biodiversity and allowing invasive species to move in [4,8,9]. Other species can benefit from urbanisation; these are often birds with a more generalist diet and life-history traits that are conducive to living in a fragmented habitat [10].

The only feral species in the records was *Columba livia*—the common pigeon, rock dove or rock pigeon. This species was recorded in the 1998 records only, most likely due to an eradication programme undertaken by the Parks and Wildlife Commission, NT, beginning in 1996, that saw the species considered eradicated from the Darwin area by 2004 [59,60]. The lack of established feral bird populations in the Darwin region could, potentially, suggest that native bird species are filling the 'urban exploiter' or 'suburban adapter' niche that is commonly filled by exotic species in other cities, but there is little or no research to show what these species may be. Additionally, a unique aspect of the Darwin region is that it is the only capital in Australia where urban fires are a regular annual occurrence. The city is located within the monsoonal tropics; a landscape dominated by tropical savannas, experiencing frequent fires during the prolonged dry season from May to October, often within just metres of suburban housing areas [61]. In addition to wildfires, many of these fires are prescribed, being undertaken by various land-management bodies as part of a yearly landscape management approach [62,63]. Whether these annual fires are in some way supporting and maintaining the local bird assemblages whilst preventing invasive species from establishing populations is an intriguing question. Research into the combined impact of urban fires and human modification of the environment on local bird populations over time is ongoing; it is hoped that the results will go some way towards answering this question.

5. Conclusions

Whilst very transient, the population of Darwin and its surrounds has grown significantly between the years 1998 and 2018, and urbanisation has expanded more than 50 km from the city centre. Despite this, this general overview of bird assemblages shows that although species numbers fluctuate, avian communities appear to be stable. The increase in the insectivorous species may be due to the establishment, maturity and vegetative composition of suburban gardens. This may potentially signify a shift in the species diversity of the area; however, further research is needed to investigate this.

Author Contributions: S.E.F. conceived, designed and performed the experiments. S.E.F. analysed the data. S.E.F. and A.C.E. wrote the paper. P.W., S.T.G. and T.G.W. provided guidance and comments on the manuscript. All authors have read and agreed to the published version of the manuscript.

Funding: This study is part of a larger Ph.D. project that is supported by Charles Darwin University and the Australian Government via the Research Training Programme scholarship.

Institutional Review Board Statement: Not applicable.

Data Availability Statement: The authors confirm that the data supporting the findings of this study are available within the article, its Appendices, and Australian Bird Data Version 1.0 (https://doi.org/10.6084/m9.figshare.1499292_D8, access date 10 September 2019) at http://www.birdsaustralia.com.au/ (access date 19 June 2019) and https://www.ala.org.au/ (access date 1 September 2019).

Acknowledgments: The authors would like to thank Andrew Silcox and BirdLife Australia for access to bird data records.

Conflicts of Interest: The authors declare no conflict of interest.

Appendix A. Landsat Satellite Specifications

Table A1. Landsat 4 and Landsat 5 Thematic Mapper (TM) 1998 imagery (Data Taken from USGS [64,65]).

Band	Wavelength (Micrometres)	Resolution (Metres)
1—Blue	0.45–0.52	30
2—Green	0.52–0.60	30
3—Red	0.63–0.69	30
4—Near Infrared	0.76–0.90	30
5—Short-wave Infrared	1.55–1.75	30
6—Thermal Infrared	10.40–12.50	120 (30) *
7—Short-wave Infrared	2.08–2.35	30

* Band 6 has thermal infrared resolution of 120 metres but is re-sampled to 30-metre pixels.

Table A2. Landsat 8 and 9 Operational Land Imager (OLI) and Thermal Infrared Sensor (TIRS) 2018 imagery (Data Taken from USGS [64,65]).

Band	Wavelength (Micrometres)	Resolution (Metres)
1—Coastal aerosol	0.43–0.45	30
2—Blue	0.45–0.51	30
3—Green	0.53–0.59	30
4—Red	0.64–0.67	30
5—Near Infrared (NIR)	0.85–0.88	30
6—SWIR 1	1.57–1.65	30
7—SWIR 2	2.11–2.29	30
8—Panchromatic	0.50–0.68	15
9—Cirrus	1.36–1.38	30
10—Thermal Infrared (TIRS) 1	10.6–11.19	100
11—Thermal Infrared (TIRS) 2	11.50–12.51	100

Appendix B. Bird Species

Table A3. Total bird species recorded.

Species Name	Year Recorded		Primary Food Source	Number of Food Sources
	1998	2018		
Australasian Figbird *(Sphecotheres vieilloti ashbyi)*	√	√	Fruit	2
Australian Magpie *(Cracticus tibicen)*	X	√	Omnivore	3+
Australian Pied Oystercatcher *(Haematopus longirostris)*	√	√	Invertebrates	2
Australian White Ibis *(Threskiornis moluccus)*	√	√	Omnivore	3+
Azure Kingfisher *(Ceyx azureus ruficollaris)*	√	X	Invertebrates	2
Bar-breasted Honeyeater *(Ramsayornis fasciatus)*	X	√	Nectar/Insects	2
Bar-shouldered Dove *(Geopelia humeralis)*	√	√	Seed	1
Bar-tailed Godwit *(Limosa lapponica)*	√	√	Invertebrates	3+
Beach Stone-curlew *(Esacus magnirostris)*	X	√	Invertebrates	1
Black Bittern *(Ixobrychus flavicollis)*	√	√	Vertebrates	1
Black Butcherbird *(Cracticus quoyi)*	√	√	Vertebrates	3+
Black Kite *(Milvus migrans)*	√	√	Raptor	2
Black-faced Cuckoo-shrike *(Coracina novaehollandiae melanops)*	√	√	Invertebrates	3+
Black-faced Woodswallow *(Artamus cinereus melanops)*	√	√	Insects	1
Black-fronted Dotterel *(Elseyornis melanops)*	√	X	Invertebrates	2
Black-necked Stork *(Ephippiorhynchus asiaticus)*	√	√	Vertebrates	2
Black-shouldered Kite *(Elanus axillaris)*	X	√	Raptor	2
Black-tailed Godwit *(Limosa limosa)*	√	√	Invertebrates	3+
Black-winged Stilt *(Himantopus himantopus)*	√	√	Invertebrates	2
Blue-faced Honeyeater *(Entomyzon cyanotis albipennis)*	√	√	Insects	3+
Blue-winged Kookaburra *(Dacelo leachii leachii)*	√	√	Vertebrates/ Invertebrates	2
Brahminy Kite *(Haliastur indus)*	√	√	Raptor	2
	1998	2018		
Broad-billed Flycatcher *(Myiagra ruficollis)*	√	X	Insects	2
Broad-billed Sandpiper *(Limicola falcinellus)*	X	√	Invertebrates	2
Brolga *(Grus rubicunda)*	√	√	Omnivore	3+
Brown Falcon *(Falco berigora)*	X	√	Raptor	1
Brown Goshawk *(Accipiter fasciatus didimus)*	√	√	Raptor	1
Brown Honeyeater *(Lichmera indistincta indistincta)*	√	√	Nectar	2
Brown Quail *(Coturnix ypsilophora)*	√	√	Seed	3+
Brown-capped Emerald-Dove *(Chalcophaps longirostris)*	√	√	Seed	2
Brush Cuckoo *(Cacomantis variolosus)*	√	√	Insects	1
Buff-banded Rail *(Hypotaenidia philippensis)*	X	√	Omnivore	3+
Bush Stone-curlew *(Burhinus grallarius)*	X	√	Invertebrates	2
Cattle Egret *(Ardea ibis)*	√	X	Insects	3+
Channel-billed Cuckoo *(Scythrops novaehollandiae)*	X	√	Fruit	3+
Chestnut Rail *(Eulabeornis castaneoventris)*	√	√	Invertebrates	1
Chestnut-breasted Mannikin *(Lonchura castaneothorax)*	√	√	Seed	2
Cicadabird *(Coracina tenuirostris)*	√	X	Insects	2
Collared Kingfisher *(Todiramphus chloris sordidus)*	√	√	Invertebrates	2
Common Greenshank *(Tringa nebularia)*	√	√	Insects	3+
Common Sandpiper *(Actitis hypoleucos)*	√	√	Insects	2
Crimson Finch *(Neochmia phaeton phaeton)*	√	√	Seed	2
Curlew Sandpiper *(Calidris ferruginea)*	√	√	Insects	2
Dollarbird *(Eurystomus orientalis)*	√	√	Insects	1
Double-barred Finch *(Taeniopygia bichenovii annulosa)*	√	√	Seed	2
Dusky Honeyeater *(Myzomela obscura obscura)*	√	√	Nectar	2
Eastern Curlew *(Numenius madagascariensis)*	√	√	Invertebrates	1

Table A3. *Cont.*

Species Name	Year Recorded		Primary Food Source	Number of Food Sources
	1998	2018		
Eastern Koel *(Eudynamys orientalis subcyanocephala)*	√	√	Fruit	3+
Eastern Reef Egret *(Egretta sacra)*	√	√	Vertebrates/ Invertebrates	2
Eurasian Coot *(Fulica atra)*	√	X	Omnivore	3+
Forest Kingfisher *(Todiramphus macleayii macleayii)*	√	√	Invertebrates	2
Galah *(Eolophus roseicapilla)*	√	√	Seed	1
Glossy Ibis *(Plegadis falcinellus)*	√	√	Invertebrates	2
Golden-headed Cisticola *(Cisticola exilis)*	√	√	Insects	2
Great Bowerbird *(Ptilonorhynchus nuchalis nuchalis)*	√	X	Omnivore	3+
Great Egret *(Ardea modesta)*	√	√	Vertebrates	2
Great Knot *(Calidris tenuirostris)*	√	√	Invertebrates	1
Greater Sand Plover *(Charadrius leschenaultii)*	√	√	Invertebrates	2
Green-backed Gerygone *(Gerygone chloronota)*	√	√	Insects	1
Grey Butcherbird *(Cracticus torquatus colletti)*	X	√	Invertebrates	3+
Grey Plover *(Pluvialis squatarola)*	√	√	Insects	3+
Grey Shrike-thrush *(Colluricincla harmonica)*	√	√	Omnivore	3+
Grey Whistler *(Pachycephala simplex simplex)*	√	√	Insects	1
Grey-crowned Babbler *(Pomatostomus temporalis rubeculus)*	√	√	Invertebrates	3+
Grey-tailed Tattler *(Tringa brevipes)*	√	√	Invertebrates	1
Helmeted Friarbird *(Philemon buceroides ammitophila)*	√	√	Nectar	3+
Horsfield's Bushlark *(Mirafra javanica)*	√	√	Seed	2
Intermediate Egret *(Ardea intermedia)*	√	√	Vertebrates	1
Jacky Winter *(Microeca fascinans pallida)*	X	√	Insects	1
Large-billed Gerygone *(Gerygone magnirostris magnirostris)*	√	√	Insects	1
Large-tailed Nightjar *(Caprimulgus macrurus)*	√	X	Insects	1
Leaden Flycatcher *(Myiagra rubecula concinna)*	√	√	Insects	1
	1998	2018		
Lemon-bellied Flycatcher *(Microeca flavigaster flavigaster)*	√	√	Insects	1
Lesser Sand Plover *(Charadrius mongolus)*	√	√	Invertebrates	1
Little Bronze-cuckoo *(Chalcites minutillus minutillus)*	√	√	Insects	1
Little Corella *(Cacatua sanguinea sanguinea)*	√	√	Seed	2
Little Curlew *(Numenius minutus)*	√	√	Insects	3+
Little Egret *(Egretta garzetta)*	√	√	Invertebrates	2
Little Friarbird *(Philemon citreogularis sordidus)*	√	√	Nectar	3+
Little Kingfisher *(Ceyx pusilla ramsayi)*	X	√	Vertebrates/ Invertebrates	2
Little Shrike-thrush *(Colluricincla megarhyncha parvula)*	√	√	Insects	2
Long-tailed Finch *(Poephila acuticauda hecki)*	√	√	Insects	1
Long-toed Stint *(Calidris subminuta)*	√	X	Insects	3+
Magpie Goose *(Anseranas semipalmata)*	√	√	Seed	2
Magpie-lark *(Grallina cyanoleuca neglecta)*	√	√	Insects	2
Mangrove Gerygone *(Gerygone levigaster levigaster)*	√	√	Insects	1
Mangrove Golden Whistler *(Pachycephala melanura melanura)*	√	X	Insects	1
Mangrove Grey Fantail *(Rhipidura phasiana)*	X	√	Insects	1
Marsh Sandpiper *(Tringa stagnatilis)*	√	√	Invertebrates	2
Masked Finch *(Poephila personata personata)*	X	√	Seed	2
Masked Lapwing *(Vanellus miles miles)*	√	√	Insects	2
Mistletoebird *(Dicaeum hirundinaceum)*	√	√	Fruit	2
Nankeen Kestrel *(Falco cenchroides)*	√	X	Raptor	2
Nankeen Night-heron *(Nycticorax caledonicus)*	√	X	Vertebrates	2
Northern Fantail *(Rhipidura rufiventris)*	√	√	Insects	1
Olive-backed Oriole *(Oriolus sagittatus affinis)*	√	√	Fruit	3+
Orange-footed Scrubfowl *(Megapodius reinwardt)*	√	√	Insects	3+

Table A3. *Cont.*

Species Name	Year Recorded		Primary Food Source	Number of Food Sources
	1998	2018		
Oriental Plover *(Charadrius veredus)*	√	√	Insects	1
Osprey *(Pandion cristatus)*	√	√	Raptor	1
Pacific Golden Plover *(Pluvialis fulva)*	√	√	Insects	3+
Peaceful Dove *(Geopelia striata placida)*	√	√	Seed	2
Pectoral Sandpiper *(Calidris melanotos)*	√	X	Insects	1
Pheasant Coucal *(Centropus phasianinus melanurus)*	√	√	Invertebrates	2
Pied Butcherbird *(Cracticus nigrogularis picatus)*	X	√	Invertebrates	3+
Pied Heron *(Egretta picata)*	√	√	Vertebrates	3+
Purple Swamphen *(Porphyrio porphyrio melanotus)*	√	X	Vertebrates/ Invertebrates	3+
Rainbow Bee-eater *(Merops ornatus)*	√	√	Insects	1
Rainbow Pitta *(Pitta iris iris)*	√	√	Invertebrates	2
Red Knot *(Calidris canutus)*	√	√	Insects	3+
Red-backed Button-quail *(Turnix maculosus)*	√	X	Seed	2
Red-backed Fairy-wren *(Malurus melanocephalus cruentatus)*	√	X	Insects	1
Red-backed Kingfisher *(Todiramphus pyrrhopygius)*	√	√	Vertebrates	2
Red-capped Plover *(Charadrius ruficapillus)*	√	√	Invertebrates	2
Red-collared Lorikeet *(Trichoglossus haematodus rubritorquis)*	√	√	Vegetation	3+
Red-headed Honeyeater *(Myzomela erythrocephala erythrocephala)*	√	√	Nectar	2
Red-kneed Dotterel *(Erythrogonys cinctus)*	√	X	Invertebrates	2
Red-necked Stint *(Calidris ruficollis)*	√	√	Omnivore	3+
Red-tailed Black-Cockatoo *(Calyptorhynchus banksii macrorhychus)*	√	√	Seed	2
Red-winged Parrot *(Aprosmictus erythropterus coccineopterus)*	√	√	Seed	3+
Restless Flycatcher *(Myiagra inquieta nana)*	√	√	Insects	2
Rock Dove *(Columba livia)*	√	X	Seed	2
Rose-crowned Fruit-Dove *(Ptilinopus regina)*	√	√	Fruit	1
	1998	2018		
Royal Spoonbill *(Platalea regia)*	√	√	Vertebrates/ Invertebrates	2
Ruddy Turnstone *(Arenaria interpres)*	√	√	Invertebrates	1
Rufous Fantail *(Rhipidura rufifrons)*	X	√	Insects	1
Rufous-banded Honeyeater *(Conopophila albogularis)*	√	√	Invertebrates	2
Rufous-throated Honeyeater *(Conopophila rufogularis)*	√	√	Invertebrates	3+
Sacred Kingfisher *(Todiramphus sanctus)*	√	√	Invertebrates	2
Sanderling *(Calidris alba)*	√	X	Invertebrates	3+
Sharp-tailed Sandpiper *(Calidris acuminata)*	√	√	Invertebrates	1
Shining Flycatcher *(Myiagra alecto melvillensis)*	√	√	Invertebrates	2
Silver-crowned Friarbird *(Philemon argenticeps argenticeps)*	√	X	Nectar	2
Sooty Oystercatcher *(Haematopus fuliginosus)*	√	√	Invertebrates	2
Spangled Drongo *(Dicrurus bracteatus baileyi)*	√	√	Omnivore	3+
Square-tailed Kite *(Lophoictinia isura)*	X	√	Raptor	1
Straw-necked Ibis *(Threskiornis spinicollis)*	√	√	Insects	2
Striated Heron *(Butorides striatus stagnatilis)*	√	√	Vertebrates	2
Striated Pardalote *(Pardalotus striatus uropygialis)*	√	√	Insects	1
Sulphur-crested Cockatoo *(Cacatua galerita fitzroyi)*	√	√	Fruit	3+
Tawny Frogmouth *(Podargus strigoides phalaenoides)*	X	√	Invertebrates	2
Terek Sandpiper *(Xenus cinereus)*	√	√	Invertebrates	1
Torresian Crow *(Corvus orru)*	√	√	Omnivore	3+
Torresian Pied Imperial-pigeon *(Ducula bicolor)*	√	√	Fruit	1
Tree Martin *(Petrochelidon nigricans)*	√	√	Insects	1
Varied Lorikeet *(Psitteuteles versicolor)*	√	√	Nectar	1
Varied Triller *(Lalage leucomela rufiventris)*	√	√	Insects	2
Weebill *(Smicrornis brevirostris flavescens)*	√	X	Insects	1

Table A3. *Cont.*

Species Name	Year Recorded		Primary Food Source	Number of Food Sources
	1998	2018		
Whimbrel *(Numenius phaeopus)*	√	√	Invertebrates	2
Whistling Kite *(Haliastur sphenurus)*	√	√	Raptor	2
White-bellied Cuckoo-shrike *(Coracina papuensis hypoleuca)*	√	√	Insects	3+
White-bellied Sea-eagle *(Haliaeetus leucogaster)*	√	√	Raptor	2
White-breasted Woodswallow *(Artamus leucorynchus)*	√	√	Insects	2
White-browed Crake *(Amaurornis cinerea)*	√	√	Invertebrates	3+
White-faced Heron *(Egretta novaehollandiae)*	X	√	Invertebrates	3+
White-gaped Honeyeater *(Stomiopera unicolor)*	√	√	Nectar	3+
White-necked Heron *(Ardea pacifica)*	√	√	Vertebrates/ Invertebrates	2
White-throated Gerygone *(Gerygone olivacea rogersi)*	√	X	Insects	1
White-throated Honeyeater *(Melithreptus albogularis)*	√	√	Insects	2
White-winged Triller *(Lalage sueurii)*	√	√	Insects	1
Willie Wagtail *(Rhipidura leucophrys picata)*	√	√	Insects	1
Yellow Oriole *(Oriolus flavocinctus flavocinctus)*	√	√	Fruit	1
Wood Sandpiper *(Tringa glareola)*	√	√	Invertebrates	1
Yellow Wagtail *(Motacilla flava tschutschensis)*	√	X	Insects	1
Yellow White-eye *(Zosterops luteus luteus)*	√	√	Insects	3+
Zitting Cisticola *(Cisticola juncidis leanyeri)*	√	√	Insects	1

Table A4. Bird species recorded in 1998 only or 2018 only.

Species Name	Primary Food Source
1998	
Azure Kingfisher *(Ceyx azureus ruficollaris)*	Invertebrates
Black-fronted Dotterel *(Elseyornis melanops)*	Invertebrates
Broad-billed Flycatcher *(Myiagra ruficollis)*	Insects
Cattle Egret *(Ardea ibis)*	Insects
Cicadabird *(Coracina tenuirostris)*	Insects
Eurasian Coot *(Fulica atra)*	Omnivore
Great Bowerbird *(Ptilonorhynchus nuchalis nuchalis)*	Omnivore
Large-tailed Nightjar *(Caprimulgus macrurus)*	Insects
Long-toed Stint *(Calidris subminuta)*	Insects
Mangrove Golden Whistler *(Pachycephala melanura melanura)*	Insects
Nankeen Kestrel *(Falco cenchroides)*	Raptor
Nankeen Night-heron *(Nycticorax caledonicus)*	Vertebrates
Pectoral Sandpiper *(Calidris melanotos)*	Insects
Purple Swamphen *(Porphyrio porphyrio melanotus)*	Vertebrates/Invertebrates
Red-backed Button-quail *(Turnix maculosus)*	Seed
Red-backed Fairy-wren *(Malurus melanocephalus cruentatus)*	Insects
Red-kneed Dotterel *(Erythrogonys cinctus)*	Invertebrates
Rock Dove *(Columba livia)*	Seed
Sanderling *(Calidris alba)*	Invertebrates
Silver-crowned Friarbird *(Philemon argenticeps argenticeps)*	Nectar
Weebill *(Smicrornis brevirostris flavescens)*	Insects
White-throated Gerygone *(Gerygone olivacea rogersi)*	Insects
Yellow Wagtail *(Motacilla flava tschutschensis)*	Insects
2018	
Australian Magpie *(Cracticus tibicen)*	Omnivore
Bar-breasted Honeyeater *(Ramsayornis fasciatus)*	Nectar/Insects
Beach Stone-curlew *(Esacus magnirostris)*	Invertebrates
Black-shouldered Kite *(Elanus axillaris)*	Raptor

Table A4. *Cont.*

Species Name	Primary Food Source
Broad-billed Sandpiper *(Limicola falcinellus)*	Invertebrates
Brown Falcon *(Falco berigora)*	Raptor
Buff-banded Rail *(Hypotaenidia philippensis)*	Omnivore
Bush Stone-curlew *(Burhinus grallarius)*	Invertebrates
Channel-billed Cuckoo *(Scythrops novaehollandiae)*	Fruit
Grey Butcherbird *(Cracticus torquatus colletti)*	Invertebrates
Jacky Winter *(Microeca fascinans pallida)*	Insects
Little Kingfisher *(Ceyx pusilla ramsayi)*	Vertebrates/Invertebrates
Mangrove Grey Fantail *(Rhipidura phasiana)* *2018*	Insects
Masked Finch *(Poephila personata personata)*	Seed
Pied Butcherbird *(Cracticus nigrogularis picatus)*	Invertebrates
Rufous Fantail *(Rhipidura rufifrons)*	Insects
Square-tailed Kite *(Lophoictinia isura)*	Raptor
Tawny Frogmouth *(Podargus strigoides phalaenoides)*	Invertebrates
White-faced Heron *(Egretta novaehollandiae)*	Invertebrates

References

1. Zhang, X.Q. The trends, promises and challenges of urbanisation in the world. *Habitat. Int.* **2016**, *54*, 241–252. [CrossRef]
2. Fund, U.N.P. World Population Trends. 2017. Available online: https://www.unfpa.org/world-population-trends (accessed on 2 March 2019).
3. Sewell, S.R.; Catterall, C.P. Bushland modification and styles of urban development: Their effects on birds in south-east Queensland. *Wildl. Res.* **1998**, *25*, 41–63. [CrossRef]
4. Cleary, G.P.; Parsons, H.; Davis, A.; Coleman, B.R.; Jones, D.; Miller, K.; Weston, M.A. Avian Assemblages at Bird Baths: A Comparison of Urban and Rural Bird Baths in Australia. *PLoS ONE* **2016**, *11*, e0150899. [CrossRef]
5. Bellocq, M.I.; Leveau, L.M.; Filloy, J. Urbanization and Bird Communities: Spatial and Temporal Patterns Emerging from Southern South America. In *Ecology and Conservation of Birds in Urban Environments*; Springer Science and Business Media LLC: New York, NY, USA, 2017; pp. 35–54.
6. Evans, K.L.; Chamberlain, D.; Hatchwell, B.J.; Gregory, R.; Gaston, K.J. What makes an urban bird? *Glob. Chang. Biol.* **2010**, *17*, 32–44. [CrossRef]
7. Garden, J.; McAlpine, C.; Peterson, A.; Jones, D.; Possingham, H. Review of the ecology of Australian urban fauna: A focus on spatially explicit processes. *Austral Ecol.* **2006**, *31*, 126–148. [CrossRef]
8. Lancaster, R.K.; Rees, W.E. Bird communities and the structure of urban habitats. *Can. J. Zool.* **1979**, *57*, 2358–2368. [CrossRef]
9. Bezzel, E. Birdlife in intensively used rural and urban environments. *Ornis Fennica* **1985**, *62*, 90–95.
10. Blair, R.B. Birds and Butterflies Along Urban Gradients in Two Ecoregions of the United States: Is Urbanization Creating a Homogeneous Fauna? In *Biotic Homogenization*; Lockwood, J.L., McKinney, M.L., Eds.; Springer US: Boston, MA, USA, 2001; pp. 33–56.
11. Wolff, P.J.; DeGregorio, B.A.; Rodriguez-Cruz, V.; Mulero-Oliveras, E.; Sperry, J.H. Bird Community Assemblage and Distribution in a Tropical, Urban Ecosystem of Puerto Rico. *Trop. Conserv. Sci.* **2018**, *11*, 1940082918754777. [CrossRef]
12. Noske, R.A. The status and ecology of the white-throated grasswren Amytornis woodwardi. *Emu* **1992**, *92*, 39–51. [CrossRef]
13. Noske, R.A. Breeding biology, demography and success of the rufous-banded honeyeater, Conopophila albogularis, in Darwin, a monsoonal tropical city. *Wildl. Res.* **1998**, *25*, 339–356. [CrossRef]
14. Franklin, D.C.; Whitehead, P.J.; Pardon, G.; Matthews, J.; McMahon, P.; McIntyre, D. Geographic patterns and correlates of the decline of granivorous birds in northern Australia. *Wildl. Res.* **2005**, *32*, 399–408. [CrossRef]
15. Crawford, D.N. Birds of Darwin Area, with Some Records from Other Parts of Northern Territory. *Emu Austral Ornithol.* **1972**, *72*, 131–148. [CrossRef]
16. Crawford, D. Waterbirds: Indices and Fluctuations in Dry-Season Refuge Areas, Northern Territory. *Wildl. Res.* **1979**, *6*, 97–103. [CrossRef]
17. Lilleyman, A.; Hensen, B.J. The occurrence of the Asian subspecies of the Gull-billed Tern ('Gelochelidon nilotica affinis') in the Darwin region, Northern Territory. *North. Territ. Nat.* **2014**, *25*, 12–17.
18. Lilleyman, A.; Franklin, D.C.; Szabo, J.K.; Lawes, M. Behavioural responses of migratory shorebirds to disturbance at a high-tide roost. *Emu Austral Ornithol.* **2016**, *116*, 111–118. [CrossRef]
19. Noske, R.A. The breeding biology of the Mangrove Gerygone, Gerygone laevigaster, in the Darwin region, with notes on brood parasitism by the Little Bronze-cuckoo, Chrysococcyx minutillus. *Emu Austral Ornithol.* **2001**, *101*, 129–135. [CrossRef]
20. Mohd-Azlan, J.; Noske, R.A.; Lawes, M. Avian species-assemblage structure and indicator bird species of mangroves in the Australian monsoon tropics. *Emu Austral Ornithol.* **2012**, *112*, 287–297. [CrossRef]

21. Mohd-Azlan, J.; Noske, R.A.; Lawes, M.J. The Role of Habitat Heterogeneity in Structuring Mangrove Bird Assemblages. *Divers* **2015**, *7*, 118–136. [CrossRef]
22. Population Division of the United Nations Department of Economic and Social Affairs. *2018 Revision of World Urbanization Prospects*; United Nations: New York, NY, USA, 2018.
23. Ridpath, M.G. Ecology in the wet-dry tropics: How different? *Proc. Ecol. Soc. Aust.* **1985**, *13*, 3–20.
24. Mittermeier, R.A.; Myers, N.; Thomsen, J.B.; Da Fonseca, G.A.B.; Olivieri, S. Biodiversity Hotspots and Major Tropical Wilderness Areas: Approaches to Setting Conservation Priorities. *Conserv. Biol.* **1998**, *12*, 516–520. [CrossRef]
25. Gaston, K.J. Global patterns in biodiversity. *Nat. Cell Biol.* **2000**, *405*, 220–227. [CrossRef] [PubMed]
26. Bureau of Meteorology. Severe Tropical Cyclone Tracy. 2020. Available online: http://www.bom.gov.au/cyclone/history/tracy.shtml (accessed on 13 January 2020).
27. National Archives of Australia, Attorney-General's Department. *Cyclone Tracy, Darwin*; Australia Government: Canberra, Australia, 2005.
28. Emlen, J.T. An Urban Bird Community in Tucson, Arizona: Derivation, Structure, Regulation. *Condor* **1974**, *76*, 184. [CrossRef]
29. Chace, J.F.; Walsh, J.J. Urban effects on native avifauna: A review. *Landsc. Urban Plan.* **2006**, *74*, 46–69. [CrossRef]
30. Conole, L.; Kirkpatrick, J. Functional and spatial differentiation of urban bird assemblages at the landscape scale. *Landsc. Urban Plan.* **2011**, *100*, 11–23. [CrossRef]
31. Northern Territory Government. Feral Animals. 2018. Available online: https://nt.gov.au/environment/animals/feral-animals (accessed on 3 October 2018).
32. Australian Bureau of Statistics. 2016 Census QuickStats—Darwin. 2016. Available online: http://quickstats.censusdata.abs.gov.au/census_services/getproduct/census/2016/quickstat/701?opendocument (accessed on 27 August 2018).
33. Black, D. *1996 Census of Population and Housing: Darwin, a Social Atlas*; Statistics, A.B.O., Ed.; Australian Bureau of Statistics: Canberra, Australia, 1997.
34. Bureau of Meteorology. Climate Statistics for Australian Locations. 2020. Available online: http://www.bom.gov.au/climate/averages/tables/cw_014015.shtml (accessed on 5 July 2020).
35. Australian Bureau of Statistics. 3218.0—Regional Population Growth, Australia, 2015–2016. 2016. Available online: http://www.abs.gov.au/ausstats/abs@.nsf/PrimaryMainFeatures/3218.0?OpenDocument (accessed on 9 May 2017).
36. Environmental Systems Research Institute (ESRI). *ArcGIS 10.4.1 for Desktop, 1999–2015*; ESRI: Redlands, CA, USA, 2017.
37. Conrad, O.; Bechtel, B.; Bock, M.; Dietrich, H.; Fischer, E.; Gerlitz, L.; Wehberg, J.; Wichmann, V.; Böhner, J. System for Automated Geoscientific Analyses (SAGA) v. 2.1.4. *Geosci. Model Dev.* **2015**, *8*, 1991–2007. [CrossRef]
38. Chen, G.; Hay, G.J.; St-Onge, B. A GEOBIA framework to estimate forest parameters from lidar transects, Quickbird imagery and machine learning: A case study in Quebec, Canada. *Int. J. Appl. Earth Obs. Geoinf.* **2012**, *15*, 28–37. [CrossRef]
39. Hay, G.J.; Castilla, G. Geographic Object-Based Image Analysis (GEOBIA): A New Name for a New Discipline. In *Object-Based Image Analysis*; Springer: Berlin/Heidelberg, Germany, 2008; pp. 75–89.
40. Fisher, R.P.; Hobgen, S.E.; Haleberek, K.; Sula, N.; Mandaya, I. Satellite Image Analysis and Terrain Modelling-A Practical Manual for Natural Resource Management, Disaster Risk and Development Planning Using Free Geospatial Data and Software. 2017. Available online: https://sagatutorials.wordpress.com/ (accessed on 17 November 2020).
41. Australia, B. Birdlife Australia. 2019. Available online: http://birdlife.org.au (accessed on 19 June 2019).
42. Atlas of Living Australia. 2019. Available online: https://www.ala.org.au/ (accessed on 1 September 2019).
43. Garnett, S.T.; Duursma, D.E.; Ehmke, G.; Guay, P.; Stewart, A.; Szabo, J.K.; Weston, M.A.; Bennett, S.; Crowley, G.M.; Drynan, D.; et al. Australian Bird Data Version 1.0, Figshare, Editor. 2015. Available online: https://doi.org/10.6084/m9.figshare.1499292_D8 (accessed on 10 September 2019).
44. Hahs, A.K.; McDonnell, M.J. Selecting independent measures to quantify Melbourne's urban–rural gradient. *Landsc. Urban Plan.* **2006**, *78*, 435–448. [CrossRef]
45. R Core Team. *R: A Language and Environment for Statistical Computing*; R Foundation for Statistical Computing: Vienna, Austria, 2019.
46. Mccrie, N.; Noske, R. *Birds of the Darwin Region*; CSIRO Publishing: Clayton, Australia, 2015.
47. Devictor, V.; Julliard, R.; Couvet, D.; Lee, A.; Jiguet, F. Functional Homogenization Effect of Urbanization on Bird Communities. *Conserv. Biol.* **2007**, *21*, 741–751. [CrossRef]
48. Öckinger, E.; Schweiger, O.; Crist, T.O.; Debinski, D.M.; Krauss, J.; Kuussaari, M.; Petersen, J.D.; Pöyry, J.; Settele, J.; Summerville, K.S.; et al. Life-history traits predict species responses to habitat area and isolation: A cross-continental synthesis. *Ecol. Lett.* **2010**, *13*, 969–979. [CrossRef] [PubMed]
49. Thomas, C.D. Dispersal and extinction in fragmented landscapes. *Proc. R. Soc. B Boil. Sci.* **2000**, *267*, 139–145. [CrossRef]
50. Slade, E.M.; Merckx, T.; Riutta, T.; Bebber, D.P.; Redhead, D.; Riordan, P.; Macdonald, D.W. Life-history traits and landscape characteristics predict macro-moth responses to forest fragmentation. *Ecology* **2013**, *94*, 1519–1530. [CrossRef]
51. Concepción, E.D.; Moretti, M.; Altermatt, F.; Nobis, M.; Obrist, M.K. Impacts of urbanisation on biodiversity: The role of species mobility, degree of specialisation and spatial scale. *Oikos* **2015**, *124*, 1571–1582. [CrossRef]
52. Parsons, H.; French, K.; Major, R. The influence of remnant bushland on the composition of suburban bird assemblages in Australia. *Landsc. Urban Plan.* **2003**, *66*, 43–56. [CrossRef]

53. Young, K.M.; Daniels, C.B.; Johnston, G. Species of street tree is important for southern hemisphere bird trophic guilds. *Austral Ecol.* **2007**, *32*, 541–550. [CrossRef]
54. Major, R.E.; Parsons, H. What do museum specimens tell us about the impact of urbanisation? A comparison of the recent and historical bird communities of Sydney. *Emu Austral Ornithol.* **2010**, *110*, 92–103. [CrossRef]
55. Kark, S.; Iwaniuk, A.; Schalimtzek, A.; Banker, E. Living in the city: Can anyone become an 'urban exploiter'? *J. Biogeogr.* **2007**, *34*, 638–651. [CrossRef]
56. McKinney, M.L. Urbanization as a major cause of biotic homogenization. *Biol. Conserv.* **2006**, *127*, 247–260. [CrossRef]
57. Aronson, M.F.J.; La Sorte, F.A.; Nilon, C.H.; Katti, M.; Goddard, M.A.; Lepczyk, C.A.; Warren, P.S.; Williams, N.; Cilliers, S.; Clarkson, B.; et al. A global analysis of the impacts of urbanization on bird and plant diversity reveals key anthropogenic drivers. *Proc. R. Soc. B Boil. Sci.* **2014**, *281*, 20133330. [CrossRef]
58. Sol, D.; Bartomeus, I.; González-Lagos, C.; Pavoine, S. Urbanisation and the loss of phylogenetic diversity in birds. *Ecol. Lett.* **2017**, *20*, 721–729. [CrossRef]
59. Parks and Wildlife Commission of the Northern Territory. *Feral Pigeon Eradication Program: Darwin 1997*; Parks and Wildlife Commission of the Northern Territory: Darwin, Australia, 1998.
60. ABC Darwin. *Darwin Eradicates Feral Pigeons*; ABC Darwin: Darwin, Australia, 2004.
61. NAFI: North Australia and Rangelands Fire Information. 2018. Available online: https://www.firenorth.org.au/nafi3/ (accessed on 8 January 2019).
62. Bowman, D.M.; Johnston, F.H. Wildfire Smoke, Fire Management, and Human Health. *EcoHealth* **2005**, *2*, 76–80. [CrossRef]
63. Johnston, F.H.; Hanigan, I.C.; Bowman, D.M.J.S. Pollen Loads and Allergic Rhinitis in Darwin, Australia: A Potential Health Outcome of the Grass-Fire Cycle. *EcoHealth* **2009**, *6*, 99–108. [CrossRef] [PubMed]
64. USGS. What are the Band Designations for the Landsat Satellites? 2020. Available online: https://www.usgs.gov/faqs/what-are-band-designations-landsat-satellites?qt-news_science_products=0#qt-news_science_products (accessed on 17 November 2020).
65. USGS. What are the Best Landsat Spectral Bands for Use in My Research? 2020. Available online: https://www.usgs.gov/faqs/what-are-best-landsat-spectral-bands-use-my-research?qt-news_science_products=0#qt-news_science_products (accessed on 17 November 2020).

diversity MDPI

Article

An Assessment of Applicability of SNP Chip Developed for Domestic Goats in Genetic Studies of Caucasian Tur (*Capra caucasica*) †

Arsen V. Dotsev [1,*], Andrey N. Rodionov [1], Veronika R. Kharzinova [1], Sergey N. Petrov [1], Dmitry G. Medvedev [2], Vugar A. Bagirov [1], Gottfried Brem [1,3] and Natalia A. Zinovieva [1,*]

1 L.K. Ernst Federal Science Center for Animal Husbandry, 142132 Podolsk, Moscow Region, Russia; rodiand@yandex.ru (A.N.R.); veronika0784@mail.ru (V.R.K.); citelekle@gmail.com (S.N.P.); vugarbagirov@mail.ru (V.A.B.); gottfried.brem@agrobiogen.de (G.B.)
2 Department of Game Management and Bioecology, Irkutsk State University of Agriculture, 664038 Molodezhny, Irkutsk Region, Russia; dmimedvedev@yandex.ru
3 Institut für Tierzucht und Genetik, University of Veterinary Medicine (VMU), A-1210 Vienna, Austria
* Correspondence: asnd@mail.ru (A.V.D.); n_zinovieva@mail.ru (N.A.Z.); Tel.: +7-496-7651104 (A.V.D.); +7-496-7651404 (N.A.Z.)

† This paper is an extended version of the paper published in the 1st International Electronic Conference on Biological Diversity, Ecology, and Evolution (BDEE 2021), Online, 15–31 March 2021.

Abstract: Caucasian tur (*Capra caucasica*) is native to Greater Caucasus Mountain Chain from Azerbaijan and Georgia in the East to Krasnodar region of Russia in the West. This species is divided into two subspecies (by some authors into species)—East-Caucasian tur and West-Caucasian tur and a subpopulation referred to as Mid-Caucasian tur. Up to date most of the genetic studies of Caucasian tur are based on mitochondrial DNA sequences and comprehensive investigation based on nuclear DNA is required for clarification of its genetic diversity and population structure. In our work, we assessed the applicability of Illumina Goat SNP50 BeadChip for genetic studies of Caucasian tur. Total of 15 specimens of *Capra caucasica* including East-Caucasian tur from Dagestan (E_TUR, $n = 5$), West-Caucasian tur from Karachay-Cherkessia (W_TUR, $n = 5$), and Mid-Caucasian tur from Kabardino-Balkaria (M_TUR, $n = 5$) were genotyped. After quality control, 5544 polymorphic loci, which were distributed all over 29 autosomes, were detected. The lowest number of SNPs was found on the 25th chromosome—68, and the highest on the 1st chromosome—348. It was shown that all the three groups of Caucasian tur clustered separately. A total of 2061 SNPs were common for all the populations, 594 were found only in W_TUR, 689 in E_TUR, and 530 in M_TUR. Individual heterozygosity ranged from 0.273 to 0.282 in W_TUR, from 0.217 to 0.253 in E_TUR, and from 0.255 to 0.283 in M_TUR. A clinal pattern of genetic variation was revealed. It was suggested to consider Caucasian tur a single species with several ecotypes. Thus, in our study we demonstrated that the Illumina Goat SNP50 BeadChip developed for domestic goats can be used as a useful tool for genetic studies of Caucasian tur.

Keywords: wild goats; single nucleotide polymorphisms; genetic diversity; population structure

Citation: Dotsev, A.V.; Rodionov, A.N.; Kharzinova, V.R.; Petrov, S.N.; Medvedev, D.G.; Bagirov, V.A.; Brem, G.; Zinovieva, N.A. An Assessment of Applicability of SNP Chip Developed for Domestic Goats in Genetic Studies of Caucasian Tur (*Capra caucasica*). *Diversity* **2021**, *13*, 312. https://doi.org/10.3390/d13070312

Academic Editor: Michael Wink

Received: 23 June 2021
Accepted: 6 July 2021
Published: 8 July 2021

1. Introduction

Caucasian tur (*Capra caucasica*) is an ungulate species from the subfamily Caprinae (Figure 1). Its habitat is limited to Greater Caucasus Mountain Chain from Azerbaijan and Georgia in the East to Krasnodar region of Russia in the West (Figure 2). Its range is one of the smallest habitats of all ungulates—around 770 km in length and 80 km in width [1]. Taxonomically, the species is divided into two subspecies: West-Caucasian tur (*C. c. severtzovi*) and East-Caucasian tur (*C. c. cylindricornis*). Some authors even considered these populations as different species [2–5]. The areas of these subspecies are united, and in the territory between them a hybrid form, which is referred to as a Mid-Caucasian tur, is distinguished [6].

Figure 1. Caucasian tur in its natural habitat. Photo by M. Andreev and A. Andreeva (http://mountaindreams.ru, accessed on 17 June 2021).

Figure 2. Map with sampling sites of Caucasian tur used in this study and the area of the species habitat. W_TUR = West-Caucasian tur, E_TUR = East-Caucasian tur, M_TUR = Mid-Caucasian tur.

The International Union for Conservation of Nature (IUCN) has listed East-Caucasian tur with the total numbers of about 31,000–32,000 animals as near threatened [7] and West-Caucasian tur with census size only around 5000 animals as endangered [8]. Of particular concern is an increasing anthropogenic pressure (poaching, habitat destruction, grazing competition with livestock, etc.,), resulting in migrations to higher mountains and adaptation to nocturnal behavior in some populations.

Little is known about the genetic diversity and evolutionary history of this animal. Only several studies based on Y-chromosome SRY gene and mitochondrial cytochrome b and control region were conducted [9–12]. To clarify the taxonomy of Caucasian tur and investigate the diversity of its populations molecular genetic studies based on nuclear DNA are needed.

Currently, one of the most effective technologies for molecular genetic research is the genome-wide analysis of single-nucleotide polymorphisms (SNPs). Recent advances in the development of high-throughput genotyping platforms (SNP chips) have turned SNPs into a powerful tool for genetic studies of domestic animals [13,14]. Although such chips are not available for most of the wild animals, the use of the SNP chips developed for their domestic relatives were found to be suitable.

Tokarska et al. [15] used the BovineSNP50 BeadChip created for cattle (*Bos taurus*) to infer the population structure of European bison (*Bos bonasus*). Kasarda et al. [16] have demonstrated suitability of the above-mentioned SNP microarray for the evaluation of the Cervidae family diversity. Kharzinova et al. [17,18] have shown the applicability of the BovineSNP50 BeadChip and BovineHD BeadChip for genome-wide studies of reindeer (*Rangifer tarandus*). The OvineSNP50 BeadChip, created for domestic sheep, was successfully used to study wild Ovis species: bighorn (*Ovis canadensis*), thinhorn (*Ovis dalli*) [19,20] and snow sheep (*Ovis nivicola*) [21].

The first SNP chip for domestic goats—Goat SNP50 BeadChip, containing more than 50,000 markers was developed by the International Goat Genome Consortium in 2011 [22]. In terms of wild species, Goat SNP50 BeadChip, so far was only used for genotyping of Bezoar goat (*Capra aegagrus*) which is considered the ancestor of domestic goats [23].

It should be noted that for non-model species the genotyping call-rate decreases approximately 1.5% per each million years of divergence time between species and the number of polymorphic sites decline exponentially leveling off after about 5 million years of divergence [20]. Therefore, this approach is not suitable for some analyses, i.e., based on linkage disequilibrium (LD), but meanwhile could be successfully used for investigation of phylogenetic relationships, genetic diversity, admixture, and introgression.

According to the web resource TimeTree (http://www.timetree.org, accessed on 15 May 2021) [24] the estimated median time of Caucasian tur and domestic goat divergence is 1.36 million years ago, which is much less than for the majority of other non-model animals. It gives a great opportunity to investigate phylogeny and genetic characteristics of *Capra caucasica*.

In this regard, the aim of our study was to assess the applicability of Illumina Goat SNP50 Bead-Chip, developed for domestic goats, for genetic studies of Caucasian tur (*Capra caucasica*).

2. Materials and Methods

A total of 15 specimens of *Capra caucasica*, including East-Caucasian tur from Dagestan (E_TUR, n = 5), West-Caucasian tur from Karachay-Cherkessia (W_TUR, n = 5), and Mid-Caucasian tur from Kabardino-Balkaria (M_TUR, n = 5) were selected for this study. The sampling sites are presented in Figure 2. All the samples were taken from trophy hunters of the Mountain Hunters Club (www.kgo-club.ru), who were licensed to hunt Caucasian tur. DNA extraction was carried out using Nexttec columns (Nexttec Biotechnology GmbH, Leverkusen, Germany) in accordance with the manufacturer's recommendations. SNP genotyping was performed using the Illumina Goat SNP50 BeadChip containing 53,347 SNPs.

SNP quality filtering was performed in PLINK 1.9 [25]. SNPs with unknown positions, located on sex chromosomes as well as those that were genotyped in less than 90% of individuals (− −geno 0.1), with a minor allele frequency (MAF) < 1% (− −maf 0.01) and with deviations from Hardy–Weinberg equilibrium (− −hwe 1e−6) were removed.

For the population genetic analyses (PCA, Admixture, f3 statistics, etc.,) SNPs in linkage disequilibrium (− −indep-pairwise 50 5 0.5) were pruned. The observed (H_O) and unbiased expected (H_E) heterozygosity [26], inbreeding coefficient (F_{IS}), and rarified allelic richness (A_R) were calculated in the R package "diveRsity" [27]. Pairwise F_{ST} genetic distances [28] were calculated in the R package "StAMPP" [29]. Principle component analysis was performed with PLINK 1.9 (− −pca 4) and visualized in the R package "ggplot2" [30]. An individual tree based on the pairwise identity-by-state (IBS) distance matrix (− −distance 1-ibs) was constructed using the Neighbor-Net algorithm implemented in SplitsTree 4.14.6 [31]. To estimate and visualize the distribution of heterozygosity at the individual level, multilocus heterozygosity (MLH) was calculated in the R package "inbreedR" [32]. Venn diagram was constructed in the R package "VennDiagram" [33]. Cluster analysis was performed in Admixture 1.3 software [34], and visualized in the R package "pophelper" [35]. f3 statistics [36] was computed using ADMIXTOOLS [37], as implemented in the R package "admixr" [38]. The map with sampling sites was created using R packages "maps" [39] and "ggplot2". As long as the Illumina Goat SNP50 BeadChip was developed for domestic goats (*Capra hircus*), the positions on chromosomes in this paper correspond to the reference genome of a domestic goat.

3. Results

From the initial set of 53,347 SNPs available in the Illumina Goat SNP50 BeadChip after quality control procedures (− −geno 0.1, − −maf 0.01) 5544 variants (10.4%) passed all the filters while 7002 variants were removed due to missing genotype data and 37,407 variants were removed due to minor allele threshold. No SNPs were removed due to Hardy–Weinberg exact test (Table S2). Total genotyping rate for the 15 samples of Caucasian tur was 94.8%. The selected polymorphic loci were distributed all over 29 autosomes. The lowest number of SNPs was found on the 25th chromosome—68, and the highest on the 1st chromosome—348 (Figure 3A). A similar pattern was observed for the three populations examined separately (Figure 3B–D), (Figure S1).

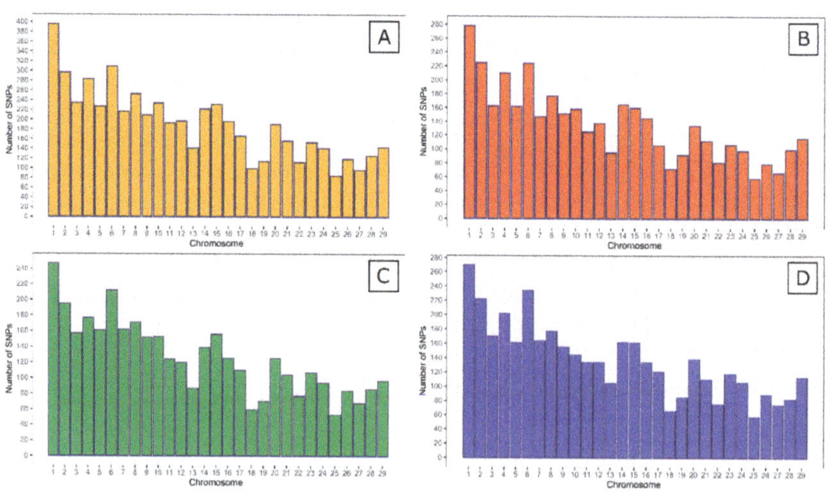

Figure 3. Number of polymorphic sites per chromosome in Caucasian tur (*Capra caucasica*). (**A**). Number of SNPs in all the studied populations. (**B**). Number of SNPs in West-Caucasian tur. (**C**). Number of SNPs in East-Caucasian tur. (**D**). Number of SNPs in Mid-Caucasian tur.

Mean minor allele frequency (MAF) computed for all the populations was 0.214 ± 0.002. About 2474 SNPs were highly informative with MAF > 0.3 and 1895 SNPs had minor allele frequency less than 0.1 (Figure S2).

While 2061 polymorphic SNPs were common for all the Caucasian tur groups, the number of unique SNPs for W_TUR, E_TUR and M_TUR was 594, 689, and 530, respectively. The most shared SNPs between two groups was found in W_TUR—M_TUR pair—869, and the least number of shared loci was observed in W_TUR—E_TUR pair—420 (Figure 4).

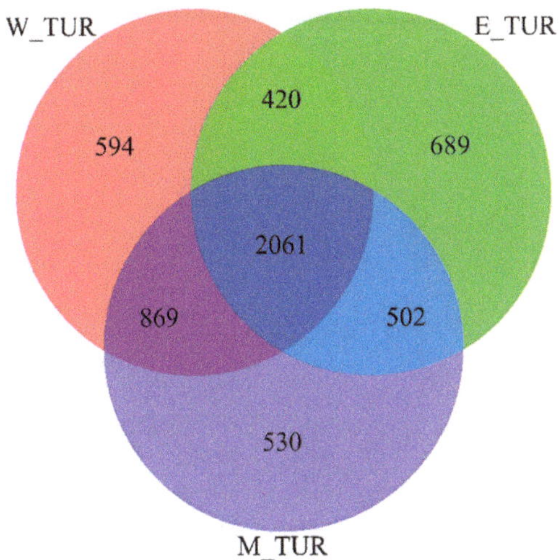

Figure 4. Venn diagram representing the number of unique and shared polymorphic SNPs in the three groups of Caucasian tur (*Capra caucasica*). W_TUR = West-Caucasian tur, E_TUR = East-Caucasian tur, M_TUR = Mid-Caucasian tur.

For all the next analyses LD pruning was performed, after which 3885 SNPs were selected.

The results of the principal component analysis (PCA) revealed that all the studied groups of Caucasian tur clustered separately (Figure 5).

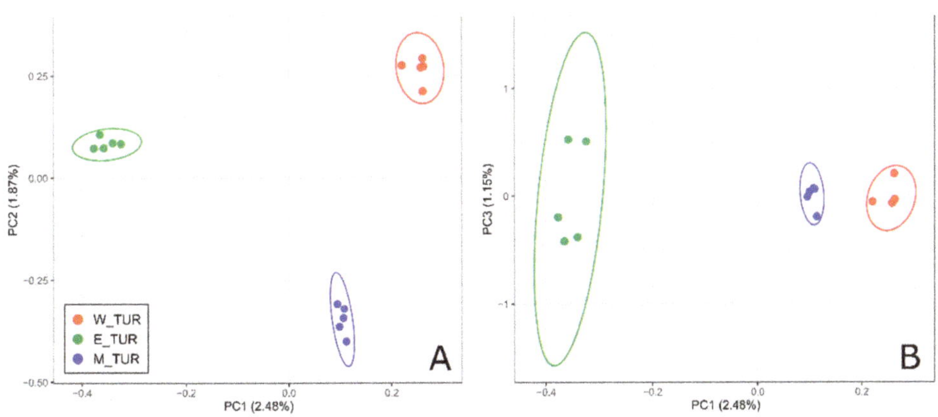

Figure 5. Principal component analysis (PCA) of the groups of Caucasian tur (*Capra caucasica*): (**A**) the first two components (PC1 and PC2), (**B**) the first and third components (PC1 and PC3). W_TUR = West-Caucasian tur, E_TUR = East-Caucasian tur, M_TUR = Mid-Caucasian tur.

The first component (PC1), which explained 2.48% of genetic variability, divided Caucasian tur into two groups. The first group consisted of E_TUR (PC1 < 0) and the second one included W_TUR and M_TUR (PC1 > 0).

The level of genetic differentiation was estimated with pairwise F_{ST} distances (Table 1).

Table 1. Pairwise F_{ST} genetic distances between the groups of Caucasian tur (*Capra caucasica*).

Group	W_TUR	E_TUR	M_TUR
W_TUR	0.000		
E_TUR	0.161	0.000	
M_TUR	0.099	0.135	0.000

Notes: W_TUR = West-Caucasian tur, E_TUR = East-Caucasian tur, M_TUR = Mid-Caucasian tur. F_{ST} values in the range 0–0.05 indicate low genetic differentiation; value between 0.05 and 0.15, moderate differentiation; values between 0.15 and 0.25, high differentiation; and values above 0.25, very high genetic differentiation [40,41].

We observed the highest F_{ST} value between W_TUR and E_TUR—0.161. M_TUR was genetically closer to W_TUR than to E_TUR, but in both cases the differentiation was moderate—0.099 and 0.135, respectively.

The above-mentioned results were also supported with Neighbor-Net dendrogram (Figure 6).

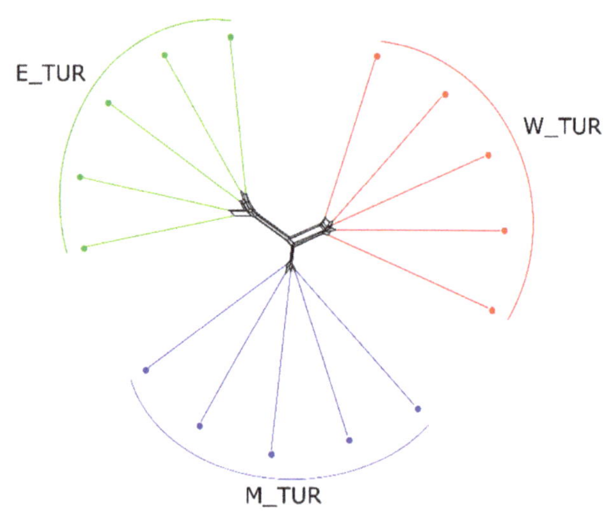

Figure 6. Neighbor-Net dendrogram of 15 Caucasian tur (*Capra caucasica*) individuals, based on IBS distances.

All the studied individuals were placed in clades according to their geographical origin. This analysis also revealed that there were no close relatives among the 15 genotyped animals.

Expected heterozygosity (H_E) and allelic richness (A_R) in W_TUR were significantly higher than in E_TUR (for both indicators t-test p-values were less than 0.0001) (Table 2). Slightly higher genetic diversity parameters were observed in M_TUR. The within population inbreeding coefficient (F_{IS}) was close to zero in W_TUR and M_TUR which showed that these groups were in a fairly stable state, and slightly higher than zero in E_TUR, which indicated weak heterozygote deficit.

Table 2. Genetic diversity in the groups of Caucasian tur (*Capra caucasica*).

Group	n	H_O (±SE)	H_E (±SE)	F_{IS} [95% CI]	A_R (±SE)
W_TUR	5	0.308 ± 0.004	0.312 ± 0.003	0.015 [0.002; 0.028]	1.775 ± 0.007
E_TUR	5	0.258 ± 0.004	0.284 ± 0.003	0.074 [0.060; 0.088]	1.730 ± 0.007
M_TUR	5	0.306 ± 0.004	0.316 ± 0.003	0.028 [0.015; 0.041]	1.790 ± 0.007

Notes: n—number of samples, H_O—observed heterozygosity, H_E—expected heterozygosity, F_{IS}—inbreeding coefficient, A_R—allelic richness, SE—standard error, CI—confidence interval, W_TUR = West-Caucasian tur, E_TUR = East-Caucasian tur, M_TUR = Mid-Caucasian tur.

Individual heterozygosity values, which were represented by multilocus heterozygosity (MLH) showed that all the E_TUR individuals had lower heterozygosity (from 0.217 to 0.253) than both W_TUR (from 0.273 to 0.282) and M_TUR (from 0.255 to 0.283) (Figure 7).

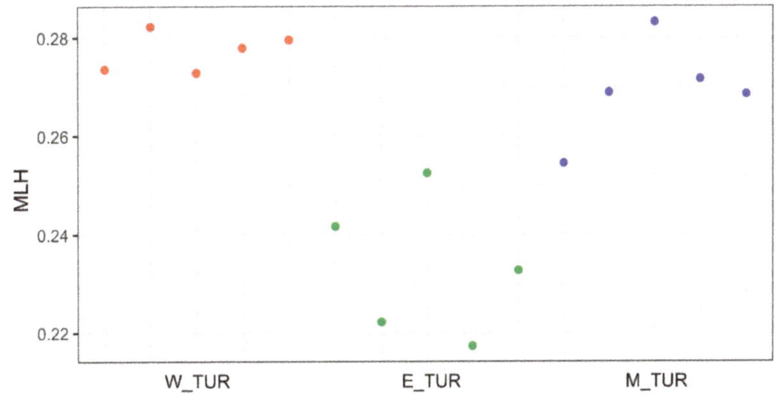

Figure 7. Individual multilocus heterozygosity (MLH) in the groups of Caucasian tur (*Capra caucasica*).

Although mean values of multilocus heterozygosity were almost equal in W_TUR—0.277 ± 0.002 and in M_TUR—0.269 ± 0.005, the range of values was wider in M_TUR than in W_TUR. In E_TUR mean MLH was significantly lower—0.233 ± 0.006.

4. Discussion

In our study we assessed the applicability of Illumina Goat SNP50 BeadChip, developed for domestic goats (*Capra hircus*), for genetic studies of Caucasian tur (*Capra caucasica*). About 15 individuals that belong to three groups of Caucasian tur (West-Caucasian tur, East-Caucasian tur, and Mid-Caucasian tur) were genotyped with 5544 polymorphic loci (3885 SNPs after LD pruning) distributed all over 29 autosomes. This number of polymorphic sites is sufficient for carrying out phylogenetic studies and assessing the genetic diversity. For instance, in the work on European bison only 960 SNPs were used [15], 1121 polymorphic loci were detected for snow sheep (*Ovis nivicola*) [21], and based on 7303 SNPs, population structure and genetic diversity of reindeer were investigated [18].

Regarding intraspecific taxonomy of Caucasian tur, there are two main questions to which there is still no answer: (1) Whether the West-Caucasian tur and the East-Caucasian tur are one species or should they be considered different species; (2) is the Mid- Caucasian tur a separate subspecies or a hybrid form. The works of 19th and 20th centuries, based on the morphological characteristics could not resolve these problems. The majority of authors such as Lydekker [2], Tsalkin [3], Geptner [4], Damm [5] suggested division into two species—*C. cylindricornis* and *C. caucasica*. Unlike the above-mentioned authors, Sokolov [42] and Tembotov [43] proposed to recognize one species—*Capra caucasica* with several subspecies *C. c. severtzovi*, *C. c. cylindricornis*, *C. c. caucasica*, and *C. c. dinniki*. Couturier [44] in 1962 suggested the presence of East West morphological cline in Caucasian tur and indicated that there is only one taxon in the Caucasus (*C. aegagrus caucasica*). Thirty years later Ayunts [45] showed a smooth transition in the variability of horns in tur, the main distinguishing feature of these populations and noted that Caucasian tur was represented by a number of slightly different groups and that the division into subspecies was not justified.

Currently in the era of molecular genetic studies, many taxonomic riddles were resolved. Clarification of the phylogeny in the subfamily Caprinae made it possible to determine the status of three subspecies of tahr, which are now considered different genera—*Hemitragus*, *Nilgiritragus*, and *Arabitragus* [46]. New subspecies of blue sheep (*Pseudois nayar*) [47] and snow sheep (*Ovis nivicola*) [21,48] were identified.

To date molecular genetic studies of Caucasian tur were based on the investigation of mitochondrial DNA and Y-chromosome. Manceau et al. [9] examined concatenated fragments of cytochrome b and control region (500 bp in length) of three samples of East-Caucasian tur and nine samples of West-Caucasian tur. Kazanskaya et al. [10] examined fragments of cytochrome b and control region of eight samples of East-Caucasian tur and nine samples of West-Caucasian tur. In the research of Zvychaynaya [11] the cytochrome b fragment (1128 bp) and Y-chromosome intron of gene SRY (1832 bp) of West-Caucasian tur (n = 6) and East-Caucasian (n =6) tur were analyzed. Kashinina and Kholodova [12] examined a 715 bp fragment of cytochrome b gene in six samples of East-Caucasian tur, 11 samples of West-Caucasian tur and eight samples of Mid-Caucasian tur. In all the above-mentioned studies strong East–West differentiation in the Caucasian tur was demonstrated and therefore all the authors supported evidence for the existence of two species—*C. caucasica* and *C. cylindricornis*. In the work of Zvychaynaya [11], some mtDNA and Y-chromosome haplotype mixture was observed in the Central Caucasus that was explained as the evidence of *C. caucasica* and *C. cylindricornis* hybridization. In all the mtDNA studies the authors emphasized that further research on nuclear markers should be performed.

Our preliminary analysis of nuclear DNA markers confirmed clear genetic differentiation between all the three studied populations. PCA analysis showed a clear division into clusters, however the percent of genetic variation was very low—2.48. Pairwise F_{ST} genetic distance between the most distant populations was only 0.161, which is too low for dividing them into species and even subspecies (e.g., in Yakut snow sheep (*Ovis nivicola lydekkeri*) pairwise F_{ST} values ranged from 0.044 to 0.144 within the subspecies, and from 0.112 to 0.205 with the proposed new subspecies [21]). Moreover, this was also confirmed by cross-validation (CV) error calculated for the cluster analysis (Figure S3), which suggested that K (optimum number of populations in our dataset) was 1 (Figure 8).

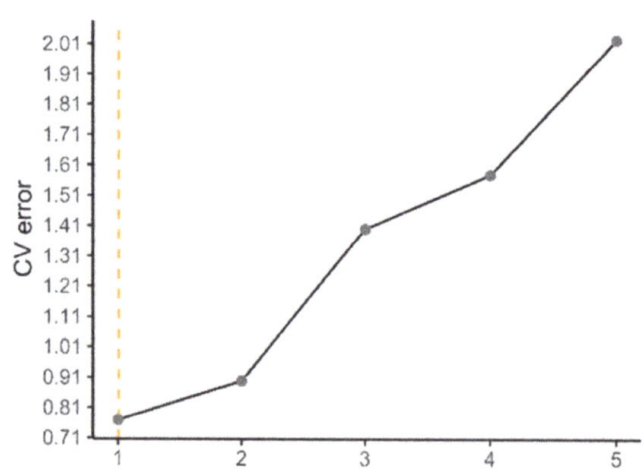

Figure 8. Admixture cross-validation (CV) error indicating the optimum number of clusters in the dataset.

Mid-Caucasian tur was genetically a little closer to West-Caucasian tur (F_{ST} 0.99) than to East-Caucasian tur (F_{ST} 0.135) but we did not find any evidence that this population arose as a result of hybridization. Heterozygosity in Mid-Caucasian tur was a little lower than in West-Caucasian tur. The number of unique SNPs was even lower than in both West-Caucasian and East-Caucasian tur. The values of f3 statistics, that is a test revealing if the population of interest is the result of admixture between two other populations, were not significant and therefore also rejected the hypothesis of hybrid origin of Mid-Caucasian tur (Table S1). Based on the above we suggest that Caucasian tur genetic variation follows a clinal pattern [49] and therefore it should be considered a single species—*Capra caucasica*, which includes several ecotypes. The division into subspecies does not seem to be justified. Thus, the results of our investigation based on nuclear molecular genetic markers agree with Ayunts's study [45] on variability of horns and confirming the hypothesis proposed by Couturier [44]. Further research with the collection of more samples will help to give a final answer regarding the intraspecific taxonomy of the Caucasian tur, reconstruct the natural history of the species formation, and assess the genetic diversity.

5. Conclusions

Here we demonstrated that the Illumina Goat SNP50 BeadChip developed for domestic goats could be used as a useful tool for genetic studies of Caucasian tur. It was shown that West-Caucasian tur, East-Caucasian tur, and Mid-Caucasian tur clustered separately from each other. Genetic diversity parameters in East-Caucasian tur were lower than in the other groups. A clinal pattern of genetic variation was revealed. It was suggested to consider Caucasian tur a single species with several ecotypes. To obtain more accurate information about population structure and genetic diversity of Caucasian tur, further studies based on Illumina Goat SNP50 Bead-Chip with larger number of samples are needed.

Supplementary Materials: The following are available online at https://www.mdpi.com/article/10.3390/d13070312/s1. Figure S1: Proportion of polymorphic sites per chromosome in Caucasian tur (*Capra caucasica*). Figure S2: Minor allele frequency (MAF) in Caucasian tur (*Capra caucasica*). Figure S3: Admixture clustering results at K = 2 and K = 3 for the three populations of Caucasian tur. Table S1: The results of f3 statistics computed for the three populations of Caucasian tur. Table S2: Hardy–Weinberg equilibrium *p*-values and minor allele frequency in SNPs selected for the study of Caucasian tur (*Capra caucasica*).

Author Contributions: Conceptualization, A.V.D. and N.A.Z., methodology, A.V.D., A.N.R., V.R.K., G.B.; software, A.N.R.; validation, A.V.D. and V.R.K.; formal analysis, A.V.D. and V.R.K.; investigation, A.V.D., A.N.R., and S.N.P.; resources, D.G.M. and V.A.B.; data curation, A.V.D., G.B., and N.A.Z.; writing—original draft preparation, A.V.D.; writing—review and editing, V.R.K. and N.A.Z.; visualization, A.V.D. and S.N.P.; supervision, N.A.Z.; project administration, A.V.D.; funding acquisition, A.V.D. All authors have read and agreed to the published version of the manuscript.

Funding: This study was supported by the Russian Scientific Foundation within Project No. 21-66-00007 (genome-wide SNP genotyping and data analysis) and the Russian Ministry of Science and Higher Education within theme No. 0445-2019-0024 (sample collecting and creation of DNA Bank).

Institutional Review Board Statement: Not applicable.

Informed Consent Statement: Not applicable.

Data Availability Statement: SNP genotypes of Caucasian tur are available on reasonable request after signing research (MTA) agreement.

Acknowledgments: We are grateful to Bendersky E.V. and the Mountain Hunters Club (www.kgo-club.ru) for providing the samples of Caucasian tur. We thank Mitya Andreev and Anya Andreeva (http://mountaindreams.ru accessed on 17 June 2021) for providing the photo of Caucasian tur.

Conflicts of Interest: The authors declare no conflict of interest.

References

1. Weinberg, P.J.; Akkiev, M.I.; Buchukuri, R.G. Tur of the Caucasus: Morphology, taxonomy and conservation strategy. Status and Protecyion of Globally Threatened Species in the Caucasus. In *CEPF Biodiversity Investments in the Caucasus Hotspot 2004–2009*; Zazanashvili, N., Mallon, D., Eds.; Contour Ltd.: Tbilisi, Georgia, 2009; pp. 83–91.
2. Lydekker, R. *Wild Oxen, Sheep, and Goats of All Lands*; Rowland Ward: London, UK, 1898; p. 318.
3. Tsalkin, V.I. Variability and systematics of tur of the West Caucasus (Caprinae, Artiodactyla). *Byulleten Mosk. Obs. Ispyt. Prir. Otd. Biol.* **1955**, *60*, 17–33. (In Russian)
4. Geptner, V.G.; Nasimovich, A.A.; Bannikov, A.G. *Mammals of the Soviet Union. Ungulates and Not Ungulates*; Higher School: Moscow, Russia, 1961; Volume 1, p. 776. (In Russia)
5. Damm, G.R.; Franco, N. *CIC Caprinae Atlas of the World*; CIC International Council for Game and Wildlife Conservation: Budakeszi, Hungary; Rowland Ward Publications: Johannesburg, South Africa, 2014.
6. Gromov, I.M.; Gureev, A.A.; Novikov, G.A.; Sokolov, I.I.; Strelkov, P.P.; Chapskii, K.K. *Mammal. Fauna of the USSR V.2.*; Academy of Sciences of the USSR: Moscow-Leningrad, Russia, 1963; pp. 1047–1050. (In Russian)
7. Lortkipanidze, B.; Weinberg, P. *Capra cylindricornis*. The IUCN Red List of Threatened Species 2020: E.T3795A91287260. [CrossRef]
8. Weinberg, P. *Capra caucasica*. The IUCN Red List of Threatened Species 2020: E.T3794A22143809. [CrossRef]
9. Manceau, V.; Després, L.; Bouvet, J.; Taberlet, P. Systematics of the genus Capra inferred from mitochondrial DNA sequence data. *Mol. Phylogenetics Evol.* **1999**, *13*, 504–510. [CrossRef] [PubMed]
10. Kazanskaya, E.Y.; Kuznetsova, M.V.; Danilkin, A.A. Phylogenetic reconstructions in the genus Capra (Bovidae, Artiodactyla) based on the mitochondrial DNA analysis. *Russ. J. Genet.* **2009**, *43*, 245–253. (In Russian) [CrossRef]
11. Zvychaynaya, E.Y. Genetic differentiation of the wild goats (genus Capra) based on the analysis of mitochondrial gene cytochrome b and fragment of nuclear gene SRY. *Galemys* **2010**, *22*, 255–276.
12. Kashinina, N.V.; Kholodova, M.V. Currunt genetic structure of Caucasian turs Capra caucasica and C. cylindricornis. In *Conference Proceedings: Topical Issues of Modern Zoology and Ecology of Animals*; Penza, Russia, 2016; p. 46.
13. Coates, B.S.; Sumerford, D.V.; Miller, N.J.; Kim, K.S.; Sappington, T.W.; Siegfried, B.D.; Lewis, L.C. Comparative performance of single nucleotide polymorphism and microsatellite markers for population genetic analysis. *J. Hered.* **2009**, *100*, 556–564. [CrossRef]
14. Kijas, J.W.; Townley, D.; Dalrymple, B.P.; Heaton, M.P.; Maddox, J.F.; McGrath, A. The International Sheep Genomics Consortium. A genome wide survey of SNP variation reveals the genetic structure of sheep breeds. *PLoS ONE* **2009**, *4*, e4668. [CrossRef] [PubMed]
15. Tokarska, M.; Marshall, T.; Kowalczyk, R.; Wójcik, J.M.; Pertoldi, C.; Kristensen, T.N.; Loeschcke, V.; Gregersen, V.R.; Bendixen, C. Effectiveness of microsatellite and SNP markers for parentage and identity analysis in species with low genetic diversity: The case of European bison. *Heredity* **2009**, *103*, 326–332. [CrossRef]
16. Kasarda, R.; Moravcikova, N.; Sidlova, V.; Trakovicka, A.; Kadlecik, O.; Pokoradi, J.; Zidek, R. Suitability of BovineSNP50 BeadChip for the evaluation of the Cervidae family diversity. *Czech. J. Anim. Sci.* **2015**, *60*, 391–399. [CrossRef]
17. Kharzinova, V.R.; Sermyagin, A.A.; Gladyr, E.A.; Okhlopkov, I.M.; Brem, G.; Zinovieva, N.A. A study of applicability of SNP chips developed for Bovine and Ovine species to whole-genome analysis of Reindeer Rangifer tarandus. *J. Hered.* **2015**, *106*, 758–761. [CrossRef]
18. Kharzinova, V.; Dotsev, A.; Solovieva, A.; Sergeeva, O.; Bryzgalov, G.; Reyer, H.; Wimmers, K.; Brem, G.; Zinovieva, N. Insight into the current genetic diversity and population structure of domestic reindeer (*Rangifer tarandus*) in Russia. *Animals* **2020**, *10*, 1309. [CrossRef]
19. Miller, J.M.; Poissant, J.; Kijas, J.W.; Coltman, D.W. The International Sheep Genomics Consortium. A genome-wide set of SNPs detects population substructure and long range linkage disequilibrium in wild sheep. *Mol. Ecol. Resour.* **2011**, *11*, 314–322. [CrossRef]
20. Miller, J.M.; Kijas, J.W.; Heaton, M.P.; McEwan, J.C.; Coltman, D.W. Consistent divergence times and allele sharing measured from cross-species application of SNP chips developed for three domestic species. *Mol. Ecol. Resour.* **2012**, *12*, 1145–1150. [CrossRef] [PubMed]
21. Dotsev, A.V.; Deniskova, T.E.; Okhlopkov, I.M.; Mészáros, G.; Sölkner, J.; Reyer, H.; Wimmers, K.; Brem, G.; Zinovieva, N.A. Genome-wide SNP analysis unveils genetic structure and phylogeographic history of snow sheep (*Ovis nivicola*) populations inhabiting the Verkhoyansk Mountains and Momsky Ridge (Northeastern Siberia). *Ecol. Evol.* **2018**, *8*, 8000–8010. [CrossRef] [PubMed]
22. Tosser-Klopp, G.; Bardou, P.; Bouchez, O.; Cabau, C.; Crooijmans, R.; Dong, Y.; Donnadieu-Tonon, C.; Eggen, A.; Heuven, H.C.; Jamli, S.; et al. Design and characterization of a 52 K SNP chip for goats. *PLoS ONE* **2014**, *9*, e86227. [CrossRef] [PubMed]
23. Bertolini, F.; Servin, B.; Talenti, A.; Rochat, E.; Kim, E.S.; Oget, C.; Palhière, I.; Crisà, A.; Catillo, G.; Steri, R.; et al. Signatures of selection and environmental adaptation across the goat genome post-domestication. *Genet. Sel. Evol.* **2018**, *50*, 57. [CrossRef]
24. Kumar, S.; Stecher, G.; Suleski, M.; Hedges, S.B. TimeTree: A Resource for Timelines, Timetrees, and Divergence Times. *Mol. Biol. Evol.* **2017**, *34*, 1812–1819. [CrossRef]
25. Purcell, S.; Neale, B.; Todd-Brown, K.; Thomas, L.; Ferreira, M.A.; Bender, D.; Maller, J.; Sklar, P.; de Bakker, P.I.W.; Daly, M.J.; et al. PLINK: A tool set for whole-genome association and population-based linkage analyses. *Am. J. Hum. Genet.* **2007**, *81*, 559–575. [CrossRef]

26. Nei, M. Estimation of average heterozygosity and genetic distance from a small number of individuals. *Genetics* **1978**, *83*, 583–590. [CrossRef]
27. Keenan, K.; McGinnity, P.; Cross, T.F.; Crozier, W.W.; Prodöhl, P.A. diveRsity: An R package for the estimation and exploration of population genetics parameters and their associated errors. *Methods Ecol. Evol.* **2013**, *4*, 782–788. [CrossRef]
28. Weir, B.S.; Cockerham, C.C. Estimating F-Statistics for the analysis of population structure. *Evolution* **1984**, *38*, 1358–1370.
29. Pembleton, L.W.; Cogan, N.O.; Forster, J.W. StAMPP: An R package for calculation of genetic differentiation and structure of mixed-ploidy level populations. *Mol. Ecol. Resour.* **2013**, *13*, 946–952. [CrossRef]
30. Wickham, H. *Ggplot2: Elegant Graphics for Data Analysis*; Springer: New York, NY, USA, 2009. [CrossRef]
31. Huson, D.H.; Bryant, D. Application of phylogenetic networks in evolutionary studies. *Mol. Biol. Evol.* **2006**, *23*, 254–267. [CrossRef] [PubMed]
32. Stoffel, M.A.; Esser, M.; Kardos, M.; Humble, E.; Nichols, H.; David, P.; Hoffman, J.I. inbreedR: An R package for the analysis of inbreeding based on genetic markers. *Methods Ecol. Evol.* **2016**, *7*, 1331–1339. [CrossRef]
33. Chen, H.; VennDiagram: Generate High-Resolution Venn and Euler Plots. R Package Version 1.6.20. 2018. Available online: https://CRAN.R-project.org/package=VennDiagram (accessed on 22 December 2020).
34. Alexander, D.H.; Novembre, J.; Lange, K. Fast model-based estimation of ancestry in unrelated individuals. *Genome Res.* **2009**, *19*, 1655–1664. [CrossRef] [PubMed]
35. Francis, R.M. pophelper: An R package and web app to analyze and visualize population structure. *Mol. Ecol. Resour.* **2017**, *17*, 27–32. [CrossRef] [PubMed]
36. Reich, D.; Thangaraj, K.; Patterson, N.; Price, A.L.; Singh, L. Reconstructing Indian population history. *Nature* **2009**, *461*, 489–494. [CrossRef]
37. Patterson, N.; Moorjani, P.; Luo, Y.; Mallick, S.; Rohland, N.; Zhan, Y.; Genschoreck, T.; Webster, T.; Reich, D. Ancient admixture in human history. *Genetics* **2012**, *192*, 1065–1093. [CrossRef]
38. Petr, M.; Vernot, B.; Kelso, J. admixr—R package for reproducible analyses using ADMIXTOOLS. *Bioinformatics* **2019**, *35*, 3194–3195. [CrossRef]
39. Becker, R.A.; Wilks, A.R.; Brownrigg, R.; Minka, T.P.; Deckmyn, A.; Maps: Draw Geographical Maps. R Package Version 3.3.0. 2018. Available online: https://CRAN.R-project.org/package=maps (accessed on 22 December 2020).
40. Wright, S. *Evolution and the Genetics of Population, Variability Within and Among Natural Populations*; The University of Chicago Press: Chicago, IL, USA, 1978.
41. Hartl, D.L.; Clark, A.G. *Principles of Population Genetics*, 3rd ed.; Sinauer Associates, Inc.: Sunderland, MA, USA, 1997.
42. Sokolov, I.I. *Fauna of the USSR: Mammals: Ungulates (Orders Perissodactyla and Artiodactyla)*; Publishing house of the Academy of Sciences of the USSR: Leningrad, Moscow, Russia, 1959; p. 265.
43. Tembotov, A.K. *Geography of North. Caucasian Mammals*; Nalchik: Elbrus, Russia, 1972; p. 151.
44. Couturier, M.A.J. *Le Bouquetin des Alpes*; Allier: Grenoble, France, 1962.
45. Ayunts, K.R. On the systematics of Caucasian tur. Bulletin of the Moscow Society of Naturalists. *Biol. Dep.* **1992**, *97*, 19–25.
46. Ropiquet, A.; Hassanin, A. Molecular evidence for the polyphyly of the genus Hemitragus (Mammalia, Bovidae). *Mol. Phylogenetics Evol.* **2005**, *36*, 154–168. [CrossRef]
47. Tan, S.; Wang, Z.; Jiang, L.; Peng, R.; Zhang, T.; Peng, Q.; Zou, F. Molecular phylogeny and phylogeography of genus Pseudois (Bovidae, Cetartiodactyla): New insights into the contrasting phylogeographic structure. *Ecol. Evol.* **2017**, *7*, 7047–7057. [CrossRef] [PubMed]
48. Dotsev, A.V.; Kunz, E.; Kharzinova, V.R.; Okhlopkov, I.M.; Lv, F.-H.; Li, M.-H.; Rodionov, A.N.; Shakhin, A.V.; Sipko, T.P.; Medvedev, D.G.; et al. Mitochondrial DNA Analysis Clarifies Taxonomic Status of the Northernmost Snow Sheep (*Ovis nivicola*) Population. *Life* **2021**, *11*, 252. [CrossRef] [PubMed]
49. Huxley, J. Clines: An Auxiliary Taxonomic Principle. *Nature* **1938**, *142*, 19–220. [CrossRef]

diversity

MDPI

Article

Comparative Study of the Genetic Diversity of Local Steppe Cattle Breeds from Russia, Kazakhstan and Kyrgyzstan by Microsatellite Analysis of Museum and Modern Samples †

Alexandra S. Abdelmanova [1,*], Veronika R. Kharzinova [1], Valeria V. Volkova [1], Arsen V. Dotsev [1], Alexander A. Sermyagin [1], Oksana I. Boronetskaya [2], Roman Yu. Chinarov [1], Evgeniya M. Lutshikhina [3], Johann Sölkner [4], Gottfried Brem [5] and Natalia A. Zinovieva [1,*]

1 L.K. Ernst Federal Research Center for Animal Husbandry, Dubrovitsy, Podolsk Municipal District, Moscow Region, 142132 Podolsk, Russia; veronika0784@mail.ru (V.R.K.); moonlit_elf@mail.ru (V.V.V.); asnd@mail.ru (A.V.D.); alex_sermyagin85@mail.ru (A.A.S.); roman_chinarov@mail.ru (R.Y.C.)
2 Museum of Livestock Named after E.F. Liskun, Timiryazev Russian State Agrarian University—Moscow Agrarian Academy, ul. Timiryazevskaya 49, 127550 Moscow, Russia; oboronetskaya@mail.ru
3 Institute of Biotechnology, National Academy of Science of Kyrgyz Republic, Bishkek 720071, Kyrgyzstan; lushihina.merinos-kg@yandex.com
4 Division of Livestock Sciences, University of Natural Resources and Life Sciences, 1180 Vienna, Austria; johann.soelkner@boku.ac.at
5 Institute of Animal Breeding and Genetics, University of Veterinary Medicine (VMU), Veterinärplatz, 1210 Vienna, Austria; gottfried.brem@vetmeduni.ac.at
* Correspondence: preevetic@mail.ru (A.S.A.); n_zinovieva@mail.ru (N.A.Z.); Tel.: +7-4967-65-11-63 (A.S.A. & N.A.Z.)

† This paper is an extended version of the paper published in the 1st International Electronic Conference on Biological Diversity, Ecology, and Evolution (BDEE 2021), Online, 15–31 March 2021.

Citation: Abdelmanova, A.S.; Kharzinova, V.R.; Volkova, V.V.; Dotsev, A.V.; Sermyagin, A.A.; Boronetskaya, O.I.; Chinarov, R.Y.; Lutshikhina, E.M.; Sölkner, J.; Brem, G.; et al. Comparative Study of the Genetic Diversity of Local Steppe Cattle Breeds from Russia, Kazakhstan and Kyrgyzstan by Microsatellite Analysis of Museum and Modern Samples. *Diversity* **2021**, *13*, 351. https://doi.org/10.3390/d13080351

Academic Editor: Michael Wink

Received: 3 July 2021
Accepted: 27 July 2021
Published: 30 July 2021

Abstract: The comparative molecular genetic study of museum and modern representatives of cattle breeds can help to elucidate the origin and maintenance of historical genetic components in modern populations. We generated the consensus genotypes for 11 microsatellite loci for 24 museum samples of Kalmyk, Kyrgyz, and Kazakh cattle, dated from the first quarter of the 20th century, and compared them with those of modern Kalmyk, Kyrgyz, and Kazakh white-headed breeds. The level of genetic diversity of the modern Kalmyk and Kyrgyz cattle (uHe = 0.771–0.778) was similar to those observed in the museum samples (uHe = 0.772–0.776), while a visible decrease in genetic variability in the modern Kazakh white-headed breed compared to museum Kazakh cattle was detected (uHe = 0.726 and 0.767, respectively). The PCA plot, F_{ST}- and Jost's D-based networks, and STRUCTURE clustering provided strong evidence of the maintenance of the historical genetic background in modern populations of Kalmyk and Kyrgyz cattle. In spite of the allele pool of Kazakh white-headed cattle having undergone great changes compared to the museum Kazakh cattle, several animals still carry the visible aspect of the historical genetic components. Our results can be used for the selection of individuals for the creation of gene banks and may significantly improve the efficiency of conservation programs aimed at preserving genetic diversity in the national genetic resources of cattle.

Keywords: cattle; Kalmyk cattle; Kyrgyz cattle; Kazakh cattle; Kazakh white-headed breed; historical specimens; microsatellites; consensus genotypes; genetic diversity

1. Introduction

Nomadic cattle husbandry across the vast territory of the south steppe has been undertaken since ancient times. The harsh climate conditions of short dry summers and cold winters have resulted in herders moving across large areas to seek better pastures for livestock [1–4]. During the short period of a plant's vegetation, cattle can restructure their body and create enough fat reserves to survive the long winters. Historically, poorly fleshed, weak livestock that are unable to find forage under the snow have died during the winter–spring period [5,6]. The similar way of life of nomadic tribes across the different parts of the south steppes over the centuries has resulted in their cattle having similar phenotypic characteristics. These animals are very hardy, of small to medium size, approximately 110–128 cm in stature, and are able to survive under poor forage conditions and be quickly fattened up with improved feeding [7–9]. Initial attempts to survey cattle inhabiting the south steppe of Russia were undertaken in the last decade of 19th century and the first decade of the 20th century, but the reports mainly described the number of cattle owned by different families and their movement routes [10]. Clear data on the exterior and particular properties of cattle of that time are not available. A comprehensive survey of the steppe cattle of Russia and the neighboring republic of former USSR countries was carried out in the first quarter of the 20th century, which resulted in the division of the entire cattle population into two groups (breeds), with a description of their properties, including coat color, exterior characteristics, and productivity [11]. The medium-sized, compact red cattle with a small head, long face, and short horns distributed in the southern regions of the European part of Russia (i.e., the Kalmykia, Rostov, Volgograd, and Astrakhan regions) and East Siberia were known as Kalmyk cattle (Supplementary Materials, Figure S1a). The origin of these cattle, officially recognized as a separate breed in 1934, was traced back to the 16th–17th centuries, when the Oirats tribe migrated from western Mongolia to the lower Volga through the territory of modern Kazakhstan, founding the Kalmyk Khanate [12,13]. At the end of the 18th century, because of disagreements with the government of the Russian Empire, part of the Oirats—now called Kalmyks—moved back to Dzungaria (the modern territories of northwest China) [14,15]. The small-sized cattle of different colors (red, black, brown, gray, white, pied, or tiger) with a relatively large head and well-developed horns that inhabited the territory of Kazakhstan and Kyrgyzstan were recognized as Kazakh or Kyrgyz cattle depending on the region of their distribution (Supplementary Materials, Figure S1b). It is well documented that the Kazakh and Kyrgyz tribes attacked Kalmyk caravans on their way to Dzungaria, capturing cattle as trophies [14,15]. This forms a basis for the suggestion that the ancient Kalmyk cattle have possibly contributed to the development of the gene pool of Kyrgyz (Kazakh) cattle. According to the cattle breeding plan of the USSR (1934), Kalmyk and Kyrgyz (Kazakh) cattle were chosen as the target breeds for further breeding in the steppe regions of the former USSR (Supplementary Materials, Figure S2) [16]. In the 1930–1940s, local steppe cattle were crossed with high-producing transboundary breeds to improve their growth capacity, meat, and carcass traits [17]. Kalmyk cattle were crossed with Brown Swiss and Simmental bulls, but most authors consider that these breeds only represent a small contribution to the development of the modern gene pool of Kalmyk cattle [8,16–19]. In Kyrgyzstan, major part of the local cattle population was crossed with Brown Swiss and Holland breeds that resulted in developing during the Soviet time several novel synthetic breeds [8,9]. The small remaining part of native Kyrgyz cattle was further developed without contribution of other breeds. The populations inhabiting the northern and western territories of Kazakhstan were crossed with Herefords followed by backcrossing of hybrid Kazakh-Hereford females with Hereford bulls over a number of generations, which resulted in the development of the Kazakh white-headed breed [17]. Additional crossing of Kazakh white-headed cows with Hereford bulls was regularly carried out during subsequent periods of breed development. Currently, the Kalmyk and Kazakh white-headed breeds are the main breeds used for meat production in the south steppe regions of Russia, Kazakhstan, and Kyrgyzstan. In Russia, the ratio of these breeds to the total number of beef cattle (2020) account for 36.6% and 11.7%, respectively [20]. Only a small number of purebred native Kyrgyz cattle are still kept by private owners in Kyrgyzstan. Because the pedigree records of modern Kyrgyz cattle are lacking, the decision on the assignment of an animal as native Kyrgyz cattle is made in most cases based on exterior characteristics and verbal information got from the owners. The origin of modern Kyrgyz cattle and its relationship with ancestral Kyrgyz cattle should be elucidated.

Native cattle breeds are an invaluable source of genetic diversity, which is necessary to ensure the sustainability of animal production systems in local geoclimatic conditions [18,21–23]. However, a drastic decline in the population size of most of the native breeds, as well as the active use of crossbreeding with high-producing commercial breeds, can lead to the irreversible loss of genetic diversity [24]. Conservation of genetic diversity is an important task in order to ensure the sustainability of the animal production system [25]. Successful conservation of local populations and breeds requires the selection of animals that are carriers of breed-specific genetic components, mostly reflecting the breeds' origin and trajectories of development and artificial selection.

A method with great potential for reconstructing breeds' origins and maintaining the ancestral genetic components in the modern populations of animal breeds is the analysis of specimens stored in museum collections [26–29]. Over the last decades, the field of ancient DNA (aDNA) heavily relied on mitochondrial DNA (mtDNA) markers. This is due to the low copy number and general difficulties associated with the recovery of nuclear DNA from archaeozoological materials [30]. However, the maternal inheritance characteristic of mtDNA has limited the ability to trace the history of cattle breeds with complex genetic background or breeds that were developed by crossing with bulls of other genetic origins. One of the most powerful tools for inferring genetic patterns of populations is microsatellite analysis [31–33]. During application of microsatellites for molecular genetic studies of farm animals over long time scales, reliable methods for analysis of these types of DNA markers have been developed [34]. Moreover, the data derived from different laboratories or different experiments can be standardized according to the Guidelines of the International Society of Animal Genetics (ISAG) [35].

The aim of our work was to compare the genetic diversity of Kalmyk, Kyrgyz, and Kazakh cattle inhabiting the steppe region of southern Russia and the neighboring republics of the former USSR in the first quarter of the 20th century with the current gene pool of the above breeds based on the analysis of microsatellite genotypes of museum and modern samples.

2. Materials and Methods

2.1. Sample Collection

The museum (historical) skulls of Kalmyk (KALM_H, n = 10), Kyrgyz (KRGZ_H, n = 11), and Kazakh (KZKH_H, n = 3) cattle dated to the first quarter of the 20th century were derived from the craniological collection of the Museum of Livestock named after E.F. Liskun (Moscow Agricultural Academy named after K.A. Timiryazev) (Figure 1).

(a) (b) (c)

Figure 1. Museum skulls of cattle breeds used for the the study dated by the first quarter of 20th century: (**a**)—Kalmyk, (**b**)—Kyrgyz, (**c**)—Kazakh.

The modern samples of Kalmyk (KALM_M, n = 28), Kyrgyz (KRGZ_M, n = 20), and Kazakh white-headed (KZWH_M, n = 30) cattle were derived from the Bioresource collection "Bank of genetic materials of domestic and wild animals and birds" of the L.K. Ernst Federal Research Center for Animal Husbandry. Tissue samples (ear skin) from modern animals were collected by trained personnel under strict veterinary rules in accordance with the guidelines for conducting laboratory research (tests) in the implementation of the veterinary control (supervision) approved by the Council Decision Eurasian Economic Commission No. 80 (10 November 2017). The 1–2 mm^2 pieces of ear skin were used for DNA extraction using Nexttec columns (Nexttec Biotechnology GmbH, Leverkusen, Germany), according to the manufacturer's instructions.

Additionally, we included modern samples of Mongolian cattle (MONG_M, $n = 41$) as a kinship breed with a similar genetic origin, which probably contributed to the development of Kalmyk cattle. Additionally, we included in the dataset samples of modern representatives of the Hereford breed (HRFD_M, $n = 26$) because of their great contribution to the development of the modern gene pool of Kazakh white-headed cattle.

2.2. DNA Extraction

All works with historical samples were performed in a dedicated facility of the L.K. Ernst Federal Research Center for Animal Husbandry [36]. The bone powder was produced from the roots of teeth using MIXER MILL MM 400 (Retsch GmbH, Haan, Germany). In total, 200 ± 5 µg of bone powder was subjected to the DNA extraction using a COrDIS Extract Decalcine Kit (GORDIZ LLC, Moscow, Russia), according to the manufacturer's instructions with modifications of the lysis conditions and volumes (lysis at 56 °C, 1200 rpm, overnight, using a 2.5× volume of lysis and washing buffers). A negative control tube ("reagent blank") containing only DNA extraction reagents without a sample was included in each batch of processed samples to trace the possible occurrence of DNA contamination.

The concentration of double-stranded DNA (dsDNA) was measured by using a Qubit™ fluorimeter (Invitrogen, Life Technologies, Waltham, MA, USA), and the purity of the DNA solutions was checked by determining the ratio of the absorption at 260 and 280 nm (OD260/280) on a NanoDrop 8000 instrument (Thermo Fisher Scientific Inc., Waltham, MA, USA). Only the samples with >1 ng/µL dsDNA concentration were used for the study.

2.3. Microsatellite Genotyping

The samples were genotyped for 11 highly polymorphic microsatellite loci (BM1818, BM2113, BM1824, ETH10, ETH225, INRA023, SPS115, TGLA53, TGLA122, TGLA126, and TGLA227) which were recommended by the International Society of Animal Genetics (ISAG) [35]. The multiplex PCRs were carried out in a final volume of 10 µL in a PCR buffer with 200 mM dNTPs, 1.0 mM $MgCl_2$, a 0.5 mM primer mix, 1 unit of Taq polymerase (Dialat Ltd., Moscow, Russia), and 1 µL of genomic DNA. The initial denaturation at 95 °C for 4 min was followed by 35 cycles of PCR amplification (95 °C, 20 s; 63 °C, 30 s; 72 °C, 1 min). Additionally, the final extension was performed at 72 °C for 10 min. The negative controls (PCR reaction without DNA template) were included in each PCR experiment to check for possible DNA contaminations. PCR fragments were separated by capillary electrophoresis on an ABI3130xl genetic analyzer (Applied Biosystems, Beverly, MA, USA) using GeneScan™-350 ET ROX as a fragment standard. The fragment lengths were determined using the Gene Mapper software v4 (Applied Biosystems, Beverly, MA, USA). Allele sizes were standardized according to an ISAG International Bovine (Bos Taurus) STR typing comparison test (2018–2019).

A modified multiple-tube approach, proposed by Mondol et al. [37] and Modi et al. [38], was used to determine the consensus genotypes as previously described [29]. Briefly, each DNA sample was analyzed in five replicates. Initially, multiplex amplification of microsatellite loci was performed in duplicates and only the samples in which at least six loci were successfully amplified (positive multiplex PCRs) were selected for further analysis. For such samples, three additional independent PCR replicates were performed using the same DNA extractions. The genotyping results of the five PCR replicates were used to calculate the "quality indices" (QIs) for each sample/locus as described by Miquel et al. [39]. For the samples with QI values less than 0.75, three additional multiplex PCRs were carried out using the same DNA extraction. Genotypes with a QI value of 0.75 and higher at each locus were considered reliable and selected for further analysis. The number of positive PCRs from the number of replicates for each locus/sample was defined as amplification failure. Those genotypes different from the most frequent genotype were defined as genotyping errors and included allelic dropout (ADO) or false alleles (FAs). The overall error rates were calculated as the ratio of genotyping errors from the total number of positive PCRs. For calculation of the ADO and FA rates, we used the protocol proposed by Broquet and Petit [40]. For the quality control of genotyping, the probability values of correct genotyping (p) for each locus were calculated according He et al. [41]. The threshold for p-values at each locus was set as <0.001.

2.4. Data Analysis

FreeNA [42] and MICRO-CHECKER 2.2.3 software [43] were used to check the presence of null alleles. We applied GenAIEx 6.5 software [44] to calculate the number and frequency of alleles. The packages diveRsity [45], adegenet [46], and ggplot2 [47] of the R software (http://cran.r-project.org, accessed on 12 May 2021) were used for calculating the main statistics, including observed heterozygosity (Ho), unbiased expected heterozygosity (uHe), unbiased inbreeding coefficient (uF_{IS}), and rarefied number of alleles (A_R) [48], performing principal component analysis (PCA), and visualizing breed relationships. Pairwise differences among studied populations for uHe and A_R estimates were tested using Wilcoxon rank-sum test [49] using Statistica 10 software (www.statsoft.com, accessed on 18 July 2021). The software environment R3.5.0 was used to prepare the data files [50].

Pairwise Jost's D genetic distances [51] and paired F_{ST} genetic distances [52] were calculated to evaluate the degree of genetic differentiation of the studied breeds. The pairwise Jost's D and F_{ST} matrices were used to construct the phylogenetic trees using the Neighbor-Net algorithm in SplitsTree 4.14.5 [53].

STRUCTURE 2.3.4 software [54] was applied to investigate the genetic structure of studied populations. We used the admixture model with the option of correlated allele frequencies. The burn-in period was set to 10,000 iterations, followed by 100,000 Markov chain Monte Carlo (MCMC) repetitions for each run. We performed 10 independent runs for a number of possible clusters (K) ranging from 2 to 8. CLUMPAK software [55], available at http://clumpak.tau.ac.il (accessed on 7 May 2021), was used to analyze multiple independent runs at a single K value. For measuring the constancy over runs, we calculated average pairwise similarity scores using CLUMPAK [55].

To infer demographic history with admixture, we applied the TreeMix software [56] to the microsatellite data. We computed the mean and variance in length at each microsatellite locus and used them to run TreeMix v1.13. We tested up to five migration events in twenty iterations per migration edge. The optimal number of migration events was determined using R package "optM" [57] based on TreeMix output files. We added the optimal number of migrations to the phylogenetic model and estimated the consistency between migration edges using TreeMix for 20 independent runs.

3. Results

3.1. Estimation of Consensus Genotypes for Museum Samples

The concentrations of dsDNA extracted from the museum specimens exceeded the set threshold of 1 ng/µL and varied from 2.88 to 38.40 ng/µL in different samples (Supplementary Materials, Table S1). Out of the 24 museum specimens, 18 samples were successfully genotyped for all 11 loci. For four samples, the consensus genotypes were obtained for 10 loci, including two samples of the Kyrgyz breed (H176, H181) and one sample each of the Kalmyk (H210) and Kazakh (H355) breeds. Two samples of Kyrgyz breed (H180 and H181) were genotyped for 9 loci (Supplementary Materials, Table S2a). No amplification failure was observed in the BM2113 locus. Among the other 10 loci, the ratio of positive PCRs varied from 99.23% in BM1824 to 72.31% in BM1818, corresponding to an amplification failure of 0.77% and 27.69%, respectively. The quality of genotyping differed between microsatellite loci and varied from QI = 0.96 in TGLA53 and TGLA122 to QI = 1.00 in

BM2113 and BM1824. We identified the allelic dropout (ADO) in 2.35% of the estimated heterozygous genotypes with variations from 0.92% in TGLA227 to 5.80% in the SPS115 locus. Three loci (BM2113, BM1824, and ETH10) were free of ADO. We found false alleles (FAs) in six genotyped loci, whereby the FA rate varied from 0.78% in ETH10 to 2.13% in BM1818. Five loci (BM2113, INRA23, TGLA126, ETH225, and BM1824) were free of FAs. The error rate varied from 0.78% in ETH10 to 4.76% in TGLA122—an average of 2.32%. No genotyping errors were observed in the BM2113 and BM1824 loci. The probability values of correct genotyping (p) for each locus met the set threshold ($p < 0.001$) (Supplementary Materials, Table S3). Therefore, all studied loci that passed quality control were selected for further analysis.

Out of 145 modern specimens, the genotypes for all 11 loci were estimated for 136 samples, while for 8 and 1 samples the microsatellite genotypes were obtained for 10 and 9 loci, respectively (Supplementary Materials, Table S2b).

3.2. Allelic Variability and Genetic Diversity of Studied Breeds

We identified 130 microsatellite alleles in eight cattle populations including 124 alleles in modern samples and 95 alleles in museum samples. The most polymorphic were the TGLA122 and TGLA53 loci (21 and 19 alleles, respectively), while the least number of alleles (7 alleles) was found in the BM1824 locus (Table 1).

Table 1. Number of identified alleles among microsatellite loci in studied cattle populations (%).

#	Locus	Entire Dataset		Museum Samples		Modern Samples	
		Observed Allele Ranges, bp	Number of Alleles	Observed Allele Ranges, bp	Number of Alleles	Observed Allele Ranges, bp	Number of Alleles
1	TGLA227	69–101	15	77–101	12	69–99	14
2	BM2113	125–143	10	125–143	8	125–143	10
3	TGLA53	152–188	19	154–184	13	152–188	18
4	ETH10	209–225	8	213–225	7	209–225	8
5	SPS115	244–262	9	248–260	5	244–262	9
6	TGLA122	137–183	21	137–173	14	137–183	20
7	INRA23	196–218	12	198–216	10	196–218	12
8	TGLA126	107–125	9	107–123	7	111–125	8
9	BM1818	256–274	10	256–274	8	258–274	9
10	ETH225	140–160	10	140–158	7	140–160	9
11	BM1824	178–190	7	178–188	4	178–190	7
In average			11.82 ± 1.38		8.63 ± 0.97		11.27 ± 1.30

Notes: locus—microsatellite locus; observed allele ranges—the limits of allelic lengths of studied microsatellite loci in historical samples analysed (a allele sizes were standardised according to International Society of Animal Genetics (ISAG) International Bovine (Bos Taurus) short tandem repeat (STR) typing comparison test 2018–2019).

Of the alleles, 83.33% found in the museum Kalmyk cattle and 81.16% in the museum Kyrgyz cattle were present in the modern representatives of these breeds. Despite long-term crossbreeding of native Kazakh cattle with Herefords, 82.93% of the alleles of museum Kazakh cattle appeared in the modern Kazakh white-headed breed. Meanwhile, 92.98% of alleles observed in Herefords appeared in Kazakh white-headed cattle. The great contribution of Mongolian cattle to the development of the gene pool of the studied breeds of steppe cattle was reflected in the high rate of common alleles with Mongolian cattle: 87.50%, 88.41%, and 92.68% in the museum Kalmyk, Kyrgyz, and Kazakh cattle, respectively. These rates decreased to 81.61% and 81.82% in the modern populations of Kalmyk and Kyrgyz cattle, respectively. The Kazakh white-headed cattle retained among other modern breeds the highest rate of alleles in common with the Mongolian cattle (93.15%) (Table 2).

Table 2. Allelic variability and number of common alleles in studied breeds.

Populations	KALM_H	KRGZ_H	KZKH_H	KALM_M	KRGZ_M	KZWH_M	HRFD_M	MONG_M
KALM_H	72							
KRGZ_H	49	69						
KZKH_H	37	30	41					
KALM_M	60	58	37	87				
KRGZ_M	63	56	35	68	88			
KZWH_M	55	53	34	63	65	73		
HRFD_M	48	40	30	49	53	53	57	
MONG_M	63	61	38	71	72	68	53	94

Notes: museum populations: KALM_H—Kalmyk, KRGZ_H—Kyrgyz, KZKH_H—Kazakh, modern populations: KALM_M—Kalmyk, KRGZ_M—Kyrgyz, KZWH_M—Kazakh White-Headed, HRFD_M—Hereford, MONG_M—Mongolian cattle; number of alleles identified in studied breeds is shown at diagonal; the number of alleles, which are common for two breeds are shown below the diagonal.

Comparing the museum and modern Kalmyk, Kyrgyz, and Kazakh cattle showed that 15 alleles, which were distributed in the museum populations, were lost in the modern cattle. By contrast, 60 novel alleles appeared in the modern populations. The greatest number of novel alleles (n = 33) was found in the Kazakh white-headed breed, reflecting the gene flow from Hereford breed. The TGLA122 and TGLA53 loci were the most altered: Three and two ancestral alleles were lost, while 11 and 9 novel alleles appeared, respectively (Supplementary Materials, Table S4). Considering the differences in sample size between museum and modern populations, it should be noted that the absence of part of alleles in historic populations could be the result of their smaller sample size.

Estimations of the genetic diversity calculated based on the genotypes for 11 microsatellite loci are summarized in Table 3.

Table 3. Summary statistics for museum and modern populations of studied breeds based on genotypes of 11 microsatellites.

Population	n	Ho (M ± SE)	uHe (M ± SE)	A_R (M ± SE)	uF_{IS} (CI, 95%)
KALM_H	10	0.671 ± 0.048	0.772 ± 0.029 *	3.635 ± 0.199 *	0.131 [0.033; 0.229]
KRGZ_H	11	0.707 ± 0.053	0.776 ± 0.035 *	3.693 ± 0.251 *	0.081 [−0.055; 0.217]
KZKH_H	3	0.818 ± 0.082	0.767 ± 0.067	3.727 ± 0.333	−0.066 [−0.174; 0.042]
KALM_M	28	0.736 ± 0.049	0.771 ± 0.034	3.664 ± 0.232 *	0.044 [−0.050; 0.138]
KRGZ_M	20	0.841 ± 0.021	0.778 ± 0.024 *	3.704 ± 0.196 *	−0.085 [−0.137; −0.033]
KZWH_M	30	0.736 ± 0.039	0.726 ± 0.030	3.336 ± 0.167 *	−0.011 [−0.063; 0.041]
HRFD_M	26	0.668 ± 0.064	0.653 ± 0.053	2.994 ± 0.211	0.005 [−0.078; 0.068]
MONG_M	41	0.672 ± 0.036	0.761 ± 0.020 *	3.531 ± 0.176 *	0.115 [0.028; 0.202]

Notes: n, number of individuals; Ho, observed heterozygosity; uHe, unbiased expected heterozygosity; A_R, rarefied allele richness; uF_{IS}, unbiased inbreeding coefficient; M, mean value; SE, standard error; CI 95%, range variation coefficient of uF is at a confidence interval of 95%; museum populations: KALM_H—Kalmyk, KRGZ_H—Kyrgyz, KZKH_H—Kazakh, modern populations: KALM_M—Kalmyk, KRGZ_M—Kyrgyz, KZWH_M—Kazakh White-Headed, HRFD_M—Hereford, MONG_M—Mongolian cattle; * $p < 0.05$ comparing to Hereford breed according to pairwise comparing Wilcoxon rank-sum test.

We did not observe significant differences in the genetic diversity between the museum and modern populations of all studied breeds. For Kalmyk and Kyrgyz cattle, the values of the unbiased expected heterozygosity and rarefied allelic richness were similar (uHe = 0.772–0.776 for the museum samples and 0.771–0.778 for the modern samples; A_R = 3.635–3.693 and 3.664–3.704, respectively). The Kazakh white-headed cattle were characterized by decreased genetic diversity compared to the museum Kazakh cattle (uHe = 0.726 vs. 0.767; A_R = 3.336 vs. 3.727). This could be associated with lower level of genetic diversity of Hereford cattle and may reflect the contribution of the limited number of Hereford bulls in formation of the Kazakh white-headed breed. A significant deficiency of heterozygotes was found in the museum Kalmyk cattle (uF_{IS} = 0.131), as well as in the modern Mongolian cattle (uF_{IS} = 0.115), while the modern Kyrgyz cattle were characterized by an excess of heterozygotes (uF_{IS} = −0.085) (Table 3).

3.3. Relationships among the Studied Breeds

The first component of the PCA explained 5.178% of the genetic variability and split the Kazakh white-headed and Hereford breeds from the other studied breeds. The second component, responsible for 3.773%, divided the museum and modern Kyrgyz cattle. The Mongolian cattle were localized in the middle of the PCA plot, explaining their contribution to the development of the studied steppe breeds. The museum Kalmyk samples did not form a clear cluster. A possible explanation could be the different genetic backgrounds of the museum specimens of Kalmyk cattle (Figure 2a).

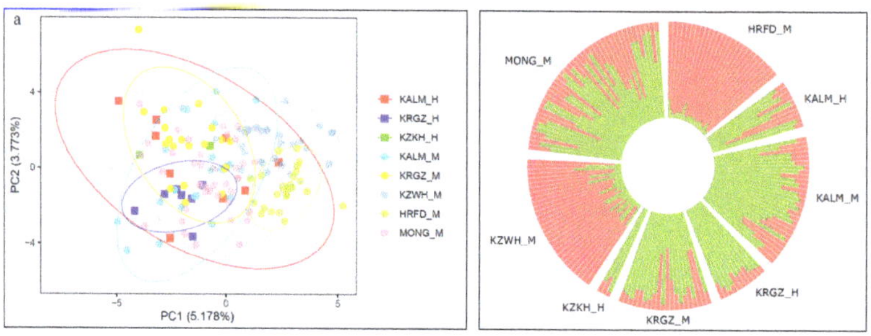

(a) (b)

Figure 2. (**a**). Principal component analysis (PCA) of the museum and modern cattle populations. X-axis, principal component 1 (PC1); Y-axis, principal component 2 (PC2); (**b**). Genetic structure of museum and modern cattle populations revealed by STRUCTURE analysis for the number of clusters K = 2. Museum populations: KALM_H—Kalmyk, KRGZ_H—Kyrgyz, KZKH_H—Kazakh, modern populations: KALM_M—Kalmyk, KRGZ_M—Kyrgyz, KZWH_M—Kazakh White-Headed, HRFD_M—Hereford, MONG_M—Mongolian cattle.

Calculation of the ΔK values for the different number of clusters (K) from one to five showed that the most probable number of ancestral populations that participated in the development of studied breeds is two. STRUCTURE analysis at K = 2 showed differences in the genetic structure of the studied breeds by the rate of native and Hereford-specific genetic components. The museum samples of the Kyrgyz and Kazakh cattle had similar genetic structures with the greatest rate of native genetic components, while part of the museum Kalmyk cattle revealed an admixed genetic structure. The modern Kalmyk and Kyrgyz cattle kept most of the native genetic components, while in the Kazakh white-headed cattle, the Hereford-specific components dominated (Figure 2b).

Analysis of F_{ST} and Jost's D genetic distances showed a lack of allelic differentiation between pairs of museum populations: Kalmyk–Kyrgyz (F_{ST} and Jost's D = 0) and Kalmyk–Kazakh ($F_{ST} = -0.020$, Jost's D = −0.013). The modern Kalmyk and Kyrgyz cattle were closest to the museum populations of their ancestral breeds (F_{ST} = 0.018 and 0.025; Jost's D = 0.017 and 0.047, respectively), while the Kazakh white-headed breed was most distant from museum populations (F_{ST} = 0.048, 0.082, and 0.096; Jost's D = 0.089, 0.175 and 0.079 for museum Kalmyk, Kyrgyz and Kazakh cattle, respectively). Museum cattle revealed a stronger Mongolian genetic background (F_{ST} = 0–0.019; Jost's D = 0–0.015) compared to the modern cattle breeds (F_{ST} = 0.026–0.052; Jost's D = 0.039–0.098) (Supplemental Materials, Table S5).

A similar allelic pattern of the museum populations was reflected in their neighbored localization on the edges of the Neighbor-Net tree constructed based on Jost's D values (Figure 3) and F_{ST} values (Supplemental Materials, Figure S3). Modern populations of the Kalmyk and Kyrgyz breeds were localized in the same cluster, but formed their own branches. This suggests that the development of these breeds was based on the historical genetic background. The Kazakh white-headed cattle formed a separate cluster with the Hereford breed, reflecting the great contribution of the latter in the development of the allele pool of the Kazakh white-headed cattle (Figure 3).

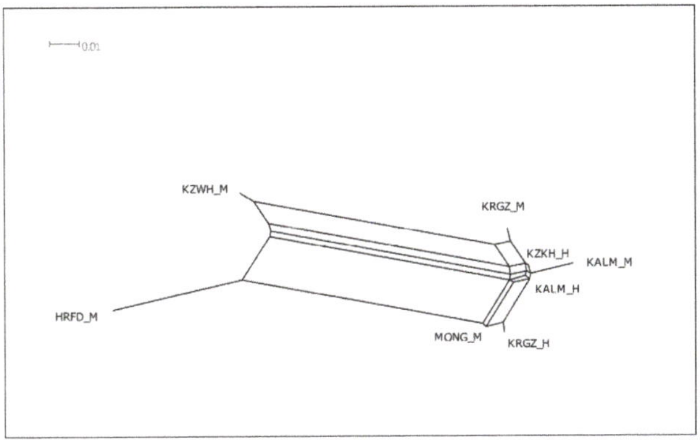

Figure 3. Neighbor-Net graphs based on Jost's D genetic distances characterizing genetic relationships between studied museum and modern cattle populations. Museum populations: KALM_H—Kalmyk, KRGZ_H—Kyrgyz, KZKH_H—Kazakh, modern populations: KALM_M—Kalmyk, KRGZ_M—Kyrgyz, KZWH_M—Kazakh White-Headed, HRFD_M—Hereford, MONG_M—Mongolian cattle.

A single migration event was found to be the most optimal for describing the demographic history of the studied populations using TreeMix. The most stable TreeMix tree showed migration from the ancestors of the museum Kazakh cattle and Herefords to the Kazakh white-headed breed that agrees with the origin and development of the last (Figure 4).

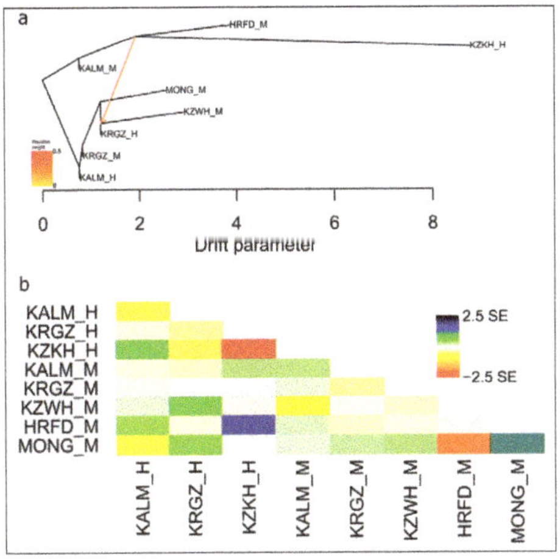

Figure 4. (**a**) Maximum likelihood tree inferred from studied cattle populations with single migration edge. Scale bar shows 10 times the average standard error (SE) of the estimated entries in the sample covariance matrix. Putative gene flow is indicated by the arrow, pointing in the direction of flow from the donor to the recipient population, and coloured in orange, proportional to the gene flow intensity. (**b**) Residual matrix plotted from a TreeMix analysis under single migration event and expressed as the number of SE of the deviation. Positive SE values between pairs of populations indicate that populations are more closely related to each other than in the modeled tree, while negative SE values show that the observed covariance is overestimated. Museum populations: KALM_H—Kalmyk, KRGZ_H—Kyrgyz, KZKH_H—Kazakh, modern populations: KALM_M—Kalmyk, KRGZ_M—Kyrgyz, KZWH_M—Kazakh White-Headed, HRFD_M—Hereford, MONG_M—Mongolian cattle.

4. Discussion

Conservation of the genetic diversity of animal genetic resources is important for food security and the ability of agricultural systems to adapt to possible future changes in production environments, including climate, market requirements, and diseases [58,59]. Predominantly using highly productive transboundary breeds in agricultural production and the associated decline in the population size of the local breeds has led to a drastic decrease in the genetic diversity of farm animal species, including cattle [59,60]. To ensure the sustainability of agricultural production, programs for the conservation of farm animal genetic resources should be developed [59,61].

One of the important points in developing such programs is the selection of the individuals to be used for conservation. The most valuable resources are individuals who are carriers of breed-specific genetic components and would allow reconstitution of that breed in case of its extinction [62]. However, in modern populations of local cattle breeds, the complex genetic backgrounds and active crossbreeding make it difficult to estimate the ancestral genetic components and infer the origin of breeds. The absence of such information would negatively affect their future development and breeding management. The study of museum and archeological samples can help in the identification of such components in the genome of modern representatives of breeds [29,63,64].

During the century-long history of the territories in the south steppe of Russia and the neighboring republics of the former Soviet Union, cattle breeds developed were well adapted to the harsh environment of this region and able to survive under poor foraging conditions [8,9]. In the present study, we were able to obtain valid genotypes for 11 microsatellite loci for the museum specimens of the native steppe Kalmyk, Kyrgyz, and Kazakh cattle dated to the first quarter of 20th century.

The level of genetic diversity of the modern Kalmyk and Kyrgyz cattle (uHe = 0.771–0.778) was similar to that determined in previous studies of these breeds (He = 0.76–0.78) [65] and comparable to those observed in the museum samples (uHe = 0.772–0.776) (Table 3). The modern populations of both of these breeds were less affected by crossing with transboundary breeds and were not subjected to artificial selection with high selection pressure [66,67]. On the contrary, we detected a visible decrease in the genetic variability of the modern Kazakh white-headed breed compared to the museum Kazakh cattle (uHe = 0.726 and 0.767; A_R = 3.727 and 3.336, respectively) (Table 3). A possible explanation is that only a part of the Kazakh cattle inhabiting the northern and western territories of Kazakhstan were used for developing the modern population of Kazakh white-headed cattle [17]. Higher selection pressure for the limited number of traits and the use of a limited number of Hereford sires for breed improvement can be considered as additional factors leading to the decline in genetic diversity in the modern population of this breed. Meanwhile, the genetic diversity of the Kazakh white-headed cattle was higher than in Herefords (uHe = 0.726 vs. 0.653; A_R = 3.336 vs. 2.994, $p < 0.05$), which is in general agreement with other studies [65,68–70] showing a lower level of genetic diversity in transboundary commercial breeds compared to local ones; He = 0.76–0.78 in Kalmyk and Kyrgyz cattle comparing to 0.70–0.71 in transboundary Brown Swiss and Holstein breeds [65]; He = 0.679–0.802 in twenty-seven Chinese indigenous yellow cattle breeds comparing to 0.661–0.697 in three commercial breeds [68]; uHe = 0.791–0.804 in local Vietnamese pigs comparing to 0.606–0.632 in commercial Yorkshire and Landrace pig breeds [69]; He = 0.72–0.81 in twenty-two Vietnamese pig breeds comparing to 0.56–0.65 in commercial Duroc, Yorkshire and Landrace pig breeds [70].

We observed a significant deviation from the Hardy-Weinberg equilibrium in heterozygote number for museum samples of the Kalmyk breed (heterozygote deficiency), which has been indicated by a positive value of uF_{IS} (0.131). Interestingly, we found the heterozygote excess beyond the expected number of heterozygotes in the modern Kyrgyz cattle (uF_{IS} = −0.085) that has the lowest population size among studied breeds. This observation is in agreement with the results of previous studies [65], indicated the heterozygote excess in two studied populations of Kyrgyz cattle (F_{IS} = 0.132–0.135). The possible explanation could be the low pressure of the artificial selection, extremely long productive life (up to 10 lactations) of the cows, and the calves' exchanges between owners.

The PCA plot (Figure 2a), STRUCTURE clustering (Figure 2b), F_{ST}-based tree (Figure S3), and Jost's D network (Figure 3) showed strong maintenance of the historical genetic backgrounds in the modern populations of the Kalmyk and Kyrgyz cattle. The allele pool of the modern Kazakh white-headed cattle has undergone the greatest changes compared to the museum Kazakh cattle, which is associated with multiple backcrossing with Hereford bulls [8,9]. This revealed in more distant localization of Kazakh white-headed breed in relation to museum samples at PCA plot (Figure 2a); occurrence of the large part of Hereford specific genetic components in the cluster structure of Kazakh white-headed cattle (Figure 2b); formation by Kazakh white-headed and Hereford breeds of the joined branch on Neighbor-Net trees (Figure 3, Supplemental materials, Figure S3). The contribution of Herefords is confirmed by more than a threefold decrease of the genetic distance with Kazakh white-headed cattle (F_{ST} = 0.049) comparing to museum Kazakh cattle (F_{ST} = 0.154) (Supplemental materials, Table S5), as well as by observed gene flow from Hereford ancestors to the ancestral population of the Kazakh white-headed breed at TreeMix tree (Figure 4a). The positive SE value observed between Hereford and Kazakh white-headed cattle at residual matrix plotted from a TreeMix (Figure 4b) indicates that these two breeds are even more closely related to each other than in the modeled tree. However, individuals that have retained a visible portion of their historical genetic components are still present among modern Kazakh white-headed cattle (Figure 2b). Including such individuals in conservation programs has great value for maintenance of the genetic diversity of the local animal genetic resources.

Thus, our research results have demonstrated that the studied cattle breeds have still retained their historical genetic background, which allows them to this day not only to successfully compete with transboundary breeds, but also to perform important functions ranging from providing foods to socio-economic, cultural and ecological roles in their breeding areas. In our opinion, the strategy for the further preservation of the genetic resources of these breeds may consist at least of preventing mass crossbreeding with transboundary breeds. In the ongoing breeding policy, special attention should be paid to the preservation of historical allele pool, since it composes the genetic uniqueness of breeds, formed by combining the ecological processes, gene flow, local breeding practices, and geographical features.

Considering the development of new powerful molecular genetics tools that enable high-throughput analysis of cattle at the genome level [18,19,22,23,71,72], including analysis of museum and archeological specimens [73,74], additional studies are required to identify and trace more precisely the historical genomic components in the modern representatives of breeds.

5. Conclusions

Using the genotypes for 11 microsatellite loci, we compared the genetic diversity and established the genetic relationships between the museum (dated to the first part of 20th century) and modern populations of Kalmyk, Kyrgyz, and Kazakh cattle, distributed in the south steppe region of the European part of Russia, Kazakhstan, Kyrgyzstan, and West Siberia. We showed that the allele pool of modern cattle populations has undergone changes that have manifested visibly, which are most significant in the Kazakh white-headed cattle. At the same time, we were able to clearly show that modern representatives of all of the studied breeds retained a portion of their historical genetic components, making them valuable national genetic resources. These research results can be used for developing sustainable programs for conservation of the genetic diversity of native cattle breeds.

Supplementary Materials: The following are available online at https://www.mdpi.com/article/10.3390/d13080351/s1, Figure S1: Representatives of Kalmyk (a) and Kyrgyz (b) cattle demonstrated on the first agricultural exhibition of breeding cattle in Moscow 1896, Figure S2: Schematic map of the zoning of cattle breeds on the territory of the USSR in 1934, Figure S3: Neighbor-Net graphs based on F_{ST} genetic distances characterizing genetic relationships between studied museum and modern cattle populations, Table S1: Quantitative and qualitative characteristics of DNA extracted from the museum specimens and number of replicates used for estimating the consensus genotypes, Table S2: Microsatellite genotypes of museum and modern samples used for the studies, Table S3: Quality of microsatellite genotyping and distribution of genotyping errors among microsatellite loci in museum samples of studied cattle populations, Table S4: Alleles of microsatellites, which were occurred in museum samples, but were lost in modern populations and novel alleles, which are appeared in modern populations, Table S5: Genetic distances between the studied populations based on F_{ST} and Jost's D indices.

Author Contributions: Conceptualization, N.A.Z.; methodology, A.S.A. and N.A.Z.; software, A.S.A. and A.V.D.; validation, A.S.A., A.V.D. and A.A.S.; investigation, A.S.A., V.R.K., V.V.V., A.A.S. and R.Y.C.; resources, O.I.B. and E.M.L.; data curation, A.S.A. and N.A.Z.; writing—original draft preparation, A.S.A. and N.A.Z.; writing—review and editing, A.S.A., A.A.S., O.I.B., J.S., G.B. and N.A.Z.; supervision, N.A.Z.; visualization, A.S.A.; supervision, N.A.Z.; project administration, N.A.Z.; funding acquisition, N.A.Z. All authors have read and agreed to the published version of the manuscript.

Funding: This research was funded by the Russian Science Foundation within project No. 19-76-20012 (study of museum samples) and the Russian Ministry of Science and Higher Education within theme No. 0445-2019-0024 (study of modern samples).

Institutional Review Board Statement: The study was approved by the Ethics Commission of the L.K. Ernst Federal Research Center for Animal Husbandry (protocol No. 3 from the 19 January 2021).

Informed Consent Statement: Not applicable.

Data Availability Statement: The data presented in this study are available in Supplementary Materials, Table S2.

Conflicts of Interest: The authors declare no conflict of interest. The funders had no role in the design of the study; in the collection, analyses, or interpretation of data; in the writing of the manuscript, or in the decision to publish the results.

References

1. Jeong, C.; Wang, K.; Wilkin, S.; Taylor, W.T.T.; Miller, B.K.; Ulziibayar, S.; Stahl, R.; Chiovelli, C.; Bemmann, J.H.; Knolle, F.; et al. A Dynamic 6,000-Year Genetic History of Eurasia's Eastern Steppe. *Cell* **2020**, *183*, 890–904.e29. [CrossRef] [PubMed]
2. Wilkin, S.; Ventresca Miller, A.; Taylor, W.T.T.; Miller, B.K.; Hagan, R.W.; Bleasdale, M.; Scott, A.; Gankhuyg, S.; Ramsøe, A.; Uliziibayar, S.; et al. Dairy pastoralism sustained eastern Eurasian steppe populations for 5000 years. *Nat. Ecol. Evol.* **2020**, *4*, 346–355. [CrossRef] [PubMed]
3. Nandintsetseg, B.; Shinoda, M.; Erdenetsetseg, B. Contributions of multiple climate hazards and overgrazing to the 2009/2010 winter disaster in Mongolia. *Nat. Hazards* **2018**, *92*, 109–126. [CrossRef]
4. Ferret, C. Mobile pastoralism a century apart: Continuity and change in south-eastern Kazakhstan, 1910 and 2012. *Cent. Asian Surv. Taylor Fr.* **2018**, *37*, 503–525. [CrossRef]
5. Suttie, J.M. *Grazing Management in Mongolia;* Suttie, J.M., Reynolds, S.J., Batelo, C., Eds.; Grasslands of the World, Plant Production and Protection Series 34; FAO-UN: Rome, Italy, 2005; pp. 265–304.
6. Nandintsetseg, B.; Shinoda, M.; Du, C.; Munkhjargal, E. Cold-season disasters on the Eurasian steppes: Climate-driven or man-made. *Sci. Rep.* **2018**, *8*, 14769. [CrossRef]
7. Leopoldov, A.V. *Cattle at the First All-Union Agricultural Exhibition of 1923;* Novaya Derevnya: Moscow, Russia, 1924; 178p, Available online: http://elib.cnshb.ru/books/free/0365/365927 (accessed on 15 June 2021).
8. Liskun, E.F. *Domestic Breeds of Cattle;* State Publishing House of Agricultural Literature: Moscow, Russia, 1949; 175p.
9. Ernst, L.K.; Dmitriev, N.G.; Paronyan, I.A. *Genetic Resources of Farm Animals in Russia and Neighboring Countries;* All-Russian Research Institute of Animal Genetics and Breeding: St. Petersburg, Russia, 1994; p. 469.
10. Eldarov, N.A. About the History of Continuous Expedition Surveys in Kalmykia. *Stat. Issues* **2006**, *6*, 63–66. Available online: http://www.lib.kalmsu.ru/text/TRUD/Eldyaeva_NA/p002.pdf (accessed on 15 June 2021).
11. Liskun, E.F. *Russian Cattle Breeds;* Novyj Agronom: Moscow, Russia, 1928; 212p.
12. Pozdneev, A.M. Astrakhan Kalmyks and their attitude to Russia before the beginning of the present century. *J. Minist. Public Educ.* **1886**, *244*, 140–171.
13. Lee, J.-Y. Were the Historical Oirats "Western Mongols"? An Examination of Their Uniqueness in Relation to the Mongols. *Études Mongoles et Sibériennes, Centrasiatiques et Tibétaines* **2016**, *47*. Available online: http://journals.openedition.org/emscat/2820 (accessed on 8 June 2021). [CrossRef]
14. Grousset, R. *The Empire of the Steppes: A History of Central Asia;* Rutgers University Press: New Brunswick, NJ, USA, 1970.

15. Kendirbai, G.T. The Politics of the Inner Asian Frontier and the 1771 Exodus of the Kalmyks. *Inner Asia* **2018**, *20*, 263–292. [CrossRef]
16. Ivanova, O.A.; Rogov, A.L.; Fedorov, V.I. *Cattle Breeding*; Selkhozgiz: Moscow, Russia, 1934; 287p.
17. Dmitriev, N.G.; Ernst, L.K. *Animal Genetic Resources of the USSR*; FAO and UNEP: Rome, Italy, 1989. Available online: http://www.fao.org/3/ah759e/AH759E07.htm (accessed on 10 February 2021).
18. Sermyagin, A.A.; Dotsev, A.V.; Gladyr, E.A.; Traspov, A.A.; Deniskova, T.E.; Kostjunina, O.V.; Reyer, H.; Wimmers, K.; Barbato, M.; Paronyan, I.A.; et al. Whole-genome SNP analysis elucidates the genetic structure of Russian cattle and its relationship with Eurasian taurine breeds. *Genet. Sel. Evol.* **2018**, *50*, 37. [CrossRef]
19. Yurchenko, A.; Yudin, N.; Aitnazarov, R.; Plyusnina, A.; Brukhin, V.; Soloshenko, V.; Lhasaranov, B.; Popov, R.; Paronyan, I.A.; Plemyashov, K.V.; et al. Genome-wide genotyping uncovers genetic profiles and history of the Russian cattle breeds. *Heredity* **2018**, *120*, 125–137. [CrossRef] [PubMed]
20. Butusov, D.V.; Ministry of Agriculture of the Russian Federation, Moscow, Russia. Animal Genetic Resources, State Support Measures, Key Tasks. Personal communication, 2021.
21. Felius, M.; Theunissen, B.; Lenstra, J. Conservation of cattle genetic resources: The role of breeds. *J. Agric. Sci.* **2015**, *153*, 152–162. [CrossRef]
22. Zinovieva, N.A.; Dotsev, A.V.; Sermyagin, A.A.; Deniskova, T.E.; Abdelmanova, A.S.; Kharzinova, V.R.; Sölkner, J.; Reyer, H.; Wimmers, K.; Brem, G. Selection signatures in two oldest Russian native cattle breeds revealed using high-density single nucleotide polymorphism analysis. *PLoS ONE* **2020**, *15*, e0242200. [CrossRef] [PubMed]
23. Zinovieva, N.A.; Sheiko, I.P.; Dotsev, A.V.; Sheiko, R.I.; Mikhailova, M.E.; Sermyagin, A.A.; Abdelmanova, A.S.; Kharzinova, V.R.; Reyer, H.; Wimmers, K.; et al. Genome-wide SNP analysis clearly distinguished the Belarusian Red cattle from other European cattle breeds. *Anim. Genet.* **2021**. [CrossRef] [PubMed]
24. Scherf, B.D.; Pilling, D. *The Second Report on the State of the World's Animal Genetic Resources for Food and Agriculture*; FAO: Rome, Italy, 2015. Available online: http://www.fao.org/3/a-i4787e/index.html (accessed on 8 June 2021).
25. Sponenberg, D.P. Conserving the Genetic Diversity of Domesticated Livestock. *Diversity* **2020**, *12*, 282. [CrossRef]
26. Rowe, K.C.; Singhal, S.; Macmanes, M.D.; Ayroles, J.F.; Morelli, T.L.; Rubidge, E.M.; Bi, K.; Moritz, C.C. Museum genomics: Low-cost and high-accuracy genetic data from historical specimens. *Mol. Ecol. Resour.* **2011**, *11*, 1082–1092. [CrossRef]
27. Billerman, S.M.; Walsh, J. Historical DNA as a tool to address key questions in avian biology and evolution: A review of methods, challenges, applications, and future directions. *Mol. Ecol. Resour.* **2019**, *19*, 1115–1130. [CrossRef]
28. Zinovieva, N.A.; Sermyagin, A.A.; Dotsev, A.V.; Boronetslaya, O.I.; Petrikeeva, L.V.; Abdelmanova, A.S.; Brem, G. Animal genetic resources: Developing the research of allele pool of Russian cattle breeds—Minireview. *Sel'skokhozyaistvennaya Biologiya* **2019**, *54*, 631–641. [CrossRef]
29. Abdelmanova, A.S.; Kharzinova, V.R.; Volkova, V.V.; Mishina, A.I.; Dotsev, A.V.; Sermyagin, A.A.; Boronetskaya, O.I.; Petrikeeva, L.V.; Chinarov, R.Y.; Brem, G.; et al. Genetic Diversity of Historical and Modern Populations of Russian Cattle Breeds Revealed by Microsatellite Analysis. *Genes* **2020**, *11*, 940. [CrossRef]
30. McHugo, G.P.; Dover, M.J.; MacHugh, D.E. Unlocking the origins and biology of domestic animals using ancient DNA and paleogenomics. *BMC Biol.* **2019**, *17*, 98. [CrossRef]
31. Putman, A.I.; Carbone, I. Challenges in analysis and interpretation of microsatellite data for population genetic studies. *Ecol. Evol.* **2014**, *4*, 4399–4428. [CrossRef]
32. Laoun, A.; Harkat, S.; Lafri, M.; Gaouar, S.B.S.; Belabdi, I.; Ciani, E.; De Groot, M.; Blanquet, V.; Leroy, G.; Rognon, X.; et al. Inference of Breed Structure in Farm Animals: Empirical Comparison between SNP and Microsatellite Performance. *Genes* **2020**, *11*, 57. [CrossRef]
33. Eusebi, P.G.; Martinez, A.; Cortes, O. Genomic Tools for Effective Conservation of Livestock Breed Diversity. *Diversity* **2020**, *12*, 8. [CrossRef]
34. Brenig, B.; Schütz, E. Recent development of allele frequencies and exclusion probabilities of microsatellites used for parentage control in the German Holstein Friesian cattle population. *BMC Genet.* **2016**, *17*, 18. [CrossRef] [PubMed]
35. *Molecular Genetic Characterization of Animal Genetic Resources*; FAO Animal Production and Health Guidelines. FAO: Rome, Italy, 2011; pp. 68–69. Available online: http://www.fao.org/3/i2413e/i2413e00.pdf (accessed on 8 June 2021).
36. Abdelmanova, A.S.; Mishina, A.I.; Volkova, V.V.; Chinarov, R.Y.; Sermyagin, A.A.; Dotsev, A.V.; Boronetskaya, O.I.; Petrikeeva, L.V.; Kostyunina, O.V.; Brem, G.; et al. Comparative study of different methods of DNA extraction from cattle bones specimens maintained in a craniological collection. *Sel'skokhozyaistvennaya Biol. [Agric. Biol.]* **2019**, *54*, 1110–1121. [CrossRef]
37. Mondol, S.; Karanth, K.U.; Kumar, N.S.; Gopalaswamy, A.M.; Andheria, A.; Ramakrishnan, U. Evaluation of non-invasive genetic sampling methods for estimating tiger population size. *Biol. Conserv.* **2009**, *142*, 2350–2360. [CrossRef]
38. Modi, S.; Habib, B.; Ghaskadbi, P.; Nigam, P.; Mondol, S. Standardization and validation of a panel of cross-species microsatellites to individually identify the Asiatic wild dog (Cuon alpinus). *Peer J.* **2019**, *7*, e7453-10. [CrossRef] [PubMed]
39. Miquel, C.; Bellemain, E.; Poillot, C.; Bessière, J.; Durand, A.; Taberlet, P. Quality indexes to assess the reliability of genotypes in studies using noninvasive sampling and multiple-tube approach. *Mol. Ecol. Notes* **2006**, *6*, 985–988. [CrossRef]
40. Broquet, T.; Petit, E. Quantifying genotyping errors in noninvasive population genetics. *Mol. Ecol.* **2004**, *13*, 3601–3608. [CrossRef]
41. He, G.; Kang, H.; Guo, S.T.; Ji, W.H.; Qi, X.G.; Ren, Y.; Jin, X.L.; Li, B.G. Evaluating the reliability of microsatellite genotyping from low-quality DNA templates with a polynomial distribution model. *Chin. Sci Bull* **2011**, *56*, 2523–2530. [CrossRef]

42. Chapuis, M.-P.; Estoup, A. Microsatellite Null Alleles and Estimation of Population Differentiation. *Mol. Biol. Evol.* **2007**, *24*, 621–631. [CrossRef]
43. Van Oosterhout, C.; Hutchinson, W.F.; Wills, D.P.M.; Shipley, P. micro-checker: Software for identifying and correcting genotyping errors in microsatellite data. *Mol. Ecol. Notes* **2004**, *4*, 535–538. [CrossRef]
44. Peakall, R.; Smouse, P.E. GenAlEx 6.5: Genetic analysis in Excel. Population genetic software for teaching and research–an update. *Bioinformatics* **2012**, *28*, 2537–2539. [CrossRef] [PubMed]
45. Keenan, K.; McGinnity, P.; Cross, T.F.; Crozier, W.W.; Prodöhl, P.A. diveRsity : An R package for the estimation and exploration of population genetics parameters and their associated errors. *Methods Ecol. Evol.* **2013**, *4*, 782–788. [CrossRef]
46. Jombart, T. adegenet: A R package for the multivariate analysis of genetic markers. *Bioinformatics* **2008**, *24*, 1403–1405. [CrossRef] [PubMed]
47. Wickham, H. *Ggplot2: Elegant Graphics for Data Analysis; Use R!* Springer: New York, NY, USA, 2009; ISBN 978-0-387-98140-6.
48. Kalinowski, S.T. Counting Alleles with Rarefaction: Private Alleles and Hierarchical Sampling Designs. *Conserv. Genet.* **2004**, *5*, 539–543. [CrossRef]
49. Wilcoxon, F. Individual Comparisons by Ranking Methods. *Biom. Bull.* **1945**, *1*, 80–83. [CrossRef]
50. R Core Team. *R: A Language and Environment for Statistical Computing*; R Foundation for Statistical Computing: Vienna, Austria. Available online: https://www.R-project.org/ (accessed on 12 May 2020).
51. Jost, L. G ST and its relatives do not measure differentiation. *Mol. Ecol.* **2008**, *17*, 4015–4026. [CrossRef]
52. Weir, B.S.; Cockerham, C.C. Estimating F-Statistics for the Analysis of Population Structure. *Evolution* **1984**, *38*, 1358–1370. [CrossRef]
53. Huson, D.H.; Bryant, D. Application of Phylogenetic Networks in Evolutionary Studies. *Mol. Biol. Evol.* **2006**, *23*, 254–267. [CrossRef] [PubMed]
54. Pritchard, J.K.; Stephens, M.; Donnelly, P. Inference of population structure using multilocus genotype data. *Genetics* **2000**, *155*, 945–959. [CrossRef]
55. Kopelman, N.M.; Mayzel, J.; Jakobsson, M.; Rosenberg, N.A.; Mayrose, I. CLUMPAK: A program for identifying clustering modes and packaging population structure inferences across K. *Mol. Ecol. Resour.* **2015**, *15*, 1179–1191. [CrossRef]
56. Pickrell, J.K.; Pritchard, J.K. Inference of population splits and mixtures from genome-wide allele frequency data. *PLoS Genet.* **2012**, *8*, e100296. [CrossRef]
57. Fitak, R.R. optM: An R package to optimize the number of migration edges using threshold models. *J. Hered.* **2019**. Available online: https://cran.r-project.org/web/packages/OptM/index.html (accessed on 21 June 2021).
58. Hoffmann, I. Climate change and the characterization, breeding and conservation of animal genetic resources. *Anim. Genet.* **2010**, *41*, 32–46. [CrossRef] [PubMed]
59. FAO. Status and trends of Animal Genetic Resources. Commission on Genetic Resources for Food and Agriculture. 2019. Available online: http://www.fao.org/3/my867en/my867en.pdf (accessed on 24 June 2021).
60. Hanotte, O.; Toll, J.; Iniguez, L.; Rege, E. Farm animal genetic resources: Why and what do we need to conserve. In *Options and Strategies for the Conservation of Farm Animal Genetic Resources: Report of an International Workshop, Montpellier, France, 7–10 November 2005*; Gibson, J., Gamage, S., Hanotte, O., Iniguez, L., Maillard, J.C., Rischkowsky, B., Semambo, D., Toll, J., Eds.; CGIAR System-Wide Genetic Resources Programme (SGRP): Rome, Italy; Bioversity International: Rome, Italy, 2006; p. 53.
61. Cao, J.; Baumung, R.; Boettcher, P.; Scherf, B.; Besbes, B.; Leroy, G. Monitoring and Progress in the Implementation of the Global Plan of Action on Animal Genetic Resources. *Sustainability* **2021**, *13*, 775. [CrossRef]
62. van Breukelen, A.E.; Doekes, H.P.; Windig, J.J.; Oldenbroek, K. Characterization of Genetic Diversity Conserved in the Gene Bank for Dutch Cattle Breeds. *Diversity* **2019**, *11*, 229. [CrossRef]
63. Edwards, C.J.; Connellan, J.; Wallace, P.F.; Park, S.D.E.; McCormick, F.M.; Olsaker, I.; Eythórsdóttir, E.; MacHugh, D.E.; Bailey, J.F.; Bradley, D.G. Feasibility and utility of microsatellite markers in archaeological cattle remains from a Viking Age settlement in Dublin. *Anim. Genet.* **2003**, *34*, 410–416. [CrossRef]
64. Gargani, M.; Pariset, L.; Lenstra, J.A.; De Minicis, E.; European Cattle Genetic Diversity Consortium; Valentini, A. Microsatellite genotyping of medieval cattle from central Italy suggests an old origin of Chianina and Romagnola cattle. *Front. Genet.* **2015**, *6*, 68. [CrossRef] [PubMed]
65. Svishcheva, G.; Babayan, O.; Lkhasaranov, B.; Tsendsuren, A.; Abdurasulov, A.; Stolpovsky, Y. Microsatellite diversity and phylogenetic relationships among East Eurasian Bos taurus breeds with an emphasis on rare and ancient local cattle. *Animals* **2020**, *10*, 1493. [CrossRef]
66. Kayumov, F.G.; Eremenko, V.K. *Kalmyk Cattle in the Conditions of South Ural and West Kazakhstan*; Gazprompechat: Orenburg, Russia, 2004; 383p.
67. Zhumakanov, K.T.; Abdurasulov, A.H.; Zhunushov, A.T. Conservation of Gene Pool of Agricultural Animals in Kyrgyzstan. *Agric. J.* **2016**, *9*, 50–54. Available online: https://cyberleninka.ru/article/n/sohranenie-genofonda-selskohozyaystvennyh-zhivotnyh-kyrgyzstana-problema-gosudarstvennogo-znacheniya (accessed on 26 June 2020).
68. Zhang, G.X.; Wang, Z.G.; Chen, W.S.; Wu, C.X.; Han, X.; Chang, H.; Zan, L.S.; Li, R.L.; Wang, J.H.; Song, W.T.; et al. Genetic diversity and population structure of indigenous yellow cattle breeds of China using 30 microsatellite markers. *Anim. Genet.* **2007**, *38*, 550–559. [CrossRef]

69. Berthouly-Salazar, C.; Thévenon, S.; Van, T.N.; Nguyen, B.T.; Pham, L.D.; Chi, C.V.; Maillard, J.C. Uncontrolled admixture and loss of genetic diversity in a local Vietnamese pig breed. *Ecol. Evol.* **2012**, *2*, 962–975. [CrossRef]
70. Ba, N.V.; Arakawa, A.; Ishihara, S.; Nam, L.Q.; Thuy, T.T.T.; Dinh, N.C.; Ninh, P.H.; Cuc, N.T.K.; Kikuchi, K.; Pham, L.D.; et al. Evaluation of genetic richness among Vietnamese native pig breeds using microsatellite markers. *Anim. Sci. J.* **2020**, *91*, e13343. [CrossRef]
71. McTavish, E.J.; Decker, J.E.; Schnabel, R.D.; Taylor, J.F.; Hillis, D.M. New World cattle show ancestry from multiple independent domestication events. *Proc. Natl. Acad. Sci. USA* **2013**, *110*, E1398–E1406. [CrossRef] [PubMed]
72. Decker, J.E.; McKay, S.D.; Rolf, M.M.; Kim, J.; Molina Alcalá, A.; Sonstegard, T.S.; Hanotte, O.; Götherström, A.; Seabury, C.M.; Praharani, L.; et al. Worldwide Patterns of Ancestry, Divergence, and Admixture in Domesticated Cattle. *PLoS Genet* **2014**, *10*, e1004254. [CrossRef] [PubMed]
73. Upadhyay, M.; Chen, W.; Lenstra, J.; Goderie, C.R.; MacHugh, D.E.; Park, S.D.; Magee, D.A.; Matassino, D.; Ciani, F.; Megens, H.J.; et al. European Cattle Genetic Diversity Consortium; RPMA Crooijmans. Genetic origin, admixture and population history of aurochs (Bos primigenius) and primitive European cattle. *Heredity* **2017**, *118*, 169–176. [CrossRef] [PubMed]
74. Verdugo, M.P.; Mullin, V.E.; Scheu, A.; Mattiangeli, V.; Daly, K.G.; Maisano Delser, P.; Hare, A.J.; Burger, J.; Collins, M.J.; Kehati, R.; et al. Ancient cattle genomics, origins, and rapid turnover in the Fertile Crescent. *Science* **2019**, *365*, 173–176. [CrossRef] [PubMed]

diversity

MDPI

Article

Biodiversity of Russian Local Sheep Breeds Based on Pattern of Runs of Homozygosity †

Tatiana Deniskova [1,*], Arsen Dotsev [1], Marina Selionova [2], Gottfried Brem [1,3] and Natalia Zinovieva [1]

1 L.K. Ernst Federal Research Center for Animal Husbandry, Dubrovitsy, Podolsk Municipal District, Moscow Region, 142132 Podolsk, Russia; arsendotsev@vij.ru (A.D.); gottfried.brem@vetmeduni.ac.at (G.B.); n_zinovieva@mail.ru (N.Z.)

2 Russian State Agrarian University—Moscow Timiryazev Agricultural Academy, 127550 Moscow, Russia; m_selin@mail.ru

3 Institute of Animal Breeding and Genetics, University of Veterinary Medicine, 1210 Vienna, Austria

* Correspondence: horarka@yandex.ru

† This paper is an extended version of the paper published in the 1st International Electronic Conference on Biological Diversity, Ecology, and Evolution (BDEE 2021), Online, 15–31 March 2021.

Abstract: Russian sheep breeds traditionally raised in specific environments are valuable parts of sociocultural heritage and economic component of the regions. However, the import of commercial breeds negatively influences the population sizes of local sheep populations and might lead to biodiversity loss. Estimation of the runs of homozygosity (ROH) in local sheep genomes is an informative tool to address their current genetic state. In this work, we aimed to address the ROH distribution and to estimate genome inbreeding based on SNP data to evaluate genetic diversity in Russian local sheep breeds. Materials for this study included SNP-genotypes from twenty-seven Russian local sheep breeds which were generated using the Illumina OvineSNP50 BeadChip ($n = 391$) or the Illumina Ovine Infinium HD BeadChip ($n = 315$). A consecutive runs method was used to calculate ROH which were estimated for each animal and then categorized in the ROH length classes. The ROH were found in all breeds. The mean ROH length varied from 86 to 280 Mb, while the ROH number ranged from 37 to 123. The genomic inbreeding coefficient varied from 0.033 to 0.106. Our findings provide evidence of low to moderate genomic inbreeding in major local sheep populations.

Keywords: sheep; single nucleotide polymorphisms; runs of homozygosity; genetic diversity; genomic inbreeding

Citation: Deniskova, T.; Dotsev, A.; Selionova, M.; Brem, G.; Zinovieva, N. Biodiversity of Russian Local Sheep Breeds Based on Pattern of Runs of Homozygosity. *Diversity* **2021**, *13*, 360. https://doi.org/10.3390/d13080360

Academic Editor: Michael Wink

Received: 18 July 2021
Accepted: 3 August 2021
Published: 4 August 2021

1. Introduction

The introduction of high throughput arrays for single nucleotide polymorphisms genotyping has led to the development of new bioinformatic approaches, which allow evaluation of genetic diversity more fully and address demographic history of the mammalian species. For example, estimation of genomic inbreeding and the analysis of patterns of distribution of runs of homozygosity regions in the genome are gaining popularity among the geneticists and are used in addition to classical methods to assess genetic processes in the populations.

Runs of homozygosity are contiguous stretches of homozygous loci that inbred offspring inherit from both parents originated from a common ancestor [1,2]. The number and length of ROH reflect individual demographic history and evaluate the homozygosity burden [3,4]. The length of ROH indicates whether inbreeding was recent or ancient in a population [1,2].

Nonetheless, livestock breeding practices often use selection schemes involving inbreeding as a tool to stabilize useful traits in farm animals. Thus, to trace the inbreeding events, the genome scanning for ROH segments was performed in various livestock species including cattle [5] (for example, Angus, Charolais, Hereford, Holstein, Simmental [6],

Russian Kholmogory and Yaroslavl breeds [7]) and small ruminant species (for example, German White-headed Mutton sheep [8] and African native goats [9]).

The patterns of ROH distribution were analyzed in local and commercial sheep breeds which were selected for various purposes, inhabit different environments, and are kept under diverse production systems. Al-Mamun et al. [10] performed a search for ROH segments in Australian populations of Border Leicester, Merino, Poll Dorset, and their crosses and found that Border Leicester sheep were characterized by a higher genome coverage by ROH. In addition, analysis of ROH distribution was used to elucidate the demographic history of the six commercial meat breeds including Belclare, Beltex, Charollais, Suffolk, Texel, and Vendeen [11]. Based on estimation of the genomic inbreeding coefficient based on ROH (FROH), He et al. [12] suggested that Chinese Merino had the lowest levels of inbreeding.

However, more precise attention was paid to estimation of genomic inbreeding based on ROH in populations of local sheep. Such populations often lack reliable pedigree information, and according to Purfield et al. [11], ROH might be recommended as a predictor of the pedigree inbreeding coefficient (correlation 0.62). Mastrangelo et al. [13] investigated the occurrence of ROH in 21 Italian sheep breeds using medium-density SNP genotypes and found that Barbaresca, Leccese and Valle del Belice breeds have been affected by recent inbreeding events. Signer-Hasler et al. [14] found a high correlation (0.95) between genomic inbreeding coefficients based on ROH (FROH) estimates from medium-density data and HD data in eight local Swiss sheep breeds. Low genomic inbreeding was observed in the Kyrgyz local sheep breeds including Alai, Aykol, Gissar, and Tien-Shan [15]. Abied et al. [16] suggested that some animals have experienced recent inbreeding events in Chinese indigenous sheep breeds. Using Illumina OvineSNP50 BeadChip, Dzomba et al. [17] analyzed ROH distribution in 400 animals from South African sheep populations representing mutton, pelt and mutton and wool dual-purpose breeds, as well as indigenous non-descript breeds and contributed to the better understanding of the genomic landscape of African sheep breeds.

After the severe crisis caused by the USSR collapse, Russian sheep breeding, which has been focused mainly on wool production for several decades, became unprofitable due to weak demand and low prices [18]. Thus, over the eighteen-year period, the share of fine wool sheep breeds decreased by 20.9%, the population number of semi-fine wool breeds reduced by 2.3 times, and the share of coarse wool breeds increased by 5.4 times [19].

Besides an increase in the population number of unproductive sheep may lead to drastic consequences in sheep farming by financial ruining of farmers and smallholders because sale prices for wool and mutton do not cover the costs for keeping sheep [18]. Considering that majority of sheep rearing farms are in the geographical areas of underdeveloped or risky farming [20], the recessions in the regional sheep industry have notable negative consequences for local people. In this regard, along with the economic aspect, sheep farming is of significant social and cultural importance in Russia.

Summarizing, contemporary Russian sheep breeding should be focused on increasing the meat production and using low-cost technologies. The import of commercial meat breeds was not beneficial because the specific harsh feeding and keeping conditions did not allow these sheep to realize full genetic potential. In addition, development of new hybrid breeds was expected to have a positive effect on the sheep rearing industry; however, this prediction did not come true [21].

Sustainable use and management of existing local breeds is a topic priority for the rising national sheep industry. Genetic resources of Russian local sheep include breeds, which were specifically selected for wool and dual-purpose production (wool and meat), and autochthonous breeds, which are adapted to extreme environments and from which all types of sheep products are used by local smallholders [22,23].

However, the levels of genetic diversity of local breeds, including the addressing of inbreeding events, should precede the changes in the selection direction to design scientific-based breeding programs. The evaluation of genetic diversity by calculating heterozygosity and effective population sizes in the most popular Russian local breeds was not a very informative tool [24].

In this regard, the aim of our present study is to address the distribution of the runs of homozygosity and to estimate genome inbreeding in Russian local sheep breeds based on SNP-genotyping data for better understanding of current levels of genetic diversity in these valuable livestock resources.

2. Materials and Methods

2.1. Samples and Genotyping

Samples for this study included SNP-genotypes of 706 individuals from twenty-seven Russian local sheep breeds which were genotyped using the OvineSNP50 BeadChip (Illumina, San Diego, CA, USA) or the Ovine Infinium HD BeadChip (Illumina, San Diego, CA, USA) [20]. The 50k SNP genotypes for 391 samples were generated in our previous research [24]. Additional 315 samples were genotyped using the Ovine Infinium HD BeadChip [25]. The details on the used data collection are given in Table 1.

Table 1. Sample collection of Russian sheep populations used in this study.

Breed	Code	n [1]	n_{50k} [2] Deniskova et al. (2018) [24]	n_{600k} [3] This Study	Region of Sampling [4]	Main Products
			Coarse wool breeds			
Andean	ANDB	17	17	-	Dagestan/North Caucasus	Meat, wool, milk
Buubei	BUUB	39	17	22	Yakutia/Far Eastern	Meat, wool
Edilbai	EDLB	44	17	27	Volgograd region/South	Meat, fat
Kalmyk	KALM	20	18	2	Kalmykia/South	Meat, fat
Karakul	KARA	41	16	25	Astrakhan region/South	Pelts, fur
Karachaev	KRCH	43	22	21	Karachay-Cherkessia/North Caucasus	Meat, wool, milk
Kuchugur	KUCH	16	16	-	Voronezh region/Central	Meat, wool
Lezgin	LEZG	41	15	26	Dagestan/South	Meat, wool, milk
Mongolian	MONG	27	-	27	Buryatia/Far Eastern	Meat
Ossetin	OSET	30	-	30	Ossetia/North Caucasus	Meat, wool, milk
Romanov	RMNV	36	26	10	Yaroslavl, Kaluga, and Tula regions/Central	Meat, skins
Tushin	TUSH	17	9	8	Dagestan/North Caucasus	Meat, wool, milk
Tuva	TUVA	41	16	25	Tyva/Siberian	Meat, wool,
			Semi-fine wool breeds			
Altai Mountain	ALTM	14	12	2	Altai region/Siberian	Wool, meat
Kuibyshev	KUIB	15	15	-	Samara region/Volga	Meat, wool
North Caucasian	NCSN	16	16	-	Stavropol region/North Caucasus	Meat, wool
Russian longhaired	RULH	32	16	16	Voronezh region/Central	Meat, wool
Tsigai	TZYG	16	16	2	Saratov region/Volga	Meat, wool

Table 1. *Cont.*

Breed	Code	n [1]	n_{50k} [2] Deniskova et al. (2018) [24]	n_{600k} [3] This Study	Region of Sampling [4]	Main Products
			Fine wool breeds			
Baikal fine-fleeced	BKFF	15	7	8	Yakutia/Far Eastern	Wool, meat
Dagestan Mountain	DAGM	16	16	-	Dagestan/North Caucasus	Meat, wool
Groznensk	GRZN	35	13	22	Stavropol region/North Caucasus	Wool
Kulundin	KLND	16	16	-	Altai region/Siberian	Wool, meat
Manych Merino	MANM	16	16	-	Stavropol region/North Caucasus	Wool
Salsk	SALS	35	16	19	Rostov region/South	Wool
Soviet Merino	SOVM	15	14	1	Stavropol region/North Caucasus	Wool
Stavropol	STAV	16	14	2	Stavropol region/North Caucasus	Wool, meat
Volgograd	VOLG	37	15	22	Volgograd region/South	Meat, wool

[1] n—total sample number; [2] n_{50k}—number of samples generated by using OvineSNP50 BeadChip (Illumina, San Diego, CA, USA) for our previous study [24]; [3] n_{600k}—number of samples generated by using Ovine Infinium HD BeadChip (Illumina, San Diego, CA, USA); [4]—the details on the origin and developmental history for each breed are available in [24].

2.2. Quality Control

Genotype quality control (QC) procedures were performed using PLINK v1.90 [26]. To consider the accuracy and efficiency of SNP genotyping, valid genotypes for each SNP were determined by setting a cut-off of 0.5 for the GenCall (GC) and GenTrain (GT) scores [27]. Samples that did not meet the quality criteria (missing genotype call rate 0.1) were eliminated from the analysis.

After merging the genotypic data from the 600K and 50K arrays, a total of 42,230 autosomal SNPs that overlapped between the two DNA chips were left in the analysis. SNPs with a call rate below 0.90, a minor allele frequency (MAF) lower than 0.05, or those which were located on sex chromosomes were eliminated from the analysis.

2.3. Principal Component Analysis (PCA)

Principal component analysis (PCA) was performed in PLINK v1.9 and visualized with the R package "ggplot2" [28]. The PCA was performed before the analysis of the runs of homozygosity distribution and showed that the individuals from the Buubei breed were divided into two groups (Buubei_1 and Buubei_2). The final sample numbers are shown in Table 2.

2.4. Runs of Homozygosity Estimation

For ROH calculation, we used a window-free method for consecutive SNP-based detection [29] implemented in the R package "detectRUNS" [30]. One SNP with a missing genotype and up to one possible heterozygous genotype was allowed in the run. The minimum ROH length was 1000 kb.

ROH were estimated for each animal and then categorized in the corresponding ROH length classes: (1–2 Mb, 2–4 Mb, 4–8 Mb, 8–16 Mb, >16 Mb). The total number of identified ROH was calculated for each length category in each of the individuals of each breed. The mean sum of ROH was computed by adding up the length of all ROH for each individual in the sheep populations and then the results were averaged per breed population.

Table 2. Mean ROH length and mean ROH number in Russian sheep populations.

Breed	Code	n [1]	ROH Length Mean	ROH Length Min [2]	ROH Length Max [3]	ROH Number Mean	ROH Number Min [2]	ROH Number Max [3]
Coarse wool breeds								
Andean	ANDB	17	190.64 ± 19.55	76.11	383.4	78.88 ± 3.82	56	116
Buubei (1)	BUUB_1	20	120.5 ± 17	60.64	379.4	58.55 ± 2.16	43	79
Buubei (2)	BUUB_2	19	105.82 ± 21.28	54.11	483.11	56.58 ± 1.67	41	76
Edilbai	EDLB	44	111.6 ± 5.06	76.69	268.69	67.25 ± 1.25	47	83
Kalmyk	KALM	20	104.83 ± 8.95	71.6	257.81	62.7 ± 1.91	44	80
Karakul	KARA	41	128.17 ± 3.39	81.06	180.04	79.34 ± 1.64	47	94
Karachaev	KRCH	43	116.8 ± 6.92	70.6	374.81	69.12 ± 1.36	48	91
Kuchugur	KUCH	16	223.81 ± 52.81	50.34	872.75	80.38 ± 5.87	37	105
Lezgin	LEZG	41	89.8 ± 5.6	48.78	282.82	59.68 ± 1.75	38	94
Mongolian	MONG	27	86.44 ± 3.19	60.62	140.96	60.11 ± 1.42	43	73
Ossetin	OSET	30	114.57 ± 13.75	64.33	463.41	64.43 ± 1.83	44	88
Romanov	RMNV	36	282.15 ± 10.46	182.03	457.36	123,14 ± 1,84	103	146
Tushin	TUSH	17	115.93 ± 11.04	72.3	208.29	63.29 ± 3.54	41	104
Tuva	TUVA	41	91.72 ± 3.33	58.03	148.89	60.85 ± 1.6	39	86
Semi-fine wool breeds								
Altai Mountain	ALTM	14	86.77 ± 7.2	61.49	149.39	37.64 ± 1.53	26	47
Kuibyshev	KUIB	15	126.09 ± 10.08	97.66	236.33	47.4 ± 1.67	38	66
North Caucasian	NCSN	16	181.81 ± 7.31	125	256.15	66.25 ± 1.51	48	74
Russian longhaired	RULH	32	257.15 ± 9.86	164.43	418.49	84.31 ± 1.89	62	104
Tsigai	TZYG	16	91.17 ± 8.9	54.92	204.21	39.19 ± 1.69	30	55
Fine wool breeds								
Baikal fine-fleeced	BKFF	15	92.82 ± 2.77	77.1	111.44	52.53 ± 1.67	42	65
Dagestan Mountain	DAGM	16	144.26 ± 7.06	105.12	194.71	64.31 ± 2.02	52	77
Groznensk	GRZN	35	109.93 ± 3.64	75,97	183.56	60.43 ± 1.39	44	80
Kulundin	KLND	16	171.31 ± 9.69	126.28	277.98	76.44 ± 1.81	63	93
Manych Merino	MANM	16	125.24 ± 6.19	84.18	181.25	63.38 ± 1.99	46	77
Salsk	SALS	35	162.7 ± 9.44	88.94	351.75	72.17 ± 1.37	52	90
Soviet Merino	SOVM	15	125.41 ± 5.05	98.45	168.69	65.33 ± 1.89	51	76
Stavropol	STAV	16	157.18 ± 20.46	89.41	419.32	64.5 ± 2.5	51	85
Volgograd	VOLG	37	174.53 ± 4.42	110.65	224.6	77.38 ± 1.63	53	98

[1] *n*—number of individuals; [2] Min—minimum values of estimations observed in individual animals within breeds; [3] Max—maximum values of estimations observed in individual animals within breeds.

2.5. Estimation of Genomic Inbreeding (FROH)

The genomic inbreeding coefficient based on ROH (FROH) was estimated as the sum of the length of all ROH per sheep as a proportion of the total autosomal SNP coverage (2.44 Gb).

3. Results

3.1. Assessment of Genetic Links between Sheep Breeds within the Wool-Type Groups

Principal Component Analysis performed for 15 coarse wool sheep populations (Figure 1A) showed that the first Principal Component (PC1) accounting for 13.34% of genetic variability clearly separated the Romanov and the Kuchugur breeds from the other populations which were clustered together. The Buubei (2) population was differentiated from the Buubei_1 group as well as the other breeds by the second Principal Component (PC2). Besides the PC2 pierced the joined cluster into two slightly traceable subgroups

(Buubei (1) + Mongolian + Edilbai + Kalmyk + Karakul + Tuva and Andean + Karachaev + Tushin + Ossetin + Lezgin).

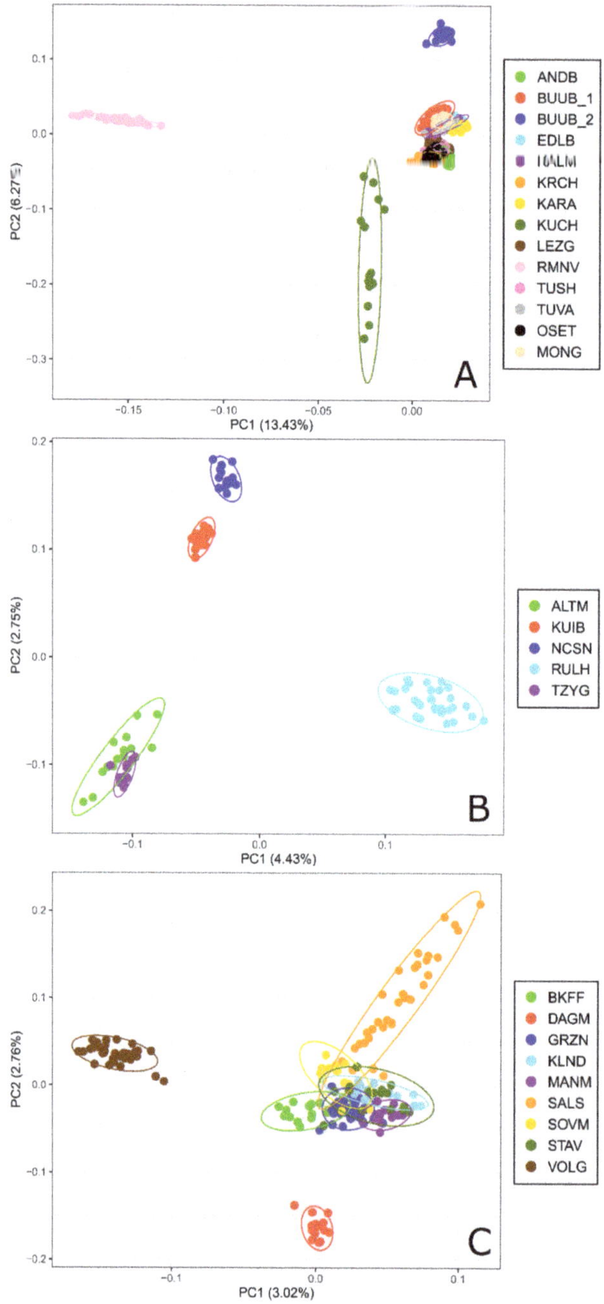

Figure 1. Principal component analysis for Russian sheep populations: (**A**) performed for coarse wool breeds; (**B**) performed for semi-fine wool breeds; (**C**) performed for fine wool breeds. For a description of the sheep breeds, see Table 1.

Based on PCA results for semi-fine wool breeds, the Russian longhaired breed was the most distant from the other breeds and located in the right down quadrant formed by the PC1 and PC2 (Figure 1B). The Kuibyshev and North Caucasian breeds occupied the left upper quadrant, while the Altai Mountain and Tsigai breeds were placed within the left down quadrant.

Within the fine wool group, the Volgograd breed was separated from the other breeds be the PC1, which accounts for 3.02% of genetic variability, while the Dagestan Mountain breed was differentiated by the PC2 (Figure 1C).

3.2. Pattern of Distribution of Runs of Homozygosity in Populations of Russian Local Sheep Breeds

The ROH segments were identified in all studied breeds on all autosomes. In all studied breeds, the highest genome coverage by ROH was found on Oar1 (10.58–12.31% in coarse wool, 9.49–12.24% in semi-fine wool, and 9.69–11.68% in fine wool group), Oar2 (8.88–12.17% in coarse wool, 10.53–12.52% in semi-fine wool, and 10.16–12.83% in fine wool group), and Oar3 (8.33–9.74% in coarse wool, 5.42–10.09% in semi-fine wool, and 8.98–10.55% in fine wool group). The Oar 26 was characterized by the lowest coverage by ROH ($\leq$2.40% in all breeds).

Mean ROH lengths varied significantly in different sheep breeds (Table 2). Mean ROH lengths ranged from 86.44 Mb in the Mongolian breed to 282.15 Mb in the Romanov breed within the group of coarse wool breeds. The Russian longhaired breed had the maximum mean ROH length (257.15 Mb), and the Altai Mountain breed showed the minimum value (86.77 Mb) within the group with semi-fine wool. Mean ROH lengths ranged from 92.82 Mb in the Baikal fine-fleeced breed to 174.53 Mb in the Volgograd breed.

The greatest mean ROH number was found in the Romanov breed (123.14) while the lowest one was detected in the Buubei_2 breed (56.58) within the group of coarse wool breeds. The mean ROH number varied from 37.64 in the Altai Mountain breed to 84.31 in the Russian longhaired breed within the group of semi-fine wool breeds. The maximum ROH number was estimated in the Volgograd breed (77.38), and the minimum was found in the Baikal fine-fleeced breed (52.53).

Among all studied breeds the maximum individual ROH length was found in the Kuchugur breed (872.75 Mb), and the minimum was identified in the Lezgin breed (48.78 Mb). Considering the individual ROH numbers, the greatest number was displayed in the Romanov breed (146) and the lowest number was detected in the Altai Mountain breed (26) (Table 2, Figure 2).

3.3. Ranging the Runs of Homozygosity by the Length Classes in Russian Local Sheep Breeds

A genetic pattern of predominance of the shortest ROH segments (1–2 Mb) was found in all studied sheep populations (Figure 3). Thus, the frequencies of the 1–2 Mb ROH segments were higher 80% (with maximums in 91% in the Lezgin and 93.59% in the Mongolian breeds (Figure 3A) within the coarse wool breeds, higher 66.82% (with maximums in 77.79% in the Baikal fine-fleeced and 78.39% in the Groznensk breeds (Figure 3B) within fine wool breeds, and higher 50.70% within semi-fine wool breeds with maximums in 68.69% in the Altai Mountain and 68.90% in the Tsigai breeds (Figure 3C). The Romanov (71.96%) and Kuchugur (67.73%) breeds had lower frequencies of the shortest ROH segments in comparison with other coarse wool populations.

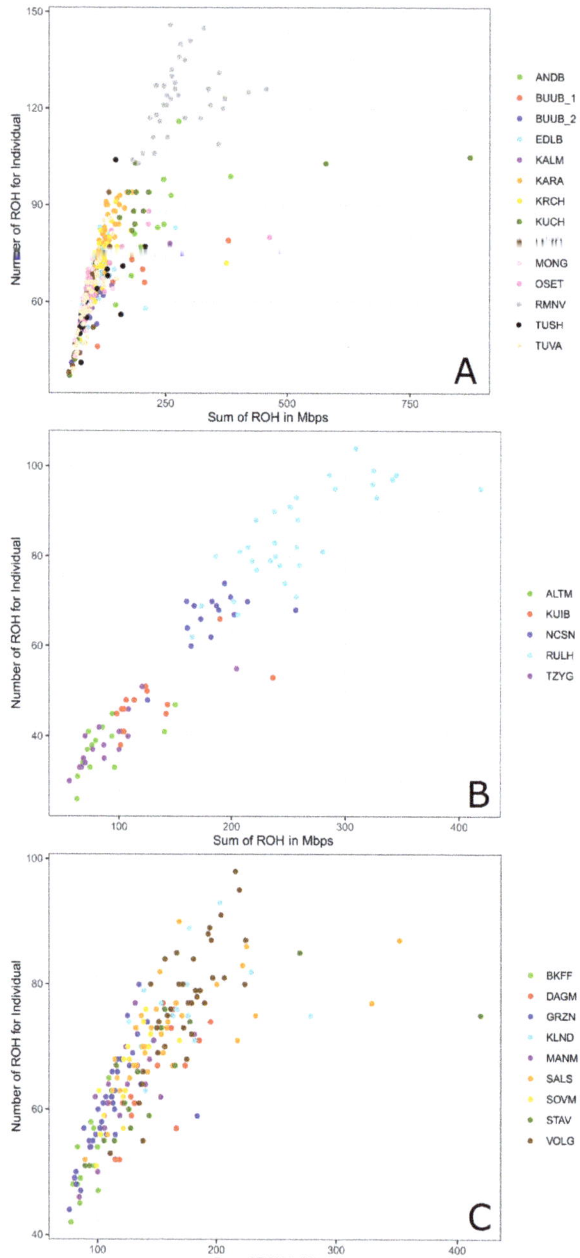

Figure 2. Genomic coverage in ROH (X-axis) and ROH number per individual (Y-axis) in Russian sheep breeds: (**A**) Genomic coverage in ROH (X-axis) and ROH number per individual (Y-axis) within the group of coarse wool breeds; (**B**) Genomic coverage in ROH (X-axis) and ROH number per individual (Y-axis) within the group of semi-fine wool breeds; (**C**) Genomic coverage in ROH (X-axis) and ROH number per individual (Y-axis) within the group of fine wool breeds. For a description of the sheep breeds, see Table 1.

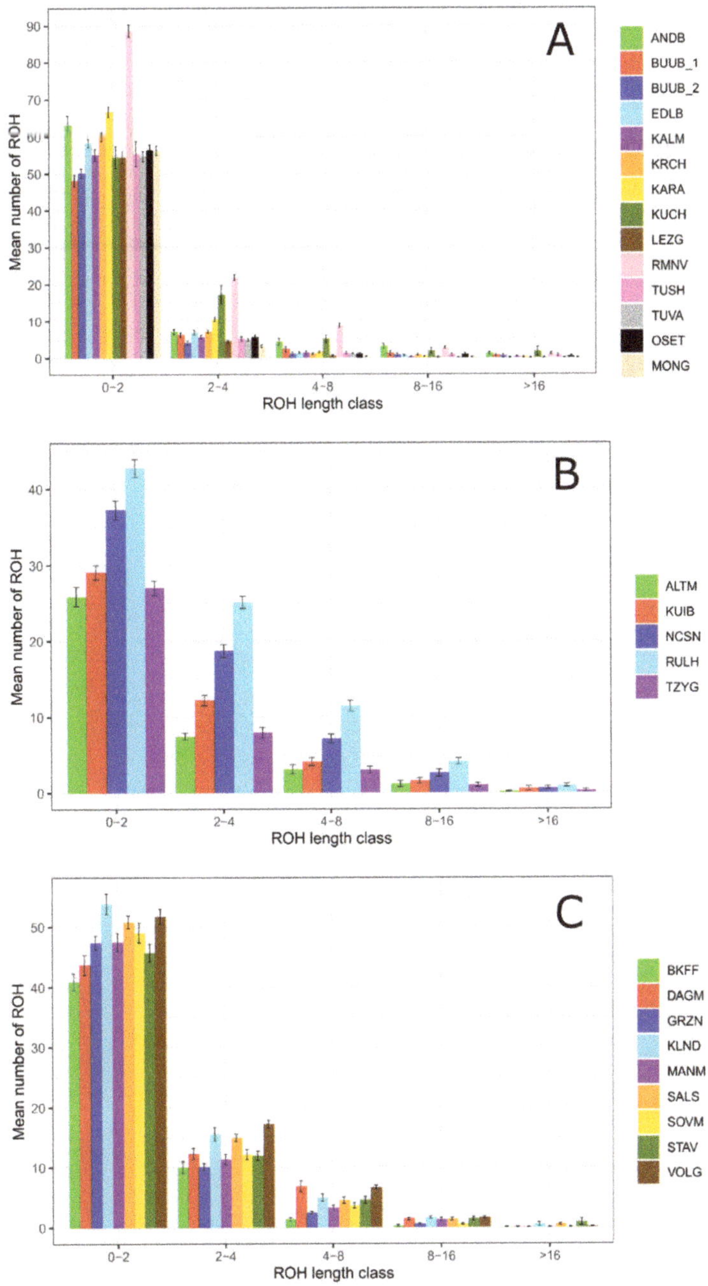

Figure 3. ROH distribution in length classes in Russian sheep populations: (**A**) Mean number of ROH by class within the group of coarse wool breeds; (**B**) Mean number of ROH by class within the group of semi-fine wool breeds; (**C**) Mean number of ROH by class within the group of fine wool breeds.

The 2–4 Mb ROH segments were more distributed in semi-fine wool (from 19.73% in the Altai Mountain to 29.76% in the Russian longhaired breeds) and in fine wool breeds (from 16.64% in the Groznensk to 22.25% in the Volgograd breeds) than in coarse wool breeds (from 5.30% in the Mongolian breed to 21.46% in the Kuchugur breeds). A similar pattern was found in the 4–8 Mb ROH length class (7.66–13.60% in semi-fine wool group and 2.54–10.59% in fine wool group versus 0.62–7.20% in coarse wool group).

The classes of the longer ROH segments (8–16 Mb and >16 Mb) were less frequent in all studied breeds. The highest share of 8–16 Mb ROH segments was found in semi-fine wool breeds (2.55% in the Tsigai to 4.86% in the Russian longhaired breeds) while the maximums in coarse wool and fine wool breeds did not exceed 3.95% and 2.24%, respectively. The frequencies of the longest ROH segments (>16 Mb) varied from 0.12% in the Tuva to 2.10% in the Kuchugur breeds in coarse wool group, from 0.38% in the Altai Mountain to 1.13% in the Kuibyshev breeds in semi-fine wool group, and from 0.05% in the Groznensk to 1.36% in the Stavropol breeds in the fine-wool group.

3.4. *Estimation of Genomic Inbreeding Coefficient Based on Runs of Homozygosity in Russian Local Sheep Breeds*

Except for the Romanov, Kuchugur, and Russian longhaired breeds, studied sheep populations regardless of the wool type had low to moderate FROH values. Thus, the FROH values varied from 0.033 (MONG) to 0.072 (ANDB) in coarse wool group (Figure 4A), and from 0.033 (ALTM) to 0.069 (NCSN) in the semi-fine wool group (Figure 4B), and from 0.035 (BKFF) to 0.066 (VOLG) in fine wool group (Figure 4C).

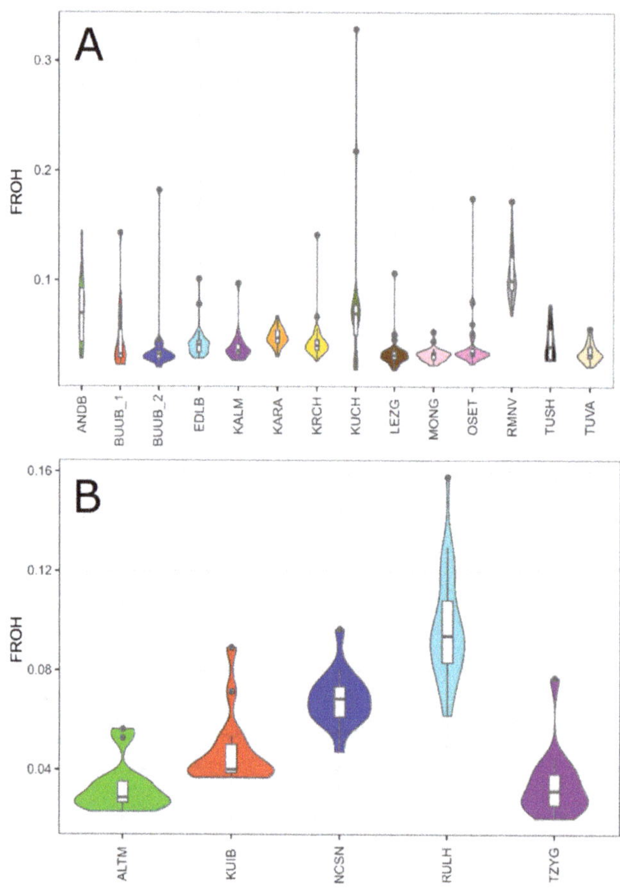

Figure 4. *Cont.*

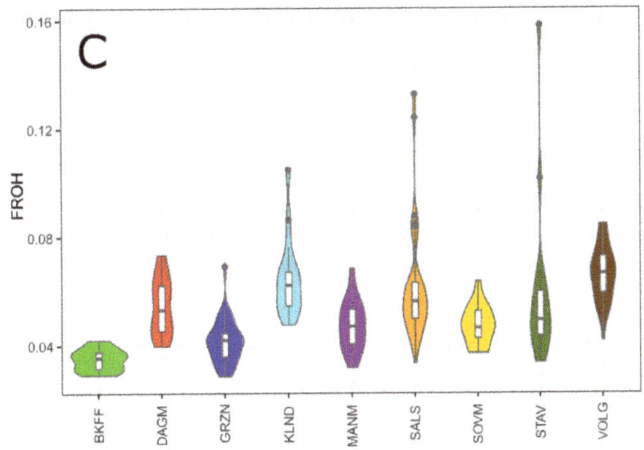

Figure 4. The genomic inbreeding coefficient based on ROH (FROH) in Russian sheep breeds: (**A**) FROH values within the group of coarse wool breeds; (**B**) FROH values within the group of semi-fine wool breeds; (**C**) FROH values within the group of fine wool breeds.

The mean genomic inbreeding coefficient values were the highest in the Romanov (FROH = 0.106), Russian longhaired (FROH = 0.097), and Kuchugur breeds (FROH = 0.084).

Among all studied animals, the maximum individual FROH value was calculated in the Kuchugur breed (FROH = 0.33) from the coarse wool group. The highest individual FROH values were found in animals from the Stavropol breed from fine wool group (FROH = 0.16), from the Russian longhaired breed from semi-fine wool group (FROH = 0.16) and from the Romanov breed from coarse wool group (FROH = 0.17).

4. Discussion

Estimation of genetic diversity in local livestock species is of special priority to prevent the steady inbreeding increase which leads to negative consequences of inbreeding depression and to endangered status. There are several approaches to address biodiversity and its dynamics in the populations of livestock species: effective population size, heterozygosity and runs of homozygosity [10].

In our previous study, we calculated and analyzed effective population sizes and heterozygosity to unlock the current state of genetic diversity in Russian local sheep breeds [24]. Nonetheless, estimation of runs of homozygosity is a useful tool to reveal the presence of long-term inbreeding in livestock populations [2,5].

A strong primary subdivision of the Russian local sheep populations according to their wool type (fine wool, semi-fine wool, and coarse wool) was reported based on using the medium density DNA arrays [24]. Therefore, in the present study, we divided Russian sheep populations corresponding to their wool type to analyze the specific patterns of distribution of the runs of homozygosity in their genomes.

The breeds in the fine wool group demonstrated a high consistency in the ROH distribution. Thus, most individuals from these breeds fit into «90 ROH number and 200 Mb sum of ROH length» pattern. These findings might be occurred because of similarities in the developmental history and of long-term underling of strong positive selection for wool production [22]. Besides Dzomba et al. [17] reported that the Merino-type breeds had similarities in ROH distribution. These results are agreed with our findings on Russian fine wool breeds which also belong to the group of Merino-derived breeds.

A common trend was not established in semi-fine wool breeds which might be divided into three groups based on ROH distribution. The first group included Kuibyshev, Altai Mountain, and Tsigai breeds. Most individuals of these breeds had 55–60 ROH numbers and 150 Mb sum ROH length. The second group represented by sheep from the North

Caucasian breed was characterized by 60–80 ROH numbers and 150–220 (250) Mb sum ROH length per individual. In addition, the individuals from the Russian longhaired had the greatest ROH numbers (≥75) and ROH length (200–350 Mb) as well as the highest genomic inbreeding coefficients.

The coarse wool group included fourteen populations collected from thirteen breeds. Based on the PCA results, animals from the Buubei breed clearly separated into two groups. This pattern was not explained by the sampling locations. The history of the Buubei breed provides a tragic lesson for future generations because the valuable gene pool of this ancient native breed was lost in the Republic of Buryatia. The contemporary Buubei breed was re-introduced into the territory of Russia from a small group of animals that had been previously imported to China and which escaped the extinction in the homeland habitat [31]. However, it might be hypothesized that some re-introduced sheep were of admixed origin which resulted in establishment of a few genetic strains within the contemporary gene pool. Nonetheless, there were no significant differences between two Buubei populations in the ROH distribution.

Most sheep in the coarse wool group had similar ROH numbers and ROH lengths which were up 105 and 250 Mb, which corresponded to previously detected patterns based on high-density genotypes [32].

However, three breeds did not fit into the genetic patterns, which were characteristic for coarse wool (Romanov, and Kuchugur) and semi-fine wool groups (Russian longhaired) and displayed the highest estimates of mean ROH length (282 Mb in Romanov; 257 Mb in Russian longhaired; 223 Mb in Kuchugur).

The contemporary gene pools of studied coarse wool breeds were formed by folk selection (somehow or other) and by required adaptions to survive in severe natural environments. Nonetheless, the Romanov breed was underling a stronger selective pressure by selection individuals, which had the best pelt traits and the highest prolificacy [22]. This could result in fixing definite genome regions that related to desirable traits and might overlap with the ROH segment. Nevertheless, this assumption should be addressed more fully with a larger sample.

However, the severity of the consequences of the recent autozygosity's events on the gene pool of these three breeds is different. Due to higher resilience to feeding and keeping conditions, the Romanov breed is reared in 26 regions. Besides, the pedigree base for the Romanov breed includes twenty breeding enterprises and multipliers [33]. Thus, a rising of the genomic inbreeding might be prevented in the future by smart choices for the unrelated rams and rotation of the founder's lines within the breed.

In comparison with the Romanov breed, the state of genetic resources of the Kuchugur and Russian longhaired breeds are more unstable. The Kuchugur breed was created by intense folk selection in the Voronezh region in the second half of the 19th century [22]. An identification of highly inbred animals (FROH = 0.33) in the Kuchugur breed was expected because this breed is in endangered status (no official census recordings are available) and most likely only a few sires are used to multiply the last existing flocks which are kept by the smallholder farmers in the Voronezh and Kursk regions. Considering Russian longhaired breeds, a rising demand for mutton has contributed to revived interest in raising breeds to produce meat (Kuibyshev, Altai Mountain, North Caucasian breed and Tsigai). However, the Russian longhaired breed has long crossbred wool (in Lincoln type), which is not in high demand currently. Thus, higher inbreeding level in this breed might correspond to the small population size (1400 heads at the end of 2019, Supplementary Material Table S1) [33] and to using of a limited number of rams. Thus, the Kuchugur and Russian longhaired breeds without proper management might be extinct in the nearest future.

Nevertheless, common genetic patterns were found in all studied Russian breeds as well. A prevalence of short ROH segments detected in all Russian local sheep populations are compatible with the relevant patterns identified in other local and cosmopolitan sheep breeds. Thus, Border Leicester, and Poll Dorset breeds predominantly had short ROH segments (1 to 5 Mb) [10] as well as Italian local sheep breeds were characterized by the highest numbers of short ROH segments (<10 Mb) [13]. Analyzing the ROH distribution in genomes of South African sheep breeds, Dzomba et al. [17] showed that 88.2% of identified ROH were in the short (1–6 Mb) category [17]. A similar pattern was reported in five Chinese sheep breeds [16] and six commercial meat breeds including Suffolk, and Texel [11].

Nonetheless, chromosome coverage in ROH varied in different sheep populations. Thus, Abied et al. [16] showed the highest coverage rate on OAR2 in Chinese sheep populations, which corresponded with our findings. Purfield et al. [11] reported the highest and the lowest percentage of the autosome residing in a ROH on OAR15 and on OAR24 in the Charollais and Suffolk populations. Dzomba et al. [17] found that South African sheep breeds were characterized by even ROH distribution amongst chromosomes.

Considering genome coverage in ROH, Russian sheep populations displayed greater mean ROH length (86.77–282.15 Mb) and higher mean ROH number (37.64–123.14) in comparison with those estimated in Italian (3.85–5.51 Mb and 10.58–44.54) [13] and in eight Swiss sheep breeds (1.88–103.25 Mb and 6.58–29.14) [14]. In addition, mean ROH length calculated in our study was a bit higher than those obtained in commercial sheep breeds (92.61–128.31 Mb [11] and 94.88–126.06 Mb [10]). However, larger variation was observed in Chinese sheep breeds for which ROH number ranged from 259 to 796, and mean ROH length varied from 15.23 Mb to 46.8 Mb with individuals values up to 273 and 984 Mb [16].

Variation of ROH lengths within the studied breeds was from moderate to high. In addition, animals with large ROH coverage were found in several breeds including Kuchugur (872.75 Mb), Romanov (457.36 Mb), Ossetin (463.41 Mb), and Buubei (2) (483.11 Mb). However, in general, maximum individual ROH length values estimated in our study were close to those obtained in Australian populations of Border Leicester, Merino, and Poll Dorset breeds (427.2, 410.5 and 396.45 Mb) [10]. Besides, the presence of several individuals, which had experienced recent autozygosity events, is typical in livestock species [13].

Although standard deviation values revealed high variability in autozygosity levels within each population, genomic inbreeding coefficients estimated per breed predominantly demonstrated a pattern of low to moderate inbreeding in Russian sheep populations (FROH from 0.033 to 0.106). Comparable results were observed in Italian (FROH from 0.016 to 0.099) [13] and Swiss local sheep breeds (FROH from 0.021 to 0.102) [14], while most South African breeds exhibited much more high inbreeding levels (FROH from 0.10 to 0.31) [17].

5. Conclusions

Here, we presented a detailed analysis of the pattern of the runs of homozygosity distribution in twenty-seven Russian local sheep breeds based on SNP profiles. The results corresponded to breed history and used production system under which populations are reared. The calculated levels of ROH reflect the inbreeding history of the studied sheep populations. Our findings provide evidence of a low to moderate genomic inbreeding in major local sheep populations. The results suggest that several animals from Kuchugur, Romanov, Ossetin, and Buubei breeds have experienced recent autozygosity events. The study results provide useful information and might contribute to designing conservation programs for local genetic resources of sheep in Russia.

Supplementary Materials: The following are available online at https://www.mdpi.com/article/10.3390/d13080360/s1, Table S1: Population numbers of breeds in breeding farm (by the end of 2019).

Author Contributions: Conceptualization, N.Z., G.B. and T.D.; methodology, A.D. and T.D.; software, A.D.; validation, T.D., A.D. and N.Z.; formal analysis, A.D.; investigation, T.D. and A.D.; resources, N.Z. and M.S.; data curation, A.D.; writing—original draft preparation, T.D.; writing—review and editing, T.D.; visualization, A.D.; supervision, N.Z.; project administration, T.D.; funding acquisition, T.D. and A.D. All authors have read and agreed to the published version of the manuscript.

Funding: The 50k SNP genotypes of the studied breeds were obtained within the framework of the Ministry of Science and Higher Education of Russia theme No. 0445-2019-0024. The 600k SNP genotypes were obtained with financial supported by the Russian Scientific Foundation (RSF) within Project No. 19-16-00070.

Institutional Review Board Statement: The study was approved by the Ethics Commission of the L.K. Ernst Federal Research Center for Animal Husbandry (protocol No. 4 from the 19 January 2021).

Informed Consent Statement: Not applicable.

Data Availability Statement: The SNP-genotypes presented in this study are available on request from the corresponding author.

Acknowledgments: The ovine HD BeadChip was developed under the auspices of the International Sheep Genomics Consortium in work underwritten by FarmIQ [34], a joint New Zealand government and industry Primary Growth Partnership programme.

Conflicts of Interest: The authors declare no conflict of interest. The Russian Scientific Foundation (RSF) and Ministry of Science and Higher Education of the Russian Federation funders had no role in the design of the study; in the collection, analyses, or interpretation of data; in the writing of the manuscript, or in the decision to publish the results.

References

1. McQuillan, R.; Leutenegger, A.L.; Abdel-Rahman, R.; Franklin, C.S.; Pericic, M.; Barac-Lauc, L. Runs of homozygosity in European populations. *Am. J. Hum. Genet.* **2008**, *83*, 359–372. [CrossRef]
2. Curik, I.; Ferencčakovicć, M.; Soölkner, J. Inbreeding and runs of homozygosity: A possible solution to an old problem. *Livest. Sci.* **2014**, *166*, 26–34. [CrossRef]
3. Kirin, M.; McQuillan, R.; Franklin, C.S.; Campbell, H.; McKeigue, P.M.; Wilson, J.F. Genomic runs of homozygosity record population history and consanguinity. *PLoS ONE* **2010**, *5*, e13996. [CrossRef]
4. Ceballos, F.C.; Joshi, P.K.; Clark, D.W.; Ramsay, M.; Wilson, J.F. Runs of homozygosity: Windows into population history and trait architecture. *Nat. Rev. Genet.* **2018**, *19*, 220–234. [CrossRef]
5. Sölkner, J.; Ferencčakovicć, M.; Gredler, B.; Curik, I. Genomic metrics of individual autozygosity, applied to a cattle population. In *Book of Abstracts of the 61st Annual Meeting of the European Association of Animal Production*; Wageningen Academic Publishers: Wageningen, The Netherlands, 2010; p. 306.
6. Purfield, D.C.; Berry, D.P.; McParland, S.; Bradley, D.G. Runs of homozygosity and population history in cattle. *BMC Genet.* **2012**, *13*, 70. [CrossRef] [PubMed]
7. Zinovieva, N.A.; Dotsev, A.V.; Sermyagin, A.A.; Deniskova, T.E.; Abdelmanova, A.S.; Kharzinova, V.R.; Sölkner, J.; Reyer, H.; Wimmers, K.; Brem, G. Selection signatures in two oldest Russian native cattle breeds revealed using high-density single nucleotide polymorphism analysis. *PLoS ONE* **2020**, *15*, e0242200. [CrossRef] [PubMed]
8. Addo, S.; Klingel, S.; Thaller, G.; Hinrichs, D. Genetic diversity and the application of runs of homozygosity-based methods for inbreeding estimation in German White-headed Mutton sheep. *PLoS ONE* **2021**, *16*, e0250608. [CrossRef] [PubMed]
9. Nandolo, W.; Mészáros, G.; Banda, L.J.; Gondwe, T.N.; Lamuno, D.; Mulindwa, H.A.; Nakimbugwe, H.N.; Wurzinger, M.; Utsunomiya, Y.T.; Woodward-Greene, M.J.; et al. Timing and Extent of Inbreeding in African Goats. *Front. Genet.* **2019**, *10*, 537. [CrossRef] [PubMed]
10. Al-Mamun, H.A.; Clark, S.A.; Kwan, P.; Gondro, C. Genome-wide linkage disequilibrium and genetic diversity in five populations of Australian domestic sheep. *Genet. Sel. Evol.* **2015**, *47*, 90. [CrossRef]
11. Purfield, D.C.; McParland, S.; Wall, E.; Berry, D.P. The distribution of runs of homozygosity and selection signatures in six commercial meat sheep breeds. *PLoS ONE* **2017**, *12*, e0176780. [CrossRef]
12. He, S.; Di, J.; Han, B.; Chen, L.; Liu, M.; Li, W. Genome-wide scan for runs of homozygosity identifies candidate genes related to economically important traits in Chinese Merino. *Animals* **2020**, *10*, 524. [CrossRef]
13. Mastrangelo, S.; Ciani, E.; Sardina, M.T.; Sottile, G.; Pilla, F.; Portolano, B.; the Bi.Ov. Ita Consortium. Runs of homozygosity reveal genome-wide autozygosity in Italian sheep breeds. *Anim. Genet.* **2018**, *49*, 71–81. [CrossRef]
14. Signer-Hasler, H.; Burren, A.; Ammann, P.; Drögemüller, C.; Flury, C. Runs of homozygosity and signatures of selection: A comparison among eight local Swiss sheep breeds. *Anim. Genet.* **2019**, *50*, 512–525. [CrossRef] [PubMed]
15. Deniskova, T.; Dotsev, A.; Lushihina, E.; Shakhin, A.; Kunz, E.; Medugorac, I.; Reyer, H.; Wimmers, K.; Khayatzadeh, N.; Sölkner, J.; et al. Population structure and genetic diversity of sheep breeds in the Kyrgyzstan. *Front. Genet.* **2019**, *10*, 1311. [CrossRef] [PubMed]
16. Abied, A.; Xu, L.; Sahlu, B.W.; Xing, F.; Ahbara, A.; Pu, Y.; Lin, J.; Berihulay, H.; Islam, R.; He, X.; et al. Genome-wide analysis revealed homozygosity and demographic history of five chinese sheep breeds adapted to different environments. *Genes* **2020**, *11*, 1480. [CrossRef] [PubMed]

17. Dzomba, E.F.; Chimonyo, M.; Pierneef, R.; Muchadeyi, F.C. Runs of homozygosity analysis of South African sheep breeds from various production systems investigated using OvineSNP50k data. *BMC Genom.* **2021**, *22*, 7. [CrossRef]
18. Yakovenko, A.M.; Aboneev, V.V.; Gorkovenko, L.G.; Marchenko, V.V. An effective method to improve the competitiveness of sheep breeding. *Ovcy Kozy Sherstyanoe Delo* **2016**, *2*, 25–27.
19. Novopashina, S.I.; Sannikov, M.Y.; Khatataev, S.A.; Kuzmina, T.N.; Khmelevskaya, G.N.; Stepanova, N.G.; Tikhomirov, A.I.; Marinchenko, T.E. *Status and Perspective Areas for Improving the Genetic Potential of Small Cattle: Scientific and Analytic Overview*; Rosinformagrotekh: Moscow, Russia, 2019.
20. Vojtyuk, M.M.; Machneva, O.P. The current state of sheep breeding in Russia. *Effektivnoe Zhivotnovodstvo* **2021**, *4*, 102–105. [CrossRef]
21. Aboneev, V.V.; Gorkovenko, L.G. Some problems of the breed-forming process in domestic sheep breeding. *Ovcy Kozy Sherstyanoe Delo* **2018**, *3*, 13–17.
22. Semyonov, S.I.; Selkin, I.I. Sheep. In *Animal Genetics Resources of the USSR*; Dmitriev, N.G., Ernst, L.K., Eds.; Food and Agriculture Organization of the United Nations: Rome, Italy, 1989.
23. Dunin, I.M.; Dankvert, A.G. *Spravochnik Porod i Tipov Sel'Skokhozyastvennykh Zhivotnykh, Razvodimykh v Rossiiskoi Federatsii*; VNIIPLEM: Moskva, Russia, 2013.
24. Deniskova, T.E.; Dotsev, A.V.; Selionova, M.I.; Kunz, E.; Medugorac, I.; Reyer, H.; Wimmers, K.; Barbato, M.; Traspov, A.A.; Brem, G.; et al. Population structure and genetic diversity of 25 Russian sheep breeds based on whole-genome genotyping. *Genet. Sel. Evol.* **2018**, *50*, 29. [CrossRef]
25. Kijas, J.W.; Lenstra, J.A.; Hayes, B.; Boitard, S.; Porto Neto, L.R.; San Cristobal, M.; Servin, B.; McCulloch, R.; Whan, V.; Gietzen, K.; et al. International Sheep Genomics Consortium Members. Genome-wide analysis of the world's sheep breeds reveals high levels of historic mixture and strong recent selection. *PLoS Biol.* **2012**, *10*, e1001258. [CrossRef] [PubMed]
26. Chang, C.C.; Chow, C.C.; Tellier, L.C.; Vattikuti, S.; Purcell, S.M.; Lee, J.J. Second-generation PLINK: Rising to the challenge of larger and richer datasets. *GigaScience* **2015**, *4*, 1–16. [CrossRef]
27. Fan, J.B.; Oliphant, A.; Shen, R.; Kermani, B.G.; Garcia, F.; Gunderson, K.L.; Hansen, M.; Steemers, F.; Butler, S.L.; Deloukas, P.; et al. Highly parallel SNP genotyping. *Cold Spring Harb. Symp. Quant. Biol.* **2003**, *68*, 69–78. [CrossRef]
28. Wickham, H. *Ggplot2: Elegant Graphics for Data Analysis*; Springer: New York, NY, USA, 2009.
29. Marras, G.; Gaspa, G.; Sorbolini, S.; Dimauro, C.; Ajmone-Marsam, P.; Valentini, A.; Williams, J.L.; Macciotta, N.P. Analysis of runs of homozygosity and their relationship with inbreeding in five cattle breeds farmed in Italy. *Anim. Genet.* **2014**, *46*, 110–121. [CrossRef] [PubMed]
30. Cran, R. Project.org: detectRUNS: Detect Runs of Homozygosity and Runs of Heterozygosity in Diploid Genomes. Available online: https://cran.r-project.org/web/packages/detectRUNS/index.html (accessed on 25 January 2021).
31. Tayshin, V.A.; Lkhasaranov, B.B. *Ecological Adaptive Qualities of the Buryat Sheep: Aboriginal Buryat Sheep*; Izd-vo BNTs SO RAN: Ulan-Ude, Russia, 1997; p. 123. (In Russian)
32. Deniskova, T.E.; Abdelmanova, A.; Dotsev, A.V.; Reyer, H.; Selionova, M.I.; Fornara, M.S.; Wimmers, K.; Brem, G.; Zinovieva, N.A. PSX-25 The distribution of runs of homozygosity in nine native Russian sheep breeds. *J. Anim. Sci.* **2020**, *98*, 456–457. [CrossRef]
33. Dunin, I.M.; Amerhanov, H.A.; Safina, G.F.; Grigoryan, L.N.; Hatataev, S.A.; Hmelevskaya, G.N.; Pavlov, M.B.; Stepanova, N.G. *Ezhegodnik po Plemennoj Rabote v Ovcevodstve i Kozovodstve v Hozyajstvah Rossijskoj Federacii (2019 god)*; FGBNU Vserossiiskii nauchno-issledovatelskii institut plemennogo dela Lesnye Poliany: Moskva, Russia, 2019.
34. FarmIQ. Available online: https://farmiq.co.nz/ (accessed on 4 August 2021).

diversity

MDPI

Article

Bee Guilds' Responses to Urbanization in Neotropics: A Case Study †

Sônia Guimarães Alves * and Maria Cristina Gaglianone

Laboratório de Ciências Ambientais, PPG-Ecologia e Recursos Naturais, Universidade Estadual do Norte Fluminense Darcy Ribeiro, CBB, Avenida Alberto Lamego 2000, Campos dos Goytacazes 28013-602, Brazil; mcrisgag@uenf.br

* Correspondence: soniagui68@gmail.com; Tel.: +55-(22)-998-001-578

† This paper is an extended version of the paper published in the 1st International Electronic Conference on Biological Diversity, Ecology, and Evolution (BDEE 2021), Online, 15–31 March 2021.

Abstract: The consequent deforestation of urban sprawl is one of the causes of the decline of wild bee communities. In this context, urban green areas (UGA) may play an important role and constitute refuge areas for bees. This study analyzed the influence of UGA conditions and their surroundings in bee guilds' responses in a medium-sized Brazilian city (Campos dos Goytacazes, RJ). The bees were sampled for 12 months (2017–2018) in 12 UGAs, and bee abundance and species richness were evaluated in guilds considering: nesting behavior, nesting site, and trophic specialization. We used as explanatory variables conditions of UGAs—the number of trees (NT), diameter at breast height (DBH), flower cover (FC), plant richness (PR), percentage of paving (PV)—and of their surroundings—paving (SPV) and the number of buildings (NB). Results showed 80% of eusocial bees, 82% nest in cavities, and 99% were generalists. FC, DBH, and NB mainly explained the responses of different guilds in study areas from all explanatory variables. Thus, this study confirms different responses associated with bee guilds' attributes. In order to conserve bee diversity, city planning must include more green areas with large flower covers and avoid long corridors of high buildings that can impact bee dispersion.

Keywords: bees; conservation; pollinators; urban management

Citation: Guimarães Alves, S.; Gaglianone, M.C. Bee Guilds' Responses to Urbanization in Neotropics: A Case Study. *Diversity* **2021**, *13*, 365. https://doi.org/10.3390/d13080365

Academic Editor: Eric Buffetaut

Received: 20 July 2021
Accepted: 4 August 2021
Published: 7 August 2021

1. Introduction

Cities in the world have faced steady growth in the last decades caused by migration from rural areas. As a result, the urban population is around 55%, and the UN estimates that the world urban population might increase by over 68% by the year 2050, which will demand faster urban expansion. The changes resulting from urbanization will affect mainly medium-sized cities (between 500 thousand and one million inhabitants), where half of the world's urban population currently lives [1]. Usually, the urbanization process in tropical and developing countries is not planned and causes drastic changes in the landscape, often irreversible, such as the increase of impervious areas (pavement, asphalt, buildings) and the destruction of wild vegetation. The great challenge for these cities is the sustainable growth that guarantees housing, transport system and energy for the population, combined with environment preservation and conservation of biodiversity [2].

The expansion of urban areas is one of the leading causes of pollinators' decline [3–5]. Among the affected pollinators are the wild bee species (Hymenoptera, Apoidea). Many bee species have a short lifecycle and high sensitivity to temperature, luminosity and humidity, responding to changes caused by urbanization [6]. Bees vary in their nesting location, build nests in the ground, in preexisting cavities in hollows in tree trunks and branches, human constructions such as walls and poles, or even exposed, out of cavities [7]. Bee species also differ in the degree of sociality, ranging from solitary to highly eusocial, and in the degree of specialization or generalization in the choice of resources [8]. Therefore, the bee community suffers direct influence from the availability of plant resources in the environment. Vegetation offers substrates such as trunks and hollows of trees for nesting, and their flowers provides resources such as nectar, pollen, oils and resin for feeding and provisioning the nests [7,9]. These different bees' biological characteristics determine how these insects interact with the environment and how they tolerate changes [10].

On the other hand, urban green areas (UGAs) such as squares, parks, urban forests and gardens are growing in importance, considering bee-friendly spaces with potential to act as a refuge for bees [11]. Urban green areas are patches of vegetation within cities, where there is a lower application of pesticides when compared to agricultural areas. The vegetation in these urban areas is composed of wild or exotic plants inserted through landscaping projects that can result in a more heterogeneous environment. In addition, these green areas' characteristics can provide nesting sites, foraging and sources of other materials for bees [12]. Therefore, even small green areas have importance and can function as ecological corridors, connecting patches of vegetation to improve bee distribution [13].

However, there is still no consensus on which characteristics of urban green areas are essential to conserving the bee community in cities [11]. One expects that pollinators respond to urbanization with positive and negative effects observed depending on taxon and environment traits. Urban expansion is not homogenous and has a significant variation in different countries and ecosystems [14]. The available data are not representative enough to guide pollinators' conservation actions globally, because most focus on North America and Western Europe with specific taxonomic groups and geographic situations. The results cannot be applied in other regions or taxa [15]. A recent systematic review of 141 peer-reviewed journals, that aimed to understand how urbanization affects pollinators' communities, showed that tropical regions remain little studied [16]. It is a significant gap that needs to be fulfilled, considering that these regions have incredible biodiversity, and many places are hotspots threatened by fast urban growth. South American countries already have an urban population of over 80%, and Brazil is at the top of the list, with more than 200 million people living in cities [1].

Here, we used a medium-sized Brazilian city to investigate which conditions of urban green areas and their surroundings can potentially contribute to shelter and conserve bees' communities. Considering biological and functional traits, we also grouped bees in guilds to provide data that can increase the efficiency of management and conservation of bee and plant species [17]. Therefore, this work aims to answer the following questions: (1) What is the community structure found in urban green areas associated with a medium-sized city in Brazil? (2) How do the different bee guilds respond to the conditions of urban green areas? Answering these questions, we intend to test two hypotheses. First, urban green areas can provide nesting sites and foraging resources to provide refuge for the bees' community. Second, bee species respond to urbanization according to their nesting behavior, nest site and trophic specialization.

2. Materials and Methods

2.1. Study Area

This study was conducted in Campos dos Goytacazes, RJ, Brazil, with approximately 4.037 km^2 and an urban area of 87.73 km^2. The total population is 507.548 thousand inhabitants, and almost 85% of the population lives in the urban area [18]. The climate is hot and humid tropical AW (Köppen-Geiger classification), with dry winter (April to September) and rainy summer (October to March), and average annual rainfall between 800 and 1200 mm [19]. The average temperature is around 26 °C in the hottest months and 19 °C in the coldest ones. Most of the vegetation found in urban areas results from public and private planting in specific places [20], modified over time. The selection of sampling points was made by marking over an aerial Google Earth image of Campos dos Goytacazes and validating the presence of vegetation in loco. Twelve urban green areas (UGAs—U1 to U12) were selected, including squares, parks and gardens. ArcGIS 10.0 was used to create location map (Figure 1).

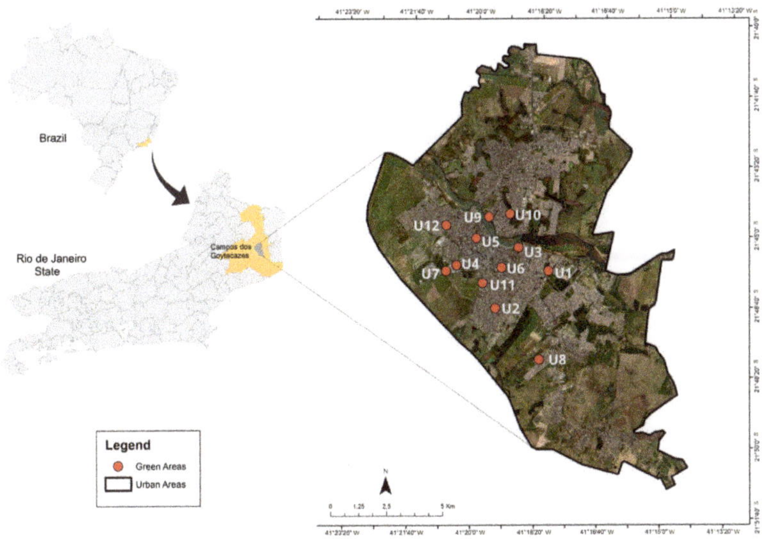

Figure 1. Red dots show the location of urban green areas (U1 to U12) evaluated in this study in the urban area of Campos dos Goytacazes, RJ, Brazil (Google image of the ArcGIS database).

2.2. Bee Sampling

We sampled the bees during their visits on flowers with an entomological net in the 12 UGAs between October 2017 and September 2018, with two samples in the rainy season and two samples in the dry season at each UGA. A sampling at each site was carried out by two collectors, who walked along the 12 UGAs searching for flowers in 3 periods of about 1 h each, between 7 am and 1 pm. Bee species were identified and deposited at the UENF Pollinating Insects Collection. They were classified by: (1) nesting site (soil or cavity), (2) nesting behavior (solitary, intermediate, eusocial) and (3) trophic specialization (generalist—collecting resources on flowers of a variety of plant species, specialist—collecting exclusively at a single plant genus). Most biological information derives from the literature [21,22], while for bee species whose functional traits were not available in the literature, we used information known for closely related taxa.

2.3. Environmental Conditions

In order to determine which environmental variables, influence the total richness and abundance of species, we measured the following traits in each study point: paved area, the richness of plants visited by bees, flower coverage, number of trees and diameter at breast height (DBH), of trees with more than 90 cm in circumference. The identification of the plant species where the bees were captured was made with the help of specialized literature [23–25] and the Flora Brazil project [26]. We measured the paved area inside the UGAs using the Google Earth Pro polygon tool (version 7.3.3.7786).

We identified the plants using photographs and exsiccates made with plant samples and deposited in Herbarium UENF. First, plant species inside the study areas were classified based on the following categories: native or exotic in Brazil, habit (herbaceous, shrubby, arboreal), nectar, pollen and/or oil as resources for pollinators [26,27]. Next, we estimated flower coverage for all flowering species on the same days of bee sampling. Finally, we calculated the area occupied by each plant with a measuring tape. From this total, we estimated the percentage covered by the flowers for each plant and we multiplied this percentage by the total plant coverage to obtain the flower coverage in square meters, we summed the results for all plants to obtain a day value and we considered the flower coverage in each UGA as the mean value for the four sampling days.

To analyze the environmental variables around the UGAs, we used the paved area around and the number of buildings with more than three floors that potentially can act as a barrier to bee dispersal, increasing isolation [28]. We measured these variables within a 500 m radius (buffers) from the center of each of the 12 UGAs. The paved area inside each buffer was measured with the polygon tool and we counted the buildings with more than three floors with the street view tool (both tools from Google Earth Pro, version 7.3.3.7786).

2.4. Data Analysis

To assess species diversity in each UGA, we calculated diversity indices of Shannon–Wiener (H′) and dominance of Berger–Parker [29]. In addition, we calculated the similarity between UGAs concerning the species composition fauna through the Bray-Curtis index.

The correlation between plant habit and species richness for each guild (related to trophic specialization, nesting behavior and nesting site) was evaluated using a Principal Component Analysis (PCA). These analyses underwent the Past 3.2 program, assuming a 95% significance level [30].

We constructed multivariate generalized linear models (GLMs) with negative binomial distribution to identify the effects of environmental variables (predictors: paved area, DBH of trees, plant richness, flower coverage, paved area around and number of buildings with more than three floors) on richness and abundance of bees in each guild. We tested the collinearity between the predictor variables using the variance inflation factor (VIF) of the car package of the R program. The best model was selected using the lowest value of the Akaike information criterion (AIC). We used the R version 4.0.5 program for these analyses, assuming a 95% significance level [31].

3. Results

3.1. Bee Community

In total, we collected 1163 bees of 39 species. Apidae was the family with the most extraordinary richness (19 species) and abundance (991 individuals). The tribes with the most remarkable species richness were Augochlorini (13) and Meliponini (5). The most abundant species were *Apis mellifera* (32% of individuals sampled), *Trigona spinipes* (26%) and *Plebeia droryana* (13%), all eusocial species. Among the non-eusocial species, the most abundant were Dialictus sp1 (4%), *Augochlora thalia* (3%) and *Xylocopa frontalis* (2%). The most remarkable species richness was found in the U1 area ($n = 20$) and the highest diversity indexes were found in U2 and U12 (H′ = 2.065 and H′ = 2.032), respectively. The highest dominance was registered in the U7 area (Berger–Parker = 0.7206) (Table 1).

The most similar areas were U7 and U8 (67%), and least similar were U7 and U4 (12%). The values of bee abundance and total species richness of sampled bees were higher in the rainy season (668 individuals and 31 species, respectively), when the average daily temperatures varied between 25 and 39 °C. In the dry season, with average daily temperatures between 18 and 38 °C, the abundance of bees was 495 and the richness was 28 species. Grouping bees in guilds, from all sampled individuals, 80% showed eusocial behavior, 11% belonged to a solitary bee and 9% had an intermediate level of sociality. Cavity bees represented 82% of the bees collected and 18% of soil-nesting bees. Considering trophic specialization, 99% of bee species were generalists. A higher richness of eusocial bees was found in the U5 and U6 areas, while intermediate and solitary species had a higher richness in the U1 area (Figure 2a,b). Solitary and soil-nesting bees composed the highest richness of sampled species and no eusocial/soil bee species were found (Figure 2c).

Table 1. Abundance of bee species collected on flowers with an entomological net at 12 Urban Green Areas (UGA) in Campos dos Goytacazes, RJ, Brazil, and respective biological characteristics: Nesting Behavior (NB)—Solitary (S), Eusocial (E), Intermediate (I); Nesting Site (NS)—Soil (SO), Cavity (C); Trophic Specialization (TS)—Generalist (G), Specialist (S).

	Urban Green Area (UGA)												Abundance		Bee Guilds		
Species	U1	U2	U3	U4	U5	U6	U7	U8	U9	U10	U11	U12	Total	Rel.%	NB	NS	TS
APIDAE																	
Apini																	
Apis mellifera L.	64	11	19	8	12	24	98	56	22	46	15	6	381	32.7	E	C	G
Centridini																	
Centris analis Lep.	2	2	0	0	0	1	0	0	0	0	0	0	5	0.4	S	C	G
Centris tarsata Sm.	1	0	0	0	0	0	0	1	0	3	1	0	6	0.5	S	C	G
Emphorini																	
Melitoma segmentaria (Fab.)	9	0	0	0	0	0	0	0	0	0	0	0	9	0.8	S	SO	S
Euglossini																	
Euglossa cordata (L.)	1	0	0	0	0	0	0	0	0	0	0	0	1	0.1	I	C	G
Euglossa sp	2	0	0	0	0	1	0	0	1	0	0	0	4	0.3	I	C	G
Eulaema flavescens (Fr.)	0	0	0	1	0	0	0	0	0	0	0	0	1	0.1	I	SO	G
Eulaema nigrita Lep.	0	1	0	0	0	2	0	0	0	0	0	0	3	0.3	I	SO	G
Exomalopsini																	
Exomalopsis analis Spinola	4	0	0	0	0	0	2	0	1	0	0	0	7	0.6	I	SO	G
Exomalopsis auropilosa Spinola	0	0	0	0	0	1	5	1	0	0	0	0	7	0.6	I	SO	G
Meliponini																	
Nannotrigona testaceicornis (Lep.)	0	0	0	0	52	0	0	0	0	0	12	0	64	5.5	E	C	G
Plebeia droryana (Fr.)	0	4	17	7	21	49	0	0	15	0	31	6	150	12.9	E	C	G
Plebeia sp	0	0	0	0	0	1	0	0	0	0	0	0	1	0.1	E	C	G
Tetragonisca angustula (Latr.)	0	15	0	0	1	0	0	0	0	0	0	0	16	1.4	E	C	G
Trigona spinipes (Fab.)	73	24	0	0	7	6	15	10	110	32	9	21	307	26.4	E	C	G
Xylocopini																	
Xylocopa frontalis Ol.	0	2	0	0	4	2	0	0	9	0	1	2	20	1.7	I	C	G
Xylocopa nigrocincta Brèthes	1	0	0	0	0	0	1	0	0	2	0	0	4	0.3	I	C	G
Xylocopa ordinaria Sm.	0	2	1	0	0	0	0	0	0	0	0	0	3	0.3	I	C	G
Xylocopa suspecta Moure & Camargo	0	0	0	0	0	0	0	0	0	0	1	0	1	0.1	I	C	G
COLLETIDAE																	
Hylaeini																	
Hylaeus tricolor (Schr.)	0	0	0	0	0	1	0	0	0	5	0	0	6	0.5	S	SO	G
Colletes sp	0	0	0	0	0	0	0	0	0	0	1	0	1	0.1	S	SO	G

Table 1. *Cont.*

	Urban Green Area (UGA)												Abundance		Bee Guilds		
Species	U1	U2	U3	U4	U5	U6	U7	U8	U9	U10	U11	U12	Total	Rel.%	NB	NS	TS
HALICTIDAE																	
Augochlorini																	
Augochlora (Oxytoglossella) thalia Sm.	4	5	11	2	0	0	0	4	0	2	4	2	34	2.9	S	SO	G
Augochlora (Augochlora) esox Vachal	3	2	0	1	0	0	0	4	0	0	0	2	12	1.0	S	SO	G
Augochlora sp3	0	0	0	0	0	0	0	0	0	1	0	3	4	0.3	S	SO	G
Augochlora sp4	1	1	0	1	0	0	0	0	2	0	2	0	7	0.6	S	SO	G
Augochlora sp5	2	1	1	3	0	1	0	1	0	0	0	4	13	1.1	S	SO	G
Augochlora sp6	1	0	0	0	0	0	0	0	0	0	0	0	1	0.1	S	SO	G
Augochlora sp7	1	3	2	1	0	0	1	1	0	1	0	3	13	1.1	S	SO	G
Augochlora sp8	3	0	0	0	0	1	1	2	0	0	0	2	9	0.8	S	SO	G
Augochlora sp9	0	0	1	0	0	1	0	0	0	1	0	0	3	0.3	S	SO	G
Augochloropsis sp1	0	0	0	0	2	0	0	0	0	2	0	0	4	0.3	I	SO	G
Augochloropsis sp2	0	0	0	0	0	0	0	1	0	0	0	0	1	0.1	I	SO	G
Augochloropsis sp3	1	0	0	0	0	0	0	0	0	0	0	0	1	0.1	I	SO	G
Augochloropsis sp4	0	0	0	0	0	0	0	0	0	1	0	0	1	0.1	I	SO	G
Halictini																	
Dialictus sp1	5	1	1	1	0	1	13	19	4	2	0	4	51	4.4	I	SO	G
MEGACHILIDAE																	
Megachilini																	
Megachile affabilis Mitchell	0	0	0	0	0	0	0	0	0	0	1	0	1	0.1	S	C	G
Megachile neoxanthoptera Cock.	3	0	0	0	0	0	0	0	0	0	0	0	3	0.3	S	C	G
Anthidiini																	
Dicranthidium seabrai Urban	0	0	0	0	0	0	0	0	0	0	2	0	2	0.2	S	C	G
Dicrantidium sp	2	0	0	0	0	0	0	0	0	4	0	0	6	0.5	S	C	G
Total sampled	183	74	53	25	99	92	136	100	164	102	80	55	1163				
Species Richness	20	14	8	9	7	14	8	11	8	13	12	11					
Diversity (Shannon H′)	1.760	2.065	1.482	1.821	1.322	1.545	1.072	1.500	1.122	1.591	1.781	2.073					
Dominance (Berger–Parker)	0.399	0.324	0.359	0.320	0.552	0.521	0.715	0.560	0.671	0.451	0.400	0.382					

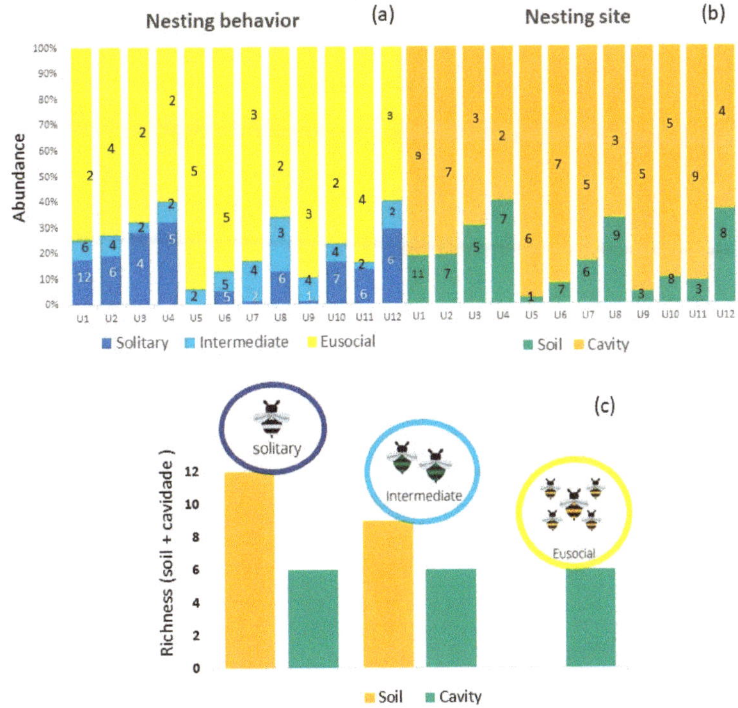

Figure 2. Relative abundance and richness (numbers inside bars) in each UGA: (**a**) nesting behavior, (**b**) nesting site, (**c**) species richness of soil- and cavity-nesting bees considering the different social groups.

3.2. Environmental Influence

Six studied UGAs presented more than 50% of their inside area as paved, and only one of the twelve UGAs had less than 30% of the surrounding paved area. The average of flower coverage in each UGA varied between 6.57 m^2 (U4) and 82.52 m^2 (U7). The most significant number of trees was registered at the U11 and the highest DBH (median = 63) at the U7 area. Three of the twelve areas studied did not have buildings higher than three floors (Table 2).

In the 12 UGAs, the bees visited 84 plant species belonging to 39 families. The average of flowering plant families among areas over the study period was 10 (ranging from 6 to 14). The families found the most in the studied UGAs were Asteraceae (10 of the 12 areas), Fabaceae (8) and Verbenaceae (8). The total plant richness per UGA ranged between 7 and 18 species. The sampling area with the highest number of plant richness was U11. The plant species with the highest frequency of occurrence in the areas were *Emilia sonchifolia* (83.3% of the areas), *Duranta erecta* (67.7%), *Tridax procumbens* (58.3%) and *Ixora coccinea* (50%). Nectar and pollen were primary floral resources for pollinators in 62 plant species, while 8 species offered only nectar, 6 species only pollen and 1 specie offered pollen and flower oil. Of the sampled bees, 55% come from flowers of plant species considered native to Brazil, and this total was mainly due to visits by Augochlorini (67% of all individuals in this tribe) and Meliponini (59%). In contrast, most bees from the tribes Apini (a species with 55% of the total number of individuals) and Halictini (a species with 58%) relate to 4 exotic plant species.

Table 2. Environmental conditions of 12 urban green areas studied in Campos dos Goytacazes/RJ. Inside UGA: percentage of paving (PV), number of trees (NT), diameter at breast height (DBH), plant richness (PR), flower coverage in m^2 (FC). Surrounding conditions measured inside a buffer of 500 m from UGA center: surrounding paved area percentage (SPV) and the number of buildings with more than three floors (NB).

	Urban Green Areas											
	U1	**U2**	**U3**	**U4**	**U5**	**U6**	**U7**	**U8**	**U9**	**U10**	**U11**	**U12**
PV	61.3	92.5	50.4	37.3	35.9	27.8	0	63.0	90.3	19.4	8.7	70.9
NT	34	36	33	22	46	78	22	00	15	63	369	17
DBH	23	27	16	22	26	35	63	30	53	22	32	32
PR	13	11	12	6	13	14	17	9	10	14	18	14
FC	60.8	47.5	10.6	6.5	42.6	33.7	82.5	22	22.7	26.7	21.3	11.7
SPV	63	96	87	71	92	97	45	51	72	84	75	89
NB	12	0	11	14	60	37	4	0	8	0	7	10

Trees' flowers received visits from 47% of the bee species, followed by herbaceous (30%) and shrubby plants (23%). Principal component analysis (PCA) showed that solitary bee correlated with herbaceous plants. Specialists, generalists, cavity and intermediate level of sociality were correlated with plants with shrubby habit, while bees from the eusocial showed correlation with trees (Figure 3).

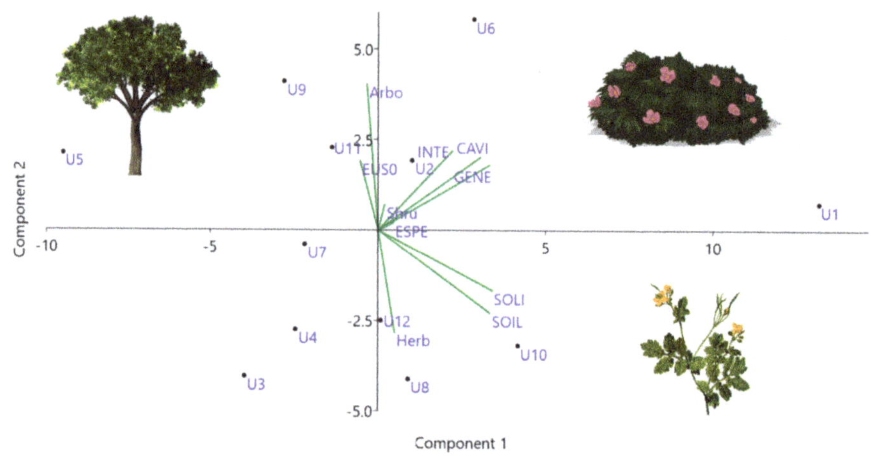

Figure 3. Principal Component Analysis (PCA) plot of correlation between bee guilds: SOIL (soil), CAVI (cavity), SPEC (specialist), GENE (generalist), SOLI (solitary), INTE (intermediate), EUSO (eusocial), and plant habits: Arbo (arboreal), Shru (shrubby) and Herb (herbaceous), U1 to U12 represent the urban green areas studied in Campos dos Goytacazes, Brazil.

Flower coverage was the predictor variable that more influenced the guilds analyzed in this study. The number of buildings with more than three floors explained the abundance of solitary and soil bees. The DBH negatively impacted the abundance of specialist and solitary bees. The paving percentage did not influence the bees' responses, and the area and number of trees were excluded from analyses because they presented collinearity (Table 3).

Table 3. Parameter estimates of a generalized linear model, explaining abundance and richness of bee guilds in urban green area (UGA) in Campos dos Goytacazes, RJ, Brazil. Inside UGA: flower cover in m^2 (FC), diameter at breast height (DBH), plant richness (PR), surrounding percentage of paving (SPV) and the number of buildings with more than three floors (NB).

Variables	Model	FC	DBH	PR	SPV	NB	Intercept	df	AICc	ΔAICc	Weight	$AdjR^2$
Abundance												
Eusocial	2		0.029				3.186	3	123.0	0.00	0.405	0.396
Intermediate	11				−0.289		4.301	3	76.8	0.00	0.318	0.477
Solitary	19		−0.052			−0.038	4.350	4	82.0	0.00	0.329	0.069
Soil	17					−0.032	3.102	3	88.4	0.00	0.355	0.436
Cavity	2	0.015					3.830	3	127.6	0.00	0.276	0.318
Specialist	4	1.447	−1.521				−50.88	4	17.8	0.00	0.217	0.993
Generalist	2	0.015					4.096	3	127.1	0.00	0.285	0.345
Richness												
Eusocial	17					0.012	0.937	3	47.3	1.77	0.134	0.151
Intermediate	3	0.101					0.851	3	49.2	1.51	0.168	0.169
Solitary	2		−0.029				2.486	3	65.8	0.00	0.179	0.264
Soil	17					−0.015	2.034	3	65.7	0.41	0.215	0.239
Cavity	129			0.079			0.661	3	57.6	0.00	0.143	0.282
Specialist	2	0.047					−46.38	3	14.6	2.27	0.095	0.250
Generalist	2	0.002					2.311	3	66.7	3.23	0.083	0.035

4. Discussion

The study data suggest that despite being a medium-sized city, the study area is at a high level of urbanization, with percentages of paving above 50% [16] and urban densification. Large paved areas have already replaced the natural areas at this stage, where only low flower covers and few potential nesting places for bees remain, making these areas harmful to the pollinator community [4]. This condition highlights the importance of urban green areas as bees' refuge, functioning as a shelter for different bee species. The lack of previous data does not allow comparisons of community parameters over time. However, the comparison between areas enables us to realize variations in bee composition. The presence of robust bees such as *X. frontalis, X. nigrocincta* and *E. nigrita* in UGA (U2, U5, U6), with a high percentile of paved surrounding (over 90%), corroborates the idea that UGAs can provide resources in scarcity situations. These bees have a long flight range and can cross large distances in the urban matrix searching for food.

Results indicated that the different guilds of bees diverged in responses to the conditions found in urban green areas and their surroundings. The structure of the bee community showed a composition mainly of eusocial and generalist bees, typical in open and highly modified environments. Other studies in Brazil have shown the tendency for eusocial bees to be more abundant in urban areas [32–35]. The fact that the colonies of eusocial species have thousands of individuals and are active throughout the year explains these results, allowing to take advantage of resources from plant species with different phenological cycles [15]. In addition, eusocial bees recruit and can collect a large number of resources quickly. Therefore, in places with little diversity of resources, these eusocial bees usually benefit compared to solitary bees.

In 50% of UGAs, the most abundant species was *Apis mellifera* (32% of individuals sampled). The more remarkable plasticity of *A. mellifera* favors the increase of its abundance in urban areas and the competitive pressure on native species, which can lead to the homogenization of the structure of the bee community in these areas [36] through the replacement of native bee species by this exotic one [37]. Although several human activities promote biotic homogenization, urbanization is the one that most favors this process [38,39]. Cities are made to attend to human needs, and therefore have a uniform nature, repeated throughout the world, with buildings, roads and houses. The construction of cities destroys the habitat of many native species, but on the other hand, creates habitats for other species, such as the exotic *A. mellifera* that can use cavities in human constructions to build nests. Consequently, more vulnerable species tend to disappear, decreasing the community species richness [38,39].

The similarity of 67% between U7 and U8 draws attention, as they are about 5 km apart. Their surrounding conditions are similar, with lower percentages of surrounding paving (41% and 51%, respectively) and low numbers of buildings with more than three floors (n = 4 and n = 0). Both are in the city borderline connected with vegetated areas. On the other hand, U7 and U4—the less similar areas, are only 700 m apart, and U4 is surrounded by a high percentage of paved areas (71%) and buildings (n = 14) and had low flower coverage (6%). These environmental conditions can compromise the movement of bees between areas and influence the composition of the bee's community since it responds in small scales to the characteristics of the environment associated with the supply of resources and nesting site [40].

The distribution of bees between native and exotic plant species did not explain the abundance and richness of the different guilds between the areas. The average flower coverage was 32%, and considering that this variable can measure the available resource, it is possible that the bees are using exotic plants as an alternative to obtain resources. This behavior is present in bees studied in an environmental gradient in the United States, aiming to examine the importance of exotic plants for native bee species [41]. In environments with anthropogenic disturbances, bees used many exotic species, and the authors found a correlation between visits to exotic plants and the low abundance of plants in the environment. This study corroborates this idea because 23% of bees sampled in our study were in shrub species, with 16% sampled in three species: *Duranta erecta*, *Ixora coccinea* and *Ruellia simplex*. All three are exotic species widely used in urban afforestation in the region. Resistant to heat and lack of water, they bloom year-round and have often been one of the few resource sources for pollinators in the UGAs.

Plant habit appears as a determinant in the distribution of bee guilds. The work highlights the importance of herbaceous plants for maintaining the solitary bees and those that nest in the soil. Among them 63% were collected in herbaceous plant species (25% solitary, 38% soil-nesting bees). Herbaceous plants were considered an essential source of resources and attractive to a large number of visitors in a study carried out in a restinga (coastal sand area) conservation unit in the state of Rio de Janeiro (Brazil) [42]. According to the authors, naturally born herbaceous/pioneers were associated with higher richness and abundance of floral visitors in restoration areas in restinga remnants. Different herbaceous plants offer various resources that can benefit wild bees when supplying their brood cells [43]. As herbaceous plants are born spontaneously in areas of exposed soil, their presence in UGAs represents potential resources of foraging and nesting places for bees that nest in the soil.

The abundance of eusocial bee appears associated with plant species of arboreal habit. Eusocial bees forage in groups and demand a greater abundance of flowers. Among the eusocial species are the stingless bees considered the primary pollinators of tropical trees with "mass" flowering [44]. Furthermore, eusocial species nest in pre-existing cavities in tree trunks, increasing the importance of tree species to native eusocial species.

Among the guilds that demonstrate greater vulnerability to the urban environment are the specialist bees. In this study, we found 1% of specialist bees in the community, similar to that described by other studies in the urban area [16]. The decline of specialist bees can be directly related to the loss of the host plant, as described in a study that found the decline of specialist bees associated with the decline of Fabaceae plants caused by the management of agricultural areas and the removal of native vegetation [45].

Analyzing the abundance of bees that nest in the soil, we found a sensitivity related to the number of buildings with more than three floors. Of the 39 species of bees sampled in this study, 16% nest in the soil. As the number of buildings increases, there is less abundance of bees from this guild in the UGAs. These high buildings are usually in large areas of paving surface that limit nesting sites as a consequence of the replacement of exposed soil areas by paved areas and the removal of small bushes or spontaneous vegetation that provide food resources [35]. However, for bees that nest in the soil, the availability of nesting sites is a decisive factor in the establishment and growth of their populations [46].

Considering that 60% of solitary bees are also soil-nesting bees, they are also influenced by a variable "number of buildings with more than three floors" on richness and abundance. A lower abundance of solitary bees appeared overall, with only 11% of the captured bees belonging to solitary species. The urban environment is composed of mosaics of vegetation and buildings that probably prevent or hinder the circulation of bees with a small flight radius, such as species of Augochlorini that we found in this study. Few individuals in this group reach long flight ranges, which increases the dependence of solitary bees on the resources available near the nests. The distance from the resource determines the ability to maintain the species [47]. According to the authors, the number of descendants generated by the species *Osmia lignaria* was more significant when the nest was close to areas with a good supply of resources. We can expect a similar situation for solitary bees in the studied areas.

The results of the solitary, soil and intermediate bees influenced by the "number of buildings with more than three floors" and by the surrounding pavement demonstrate the importance of creating green corridors that increase the possibility of bees' movement, avoiding isolation. According to a recent study [48], it is not urban environments that negatively influence the response of bee communities, but the way cities are organized act as filters that limit the presence of some species. Our results corroborate this affirmation as they indicated that the green areas and surroundings do not consider the necessities of different bee species.

5. Conclusions

We suggest that decisions about management of urban green areas must be transdisciplinary in decision making and planning. We need engineers and architects that comprehend the importance of including green areas in their projects, and of keeping native vegetation. We need public managers that consider the creation of parks and squares a priority in the city. We must also make residents understand the importance of trees and vegetation for a healthy city and to not exchange vegetation for paving. A bee conservation program in urban areas should include the management of plants that provide resources for bees and other pollinators in abundance all year. Creating and maintaining of urban parks with more vegetation cover is important to maintain the milder climate and increase the diversity of essential animals, such as pollinators and seed dispersers, in the urban area. This vision constitutes two of the UN Sustainable Development Goals (SDGs): item 15, which deals with terrestrial life and aims to protect, recover and promote the sustainable use of terrestrial ecosystems, sustainably manage forests, combat desertification, stop and reverse the degradation of the Earth and stop the loss of biodiversity, and item 11, which deals with cities and sustainable communities to make urban spaces and human settlements inclusive, safe, resilient and sustainable. This work considers that the planned urban policies, considering other organisms besides human beings, can lessen the negative impacts of rapid urbanization. A new way of looking at the city is the one that considers man as part of the ecosystem and that by conserving other species will be conserving its quality of life in the urban environment.

Author Contributions: M.C.G. and S.G.A. conceived the ideas and designed the methodology; S.G.A. collected the data; M.C.G. and S.G.A. analyzed the data; S.G.A. and M.C.G. wrote the paper. All authors contributed critically to the drafts. All authors have read and agreed to the published version of the manuscript.

Funding: This research was funded by CNPq (303894/2018-0) and FAPERJ (203.321/2017), and SGA was funded by Coordenação de Aperfeiçoamento de Pessoal de Nível Superior-Brasil (CAPES)—Finance Code 001.

Institutional Review Board Statement: Fieldwork and laboratory procedures were in compliance with CEUA—Ethical Committee for the use of animals' law n° 11.794/08, invertebrates don't require authorization.

Informed Consent Statement: Not applicable.

Data Availability Statement: The data available in this study are available on request from the corresponding author. The data are not publicly available due to needing further use.

Acknowledgments: We thank the graduation Program of Ecology and Natural Resources from Universidade Estadual do Norte Fluminense Darcy Ribeiro. We also thank Campos dos Goytacazes city hall for the permission to study in public areas, Janie Mendes Jasmim for the help in plant identification, Caique Barcelos da Silva for field help, Camila Priante for making the location map and Lucas Carneiro e Lazaro Carneiro for statistical analysis help. This study was financed in part by the Coordenação de Aperfeiçoamento de Pessoal de Nível Superior-Brasil (CAPES)—Finance Code 001. M.C.G. thanks CNPq (303894/2018-0) and FAPERJ (203.321/2017) for financial support.

Conflicts of Interest: The authors declare no conflict of interest.

References

1. United Nations, Department of Economic and Social Affairs. *Population Division;* World Population Prospects 2019: Press Release. (ST/ESA/SER.A/352); New York, NY, USA, 2018; p. 27. Available online: https://www.un.org/development/desa/en/news/population/2018-revision-of-world-urbanization-prospects.html (accessed on 13 April 2020).
2. McKinney, M.L. Urbanization, biodiversity, and conservation. *Bioscience* **2002**, *52*, 883–8902. [CrossRef]
3. Vanbergen, A.J.; Garratt, M.P. the Insect Pollinators Initiative Threats to an ecosystem service: Pressures on pollinators. *Front. Ecol. Environ.* **2013**, *11*, 251–259. [CrossRef]
4. Fortel, L.; Henry, M.; Guilbaud, L.; Guirao, A.L.; Kuhlmann, M.; Mouret, H.; Vaissière, B.E. Decreasing abundance, increasing diversity and changing structure of the wild bee community (Hymenoptera: Anthophila) along an urbanization gradient. *PLoS ONE* **2014**, *9*, e104679. [CrossRef] [PubMed]
5. Ollerton, J.; Erenler, H.; Edwards, M.; Crockett, R. Extinctions of Aculeate pollinators in Britain and the role of large-scale agricultural changes. *Science* **2014**, *346*, 1360–1362. [CrossRef] [PubMed]
6. Hamblin, A.L.; Youngsteadt, E.; Frank, S.D. Wild bee abundance declines with urban warming, regardless of floral density. *Urban Ecosyst.* **2018**, *21*, 419–428. [CrossRef]
7. Roubik, D.W. *Ecology and Natural History of the Tropical Bees;* Cambridge University: New York, NY, USA, 1989; p. 514.
8. Cane, J.H.; Sipes, S. Characterizing floral specialization by bees: Analytical methods and a revised lexicon for oligolecty. In *Plant-Pollinator Interactions: From Specialization to Generalization;* Waser, N.M., Ollerton, J., Eds.; University of Chicago Press: Chicago, IL, USA, 2006; pp. 99–122.
9. Morato, E.F.; Martins, R.P. An overview of proximate factors affecting the nesting behavior of solitary wasps and bees (Hymenoptera: Aculeata) in preexisting cavities in wood. *Neotrop. Entomol.* **2006**, *35*, 285–298. [CrossRef]
10. Steneck, R.; Dethier, M. A Functional Group Approach to the Structure of Algal-Dominated Communities. *Oikos* **1994**, *69*, 476–498. [CrossRef]
11. Threlfall, C.G.; Walker, K.; Williams, N.S.G.; Hahs, A.K.; Mata, L.; Stork, N.; Livesley, S.J. The conservation value of urban green space habitats for Australian native bee communities. *Biol. Conserv.* **2015**, *187*, 240–248. [CrossRef]
12. McFrederick, Q.S.; LeBuhn, G. Are urban parks refuges for bumble bees *Bombus* spp. (Hymenoptera: Apidae)? *Biol. Conserv.* **2006**, *129*, 372–382. [CrossRef]
13. Dearborn, D.C.; Kark, S. Motivations for Conserving Urban Biodiversity. *Conserv. Biol.* **2010**, *24*, 432–440. [CrossRef] [PubMed]
14. Seto, K.C.; Güneralp, B.; Hutyra, L.R. Global forecasts of urban expansion to 2030 and direct impacts on biodiversity and carbon pools. *Proc. Natl. Acad. Sci. USA* **2012**, *109*, 16083–16088. [CrossRef]
15. De Palma, A.; Abrahamczyk, S.; Aizen, M.A.; Albrecht, M.; Basset, Y.; Bates, A.; Purvis, A. Predicting bee community responses to land-use changes: Effects of geographic and taxonomic biases. *Sci. Rep.* **2016**, *6*, 31153. [CrossRef]
16. Wenzel, A.; Grass, I.; Belavadi, V.V.; Tscharntke, T. How urbanization is driving pollinator diversity and pollination—A systematic review. *Biol. Conserv.* **2020**, *241*, 108321. [CrossRef]
17. Normandin, É.; Vereecken, N.J.; Buddle, C.M.; Fournier, V. Taxonomic and functional trait diversity of wild bees in different urban settings. *PeerJ* **2017**, *5*, e3051. [CrossRef]
18. Instituto Brasileiro de Geografia e Estatística. IBGE. Available online: http://ibge.gov.br (accessed on 20 May 2020).
19. Köppen, W.; Geiger, R. *Klimate der Erde;* Verlag Justus Perthes: Gotha, Germany, 1928.
20. Pedlowski, M.A.; Carneiro da Silva, V.A.; Adell, J.J.C.; Heynen, N.C. Urban forest and environmental inequality in Campos dos Goytacazes, Rio de Janeiro, Brazil. *Urban Ecosyst.* **2002**, *6*, 9–20. [CrossRef]
21. Michener, C.D. *The Bees of the World*, 2nd ed.; Johns Hopkins University: Baltimore, MD, USA, 2007; p. 953.
22. Martins, A.C.; Goncalves, R.B.; Melo, G.A.R. Changes in wild bee fauna of a grassland in Brazil reveal negative effects associated with growing urbanization during the last 40 years. *Zoologia* **2013**, *30*, 157–176. [CrossRef]

23. Lorenzi, H. *Árvores brasileiras: Manual de Identificação e Cultivo de Plantas Arbóreas Nativas do Brasil*, 2nd ed.; Editora Plantarum: Nova Odessa, SP, Brasil, 1992; Volume 1, p. 384.
24. Lorenzi, H. *Árvores brasileiras: Manual de Identificação e Cultivo de Plantas Arbóreas Nativas do Brasil*, 2nd ed.; Editora Plantarum: Nova Odessa, SP, Brasil, 2002; Volume 2, p. 368.
25. Pereira, M.E.C.; Jasmim, J.M.; Pereira, M.H. *Guia de Vegetação de Praças*; Editora; EdUENF: Campos dos Goytacazes, Brasil, 2021; p. 70.
26. Brazil Flora, G. Brazilian Flora 2020 Project—Projeto Flora do Brasil 2020. Version 393.292. Instituto de Pesquisas Jardim Botanico do Rio de Janeiro. Checklist Dataset. Available online: https://doi.org/10.15468/1mtkaw accessed via GBIF.org (accessed on 19 July 2021). Available online: http://floradobrasil.jbrj.gov.br/ (accessed on 15 April 2019).
27. RCPol. Rede de Catálogos Polínicos Online. Available online: http://chave.rcpol.org.br/ (accessed on 24 May 2019).
28. Geslin, B.; Gauzens, B.; Thébault, E.; Dajoz, I. Plant pollinator networks along a gradient of urbanisation. *PLoS ONE* **2013**, *8*, e63421. [CrossRef] [PubMed]
29. Magurram, A.E. *Ecological Diversity and its Measurements*; Princeton University Press: Princeton, NJ, USA, 2004; p. 99.
30. Hammer, Q.; Harper, D.A.T.; Ryan, P.D. PAST: Paleontological Statistics software package for education and data analysis. *Paleontol. Electron.* **2001**, *4*, 9.
31. R Development Core Team. *R: A Language and Environment for Statistical Computing*; R Foundation for Statistical Computing: Vienna, Austria, 2021; Available online: http://www.R-project.org (accessed on 25 March 2020).
32. Taura, H.M.; Laroca, S. A associação de abelhas silvestres de um biótopo urbano de Curitiba (Brasil), com comparações espaço temporais: Abundância relativa, fenologia, diversidade e exploração de recursos. *Acta Biológica Parana.* **2001**, *30*, 135–137. [CrossRef]
33. Zanette, L.R.S.; Martins, R.P.; Ribeiro, S.P. Effects of urbanization on neotropical wasp and bee assemblages in a Brazilian metropolis. *Landsc. Urban Plan.* **2005**, *71*, 105–121. [CrossRef]
34. Taura, H.M.; Laroca, S.; Barbosa, J.F.; Rodrigues, J. Melissocenótica (Hymenoptera, Anthophila) no Parque Florestal dos Pioneiros, Maringá, PR. (sul do Brasil)—I. Abundância relativae diversidade. *Acta Biológica Parana.* **2007**, *36*, 47–65. [CrossRef]
35. Cardoso, M.C.; Gonçalves, R.B. Reduction by half: The impact on bees of 34 years of urbanization. *Urban Ecosyst.* **2018**, *21*, 1–7. [CrossRef]
36. McKinney, M.L. Urbanization as a major cause of biotic homogenization. *Biol. Conserv.* **2006**, *127*, 247–260. [CrossRef]
37. McKinney, M.L.; Lockwood, J.L. Biotic homogenization: A few winners replacing many losers in the next mass extinction. *Trends Ecol. Evol.* **1999**, *14*, 450–453. [CrossRef]
38. McKinney, M.L. Do exotics homogenize or differentiate communities? Roles of sampling and exotic species richness. *Biol. Invasions.* **2004**, *6*, 495–504. [CrossRef]
39. Horsák, M.; Lososová, Z.; Čejka, T.; Juřičková, L.; Chytrỳ, M. Diversity and biotic homogenization of urban land-snail faunas in relation to habitat types and macroclimate in 32 central European cities. *PLoS ONE* **2013**, *8*, e71783. [CrossRef]
40. Theodorou, P.; Radzevičiūtė, R.; Kahnt, B.; Soro, A.; Grosse, I.; Paxton, R.J. Genome-wide single nucleotide polymorphism scan suggests adaptation to urbanization in an important pollinator, the red-tailed bumblebee (*Bombus lapidarius* L.). *Proc. R. Soc. B Biol. Sci.* **2010**, *205*, 20172806. [CrossRef] [PubMed]
41. Williams, N.M.; Cariveau, D.; Winfree, R.; Kremen, C. Bees in disturbed habitats use, but do not prefer, alien plants. *Basic Appl. Ecol.* **2011**, *12*, 332–341. [CrossRef]
42. Depra, M.S.; Evans, D.M.; Gaglianone, M.C. Pioneer herbaceous plants contribute to the restoration of pollination interactions in restinga habitats in tropical Atlantic Forest. *Restor. Ecol.* under review.
43. Rollin, O.; Bretagnolle, V.; Decourtye, A.; Aptel, J.; Michel, N.; Vaissière, B.E.; Henry, M. Differences of floral resource use between honey bees and wild bees in an intensive farming system. *Agric. Ecosyst. Environ.* **2013**, *179*, 78–86. [CrossRef]
44. Ramalho, M. Stingless bees and mass flowering trees in the canopy of Atlantic Forest: A tight relationship. *Acta. Bot. Brasilica.* **2004**, *18*, 37–47. [CrossRef]
45. Scheper, J.; Reemer, M.; van Kats, R.; Ozinga, W.A.; van der Linden, G.T.; Schaminee, J.H.; Siepel, H.; Kleijn, D. Museum specimens reveal loss of pollen host plants as key factor driving wild bee decline in The Netherlands. *Proc. Natl. Acad. Sci. USA* **2014**, *111*, 17552–17557. [CrossRef]
46. Senapathi, D.; Goddard, M.A.; Kunin, W.E.; Baldock, K.C.R. Landscape impacts on pollinator communities in temperate systems: Evidence and knowledge gaps. *Funct. Ecol.* **2017**, *31*, 26–37. [CrossRef]
47. Williams, N.M.; Kremen, C. Resource distributions among habitats determine solitary bee offspring production in a mosaic landscape. *Ecol. Appli.* **2007**, *17*, 910–921. [CrossRef] [PubMed]
48. Buchholz, S.; Gathof, A.K.; Grossmann, A.J.; Kowarik, I.; Fischer, L.K. Wild bees in urban grasslands: Urbanisation, functional diversity and species traits. *Landsc. Urban Plan.* **2020**, *196*, 103731. [CrossRef]

diversity MDPI

Article

A GIS Modeling Study of the Distribution of Viviparous Invasive Alien Fish Species in Eastern Europe in Terms of Global Climate Change, as Exemplified by *Poecilia reticulata* Peters, 1859 and *Gambusia holbrooki* Girarg, 1859

Oksana Nekrasova [1,2], Volodymyr Tytar [1], Mihails Pupins [2], Andris Čeirāns [2], Oleksii Marushchak [1,*] and Arturs Skute [2]

[1] I. I. Schmalhausen Institute of Zoology, National Academy of Sciences of Ukraine, 01030 Kyiv, Ukraine; oneks22@gmail.com (O.N.); vtytar@gmail.com (V.T.)
[2] Department of Ecology, Institute of Life Sciences and Technologies, Daugavpils University, LV5400 Daugavpils, Latvia; mihails.pupins@gmail.com (M.P.); andris.ceirans@lu.lv (A.Č.); arturs.skute@du.lv (A.S.)
* Correspondence: ecopelobates@gmail.com; Tel.: +380-96-488-2670

Citation: Nekrasova, O.; Tytar, V.; Pupins, M.; Čeirāns, A.; Marushchak, O.; Skute, A. A GIS Modeling Study of the Distribution of Viviparous Invasive Alien Fish Species in Eastern Europe in Terms of Global Climate Change, as Exemplified by *Poecilia reticulata* Peters, 1859 and *Gambusia holbrooki* Girarg, 1859. *Diversity* **2021**, *13*, 385. https://doi.org/10.3390/d13080385

Academic Editor: Michael Wink

Received: 8 July 2021
Accepted: 11 August 2021
Published: 17 August 2021

Abstract: The potential distribution of tropical fish species in Eastern Europe—*Gambusia holbrooki* (introduced for biological control) and *Poecilia reticulata* (aquarium species, found in waste waters of big cities)—tend to be of particular interest in terms of global climate change. After GIS modeling of our own data and findings listed in the GBIF databases (2278 points for *G. holbrooki* and 1410 points for *P. reticulata*) using the Maxent package and 'ntbox' package in R, 18 uncorrelated variables of 35 Bioclim climatic parameters from CliMond dataset, it was found out that by 2090 guppies will appear in the south of Ukraine (Danube river's estuary, as well as in several places in the Caucasus and Turkey with habitat suitability > 0.3–0.5). *G. holbrooki* will also slightly expand its range in Europe. Limiting factors for *G. holbrooki* distribution are: bio1 (Annual mean temperature, optimum +12–+24 °C) and bio19 (Precipitation of coldest quarter (mm). Limiting factors for *P. reticulata* are: bio1 (optimum +14–+28 °C), bio4 (Temperature seasonality), bio3 (Isothermality). Unlike *G. holbrooki*, guppies prefer warmer waters. Such thermophilic fish species do not compete with the native ichthyofauna, but they can occupy niches in anthropogenically transformed habitats, playing an important role as agents of biological control.

Keywords: ecological niche model; distribution; aquaculture; mosquitofish; climate change; expansion

1. Introduction

Considering the prospects associated with the appearance of invasive species of animals at a new continent, some of the possible consequences are usually missed. It is often expected that the distribution and naturalization of invasive alien species can be controlled. However, this cannot always be possible since in open biosystems (natural and anthropogenic), in addition to other human activities leading to the spread of invasive alien species, the influence of changing climatic conditions, as the most global consequence of human activity takes place. However, the influence of all these negative factors (moving species between countries, uncontrolled release, wrong usage as agents of bimethod, accidental introduction, hunting, disbalancing of local ecosystems, etc.) as a whole can lead to the suppression of native animal species [1] and the emergence of species new to the local environment [2–7]. These factors can actively displace not only representatives of the ichthyofauna, but also batrachofauna [8]. Therefore, the main questions that we must ask ourselves are as follows: (1) What is the preferable (optimal) environment for an invasive species and in which regions naturalization and appearance of the species is possible; (2) what consequences of this appearance could be for native species and for the environment in general; (3) where do invasive species come from and what are the possible ways to avoid this; 4) what can influence invasive species as a limiting factor (besides climate conditions)—predators, parasites, illnesses. Of great interest is the appearance (or prospects of appearance) in temperate latitudes of thermophilic poikilothermic animals with their own strategy of viviparity, which gives them an opportunity to reproduce even in non-optimal conditions (for example, in human transformed territories). They also thrive and reproduce in both fresh and brackish water. Representatives of the genera *Gambusia* [9] and *Poecilia* [10] came to Europe as agents of a biological method for controlling the malaria mosquitos at the beginning of 20th century (they were released into the reservoirs of Western Europe). In addition, previously they also were bred in quantities in aquarium farms as aquarium fish. Two closely related species of this genus were introduced in Europe for the above mentioned purpose: *Gambusia affinis* (Baird and Girard, 1853) and *Gambusia holbrooki* (Girard, 1859), but the latter species is the most widespread and occurs in Ukraine.

G. holbrooki is a viviparous freshwater fish species (Poeciliidae family) originating from southern areas of North America. The species demonstrates great plasticity in the preference of comfortable water temperatures, thriving at +31–+38 °C, but being able to survive beyond these values [10]. This planktivorous species was used as an agent of biological control and was introduced in 1921 to the Iberian Peninsula for the first time to combat malaria. Later on the fish expanded an area of its invasion to Italy in 1922 and other Mediterranean countries, like Greece, Croatia, Spain etc. [11]. Currently this invasive alien species is well known in about 50 countries worldwide [9,12]. Besides being used to control mosquito populations, this species is known to have negative effects on local populations of aboriginal amphibian and fish species [13,14], which necessitates the study of its potential distribution, in terms of global climate change of particular importance in order to preserve local biodiversity.

P. reticulata is another viviparous freshwater fish species from North America that became invasive in Europe over recent decades. The wild form of the guppy was introduced to Europe in the 19th century. It was also used to fight malaria; thus it was introduced to many places. Being a popular polymorphic aquarium species, and due to their better resistance to colder water (up to +12 °C), guppies became invasive as a result of many accidental releases from aquaria [10,15]. Their ability to store sperm for months made it possible even for a one single gravid female to start a new population [16]. A wild population of guppies permanently lives in the Moskva river in the area of warm water discharge in Lyubertsy (Kuryanovsk drains) and in other places of this river [17,18]. Recently, in the Upper Volga basin, numerous self-reproducing populations of guppies have been noted in the regions of large cities (Tver, Yaroslavl, Rybinsk) in the areas of heated water discharge, as well as in settling ponds in facilities for the purification of domestic wastewater [19,20]. In their homeland, the island guppy populations live in brackish and seawater; they are bright in color and large in size. Optimal conditions for guppies include: clean water with a temperature of about +24 °C (the range being +16–+30 °C), the presence of zones with vegetation and free for swimming, a varied diet with a substantial proportion of live food. There are no data on the biology of this species from the Moskva river. It is known that there it reaches a high number and can be caught with a net in large quantities. For water bodies of the Upper Volga, there is an indication that it does not occur in water bodies with a temperature under +17 °C [20].

Such assumptions about appearance and distribution of these species in Eastern Europe are most relevant in connection with their potential usage as agents of biological control against the emergence of new carriers of various diseases—blood-sucking insects [21]. Therefore, the purpose of our work was to study both climatic indicators of environmental optima for these invasive alien species, and possible spread of these species over Eastern Europe in space and time.

2. Materials and Methods

2.1. Occurrence Data Collection

Occurrence data were collected from the original datasets [22], collection materials (I. I. Schmalhausen Institute of Zoology, National Academy of Sciences of Ukraine, Kyiv), GBIF databases [23,24]—all non-duplicate. We also studied the typical biotopes generally occupied by these fish species in Eastern Europe (Ukraine and Turkey) within 2012–2021 period of conducted field research (20 expeditions) (Figure 1). Fish were registered visually or caught by a manual fishing net (d = 2 mm) with subsequent species determination and release at the place of capture; the biotopes were photographed. To account for sampling bias, we used the nearest neighbor distance ('ntbox' package in R [25]) method to thin the data, where occurrence points that were $\leq$0.1 units away from each other were removed to avoid errors due to spatial autocorrelation. As a result, the number of points has significantly decreased; from 26,140 total points to 2278 for *G. holbrooki* and from 4200 points to 1410—for *P. reticulata.*

Figure 1. Biotope occupied by (**A**)—*P. reticulata* along the Bortnychi sewage drain (Ukraine, localities are marked with red circles); (**B**)—*G. holbrooki* is found in roadside fresh water bodies (Turkey).

2.2. Environmental Data

We used bioclimatic variables from the CliMond dataset (https://www.climond.org/ (accessed on 27 December 2020), following A1B climate change prediction scenario, of MIROC H global climate model. The above mentioned variables represent annual trends (e.g., mean annual temperature, annual precipitation) and extreme limiting environmental factors (e.g., temperature of the warmest month, precipitation in the wettest quarter), that are predicted for the whole planet (or its particular regions) and are known to influence species distributions [26]. Of 35 bioclimatic variables, highly correlated (r > 0.7) predictors were chosen using the 'virtualspecies' package in R, resulting in a selection of 18 variables predicted for 1975 (1970–2000) and 2090 (2081–2100) years. The following bioclimatic variables (CliMond) were chosen to be used in the analysis: bio01 Annual mean temperature (°C), bio02 Mean diurnal temperature range (mean (period max-min)) (°C), bio03 Isothermality (Bio02–Bio07), bio04 Temperature seasonality (C of V), bio14 Precipitation of driest week (mm), bio06 Min temperature of coldest week (°C), bio07 Temperature annual range (Bio05–Bio06) (°C), bio08 Mean temperature of wettest quarter (°C), bio10 Mean temperature of warmest quarter (°C), bio11 Mean temperature of coldest quarter (°C), bio12 Annual precipitation (mm), bio14 Precipitation of driest week (mm), bio15 Precipitation seasonality (C of V), bio18 Precipitation of warmest quarter (mm), bio25 Radiation of driest quarter (W m^{-2}), bio28 Annual mean moisture index, bio31 Moisture index seasonality (C of V), bio34 Mean moisture index of warmest quarter.

2.3. Model Building

In order to illustrate the niche-biotope dualism (known as Hutchison's duality [27])—the fact that the two invasive fish species can occupy ecological niches in different biotopes that have different combination of bioclimatic variables and make an ecological niche modelling (ENM), we used niche clustering ('ntbox' package in R [25]), which provides the opportunity to perform k-means clustering and project the results in the geographic and environmental spaces in a form of a world's map. The map shows the world's areas where the two fish species were found in biotopes that differ from those within their home range (different combinations of variables in places of species findings are market with differently colored circles).

Species distribution modelling (creation of standard distribution models or SDMs) was used to determine the potential change of distribution ranges of invasive alien species in new environments with time (MaxEnt [28] with 35 replicates, DivaGis (version 7.5; using CliMond dataset of current and predicted variables)). The Maxent software (version 3.4.1, [28,29]) was utilized for modelling, using the default settings. Maxent, unlike other distributional modelling techniques, uses only presence (registration points) data instead of presence and absence data. The SDMs were shown as maps where the areas of the highest habitat suitability (r > 0.3–0.5) are colored in red and areas of the lowest (r < 0.2)—in blue when visualized in SagaGis (System for Automated Geoscientific Analyses, version 7.6.0). The evaluation metrics for the obtained SDMs (performance) included: Partial receiver operating characteristic (ROC) [29], binomial tests [30], and the confusion matrix [31]. The ROC area under the receiver-operator curve (AUC) was used for assessing the discriminatory capacity of the models: AUC > 0.9 is considered excellent. The true skills statistics (TSS) was used to make a post-modelling check of obtained SDMs based on the standard confusion matrix that represents matches and mismatches between real observations of species (collected data) and predicted theoretical points within areas of high habitat suitability according to the created SDMs [26]. Binomial test mentioned earlier, as one of standard ways for evaluating obtained SDMs' quality, allowed to determine whether test points (from obtained SDMs' .acs files) fall into regions of predicted presence more often than expected by chance, given the proportion of map pixels predicted present by the model and showed on the world's map) [30]. This also helped to visualize the difference in predicted world areas occupied by the two fish species. GIS-modelling was accomplished using visualization in SagaGis, DivaGis, QGIS (version 3.16.9) [26]. Statistical processing of the obtained data was carried out using Statistica for Windows v.10.

3. Results

3.1. Limiting Factors for *P. reticulata* Distribution

MaxEnt GIS modeling revealed the following limiting and important factors (percent contribution) for the distribution of guppies: bio4 (Temperature seasonality)—21.9%, bio3 (Isothermality)—10.6%. In this "ecological envelope" 68.1% of guppies are found in the temperature range of +10–+28 °C, mean—22 °C (bio1, Figure 2, [28]). The temperature of +10 °C may have appeared as a result of extrapolation to northern areas, where warm drains are located (in large cities, etc.). According to our observations of "wild" populations of guppies in warm water bodies of the Bortnychi aeration station in Kyiv (2011–2020; 50.3837° N, 30.6642° E, Ukraine), these fish are quite unpretentious in terms of water quality and are generally demanding to the water temperature. The total body length in males is 1.85–3.24 cm, and in females—2.00–5.50 cm. Guppies breed all year round and even in winter in warm sewage waters; in December 2020 (water temperature +16 °C) mainly juvenile individuals were registered (Figure 1A). And owing to the high water temperature, they reproduce all year round. Therefore, these fish are more synanthropic, and are more likely to get along in warm urban drains in Eastern Europe. Consequently, annual mean temperature (bio1) within the native range has optimum—+14–+28 °C, mean—24 °C (Figure 2, [28]), and this species is rather unpretentious to water quality (Figure 1).

3.2. Limiting Factors for *G. holbrooki* Distribution

On the contrary, mosquitofish can coexist in open water bodies outside the warm wastewaters of large cities and anthropogenic areas, in roadside water bodies (Figure 1B) in western and southern parts of Europe, as well as in Ukraine—in the Odessa region and the Crimea [32]. As a result of GIS modeling, it was revealed that the most important factors (percent contribution) limiting the distribution of *G. holbrooki* in Europe are: bio21 Highest weekly radiation (W m^{-2})—26.2%, bio19 Precipitation of coldest quarter (mm)—18.6%, bio1 annual mean temperature—14.7% has the optimum—+12–+24 °C (mean—+16.5 °C), within the native range of +15–+24 °C (mean—+19.5 °C, Figure 2, [28]). This fish is able to reproduce throughout the year as long as the water temperature is above +15 °C (April–November). An interesting observation is that at a water temperature below +10 °C *G. holbrooki* burrows into the silt and falls into a suspended anabiosis [33]. During severe winters when reservoirs are bound with ice, *G. holbrooki* perishes in great numbers. For instance, this species died en masse as a result of cold winter in Sochi in 2020 [34,35].

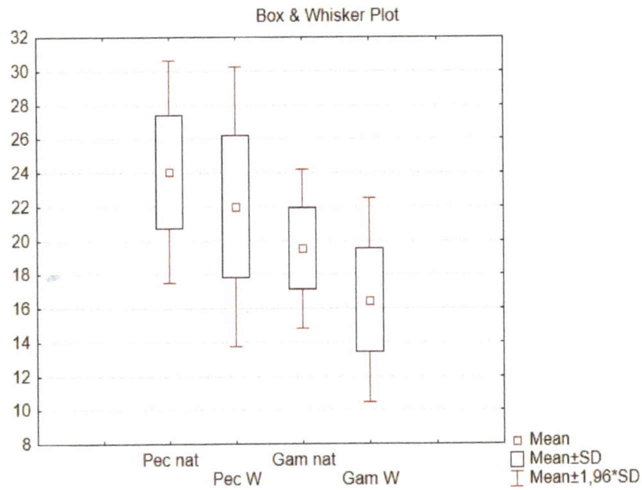

Figure 2. Plot variable bio1 Annual mean temperature (°C, CliMond, 1975 (1970–2000)) from: Pec—*P. reticulata*; Gam—*G. holbrooki*; nat—native range; W—World areas.

3.3. Niche Clustering

Using the clustering algorithms (a statistical tool which explains the difference between the variables that describe climatic conditions) for visualizing the Hutchison's duality (K-means clustering, Colwell and Rangel 2009) it was shown that the native range of American alien fish species differs in bioclimatic parameters (Figure 3). Therefore, understanding such perspectives and conditions is very important for interpreting expansion of the species range in future. Therefore, the preference for a warmer temperature regimen (Figure 2), unpretentiousness, the possibility of live birth and great popularity among fans of exotic aquacultures make it possible to "expect" more favorable conditions during climate warming. And it is even possible to adapt to new conditions.

3.4. Ecological Niche Modeling

To study the distribution of alien fish species in the present and in the future, we created two models (Maxent, CliMond 1975 and 2090) for each species (Figures 4 and 5). As a result, it was shown that taking into account tendencies of global warming till 2090, the range will increase by 1.2 times for guppies and by 1.07 times for mosquitofish. Despite the fact that these American viviparous fish species are thermophilic, they occupy different ecological niches (Figures 2 and 3), which certainly do not coincide with native northern fish species. To compare the SDM models (Maxent, CliMond 1975) of two species of American fish we used the regression module in SagaGis; the resulting coefficient of correlation was only r = 0.18, with global warming by 2090, the rate increases: r = 0.33. The similarity will become greater over time, thus suggesting that with global warming these fish species can occupy similar habitats in Eastern Europe. Nevertheless, the tendencies for the expansion of the range of these two invasive alien species in Eastern Europe (as well as in the south of Ukraine, the Caucasus) are quite realistic in the future (by 2090). Evaluating the performance of the SDM by the threshold-independent receiver operating characteristic (ROC) approach, where the calculated area under the ROC curve (AUC) is considered as a measure of prediction success, we came to the conclusion that the indicators are quite high—AUC > 0.95, TSS > 0.62.

To calculate the significance of the model, we used all the American fish meeting points (see Materials and Methods, [23,24]) and the Maxent model (CliMond 2090, Figure 6 [30]). The check was done using a binomial test using the cumulative binomial probability of success of predicting correctly an occurrence given the validation data and the proportional area predicted as present in the niche model. Therefore, we can conclude that the mosquitofish has great prospects for distribution in the south of Eastern Europe, since it is less demanding on temperature factors. (Figures 2 and 6).

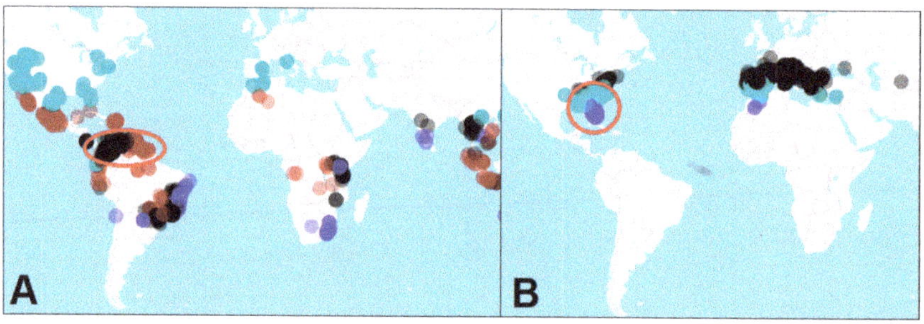

Figure 3. Niche clustering (Geographic space, CliMond 1975 (1970–2000)) from: (**A**)—*P. reticulata*; (**B**)—*G. holbrooki*, Differently coloured spots represents biotopes where the fish species was registered according to data collected from GBIF [23,24], where the species was registered and which has combination of climatic factors different from those in their native distribution area (marked with red circles).

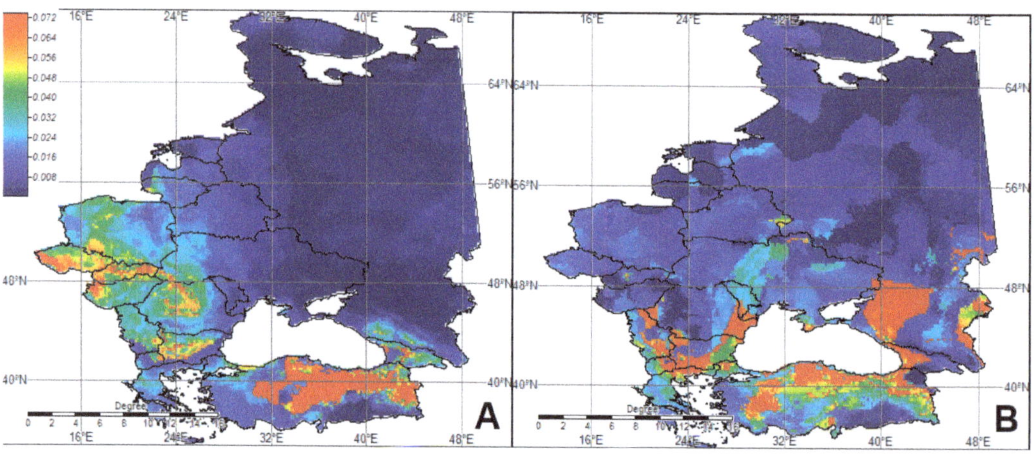

Figure 4. Potential (probabilistic) model of *P. reticulata* expansion built in the Maxent program based on the CliMond: (**A**)—1975 (1970–2000); (**B**)—2090 (2081–2100)) climatic data and GBIF data (2021). Areas of the highest habitat suitability (r > 0.3–0.5) are colored in red and areas of the lowest (r < 0.2)—in blue (SagaGis).

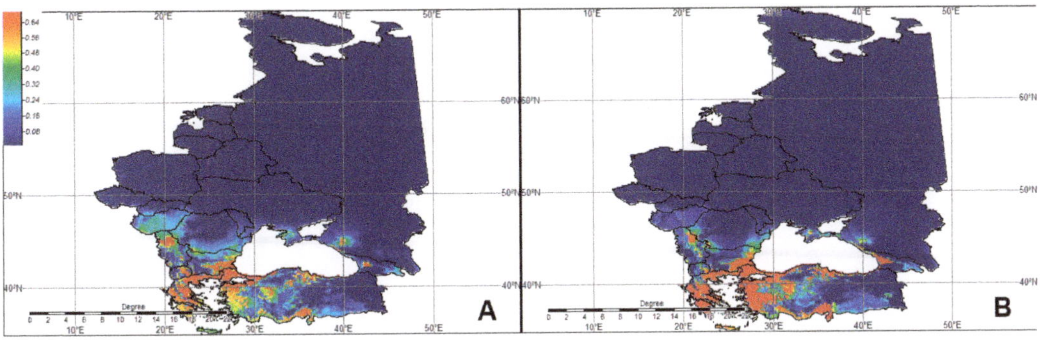

Figure 5. Potential (probabilistic) model of *G. holbrooki* world expansion built in the Maxent program based on the CliMond: (**A**)—1975 (1970–2000); (**B**)—2090 (2081–2100)) climatic data and GBIF data (2021). Areas of the highest habitat suitability (r > 0.3–0.5) are colored in red and areas of the lowest (r < 0.2)—in blue (SagaGis).

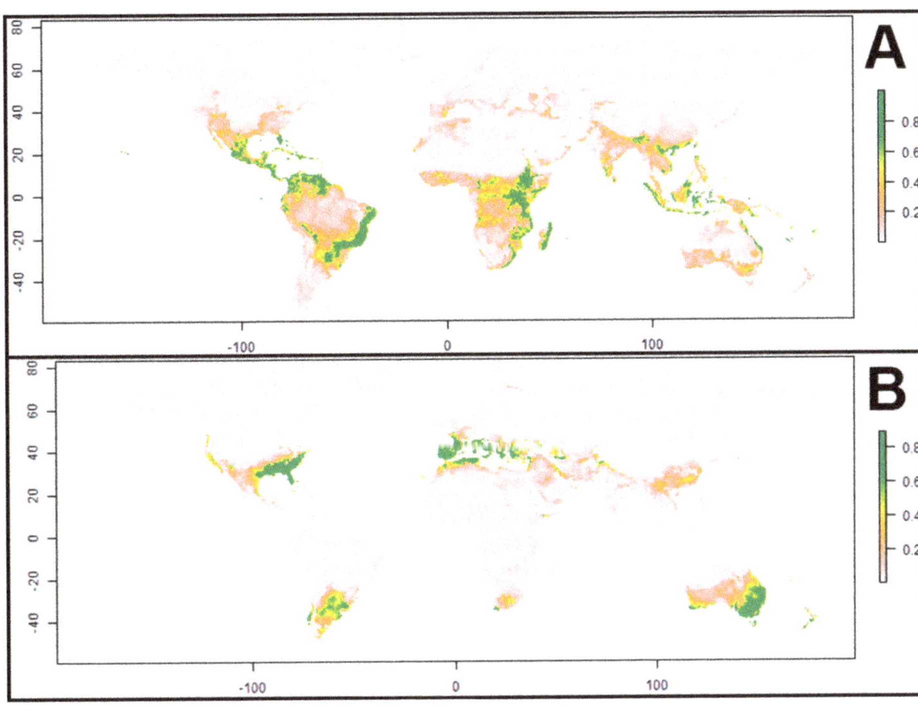

Figure 6. Result of the analysis of binomial tests (CliMond 2090 (2081–2100) [30]): (**A**)—*P. reticulata*; (**B**)—*G. holbrooki*.

4. Discussion

As a result of GIS modeling, climatic factors have been identified that allow new invasive alien species to inhabit southern Europe in the future. The new perspectives in Eastern Europe are very different from the conditions of the ecological niche of the native range of these fish species (Figure 3). However, due to increasing effects of climate change (to 2090; Figures 4B and 5B) and the emergence of "synanthropic ecological niches" in anthropogenic territories [36], these new perspectives provide great opportunities for the naturalization of invasive alien thermophilic poikilothermic animals with their own viviparity strategy. Under such conditions, not all native species will be able to survive in the changing climatic conditions under the influence of anthropogenic pressure [37]. In the more northern regions of Eastern Europe in a changing climate such invasive alien thermophilic fish species will not compete directly with the native ichthyofauna in the nearest future, as in very warm drains in anthropogenic territories or during mass summer fish kills, most of the native fish species die [1]. Instead, they can occupy their niche in new habitats, having an important role in ecosystems and for biological control (especially in anthropogenic areas) in a changing climate. It should be noted that in the so-called "warm countries" of Europe and other parts of the world where these two species are invasive, they can pose a threat, particularly mosquitofish, by consuming eggs of fish and amphibians of the 'Critically Endangered (CR)' yellow-spotted tree frog (see *Litoria castanea* in IUCN Red List of Threatened Species); the 'Endangered (EN)' green and gold frog (see *Litoria raniformis* in IUCN Red List of Threatened Species); and the 'Vulnerable (VU)' golden bell frog (see *Litoria aurea* in IUCN Red List of Threatened Species) [38]. Taking into account mouth size and ability to switch to a more zooplanktivorous diet under certain conditions [12,39–41], it is quite likely that in the northern regions such species can possibly become an additional threat for amphibian species native for Northern Europe, feeding on eggs and tadpoles/larvae of oriental tree frogs *Hyla orientalis* (Bedriaga, 1890), fire-bellied toads *Bombina bombina* (Linnaeus, 1761), crested newts *Triturus cristatus* (Laurenti, 1768) and common newts *Lissotriton vulgaris* (Linnaeus, 1758). It was confirmed by some studies [42] that in countries with warmer climate these alien invasive fish species can become a threat to native ichthyofauna, competing with aboriginal species for food resources, feeding on phyto- and zooplankton (cladocerans, copepods, ostracods, rotifers and insects) decreasing feeding base for native species and damaging already established food chains. Therefore, their potential impact on aboriginal biota in terms of changing climate conditions should not be underestimated.

5. Conclusions

We determined the bioclimatic variables influencing the expansion of the area of these two invasive alien fish species, which will allow them to occupy their special niche in anthropogenic territories of Southern Europe. As a result of GIS modeling of collected data, it was found out that by 2090 expansion of these species to Eastern Europe is highly likely. The appearance of guppies in the south of Ukraine is possible (the estuary of the Danube river, as well as in several parts of the Caucasus, Turkey, lower part of the Don river (>0.3–0.5, Figure 4). *G. holbrooki* will also slightly expand its range in Europe, the Caucasus as well as in Ukraine, namely in the Crimean peninsula, western part of the Black Sea region and the Danube estuary (Figure 5). The results obtained showed that *P. reticulata* are more attached to warm drains than *G. holbrooki* and the quality of water seems not affecting its distribution.

Author Contributions: Conceptualization, O.N., M.P. and V.T.; data curation, O.N., M.P., A.S. and A.Č.; Formal analysis, O.N., V.T. and O.M.; funding acquisition, M.P., A.S. and A.Č.; investigation, O.N., M.P., O.M. and A.Č.; methodology, O.N. and V.T.; project administration, A.S., O.N., A.Č. and M.P.; resources, O.N., M.P. and A.Č.; Software, O.N. and V.T.; supervision, O.N., V.T. and M.P.; validation, O.N., V.T., A.Č. and M.P.; visualization, O.N., V.T. and M.P.; writing—original draft and writing—review and editing, author: O.N., V.T., M.P., A.Č., O.M. and A.S. All authors have read and agreed to the published version of the manuscript.

Funding: This research was funded by the project "Pond aquaculture production and ecosystem service innovative research with modeling of the climate impact to tackle horizontal challenges and improve aquaculture sustainability governance in Latvia" (lzp–2020/2–0070) financed by Fundamental and Applied Research Projects (FLPP).

Institutional Review Board Statement: Not applicable.

Data Availability Statement: The data presented in this study are openly available in [*Gambusia holbrooki* Girard, 1859 in GBIF Secretariat (2021); GBIF.org (2 April 2021) GBIF Occurrence Download at https://doi.org/10.15468/dl.d5e6w9, accessed on 2 April 2021], reference number [23] and *Poecilia reticulata* Peters, 1859 in GBIF Secretariat (2021); GBIF.org (16 June 2021) GBIF Occurrence Download at https://doi.org/10.15468/dl.tuwazq, accessed on 16 June 2021, reference number [24].

Acknowledgments: Special thanks are due to Alex Volkov, freelance translator, for editing the manuscript.

Conflicts of Interest: The authors declare no conflict of interest.

References

1. Kuybida, V.V.; Nekrasova, O.D.; Kutsokon, Y.K.; Lopatynska, V.V. Summer fish kills in the Kaniv Reservoir. *Hydrobiol. J.* **2019**, *55*, 103–106. [CrossRef]
2. Ruchin, A.B.; Osipov, V.V.; Fayzulin, A.I.; Bakin, O.V.; Tselishcheva, L.G.; Bayanov, N.G. Chinese sleeper (*Perccottus glenii* Dybowski, 1877) (Pisces, Odontobutidae) in the reserves and National Parks of the middle and lower Volga (Russia): Mini-review. *AACL Bioflux* **2019**, *12*, 1114–1124.
3. Takács, P.; Abonyi, A.; Bánó, B.; Erős, T. Effect of non-native species on taxonomic and functional diversity of fish communities in different river types. *Biodivers. Conserv.* **2021**, *30*, 2511–2528. [CrossRef]
4. Clusa, L.; Garcia-Vazquez, E.; Fernández, S.; Meyer, A.; Machado-Schiaffino, G. Nuisance species in lake constance revealed through eDNA. *Biol. Invasions* **2021**, *23*, 1619–1636. [CrossRef]
5. Radočaj, T.; Špelić, I.; Vilizzi, L.; Povž, M.; Piria, M. Identifying threats from introduced and translocated non-native freshwater fishes in Croatia and Slovenia under current and future climatic conditions. *Glob. Ecol. Conserv.* **2021**, *27*, e01520. [CrossRef]
6. Iftime, A.; Iftime, O. Alien fish, amphibian and reptile species in Romania and their invasive status: A review with new data. *Trav. Mus. Natl. d'Hist. Nat. "Grigore Antipa"* **2021**, *64*, 131–186. [CrossRef]
7. Kestemont, B. The bottom-up assessment of threatened species. *Nat. Conserv. Res.* Заповедная наука **2019**, *4*, 93–106. [CrossRef]
8. Pupina, A.; Pupins, M.; Nekrasova, O.; Tytar, V.; Kozynenko, I.; Marushchak, O. Species distribution modelling: *Bombina bombina* (Linnaeus, 1761) and its important invasive threat *Perccottus glenii* (Dybowski, 1877) in Latvia under global climate change. *J. Environ. Res. Eng. Manag.* **2018**, *74*, 79–86. [CrossRef]
9. Vidal, O.; García-Berthou, E.; Tedesco, P.A.; Garcia-Marin, J.-L. Origin and genetic diversity of mosquitofish (*Gambusia holbrooki*) introduced to Europe. *Biol. Invasions* **2010**, *12*, 841–851. [CrossRef]
10. Padilla, D.K.; Williams, S.L. Beyond ballast water: Aquarium and ornamental trades as sources of invasive species in aquatic ecosystems. *Front. Ecol. Environ.* **2004**, *2*, 131–138. [CrossRef]
11. Kats, L.B.; Ferrer, R.P. Alien predators and amphibian declines: Review of two decades of science and the transition to conservation. *Divers. Distrib.* **2003**, *9*, 99–110. [CrossRef]
12. Krumholz, L.A. Reproduction in the western mosquitofish, *Gambusia affinis affinis* (Baird & Girard), and its use in mosquito control. *Ecol. Monogr.* **1948**, *18*, 1–43.
13. Garcı´a-Berthou, E.; Alcaraz, C.; Pou-Rovira, Q.; Zamora, L.; Coenders, G.; Feo, C. Introduction pathways and establishment rates of invasive aquatic species in Europe. *Can. J. Fish. Aquat. Sci.* **2005**, *62*, 453–463. [CrossRef]
14. Alcaraz, C.; Bisazza, A.; García-Berthou, E. Salinity mediates the competitive interactions between invasive mosquitofish and an endangered fish. *Oecologia* **2008**, *155*, 205–213. [CrossRef] [PubMed]
15. Strecker, A.L.; Campbell, P.M.; Olden, J.D. The aquarium trade as an invasion pathway in the Pacific Northwest. *Fisheries* **2011**, *36*, 74–85. [CrossRef]
16. Evans, J.P.; Magurran, A.E. Multiple benefits of multiple mating in guppies. *Proc. Nat. Acad. Sci. USA* **2000**, *97*, 10074–10076. [CrossRef] [PubMed]
17. Sokolov, L.I.; Sokolova, E.L.; Pegasov, V.A. Ichthyofauna of Moskva river within the city of Moscow and some data on its condition. *Quest. Ichthyol.* **1994**, *34*, 634–641.
18. Reshetnikov, Y.S.; Bogutskaya, N.G.; Vasiljeva, E.D. List of fish-like animals and fish of fresh water of Russia. *Quest. Ichthyol.* **1997**, *37*, 723–771.
19. Slynko, Y.V.; Kyiashko, V.N.; Yakovlev, V.N. List of fish-like animals and fish of Volga river basin. In *Catalogue of Plants and Animals of Volga River Basin*; ИБВВ РАН: Yaroslavl, Russian, 2000; pp. 252–277.
20. Yakovlev, V.N.; Slynko, Y.V.; Kyiashko, V.N. *Annotated List of Cyclostomata and Fish of Water Bodies of Upper Volga. Ecological Problems of Upper Volga*; ИБВВ РАН: Yaroslavl, Russian, 2001; pp. 52–69.
21. Kozynenko, I.I.; Tytar, V.M. Bioclimatic modeling of the European distribution of the invasive Asian tiger mosquito, *Aedes (Stegomyia) albopictus* (Skuse, 1895), with special reference to Ukraine. *Rep. Natl. Acad. Sci. Ukr.* **2020**, *3*, 88–93. [CrossRef]
22. Kutsokon, Y.; Nekrasova, O.; Shkamerda, V.; Loparev, S. The spread of guppy (*Poecilia reticulata* Peters, 1859) in the Bortnychi aeration station channel of Kyiv City. In *Biodiversity Dynamics 2012: In the Abstract of Scientific Materials*; Zagorodniuk, I., Ed.; Lugansk, Ukraine, 2012; pp. 94–95. Available online: http://www.terioshkola.org.ua/library/conf2012-biodiv/DBD2012-Dynamics_of_Biodiversity-all.pdf (accessed on 12 May 2021).

23. *Gambusia holbrooki* Girard, 1859 in GBIF Secretariat (2021). GBIF.org (2 April 2021) GBIF Occurrence Download. Available online: https://doi.org/10.15468/dl.d5e6w9 (accessed on 2 April 2021).
24. *Poecilia reticulata* Peters, 1859 in GBIF Secretariat (2021). GBIF.org (16 June 2021) GBIF Occurrence Download. Available online: https://doi.org/10.15468/dl.tuwazq (accessed on 16 June 2021).
25. Osorio-Olvera, L.; Lira-Noriega, A.; Soberón, J.; Peterson, A.T.; Falconi, M.; Contreras-Díaz, R.G. ntbox: An r package with graphical user interface for modelling and evaluating multidimensional ecological niches. *Methods Ecol. Evol.* **2020**, *11*, 1199–1206. [CrossRef]
26. Kriticos, D.J.; Jarošik, V.; Ota, N. Extending the suite of Bioclim variables: A proposed registry system and case study using principal components analysis. *Methods Ecol. Evol.* **2014**, *5*, 956–960. [CrossRef]
27. Colwell, R.K.; Rangel, T.F. Hutchinson's duality: The once and future niche. *Proc. Natl. Acad. Sci. USA* **2009**, *106*, 19651–19658. [CrossRef]
28. Phillips, S.J. A brief tutorial on Maxent. *AT&T Res.* **2005**, *190*, 231–259.
29. Peterson, A.T.; Papes, M.; Soberón, J. Rethinking receiver operating characteristic analysis applications in ecological niche modeling. *Ecol. Model.* **2008**, *213*, 63–72. [CrossRef]
30. Anderson, R.P.; Lew, D.; Peterson, A.T. Evaluating predictive models of species' distributions: Criteria for selecting optimal models. *Ecol. Model.* **2003**, *162*, 211–232. [CrossRef]
31. Fielding, A.H.; Bell, J.F. A review of methods for the assessment of prediction errors in conservation presence/absence models. *Environ. Conserv.* **1997**, *24*, 38–49. [CrossRef]
32. Kvach, Y.; Kutsokon, Y. The non-indigenous fishes in the fauna of Ukraine: A potentia ad actum. *BioInvasions Rec.* **2017**, *6*, 269–279. [CrossRef]
33. National Geographic Russia. Available online: https://nat-geo.ru/nature/v-sochi-ozhidayut-nashestviya-malyarijnyh-komarov-iz-za-massovoj-gibeli-ryby-gambuzii/ (accessed on 11 June 2020).
34. Nekrasova, O.; Tytar, V.; Pupins, M.; Čeirāns, A.; Marushchak, O. Distribution of Viviparous American Fish Species in Eastern Europe on the Example of *Gambusia holbrooki* Girarg, 1859 and *Poecilia reticulata* Peters, 1859 in the Context of Global Climate Change. In Proceedings of the 1st International Electronic Conference on Biological Diversity, Ecology and Evolution, (E-meeting address: www.sciforum.net), Online, 15–31 March 2021; Volume 68. [CrossRef]
35. Nekrasova, O.D.; Tytar, V.M.; Kuybida, V.V. *GIS Modeling of Climate Change Vulnerability of Amphibians and Reptiles in Ukraine*; Shmalgausen Institute of Zoology NAS: Kyiv, Ukraine, 2019; 204p, ISBN 978-966-02-8956-7.
36. Nekrasova, O.; Marushchak, O.; Pupins, M.; Skute, A.; Tytar, V.; Čeirāns, A. Distribution and Potential Limiting Factors of the European Pond Turtle (*Emys orbicularis*) in Eastern Europe. *Diversity* **2021**, *13*, 280. [CrossRef]
37. Global Invasive Species Database: *Gambusia holbrooki*. Available online: http://issg.org/database/species/ecology.asp?si=617&fr=1&sts=sss&lang=EN (accessed on 11 June 2020).
38. Dussault, G.V.; Kramer, D.L. Food and feeding behavior of the guppy, *Poecilia reticulata* (Pisces: Poeciliidae). *Can. J. Zool.* **1981**, *59*, 684–701. [CrossRef]
39. Lawal, M.O.; Edokpayi, C.A.; Osibona, A.O. Food and feeding habits of the guppy, *Poecilia reticulata*, from drainage canal systems in Lagos, Southwestern Nigeria. *West Afr. J. Appl. Ecol.* **2018**, *20*, 1–9.
40. Gkenas, C.; Oikonomou, A.; Economou, A.; Kiosse, F.; Leonardos, I. Life history pattern and feeding habits of the invasive mosquitofish, *Gambusia holbrooki*, in Lake Pamvotis (NW Greece). *J. Biol. Res.* **2012**, *17*, 121–136.
41. Blanco, S.; Romo, S.; Villena, M.-J. Experimental study on the diet of mosquitofish (*Gambusia holbrooki*) under different ecological conditions in a shallow lake. *Int. Rev. Hydrobiol.* **2004**, *89*, 250–262. [CrossRef]
42. Singh, N.; Gupta, P.K. Food and feeding habits of an introduced mosquitofish, *Gambusia holbrooki* (Girard) (Poeciliidae) in a subtropical lake, lake Nainital, India. *Asian Fish. Sci.* **2010**, *23*, 355–366.

diversity

MDPI

Article

Mammal Species Richness at a Catena and Nearby Waterholes during a Drought, Kruger National Park, South Africa

Beanélri B. Janecke

Animal, Wildlife & Grassland Sciences, University of the Free State, 205 Nelson Mandela Road, Park West, Bloemfontein 9301, South Africa; janeckbb@ufs.ac.za; Tel.: +27-51-401-9030

Abstract: Catenas are undulating hillslopes on a granite geology characterised by different soil types that create an environmental gradient from crest to bottom. The main aim was to determine mammal species (>mongoose) present on one catenal slope and its waterholes and group them by feeding guild and body size. Species richness was highest at waterholes (21 species), followed by midslope (19) and sodic patch (16) on the catena. Small differences observed in species presence between zones and waterholes and between survey periods were not significant ($p = 0.5267$ and $p = 0.9139$). In total, 33 species were observed with camera traps: 18 herbivore species, 10 carnivores, two insectivores and three omnivores. Eight small mammal species, two dwarf antelopes, 11 medium, six large and six mega-sized mammals were observed. Some species might not have been recorded because of drought, seasonal movement or because they travelled outside the view of cameras. Mammal presence is determined by food availability and accessibility, space, competition, distance to water, habitat preferences, predators, body size, social behaviour, bound to territories, etc. The variety in body size and feeding guilds possibly indicates a functioning catenal ecosystem. This knowledge can be beneficial in monitoring and conservation of species in the park.

Keywords: catena ecosystem; ephemeral mud wallows; habitat use; mammal variety; Skukuza area; species presence; Stevenson-Hamilton supersite

Citation: Janecke, B.B. Mammal Species Richness at a Catena and Nearby Waterholes during a Drought, Kruger National Park, South Africa. *Diversity* **2021**, *13*, 387. https://doi.org/10.3390/d13080387

Academic Editor: Michael Wink

Received: 10 July 2021
Accepted: 31 July 2021
Published: 18 August 2021

1. Introduction

The habitat of many species is often distributed in patches across the landscape. Species may occupy several patches or only a few [1]. The number of patches available will influence the type of species and total number of individuals (population sizes) of a species present in a particular landscape. This is related to the distribution and abundance of the species [1]. A landscape is defined by Gertenbach [2] (p. 9) as, "an area with a specific geomorphology, climate, soil and vegetation pattern together with the associated fauna". Animals use certain habitat patches in a landscape based on several factors, such as spatial distribution of resources, ability of animals to use the resources and habitat requirements [3]. Species that are widespread are in general more abundant than species with a restricted occurrence [4,5]. Abundance information can give an indication of the number of individual animals of a certain species that occur in a population. This can provide a basis when making decisions for practical conservation [6]. It can be quite expensive or difficult to obtain abundance information of species at a local scale, depending on the landscape and type of species studied. Usually, being able to predict abundances on a fine-scale from the presence-absence data on a larger scale (that is generally more readily available) can have a considerable application in conservation biology [6]. This current study focussed on small, local scale presence of mammal species at specific habitat patches that can be used as a basis for large scale predictions on the distribution and abundance of mammals in similar areas in the landscape.

The Kruger National Park (KNP) is a savanna ecosystem where heterogeneity and complexity are studied, acknowledged and applied in management practices [7]. Mammal herbivores heavier than 5 kg represent 30 species in KNP, and that is excluding the smaller herbivores, rodents, bats, carnivores, primates, etc. [8]. A defining characteristic of savannas is the continuous herbaceous layer and a discontinuous tree and shrub stratum in the same landscape [7]. Heterogeneity in the landscape influences vegetation patterns and thus also community assemblages, densities of herbivores and predators at different spatial scales [8]. Different scales and factors exist to determine the heterogeneity of savannas, such as variation in soil conditions and properties, topography, rainfall, fire regimes and distribution of surface water [9]. Some abiotic factors, such as distance to water and steep slopes, can limit grazing and browsing in certain areas, while wind and presence of shade will affect where animals can rest and where they will graze [10]. This will impact on mammal presence in a specific area. Biotic factors such as plant productivity, phenology, species composition and forage quality have an impact on the distribution of forage. The time spent by large herbivores in different habitats is usually based on resource levels and its accessibility in that area [10]. This heterogeneity, expressed as for example diversity of habitat types created by differing abiotic and biotic components, is important for the distribution and utilisation of habitat patches by large mammals [9], which will ultimately affect their presence or absence in a local area.

The landscape in the south of Kruger National Park has a granitic geology and is dissected by a large density of streams and slightly undulating catenas [11]. A catena is a hillslope with different soil types and properties arranged in zones from its crest to its footslope. Particles travel downslope creating an environmental gradient in which different vegetation types are associated with the different soil types and soil properties of the zones [12,13]. There is a clear turnover in nutrients and vegetation structure from the uplands of a catenal sequence (nutrient-poor soils associated with broad-leaved savanna type) to the bottomlands (clay-rich, nutrient rich soils associated with fine-leaved savanna type) [7,14].

It was hypothesised in this study that some mammal species might use certain zones on the catena with higher frequency than others, since the diverse vegetation types across a catena create different habitat patches for mammals (food plants available, vegetation structure, space, cover, etc.) The aims of this study were to (1) provide a basic list of mammal species (>mongoose) that are present in the different zones of one catenal slope and at three waterholes in close vicinity, (2) indicate the feeding guild and body size of mammals present and (3) compare the mammal species richness between three short survey periods during a drought. Factors that influence mammal presence or absence were also included from the literature. This study formed part of a larger multidisciplinary project [15] where certain abiotic and biotic components of this catenal ecosystem were investigated. The study area was deliberately kept small to enable the focus of the larger project to be on providing detail of each of these ecosystem components and to find links between them on a small-scale [16]. This proved to be an important multidisciplinary project that can be expanded to the granite supersite on a landscape scale. This specific study covered here forms the foundation for numerous future research possibilities specifically on mammal diversity and how it might change under different environmental conditions.

2. Methods

2.1. Study Area

The park falls in the Savanna Biome and the study area in the Granite Lowveld vegetation type [17], consisting mainly of mixed woodland and thorn thickets [11,18]. The climate is described as semi-arid subtropical with two distinct seasons: a growing season that is hot and occasionally wet and a dormant season that is warm and dry. The average minimum temperature is 5.7 °C in June, and the average maximum temperature is 32.6 °C in January [2,7]. Topography is an undulating landscape derived from granite that covers the sequence of terrain morphology from the bottomlands to the uplands. Uplands are characterized by sandy, coarse, shallow soil overlying rock, while the bottomlands below the seepline are characterized by deep duplex soil—this is a typical catenal pattern. Sodic areas that are found between the crest and drainage line are composed of shallow sand, with high pH and reduced hydraulic conductivity [13,19].

The study was done on one catena/hillslope in the undulating landscape, its closest outcrops and waterholes in the Southern Granite Supersite [11] of Kruger National Park, South Africa. The three waterholes studied had a permanent supply of water throughout the year. Water is extracted from boreholes using either windmill or solar powered pumps and pumped into 2.5 m tall concrete reservoirs with an open top which feeds into a nearby ground-level water trough/artificial waterhole [20]. The catena was divided into four zones based on differences in vegetation and soil types and its position on the hillslope [16]: crest and midslope, sodic patch with small, wet seep area, footslope shrub veld and a riparian area (around the dry drainage line of the Sabi River near Skukuza). In the study area, the pH of the crest was 5, the sodic patch was 8 and the drainage line 6 [16]. The length of the catena from crest to drainage line is about 500 m (0.3 mile), and the width is about 2 km (1.2 mile) on its broadest part, but this is not a true representation of the surface area of the catena since the zones differ in size along its length. The normal annual rainfall of the area is 560 mm. However, a drought was experienced during the study period and according to Skukuza Weather Data only 194 mm rainfall was measured per year during the study period [15]. It was reported as being one of the most severe droughts, due to an El Niño Southern Oscillation event that reached its peak during 2015–2016 [21].

2.2. Procedure

A total of 30 camera traps were distributed randomly on the catena to give the best cover of each zone. This included the mud wallows on the catena, the nearby granite-boulder outcrops (inselbergs) and the three closest permanent waterholes. Bushnell (23), Cuddeback (4), Scoutguard (2) and Little Acorn (1) models with similar trigger response times and field of view ranges were used [15]. Cameras were operational in the growing season for a period of two weeks each during September 2015, March 2016 and March–April 2017 in order to compile the basic list of mammals present and to compare the data of similar survey periods spanning three years. All cameras were checked after one week of operation, and memory cards were replaced. The same positions were used for the three survey periods. Two photographs were taken when a camera was triggered, with a 5 s rest interval in-between.

Cameras on the catena were positioned in such a way to be able to include small (>mongoose) to mega sized mammals at heights ranging from approximately 0.5 m to 1.7 m (higher cameras pointing slightly downwards) depending on how open the view of the camera was [15]. Cameras were facing in a southerly direction as far as possible. Visual surveys of the area were used to determine the positioning of cameras based on recommendations of a camera trapping specialist. Game trials, open clearings, areas with longer distance view, termite mounds and areas where obvious animal activity was noted were selected. To ensure that the entire study area was represented, some areas with denser vegetation were also included. One camera was placed several meters from each waterhole at an average height of 2 m, facing down to include the entire waterhole in the view. This was done to prevent animal tampering that can cause cameras to malfunction (viz. [20]). The waterhole cameras were set to take photos at 5 min intervals, and this was combined with taking photos when the camera was triggered by animal movement. All cameras used infrared flashes to take images during the night, but the cameras were too far for the flash to capture animals at the waterhole during the dark part of the night. Only larger animals in the vicinity of the camera were captured by the waterhole cameras at night. There were also additional cameras placed in vicinity of the waterholes but not looking directly at the waterhole.

2.3. Data Analyses

Data of all the cameras that were positioned in the same catenal zone were combined to determine animal presence in that zone. The waterholes, mud wallows and granite-boulder outcrops were treated as three additional zones/areas. For the purpose of this study, a trigger event is described as the image/s recorded of an individual or group of a mammal species triggering the motion sensor of the camera at a specific time. If the species could not be identified due to blurred photos, it was indicated as Unknown. The date and time stamps on the photographs were used to determine separate events. A new event was recorded when there was an absence of the group or individual for at least 30 min, approximately.

The type of feeder and body size of each species present were confirmed by Estes [22] and tabulated. The total number of trigger events (in other words the number of observations) of each species was calculated per zone. The data were graphically presented as a colour gradient in different blocks—white blocks indicate no observation of that species, while the darker the colour, the higher the total number of events/observations (range) that were noted for that species. The three survey periods were indicated on this colour figure for each zone (including mud wallows, waterholes and granite outcrops, separately). The number of events was counted for all mammal species observed at each of the three waterholes during a survey period and that was totalled per species. The maximum number of individuals of a species that occurred together in a group at the waterholes during one event was indicated in a table for that survey period. Other smaller groups and solitary individuals were observed for the majority of the species visiting waterholes, but only the maximum group size was tabulated.

The Shapiro–Wilk test was used to test for statistical normality of the data. Since data were not normally distributed, the Kruskal–Wallis test was used to determine significant differences between mammal presence (indicated with a 1 if present and 0 if absent) in each catenal zone and between different survey periods, respectively. A 5% level of significance was used. The species richness was measured by the number of different species observed in each zone during a survey period, while the Shannon–Weiner Index measured the diversity and evenness. The zones were included separately in the calculations and the sum totals of the number of events per species were included for the catena and the waterholes including the surrounding areas. A two-way ANOVA was performed on these values to determine significant differences in number of events between zones and survey periods. The vegan package in R was used for the analyses [23].

3. Results

The species richness was rather high for the nutrient poor granites, with 33 mammal species observed in total. These species were subdivided into the following feeding guilds (Table 1): 18 herbivore species (seven grazers, six browsers, two mixed feeders and three general vegetarians), 10 carnivores, two insectivores, and three omnivore species. Eight species are classified as small mammals (mongoose, squirrel, etc.), two as dwarf antelopes (steenbok, duiker), 11 as medium (impala, leopard, etc.), six as large (zebra, lion, etc.) and six as mega-sized mammals (elephant, rhino, etc.)

Due to the global rhino poaching crisis, all detail of rhinoceros species that can be used to determine their exact location was removed from the figures and text. For the remainder of the species, their presence in each zone on the catena (Figure 1a) (including the additional zones/areas—Figure 1b,c) was indicated for each survey period based on number of events/observations. Species with darker colour blocks on Figure 1 were observed with higher frequency (during different events/observations) on the cameras than species with lighter colour blocks. The common species found in all zones, were buffalo, elephant, greater kudu, grey duiker, impala and lion, while the other common species that were only absent from the granite-boulder outcrop area (Figure 1c) were blue wildebeest, giraffe, plains zebra and spotted hyena. Steenbok (water-independent species, like the grey duiker) was only absent at the mud wallows and waterholes (Figure 1b). Mammal species richness was the highest at waterholes (21 species), followed by the midslope (19) and sodic patch (16) on the catena. Four species were found only in the vicinity of a waterhole and not on the catena, namely, banded mongoose, side-striped jackal, vervet monkey and waterbuck.

Table 1. Total list of mammal species, including their scientific names, observed on the catena, granite outcrops and nearby waterholes. Their feeding guild and body sizes relative to each other are also indicated.

Common Name	Scientific Name	Feeding Guild	Size
Aardvark	*Orycteropus afer*	Insectivore	Medium
African wild dog	*Lycaon pictus*	Carnivore	Medium
Banded mongoose	*Mungos mungo*	Carnivore	Small
Black rhinoceros	*Diceros bicornis*	Browser	Mega
Blue wildebeest	*Connochaetes taurinus*	Grazer	Large
Buffalo (Cape)	*Syncerus caffer*	Grazer	Mega
Bushbuck	*Tragelaphus scriptus*	Browser	Medium
Chacma baboon	*Papio ursinus*	Omnivore	Medium
Civet	*Civettictis civetta*	Omnivore	Medium
Dwarf mongoose	*Helogale parvula*	Carnivore	Small
Elephant (African)	*Loxodonta africana*	Mixed feeder	Mega
Genet species	*Genetta* species	Carnivore	Small
Giraffe (South African)	*Giraffa giraffa* (*G. camelopardalis*—old name)	Browser	Mega
Greater kudu	*Tragelaphus strepsiceros*	Browser	Large
Grey/Common duiker	*Sylvicapra grimmia*	Browser	Dwarf
Hippopotamus	*Hippopotamus amphibius*	Grazer	Mega
Impala	*Aepyceros melampus*	Mixed feeder	Medium
Leopard	*Panthera pardus*	Carnivore	Medium
Lion (African)	*Panthera leo*	Carnivore	Large
Plains zebra	*Equus quagga*	Grazer	Large
Porcupine	*Hystrix africaeaustralis*	Vegetarian	Medium
Scrub hare	*Lepus saxatilis*	Vegetarian	Small
Serval	*Leptailurus serval*	Carnivore	Medium
Side-striped jackal	*Canis adustus*	Carnivore	Medium
Slender mongoose	*Galerella sanguinea*	Carnivore	Small
Spotted hyena	*Crocuta crocuta*	Carnivore	Large
Steenbok	*Raphicerus campestris*	Browser	Dwarf
Tree squirrel	*Paraxerus cepapi*	Vegetarian	Small
Vervet monkey	*Chlorocebus pygerythrus*	Omnivore	Small
Warthog	*Phacochoerus africanus*	Grazer	Medium
Waterbuck	*Kobus ellipsiprymnus*	Grazer	Large
White rhinoceros	*Ceratotherium simum*	Grazer	Mega
White-tailed mongoose	*Ichneumia albicauda*	Insectivore	Small

Mammal species observed at each of the waterholes in closest vicinity to the catena could be distinguished more easily (Table 2). During the 2015 survey period, the De la Porte waterhole was closed for maintenance and no data is available. The herd sizes of elephants visiting waterholes differed, including herds of 6, 9, 12, 18, 21 and 30, while impala herd sizes were 5, 12, 20, 40 and 70 over the different survey periods. Species observed at all three waterholes were baboon, buffalo, elephant, giraffe, impala and warthog. Some species that were also present on the catena were observed at only one of the waterholes, such as blue wildebeest, grey duiker, greater kudu and plains zebra. The larger predators were also only observed at one of the three waterholes, i.e., lion, leopard, wild dog and spotted hyena (latter observed at two waterholes), but these results can be biased since accurate night observations were impossible because of the way the cameras were set-up at waterholes. Vervet monkey, bushbuck, waterbuck and hippopotamus were also only captured at one waterhole, albeit different waterholes (Table 2). The differences that are evident between data presented in Figure 1 and Table 2 (i.e., banded mongoose, civet, dwarf mongoose, side-striped jackal, etc.) are because of the additional cameras that were placed at waterholes to cover the vicinity of the waterhole as well, and these data were included in Figure 1 but not in Table 2 where only data from the waterhole itself are presented.

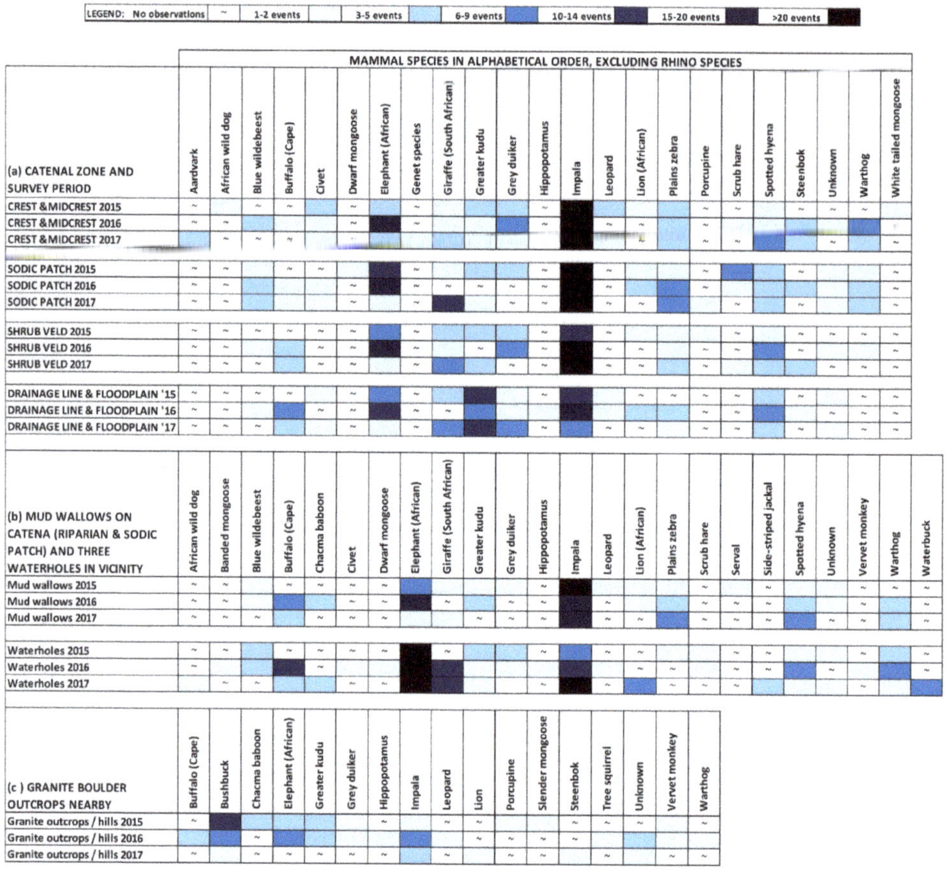

Figure 1. Mammal species (excluding rhino species) present in different zones during three survey periods: (**a**) catenal zones; (**b**) mud wallows, waterholes and (**c**) granite-boulder outcrops. The colour gradient is used to indicate number of events (observations) of each species where increasing numbers were represented by darker colours.

Table 2. Number of events (observations) of the mammal species (excluding rhinos) and the maximum group size observed at each of the three waterholes during that survey period.

Species	Number of Events			Total Events	Max Number of Individuals		
	2015	2016	2017		2015	2016	2017
	De la Porte waterhole						
Baboon	0	1	2	**3**	0	2	2
Buffalo	0	0	1	**1**	0	0	1
Elephant	0	10	25	**35**	0	14	18
Giraffe	0	3	0	**3**	0	11	0
Hippopotamus	0	2	0	**2**	0	5	0
Spotted hyena	0	4	0	**4**	0	1	0
Impala	0	0	4	**4**	0	0	8
Scrub hare	0	1	0	**1**	0	1	0
Warthog	0	1	0	**1**	0	1	0

Table 2. *Cont.*

Species	Number of Events			Total Events	Max Number of Individuals		
	2015	2016	2017		2015	2016	2017
	Kwaggaspan waterhole						
Baboon	0	1	0	**1**	0	2	0
Buffalo	0	4	3	**7**	0	22	30
Bushbuck	0	1	0	**1**	0	1	0
Elephant	14	30	5	**49**	16	21	6
Giraffe	1	1	0	**2**	0	3	1
Grey duiker	4	0	0	**4**	1	0	0
Impala	1	0	2	**3**	1	0	3
Leopard	0	1	0	**1**	0	1	0
Plains zebra	1	0	0	**1**	5	0	0
Scrub hare	1	0	0	**1**	1	0	0
Warthog	4	1	0	**5**	2	1	0
Waterbuck	0	0	9	**9**	0	0	7
	Renosterkoppies waterhole						
Baboon	0	1	1	**2**	0	2	1
Blue wildebeest	1	4	0	**5**	1	3	0
Buffalo	0	5	0	**5**	0	3	0
Elephant	13	16	3	**32**	30	18	1
Giraffe	0	5	10	**15**	0	1	1
Greater kudu	3	1	0	**4**	5	10	0
Impala	3	5	56	**64**	2	70	90
Lion	0	0	7	**7**	0	0	3
Spotted hyena	1	0	0	**1**	1	0	0
Vervet monkey	0	2	0	**2**	0	1	0
Warthog	1	5	1	**7**	0	4	3
Wild dog	0	0	2	**2**	0	0	9

The total number of events and the total number of species per zone were summarised in Table 3. The diversity index values indicating the species richness and evenness in each zone from the number of events were also included, for the sake of completeness. The riparian area (2016 and 2017), granite outcrops (2016) and crest–midslope (2017) had the highest diversity indexes. ANOVA (number of events per species) revealed that all main (Zone and Survey period, respectively: F(10,10) = 2.077, p = 0.132 and F(1,10) = 0.287, p = 0.604) and interaction effects (F(10,10) = 0.324, p = 0.955) were not significant. Small differences were also noted (Figure 1) in presence/absence of some species in the different zones (H = 2.227, n = 96, p = 0.5267) and between survey periods (H = 0.180, n = 99, p = 0.9139), but they were not statistically significant. Consequently, there was not enough evidence to test the hypothesis that species may be associated with specific zones. A longer time period is needed for the study and the zones are probably too small, relatively, to limit the mammals to one specific zone.

Table 3. Summary of the sum total of events (observations), species richness (number of species) and the Shannon–Weiner Diversity Index values per survey period for each zone. Total values of the catena and the waterholes (including their surrounding areas) are also indicated.

Zone	Survey Period	Total Events	Species Richness	Shannon Index
Crest and midslope	2015	54	12	1.91
	2016	97	14	1.74
	2017	58	12	2.01
Sodic patch	2015	93	12	1.61
	2016	136	11	1.34
	2017	86	11	1.67
Shrub veld	2015	33	9	1.85
	2016	61	9	1.69
	2017	52	9	1.76
Riparian area	2015	43	8	1.75
	2016	68	13	2.18
	2017	50	9	1.97
Catena total	2015	209	17	1.89
	2016	335	17	1.67
	2017	248	17	2.08
Mud wallows	2015	40	11	1.65
	2016	59	9	1.84
	2017	45	7	1.66
Granite outcrops	2015	28	9	1.83
	2016	36	11	2.03
	2017	9	4	1.15
De la Porte Waterhole	2015	0	0	0.00
	2016	17	7	1.20
	2017	32	4	0.73
Kwaggaspan Waterhole	2015	26	7	1.41
	2016	38	7	0.81
	2017	19	4	1.23
Renosterkoppies Waterhole	2015	22	6	1.28
	2016	44	9	1.89
	2017	80	7	1.05
Waterhole total	2015	73	10	1.38
	2016	150	17	1.98
	2017	191	11	1.36

4. Discussion

4.1. Species Richness and Habitat Description

Mammal species richness on the catena was 23 in total (Table 3) and highest at the crest—midslope (19 species) followed by sodic patch (16), riparian area (15) and shrub veld (14). Similar numbers of species in different zones and between survey periods represent different species that make up the total (Figure 1). In the context of this study, species richness can be defined as the number of different mammal species present in the ecological community [24] associated with the catena and its nearest waterholes. It is merely a count of species and does not take their abundance or distribution in KNP into consideration.

There were little differences (not statistically significant, $p = 0.53$) between species observed across the catenal zones, but a few species could be linked to specific areas outside the catena (Figure 1). The granite-boulder rock outcrops and waterholes delivered some species that were absent in the zones on the catena. A higher-than-expected number of species were noted in vicinity of the outcrops (16 species, Figure 1c), possibly because mammals pass the nearby outcrops en route to one of the waterholes. Rock hyrax and klipspringer (name literary means rock-jumper) are known to be in the area [18] and were expected to be present at these outcrops because they are adapted to such a xeric habitat, but they were not captured on the cameras and thus not included in the results.

Common species found at the catena in this current study were buffalo, elephant, greater kudu, grey duiker, impala, lion, blue wildebeest, giraffe, plains zebra, spotted hyena and steenbok. Siebert and Scogings [7] list similar common species on the catena approximately 40 km away (adjacent to the Sabi River), namely, impala, elephant, hippo, black and white rhino, blue wildebeest, Cape buffalo, plains zebra, greater kudu, steenbok, giraffe and scrub hare. Gertenbach [2] and Sutherland et al. [20] also list the following common species in the larger landscape: impala, greater kudu, steenbok, giraffe, elephant, buffalo, plains zebra, wildebeest, warthog, hippo (at the rivers), lion, leopard, wild dog and spotted hyena.

The zones of the catena, from the crest to the drainage line, consist of different plant communities composed of different densities of trees, shrubs and herbaceous plant species [25] that create different habitat patches for mammals. A large proportion of these plant species are palatable [18] and will most probably be used in different degrees by the herbivore species present, based on their feeding guild. The sodic patch in the study area is reported to be more fertile than the surrounding catenal zones [14], but the grass cover is relatively low, i.e., the sodic patch had a 16% grass cover and the shrub veld zone 32%. The riparian zone at the drainage line had the highest tree canopy cover (33.6% on a 100 m line transect) and the sodic patch the lowest cover (2.7%) [25]. Sodic soils are regarded as a stressful environment for vegetation that is usually sparse but more attractive than the upland vegetation [7]. Sodic areas are attractive to herbivores because of more nutritious vegetation than surrounding areas, predator vigilance, dietary salts or anti-acidosis minerals [13], water in ephemeral depressions and green foliage at wet seep areas. These sodic areas with higher nutrients usually form grazing lawns that are maintained by grazers in a "short-cropped state of high nutritional value", according to Martin et al. [26] (p. 2). However, in comparison to other areas in the KNP, gabbro substrates in the west-central parts of KNP and basalt plains in the eastern parts generate soils of higher fertility leading to more nutritious grazing lawns, while the sandy granitic soils (study area) are less fertile supporting predominantly sourveld [18,26]. The distribution of herbivores in the park is determined by this variation, with higher densities in general in the eastern basaltic soils than on the western granitic soils [19].

It is known that herbivores generally concentrate their foraging in zones that shift through the seasonal cycle up and down with the catenary gradient formed from the crests downward. They move progressively down the slope of the catena in the dry season when availability of nutritious and moist, green food declines upslope, while moving up the profile again during the wet season [8,27]. This catenary movement was not recorded down to the river in this small-scale study—the last part of the catena studied was the third order, dry drainage line. However, the Shannon Diversity Index (H = ~2) was the highest at this drainage line (riparian area) for two survey periods during the drought compared to all the other zones in the study (Table 3). Areas that are closer to rivers have valleys with more drainage lines, and thus, these areas have more moisture than areas on the catenary landscape that are further from rivers, representing the drier crests of the catena [27]. Consequently, the association patterns with rivers suggest that KNP's browsers and mixed feeders prefer these areas closer to the drainage lines in the dry season for the forage it provides [27].

Areas occupied for longer periods contribute to the utilization distribution of the animal and can be connected to activities performed and thus benefits derived, while predator avoidance strategies and accessibility can result in limited use of other areas [28]. A home range is generally defined as a larger general use area traversed by an animal species during routine activities (find food, water, shelter and a mate, avoid predators, etc.) that includes the areas used during different seasons [29,30]. Some areas in the demarcated home range may be occupied simply in transit, i.e., to reach surface water for drinking and seasonal food patches or for escaping predators. Availability of resources and habitat requirements will determine the extent of the area that needs to be traversed in order to fulfil the animal's needs [30]. These among other factors will impact on the species' absence or presence in certain locations and/or zones.

4.2. Waterholes and Ephemeral Mud Wallows

The waterholes had a high mammal species richness in total (21 species), but separately, each waterhole had low species richness and diversity (Table 3). This total number can probably be higher since the smaller and nocturnal mammals might have been missed because of the way the camera traps were set up to cover the entire waterhole and some of the surrounding area. Sutherland et al. [20] reported 26 mammal species at the same three waterholes but also including an additional fourth waterhole during a similar time period. They indicated that carnivores and small herbivores were observed during the night at these four waterholes, while large herbivores and primates visited mostly during the day. Some species, such as elephants and large predators, may monopolise water sources and prevent other species from drinking [5]. This was not observed during the short period of this study, as large numbers and a variety of mammal species were observed at the waterholes, even at the peak of the drought. The three permanent waterholes are located on granitic soils in close vicinity to the study area. They constantly provided drinking water in the drought, since they are filled with underground water by solar pumps and windmills, while ephemeral areas dried up at the beginning of the dry season.

Common species found at the catena were also captured at the waterholes (Figure 1 and Table 2). The black and white rhino were captured at some of the waterholes, but this information is classified. Results from Sutherland et al. [20] correspond roughly to results from this study, i.e., they found an average of 6–7 species at De la Porte for the dry and wet seasons; 3–8 species at Kwaggaspan; and 10–11 species at Renosterkoppies waterholes (compare to Table 2). It is interesting that they reported total numbers of 3 for blue wildebeest, 5 for hippo and 1 warthog that correspond exactly with group sizes observed in this current study. They also listed 3 grey duiker, 2 leopards, 7 scrub hares and 5 side-striped jackals observed in total, but these might be different events and not different individuals. These species were also noted with lower numbers in this study but mostly solitary or in pairs. Species listed by Sutherland et al. [20] that were not captured at the waterholes in this study were cheetah, black-backed jackal, honey badger, southern springhare, steenbok, serval, large spotted genet and white-tailed mongoose that were all captured during the dark part of the night. Some of these species listed were found at the catena or granite outcrops (Figure 1), and only the first four species were completely absent from our results. It is not mentioned at which waterhole these animals were captured, and it may be that these species are present at the N'waswitshaka waterhole included by these authors or that we did not record them during the night.

Based on number of events (Table 2), blue wildebeest, impala and greater kudu seemed to prefer Renosterkoppies waterhole which is the nearest to the catena studied (3.3 km in a straight line), while plains zebra was only captured at the Kwaggaspan waterhole (4.6 km in straight line). Hippopotamuses depend on water-related landscapes but are also known to graze several kilometres from water [18,22,31]. In the study area, hippos were observed quite far from the perennial water sources (i.e., the closest point to the Sabi River is 13.2 km in a straight line) and were found in the drainage line (riparian zone) (Figure 1) and at two of the nearest waterholes only during 2016 at the peak of the drought. Elephant are also known to occur 15–24 km from water in drought conditions [31]. Hippo was captured in 2016 at the De la Porte waterhole, which was the furthest from the catena studied, namely 6.2 km in a straight line, while waterbuck only occurred at the Kwaggaspan waterhole during 2017 at the end of the drought. These latter two species have water-related habitat requirements and are known to stay in close proximity of surface water. The grey duiker is known to be water independent and was not captured while drinking at the Kwaggaspan waterhole, it was just passing by the camera probably en route to reach specific forage patches in its territory.

Water-dependent species require water regularly and can range 5–6 km (non-mobile species: impala, bushbuck, warthog) or up to 10 km distances (mobile species: buffalo, zebra, blue wildebeest) from water [31,32]. Some of the listed species (Table 2) prefer to stay close to surface water, e.g., waterbuck, and they were not expected at the catena which is further away. Buffalo and waterbuck were found farther from surface water where lower forage quantity was available in dry years, while zebra, wildebeest, elephant and impala were found farther with lower forage quality [5,28,33,34]. In the nutrient poor granites (compared to the basalts), the quality and quantity of food are lower, and animals have to travel further from water than in basalt areas where their forage and water needs can be satisfied closer to the waterholes [27].

Irregular availability of water in semi-arid savannas can affect the distribution of mammals, together with the distribution of forage and certain habitat effects. Animals often migrate long distances in response to food and water availability [9,27,35]. In areas where permanent water is provided artificially, the significance of water as a limiting resource or as driver for migration patterns in dry periods is suppressed, as is the case in KNP [5]. If herbivores aggregate near waterholes, predators may also concentrate in the vicinity [26]. Grazers occurred closer to water than expected by Smit et al. [27], while mixed feeders and browsers occurred at high densities close to the rivers but at low densities close to waterholes. Waterbuck and elephant are very water-dependent and tend to be more associated with rivers than with artificial waterholes, while buffalo were found to use both rivers and waterholes. Rivers provide water, habitat and forage to these species [27]. The three waterholes are less than 6 km from the centre of the study area and the closest point to the perennial Sabi River is 13.2 km from the study area, so the area is still within reach of many of the water-dependent species, especially the more mobile species. According to Smit and Grant [5] (p. 69), "An increase in distance to the closest water source implies higher costs in terms of energy spent during travelling, as well as energy "not gained" due to time spent travelling instead of foraging." Additionally, they also state that a larger distance travelled may increase the chances of encountering predators and increase calf mortality due to exhaustion.

The distance to permanent water sources together with many other factors plays a role in the use of ephemeral sources by a mammal species, including the number of animals that can be supported by the ephemeral source, quality and accessibility of the water, other specific conditions that can be limiting and the period of time that the source contains water [31]. Ephemeral water source location depends on the topography and geology of an ecosystem together with rainfall fluctuations in that area. These water sources differ over a range of temporal scales, from small pans or mud wallows formed by rainstorms during the dry season that only contain water for a few days or weeks to pools formed in seasonal rivers that are available throughout the dry season [31,34].

Mud wallows in the study area are located on the sodic patch (with shallow rock beds) and in the riparian area. It contained water during the 2015 and 2016 survey periods but not during 2017. Mud wallows are depressions (small pans) in the ground that temporarily fill with rainwater, hold the water for a certain period thereafter and are maintained by wallowing of animals [15]. These wallows were favoured by species that cover their bodies with mud (i.e., buffalo, elephant, warthog, black and white rhino), while most of the mammals also quenched their thirst at the larger temporary holes on the catena (15 species in total were observed using mud wallows).

Other authors found that herbivores dissipate in KNP and use waterholes less after the first rains when pans and other ephemeral depressions fill with water [20,36]. This enables animals to use areas slightly further away from permanent water. If cover and forage are abundant, animals will probably stay close to the perennial water sources and minimize predation risk by travelling to satisfy water requirements, while the scarcity of food and cover in these areas as result of drought can result in animals dispersing to areas with ephemeral water sources [31]. This will depend on the animal's body size and water dependence. Redfern et al. [31] hypothesized that the distribution of large and mobile water dependent herbivores will be more influenced by ephemeral water sources than perennial sources, while distribution of smaller herbivores that have a fixed territory or specific water related habitat requirements will be limited by perennial water sources.

4.3. Feeding Guild and Body Size

Foraging commonly makes up the largest portion of an animal's mobile activity, especially for large mammals, and basically determines their space occupation [30]. The mammals listed in this study were therefor divided into two main categories, namely, feeding guild and body size. Activity patterns of the mammals observed during the study period were covered in another study [15].

Different feeding guilds were represented in the data, from herbivores to carnivores, scavengers (i.e., jackal, hyena, etc.), insectivores and omnivores (Table 1). These different feeding guilds can indicate a working ecosystem on the catena, especially if it can be linked to different habitats in the zones that include certain biotic and abiotic factors of the catena ecosystem. Herbivores were subdivided into the following dietary classes: grazers that feed mostly on grass and herbaceous material; browsers that feed mostly on leaf material from woody plants, including the concentrate selectors (steenbok and duiker) that select the more nutritious plant parts and fruit; mixed feeders that feed on grass and browse material; and general vegetarians that feed on plant material, such as roots, geophytes, bark, fruits, nuts, etc. Insectivores that feed mostly on insects such as ants, termites and other invertebrates; and omnivores (feed on plant and animal material) [22,37] were also noted.

The order Carnivora contains predators, scavengers and animals that also include large proportions of insects in their diet, such as civet, genet, banded mongoose, dwarf mongoose, slender mongoose and white-tailed mongoose found in this study. Few species are strictly carnivorous; most supplement their diet with fruits, bone, carrion, insects and other invertebrates making it difficult to group them into insectivores, omnivores or carnivores [22]. The majority of predators use the resource 'live prey', but it can be classified into various prey classes with the following characteristics: prey type, size, sex, age, activity periods and habitat needs [38]. Ambush predators (lion and leopard) prefer to hunt in areas with more cover and they are usually more successful during the night [26]. Cursorial predators (wild dog and hyena) need more open areas to run the prey down to exhaustion and then kill it. Some predators (serval, jackal) specialize in rodents and birds and use pouncing and jumping techniques to hunt [22]. All of these carnivores were present in the study area but not limited to specific zones.

Mammalian herbivores that feed on different plant types and parts can optimally utilize the diversity in vegetation resources in the same space [37]. Selective preferences for specific food resource types can generally be expressed by the herbivore either staying in profitable areas for longer times and/or by returning frequently to such areas [26]. Interspecific competition mostly occurs in the dry season when adequate quality forage becomes depleted due to most plants being dormant [28].

Mammal species that are less common generally specialize on a narrow range of resource types, while species that are more abundant usually exploit a wider range of resource types and habitat conditions, according to the niche breadth concepts [3]. Du Toit [8] and Macandza et al. [39] list several authors that describe the various scales at which distinctions in habitat occupation (presence or absence of mammal species) can be recognized, from landscape or biome to vegetation types, habitat to plant species composition of local patches and specific plant parts. Animal factors such as social and behavioural aspects, home range sizes, high- or low-density species, water-dependence, predation, age, sex, among others, also play a role in areas occupied [15]. At landscape scale, some features of resource heterogeneity include vegetation structure of woody plants, soil fertility, distance to water, topography, geomorphology, etc. [39]. Thus, according to Macandza et al. [39] (p. 176), "coexistence among large herbivores may be enabled by distinctions in resource use at one or more of these scales, underlain by differences in body size and morphological adaptations".

A large variety and range in body sizes were indicated in Table 1, from the smallest ruminant antelope, the steenbok (0.5 m shoulder height, 11 kg), to the largest land mammal, the elephant (4 m, 5000 kg), is present in the study area, spanning three orders of magnitude in body mass [8], while carnivores range from dwarf mongoose (0.07 m, 0.27 kg) to lion (1.2 m, 190 kg). There are various mechanisms that regulate food intake that can be connected to body size. Small herbivores usually require less feeding time, and they can spend relatively more time to search for higher quality food items, giving them a patchy distribution. Larger herbivores need to maintain their intake when forage is limited (quantity) and may be forced to consume a lower quality diet resulting in a more even distribution over larger areas in general [10,31,40]. Small mammal distribution may be influenced more by location of cover than distribution of larger species [31], but they may competitively displace the larger guild members away from feeding sites in a horizontal plane for grazers and vertical plane for browsers [8]. However, distribution of large and small species may actually overlap in areas that are resource rich [28,40]. The granite landscape is not nutrient rich compared to other areas in the KNP and only the sodic patch, drainage lines and termite mounds [19] have more nutritious vegetation than the surrounding study area (as explained earlier), but still a variety of mammal species and body sizes were recorded. These species may occupy the same local area because they use different food sources at different heights (based on their body size, morphological and digestive adaptations), and they are not limited to this area only, since they can include other areas in the larger landscape (distribution differences) at different times to meet their nutritional needs and so reduce competition.

4.4. General Discussion

There were small differences between observations in the different survey periods (Figure 1). The Pareto charts for each survey period included in Janecke and Bolton [15] clearly indicate the differences in species presence and frequency of observations. Most of the differences can probably be ascribed to the extreme drought that reached its peak during 2016 and the beginning of 2017 in this area. Some of the smaller mammals were not observed during this period or with low numbers of events (four mongoose species, tree squirrel, scrub hare), while other species might have already been absent at the start of the drought before the study commenced. Vegetation was sparse and the ground bare during the drought. As a result, it could not provide enough cover for mammals, while food resources became limited [15,25,41]. There was variation found in the use of grass communities by the grazers in KNP during this specific drought that was driven by erratic rainfall in different areas of the park, an increase in nutritional stress and in the perceived predation risk and by large-scale variation in the severity of the drought in different areas [21]. Donaldson et al. [21] reported that grazers used the nutritious basalt grazing lawns (Satara area of KNP) much more during the early drought phases in 2014–2015 but that they left in the second drought season (2015–2016) and did not return even after rainfall events.

Food selection, movement rates, dispersion of larger herbivores and other mechanisms that happen at small scales can contribute to explain foraging patterns observed across the landscape at a larger scale. Selection of specific patches and feeding sites (collection of patches in a spatial area that animals forage in during one feeding bout) may play a role in grazing distribution patterns [10]. This may also explain some of the mammal presence or absence in the study area. In heterogeneous habitats, herbivores tend to select nutrient-rich sites more often than sites that are less productive. Large herbivores have spatial memories that are quite accurate, and they will avoid areas with little or no food, as well as patches where food has been depleted based on their memory [10].

Kruger National Park is a large, open park (almost 20,000 km^2—[18]) where the movement of mammals through the park is not limited, except if mammals are bound by their own intraspecies territorial boundaries or by available space [29,42] and food (due to high numbers of animals, geomorphological features or accessibility of food sources) or if they are habitat specific (meaning they can only survive in a specific environment, or vegetation type [8,39]) or influenced by predation risk, to name a few. Thus, a large variety of species that are present in the bigger park can also be present on a small scale at the granite catenas and vice versa. Movement of animals is a key mechanism that allows them to cope with the highly dynamic resource productivity in environments that are spatially heterogeneous [26,35].

5. Conclusions

A total of 33 mammal species were observed in the study area, including the two rhinoceros species. A variety of mammals were present in the study area, but the basic list of mammals provided here can be expanded if a larger area in the same supersite can be included, more camera traps (especially at the waterholes) and longer survey periods (including seasonal and climatic differences) can be used. On the one granitic catena (hillslope) alone, 23 mammal species were found across all four zones, 21 species at the waterholes and their surrounding areas, 16 species at the granite-boulder outcrops and 13 species at the mud wallows, excluding the two rhino species. Some of these species were similar between different zones and areas. A few species only occurred at the outcrops and waterholes but were not observed on the catena. The hypothesis, that specific mammal species might frequent or associate with certain zone/s on the catena, could not be investigated properly in this current study. This could be due to zones being too small to limit mammals to a specific zone in such an open, natural system. Most mammals also need to pass through certain zones to reach waterholes located outside the catena studied. Some species stay close to the waterhole, while others move larger distances away to feed. The two closer waterholes were less than 4.5 km from the centre of the study area and that is close enough for non-mobile water dependent species to reach, while the furthest one was 6 km away and can even be reached by mobile water-dependent species.

If more than one catena can be included in future studies, it could provide a better understanding of how these mammals use similar zones on other hillslopes in the area and how the various feeding guilds interact with the environment on a landscape scale. Small differences were found in mammal species presence between the three survey periods but the extreme drought possibly limited species richness. The presence of a variety of different sized mammals (small to mega-sized) from different feeding guilds (herbivores to carnivores and omnivores) and dietary classes (grazer to browser and mixed feeder) most probably indicate a functioning ecosystem consisting of various interlinked trophic levels.

Some mammal species may not have been recorded because of the extreme drought during the study period (lack of predator cover and food availability forced them to leave the area), normal movement or migration (they are only present during a certain season or opportunistically arrive when conditions are right—which might have been outside the survey periods) or simply because they travelled outside the view of the cameras in the study area. There is thus scope for future studies to add to the basic list of mammals observed during this study to make it more complete, such as to include a longer time period, a wetter period with normal precipitation, seasonal variation, other types of animals (i.e., birds, reptiles, amphibians, small rodents, invertebrates), etc. It is also known that the numbers of certain mammal species (group sizes) can fluctuate seasonally due to availability of food—in summer they aggregate into large groups, while in winter or when conditions become less favourable; they break up into smaller groups and disperse more widely in search of food. All of these factors and more can contribute to the total variety and species richness of catenas inside the Southern Granite Supersite during different climatic, seasonal or environmental conditions. All the knowledge from this study can be beneficial in monitoring and conservation of species in Kruger National Park.

Funding: This research was partially funded by the University of the Free State (UFS) Strategic Research Fund and the NRF Thuthuka Grant (TTK14052267732). The UFS Central Research Fund provided the funds to publish this article.

Institutional Review Board Statement: The study was conducted according to the guidelines of the Declaration of Helsinki and approved by the University of the Free State Interfaculty Animal Ethics Committee (UFS-AED2019/0121 on 22 August 2019).

Informed Consent Statement: Not applicable.

Data Availability Statement: There is no data available from this specific study, but some of the larger multidisciplinary project's data are available at the SANParks Scientific Services repository.

Acknowledgments: Jeremy Bolton, camera trapping specialist from Bushcam Consulting is acknowledged for his part in camera trapping procedures and assistance with data processing of some results. Errol Cason from Department of Animal Science, UFS, is thanked for his assistance with Table 3. The multidisciplinary team of the UFS is acknowledged. Staff at Scientific Services in Skukuza and the game guards from SANParks are also thanked for all their assistance.

Conflicts of Interest: The author declares no conflict of interest. The founding sponsors had no role in the design of the study; in the collection, analyses or interpretation of data; in the writing of the manuscript or in the decision to publish the results.

References

1. Maurer, B.A. The relationship between distribution and abundance in a patchy environment. *Oikos* **1990**, *58*, 181–189. [CrossRef]
2. Gertenbach, W.P.D. Landscapes of the Kruger National Park. *Koedoe* **1983**, *26*, 9–121. [CrossRef]
3. Brown, J.H. On the relationship between abundance and distribution of species. *Am. Nat.* **1984**, *124*, 255–279. [CrossRef]
4. Gaston, K.J. The multiple forms of the interspecific abundance-distribution relationship. *Oikos* **1996**, *76*, 211–220. [CrossRef]
5. Smit, I.P.J.; Grant, C.C. Managing surface-water in a large semi-arid savanna park: Effects on grazer distribution patterns. *J. Nat. Cons.* **2009**, *17*, 61–71. [CrossRef]
6. Tosh, C.A.; Reyers, B.; Van Jaarsveld, A.S. Estimating the abundances of large herbivores in the Kruger National Park using presence-absence data. *Anim. Cons.* **2004**, *7*, 55–61. [CrossRef]
7. Siebert, F.; Scogings, P. Browsing intensity of herbaceous forbs across a semi-arid savanna catenal sequence. *S. Afr. J. Bot.* **2015**, *100*, 69–74. [CrossRef]
8. Du Toit, J.T. Large herbivores and savanna heterogeneity. In *The Kruger Experience: Ecology and Management of Savanna Heterogeneity*; Du Toit, J.T., Biggs, H.C., Rogers, K.H., Eds.; Island Press: Washington, DC, USA, 2003; pp. 292–309.
9. Bergstrom, R.; Skarpe, C. The abundance of large wild herbivores in a semi-arid savanna in relation to seasons, pans and livestock. *Afr. J. Ecol.* **1999**, *37*, 12–26. [CrossRef]
10. Bailey, D.W.; Gross, J.E.; Laca, E.A.; Rittenhouse, L.R.; Coughenour, M.B.; Swift, D.M.; Sims, P.L. Mechanisms that result in large herbivore grazing distribution patterns. *J. Range Manag.* **1996**, *49*, 386–400. [CrossRef]
11. Smit, I.P.J.; Riddell, E.S.; Cullum, C.; Petersen, R. Kruger National Park research supersites: Establishing long-term research sites for cross-disciplinary, multiscaled learning. *Koedoe* **2013**, *55*, Art. #1107, 7p. Available online: https://koedoe.co.za/index.php/koedoe/article/view/1107/1489 (accessed on 12 February 2021). [CrossRef]
12. Weil, R.R.; Brady, N.C. *The Nature and Properties of Soils*, 15th ed.; Pearson Education Limited: Harlow, UK, 2016.
13. Khomo, L.M.; Rogers, K.H. Proposed mechanism for the origin of sodic patches in Kruger National Park, South Africa. *Afr. J. Ecol.* **2005**, *43*, 29–34. [CrossRef]
14. Bouwer, D.; Le Roux, P.A.L.; Van Tol, J. Identification of hydropedological flowpaths in Stevenson-Hamilton catena from soil morphological, chemical and hydraulic properties. *Koedoe* **2020**, *62*, a1584. Available online: https://koedoe.co.za/index.php/koedoe/article/view/1584 (accessed on 12 February 2021). [CrossRef]
15. Janecke, B.B.; Bolton, J.G. Variation in mammal diversity and habitat affect heterogeneity and processes of a granite catena. *Koedoe* **2020**, *62*, a1592. Available online: https://koedoe.co.za/index.php/koedoe/article/view/1592 (accessed on 12 February 2021). [CrossRef]
16. Janecke, B.B.; Van Tol, J.; Smit, I.P.J.; Van Aardt, A.C.; Riddell, E.S.; Seaman, M.T.; Swart, W.J.; Du Preez, P.J.; Le Roux, P.A.L. Biotic and abiotic connections on a granitic catena: Framework for multidisciplinary research. *Koedoe* **2020**, *62*, a1600. [CrossRef]
17. Mucina, L.; Rutherford, M.C. (Eds.) *The Vegetation of South. Africa, Lesotho and Swaziland*; Strelizia 19; South African National Biodiversity Institute: Pretoria, South Africa, 2006.
18. Hilton-Barber, B.; Berger, L.R. *Prime Origins Guide to Exploring Kruger. Your Key to Unlocking Kruger's Wildlife Treasure*; Prime Origins: Cape Town, South Africa, 2007.
19. Grant, C.C.; Scholes, M.C. The importance of nutrient hot-spots in the conservation and management of large wild mammalian herbivores in semi-arid savannas. *Biol. Conserv.* **2006**, *130*, 426–437. [CrossRef]

20. Sutherland, K.; Ndlovu, M.; Perez-Rodriguez, A. Use of artificial waterholes by animals in the southern region of the Kruger National Park. *Afr. J. Wildl. Res.* **2018**, *48*, 1–14. [CrossRef]
21. Donaldson, J.E.; Parr, C.L.; Mangena, E.H.; Archibald, S. Droughts decouple African savanna grazers from their preferred forage with consequences for grassland productivity. *Ecosystems* **2020**, *23*, 689–701. [CrossRef]
22. Estes, R.D. *The Behavior Guide to African Mammals: Including Hoofed Mammals, Carnivores, Primates*, 20th ed.; University of California Press: Berkeley, South Africa, 2012.
23. Oksanen, J.; Blanchet, F.G.; Kindt, R.; Legendre, P.; Minchin, P.; O'Hara, B.; Simpson, G.; Solymos, P.; Stevens, H.; Wagner, H. Vegan: Community Ecology Package; R Package 2.2-1. 2. 1-2. 2018. Available online: https://www.researchgate.net/publication/313502495_Vegan_Community_Ecology_Package (accessed on 12 February 2021).
24. Colwell, R.K. Biodiversity: Concepts, Patterns and Measurement. In *The Princeton Guide to Ecology*; Simon, A.L., Ed.; Princeton University Press: Princeton, NJ, USA, 2009; pp. 257–263.
25. Janecke, B.B. Vegetation structure and spatial heterogeneity in the Granite Supersite, Kruger National Park. *Koedoe* **2020**, *62*, a1591. Available online: https://koedoe.co.za/index.php/koedoe/article/view/1591 (accessed on 12 February 2021). [CrossRef]
26. Martin, J.; Benhamou, S.; Yoganand, K.; Owen-Smith, N. Coping with Spatial Heterogeneity and Temporal Variability in Resources and Risks: Adaptive Movement Behaviour by a Large Grazing Herbivore. *PLoS ONE* **2015**, *10*, e0118461. [CrossRef]
27. Smit, I.P.J.; Grant, C.C.; Devereux, B.J. Do artificial waterholes influence the way herbivores use the landscape? Herbivore distribution patterns around rivers and artificial water sources in a large African savanna park. *Biol. Conserv.* **2007**, *136*, 85–99.
28. Redfern, J.V.; Ryan, S.J.; Getz, W.M. Defining herbivore assemblages in the Kruger National Park: A correlative coherence approach. *Oecologia* **2006**, *146*, 632–640. [CrossRef] [PubMed]
29. Carnaby, T. *Beat About the Bush—Mammals*, 2nd ed.; Jacana Media: Johannesburg, South Africa, 2010.
30. Owen-Smith, N.; Martin, J. Identifying space use at foraging arena scale within the home ranges of large herbivores. *PLoS ONE* **2015**, *10*, e0128821. [CrossRef]
31. Redfern, J.V.; Grant, C.C.; Gaylard, A.; Getz, W.M. Surface water availability in the management of herbivore distributions in an African savanna ecosystem. *J. Arid Environ.* **2005**, *63*, 406–424. [CrossRef]
32. Grossman, D.; Holden, P.L.; Collinson, R.F.H. Veld management on the game ranch. In *Veld management in South Africa*; Tainton, N.M., Ed.; University of Natal Press: Pietermaritzburg, South Africa, 1999; pp. 261–279.
33. Redfern, J.V.; Grant, R.; Biggs, H.; Getz, W.M. Surface-water constraints on herbivore foraging in the Kruger National Park, South Africa. *Ecology* **2003**, *84*, 2092–2107. [CrossRef]
34. Gaylard, A.; Owen-Smith, N.; Redfern, J. Surface water availability: Implications for heterogeneity and ecosystem processes. In *The Kruger Experience: Ecology and Management of Savanna Heterogeneity*; Du Toit, J.T., Biggs, H.C., Rogers, K.H., Eds.; Island Press: Washington, DC, USA, 2003; pp. 171–188.
35. Fryxell, J.M.; Sinclair, A.R.E. Causes and consequences of migration by large herbivores. *Trends Ecol. Evol.* **1988**, *3*, 237–241. [CrossRef]
36. Thrash, I.; Theron, G.; Bothma, J.D.P. Dry season herbivore densities around drinking troughs in the Kruger National Park. *J. Arid. Environ.* **1995**, *29*, 213–219. [CrossRef]
37. Bothma, J.d.P.; Van Rooyen, N.; Van Rooyen, M.W. Using diet and plant resources to set wildlife stocking densities in African savannas. *Wildl. Soc. Bull.* **2004**, *32*, 840–851. [CrossRef]
38. Volmer, R.; Hertler, C. The effect of competition on shared food resources in carnivore guilds. *Q. Int.* **2016**, *413*, 32–43. [CrossRef]
39. Macandza, V.A.; Owen-Smith, N.; Cain, J.W. Habitat and resource partitioning between abundant and relatively rare grazing ungulates. *J. Zool.* **2012**, *287*, 175–185. [CrossRef]
40. Du Toit, J.T.; Owen-Smith, N. Body size, population metabolism, and habitat specialization among large African herbivores. *Am. Nat.* **1989**, *113*, 736–740. [CrossRef]
41. Malherbe, J.; Smit, I.; Wessels, K.; Beukes, P. Recent droughts in the Kruger National Park as reflected in the extreme climate index. *Afr. J. Range For. Sci.* **2020**, *37*, 1–17. [CrossRef]
42. Joubert, S.C.J. Animal behavior. In *Game Ranch Management*, 6th ed.; Bothma, J.d.P., Du Toit, S.G., Eds.; Van Schaik Publishers: Pretoria, South Africa, 2016; pp. 385–391.

diversity

MDPI

Article

Integrative Descriptions of Two New Tardigrade Species along with the New Record of *Mesobiotus skorackii* Kaczmarek et al., 2018 from Canada

Pushpalata Kayastha [1], Milena Roszkowska [1,2], Monika Mioduchowska [3,4], Magdalena Gawlak [5] and Łukasz Kaczmarek [1,*]

1 Department of Animal Taxonomy and Ecology, Faculty of Biology, Adam Mickiewicz University in Poznań, Uniwersytetu Poznańskiego 6, 61-614 Poznań, Poland; pushpalata.kayastha@gmail.com (P.K.); mil.roszkowska@gmail.com (M.R.)
2 Department of Bioenergetics, Faculty of Biology, Adam Mickiewicz University in Poznań, Uniwersytetu Poznańskiego 6, 61-614 Poznań, Poland
3 Department of Genetics and Biosystematics, Faculty of Biology, University of Gdańsk, Wita Stwosza 59, 80-308 Gdańsk, Poland; monika.mioduchowska@ug.edu.pl
4 Department of Marine Plankton Research, Institute of Oceanography, University of Gdańsk, Marszałka Piłsudskiego 46, 81-378 Gdynia, Poland
5 Institute of Plant Protection—National Research Institute, Władysława Węgorka 20, 60-318 Poznań, Poland; m.gawlak@iorpib.poznan.pl
* Correspondence: kaczmar@amu.edu.pl

ZooBank: http://zoobank.org/urn:lsid:zoobank.org:act:F4ACE4EB-4872-43C8-8997-BD40CCBE17EE.
ZooBank: http://zoobank.org/urn:lsid:zoobank.org:act:5617E45B-865A-42DA-A70A-DB6FE1A52163.

Citation: Kayastha, P.; Roszkowska, M.; Mioduchowska, M.; Gawlak, M.; Kaczmarek, Ł. Integrative Descriptions of Two New Tardigrade Species along with the New Record of *Mesobiotus skorackii* Kaczmarek et al., 2018 from Canada. *Diversity* **2021**, *13*, 394. https://doi.org/10.3390/d13080394

Academic Editor: Michael Wink

Received: 20 July 2021
Accepted: 19 August 2021
Published: 22 August 2021

Publisher's Note: MDPI stays neutral with regard to jurisdictional claims in published maps and institutional affiliations.

Abstract: Two new tardigrade species from a moss sample collected in Canada, one representing *Macrobiotus hufelandi* complex and the second one belonging to the genus *Bryodelphax*, are described. Integrative analysis was undertaken based on morphological and morphometric data (using both light and scanning electron microscopy (SEM)) combined with multilocus molecular analysis (nuclear sequences, i.e., 18S rRNA, 28S rRNA and ITS-2 as well as mitochondrial COI barcode sequences). Based on COI sequences, *Macrobiotus birendrai* sp. nov. is most similar to *Mac. canaricus* (p-distance 17%), whereas *Bryodelphax mareki* sp. nov. is most similar to *Bry. parvulus* (p-distance 16%). Both species differ also from their congeners in some morphological and morphometric characters of adults and/or details of egg chorion. Additionally, a large population of *Mesobiotus skorackii* was found in the sample and this is the first report of this species outside its terra typica in Kirghizia. The original description of this species was prepared based solely on the morphology and morphometry, therefore, here we provide updated data for this species enclosing morphometric and molecular data for the Canadian population.

Keywords: *Bryodelphax mareki* sp. nov.; DNA barcoding; Eutardigrada; Heterotardigrada; *Macrobiotus birendrai* sp. nov.; water bears

1. Introduction

Canada is the second largest country in the world which extends its longitude from approximately 52° to 141° W to latitude approximately 42° to 83° N. It has such a distance that spans in six time zones and has a wide variety of climates. The highest peak in Canada which is Mount Logan reaches 5959 m asl and the country's landform structure can be considered a vast basin. Additionally, people living in two-thirds of the area experience very cold winters and short, cool summers. However, the interior plains of central southern area come with very cold winters, hot summers, and relatively sparse precipitation. Nonetheless, climate with hot, humid summers and cold, snowy winters also prevails in Southern Ontario and Quebec. Except for the west coast, all of Canada has a winter season with average temperatures below freezing and with continuous snow cover (https://www.britannica.com/place/Canada (accessed on 18 June 2021).

Tardigrada, also commonly known as water bears, inhabit in terrestrial and aquatic (freshwater and marine) environments. They can be found on aquatic plants and/or in lichens, leaf litter, mosses, soil, sediments [1–3]. To date, more than ca. 1300 species of tardigrades have been described throughout the world [4–7]. The genus *Bryodelphax* [8] is unique amongst Echiniscidae with some peculiar apomorphies like presence of 10 peribuccal papulae and plesiomorphies like ancestral type of the buccal apparatus, which makes *Bryodelphax* a good example of mosaic evolution in tardigrades [9,10]. Moreover, it is characterized by the presence of median plates 1 and 2 divided, median plate 3 not divided, and absence of notches on terminal plate. Up to now, 26 species were attributed to this genus [7]. The genus *Macrobiotus* [11] is one of the most species-rich and widespread genus in the phylum being also, the first formally described tardigrade genus. It is characterized by the presence of a rigid buccal tube with a straight ventral lamina lacking a ventral hook, 10 peribuccal lamellae, pharynx with two macroplacoids and microplacoid, symmetrical diploclaws and freely laid ornamented eggs [12]. Up to now, 118 species were attributed to this genus [7].

Tardigrade fauna of Canada is rather poorly known and up to now only 121 species have been reported from this region [13,14]. In this study, we applied integrative taxonomy for description of two new species from Canada belonging to the genus *Bryodelphax* and the *Macrobiotus hufelandi* complex. Moreover, we enriched this paper in additional molecular and morphometric data of the Canadian record of *Mesobiotus skorackii* Kaczmarek, Zawierucha, Buda, Stec, Gawlak, Michalczyk and Roszkowska [15], as the original description of this species was prepared based solely on the morphology and morphometry.

2. Materials and Methods

2.1. Sampling

A moss sample was collected in Banff National Park (AB, Canada) in March 2019. It was then packed in a paper envelope, dried at a temperature of ca. 20 °C and delivered to the Department of Animal Taxonomy and Ecology at the Faculty of Biology, Adam Mickiewicz University in Poznań (Poland). The tardigrade collection, extraction and mounting techniques followed the protocol of Stec et al. [16].

2.2. Microscopy and Imaging

In total, 163 animals (74 *Bryodelphax mareki* sp. nov. + 45 *Macrobiotus birendrai* sp. nov. + 44 *Mesobiotus skorackii*) and 31 eggs (12 *Macrobiotus birendrai* sp. nov + 19 *Mesobiotus skorackii*) were mounted on microscope slides in the Hoyer's medium, and then examined under Olympus BX41 Phase Contrast light Microscope (PCM) associated with Olympus SC50 digital camera (Olympus Corporation, Shinjuku-ku, Japan). The 44 animals and 8 eggs were prepared for scanning electron microscopy (SEM) analysis according to the protocol in Roszkowska et al. [17] and examined under high vacuum in Hitachi S3000N SEM. Thirty-one specimens were prepared for genotyping.

All figures were assembled in Corel Photo-Paint 2017. For deep structures that could not be fully focused in a single photograph, a series of 2–10 images were taken every ca. 0.5 µm and then manually assembled into a single deep-focus image in Corel Photo-Paint 2017.

2.3. Morphometrics and Morphological Nomenclature

All measurements are given in micrometers (µm]). Structures were measured only if their orientation was suitable. Body length was measured from the anterior extremity to the end of the body, excluding hind legs. The *sp* index in *Bryodelphax* is the ratio of the length of a given structure to the length of the scapular plate expressed as a percentage (length of structure × 100/length scapular plate) [18] and later independently proposed as the *psc* index by Fontoura and Morais [19]. Ventral plates configuration in *Bryodelphax* is given according to Kaczmarek et al. [20]. The types of bucco-pharyngeal apparatuses and claws of Macrobiotidae were classified according to Pilato and Binda [21]. All measurements and terminology of adults and eggs of Macrobiotidae were prepared according to Kaczmarek and Michalczyk [22] and Kaczmarek et al. [23]. Terminology describing the oral cavity armature (OCA) in *Macrobiotus* and *Mesobiotus* follows Michalczyk and Kaczmarek [24] and OCA morphotypes are given according to Kaczmarek and Michalczyk [22]. The macroplacoid length sequence in *Macrobiotus* and *Mesobiotus* was indicated according to Kaczmarek et al. [25]. The *pt* ratio is the ratio of the length of a given structure to the length of the buccal tube expressed as a percentage [26]. The terminology of cuticular bars in macrobiotid legs follows Kiosya et al. [27]. The classification of type of egg process, sculpture on egg processes, egg processes bases and egg shell surface between processes are given according to Kaczmarek et al. [23]. Genus abbreviations follow Perry et al. [28].

Morphometric data were handled using the "Parachela" ver. 1.8 and "Echiniscoidea" ver. 1.4 template available from the Tardigrada Register [29].

2.4. Comparative Material

For identification and differentiation of the new species, the key by Gąsiorek et al. [30] for the genus *Bryodelphax* and the key by Kaczmarek and Michalczyk [22] for the genus *Macrobiotus* were used. We also compared our new species with the type material of *Bry. aaseae* Kristensen, Michalczyk and Kaczmarek [9], *Bry. asiaticus* Kaczmarek and Michalczyk [31], *Bry. brevidentatus* Kaczmarek, Michalczyk and Degma [32], *Bry. olszanowskii* Kaczmarek, Parnikoza, Gawlak, Esefeld, Peter, Kozeretska and Roszkowska [33], *Bry. parvuspolaris* Kaczmarek, Zawierucha, Smykla and Michalczyk [20], *Mac. dulciporus* Roszkowska, Gawlak, Draga and Kaczmarek [34], *Mac. kazmierskii* Kaczmarek and Michalczyk, [35], *Mac. marlenae* Kaczmarek and Michalczyk [36], *Mac. paulinae* Stec, Smolak, Kaczmarek and Michalczyk [16], *Mac. pisacensis* Kaczmarek, Cytan, Zawierucha, Diduszko and Michalczyk [25], *Mac. polonicus* Pilato, Kaczmarek, Michalczyk and Lisi [37], *Mac. polypiformis* Roszkowska, Ostrowska, Stec, Janko and Kaczmarek [38], *Mac. porifini* Kuzdrowska, Mioduchowska, Gawlak, Bartylak, Kepel, Kepel and Kaczmarek [39], *Mac. sottilei* Pilato, Kiosya, Lisi and Sabella [40] and *Mac. wandae* Kayastha, Berdi, Miaduchowska, Gawlak, Łukasiewicz, Gołdyn and Kaczmarek [41].

2.5. Genotyping

Prior to DNA extraction, individual tardigrades from the three species were preliminarily identified in vivo using light microscopy (LM). Genomic DNA was extracted using a Chelex®® 100 resin (Bio-Rad, Hercules, CA, USA) extraction method [42] with modification in order to obtain voucher specimens, i.e., tardigrade exoskeletons [43]. After DNA extraction we performed morphological analysis following the protocol of Kaczmarek et al. [43]. Then, all exoskeletons were deposited in the collection of the Department of Animal Taxonomy and Ecology, Faculty of Biology, Adam Mickiewicz University in Poznań.

In total, four molecular markers were amplified: one mitochondrial gene, i.e., COI—the cytochrome oxidase subunit I; three nuclear markers, i.e., 18S rRNA—the small ribosome subunit and 28S rRNA—the large ribosome subunit as well as ITS-2—the internal transcribed spacer-2. The polymerase chain reaction (PCR) amplification was performed according to Kaczmarek et al. [44]. The sequences of primers applied to amplify molecular markers are listed in Table 1. All PCR reactions were conducted in a Biometra TProfessional thermocycler. Prior to the sequencing, the PCR products were treated with the FastAP Alkaline Phosphatase and thermosensitive Exonuclease I (Fermentas, Thermo Scientific, Waltham, MA, USA) according to the manufacturer's guidelines. Sanger DNA sequencing in both directions was performed by Macrogen (Amsterdam, The Netherlands).

Table 1. Primers used for PCR amplification of four DNA molecular markers of *Bryodelphax mareki* sp. nov., *Macrobiotus birendrai* sp. nov. and *Mesobiotus skorackii*.

DNA Fragment	Primer Name/Direction	Primer Sequence (5′-3′)	Source
COI	LCO1490/forward HCO2198/reverse	GGTCAACAAATCATAAAGATATTGG TAAACTTCAGGGTGACCAAAAAATCA	Folmer et al. [45]
18S rRNA	SSU01_F/forward SSU82_R/reverse	AACCTGGTTGATCCTGCCAGT TGATCCTTCTGCAGGTTCACCTAC	Sands et al. [46]
28S rRNA	28SF0001/forward 28SR0990/reverse	ACCCvCynAATTTAAGCATAT CCTTGGTCCGTGTTTCAAGAC	Mironov et al. [47]
ITS-2	ITS3/forward ITS4/reverse	GCATCGATGAAGAACGCAGC TCCTCCGCTTATTGATATGC	White et al. [48]

2.6. Comparative Genetic Analysis

Obtained mtDNA and nrDNA sequences were quality checked and consensus sequences were created for individual tardigrades in BioEdit v. 7.2.5 [49]. All COI sequences were translated into amino acid sequences to check against pseudogenes using the EMBOSS-TRANSEQ application [50,51]. The translation was performed with the invertebrate mitochondrial codon table. To verify the homology of the amplified DNA region, Basic Local Alignment Search Tool [52] searches at the National Centre for Biotechnology Information NCBI were applied.

All obtained sequences were deposited in GenBank and the accession numbers are listed in Table 2.

Table 2. The GenBank accession numbers of obtained molecular markers of three tardigrade species.

Species	GenBank Accession Numbers (Voucher Numbers of Specimens)			
	COI	**18S rRNA**	**28S rRNA**	**ITS-2**
***Bryodelphax mareki* sp. nov.**	MW655785-87 (CN8.17/S, CN8.25/S, CN8.28/S)	MW680639-40 (CN8.21/S, CN8.25/S)	MW680637-38 (CN8.21/S, CN8.22/S)	NA
***Macrobiotus birendrai* sp. nov.**	MW656266 (CN8.101/S)	MW680641 (CN8.101/S)	MW680644 (CN8.101/S)	MW680418 (CN8.101/S)
Mesobiotus skorackii	MW656257 (CN8.115/S)	MW680642-43 (CN8.22/S, CN8.115/S)	MW680636 (CN8.115/S)	NA

For molecular comparisons, all sequences of the mtDNA and nrDNA fragments of the genera *Bryodelphax*, *Macrobiotus* and *Mesobiotus* were downloaded from the GenBank and trimmed to the same length in BioEdit v. 7.2.5. The COI sequences could be unambiguously aligned without inserting gaps. In turn, the sequences of nrDNA were aligned using ClustalW Multiple Alignment tool [53] implemented in BioEdit v. 7.2.5. with default settings. Uncorrected pairwise distances were calculated using MEGA X [54].

3. Results

Taxonomic Account

Phylum: Tardigrada Doyère, 1840 [55]

Class: Heterotardigrada Marcus, 1927 [56]
Order: Echiniscoidea Richters, 1926 [57]
Family: Echiniscidae Thulin, 1928 [8]
Genus: *Bryodelphax* Thulin, 1928 [8]
***Bryodelphax mareki* sp. nov.**
(Table 3, Figures 1–4)

LSID http://zoobank.org/urn:lsid:zoobank.org:act:F4ACE4EB-4872-43C8-8997-BD40CCBE17EE

Table 3. Measurements (in µm) and *sc* values of selected morphological structures of adult females of ***Bryodelphax mareki* sp. nov.** mounted in Hoyer's medium (N—number of specimens/structures measured; RANGE refers to the smallest and the largest structure among all measured specimens; SD—standard deviation).

Character	N	Range						Mean		SD		Holotype	
		µm			*sc*			µm	*sc*	µm	*sc*	µm	*sc*
Body length	16	122	–	181	*577*	–	*780*	141	*689*	15	*60*	146	*749*
Scapular plate length	16	18.4	–	24.0		–		20.4	–	1.7	–	19.5	–
Head appendages lengths													
Cirrus *internus*	16	5.3	–	7.3	*26.8*	–	*34.6*	6.2	*30.6*	0.5	*2.7*	5.3	*27.0*
Cephalic papilla	15	4.1	–	5.9	*21.7*	–	*29.1*	4.9	*24.1*	0.5	*2.3*	4.3	*22.0*
Cirrus *externus*	16	11.1	–	13.1	*53.4*	–	*63.0*	11.9	*58.7*	0.7	*3.2*	11.1	*57.0*
Clava	12	4.3	–	6.2	*20.4*	–	*33.0*	5.2	*26.1*	0.5	*3.7*	4.8	*24.6*
Cirrus *A*	15	35.2	–	46.7	*174.1*	–	*212.9*	39.6	*193.6*	3.2	*10.7*	36.5	*186.9*
Cirrus *A*/body length ratio	15	25%	–	32%		–		28%	–	2%	–	25%	–
Cirrus *int*/*ext* length ratio	16	47%	–	56%		–		52%	–	3%	–	47%	–
Body appendages lengths													
Papilla on leg IV length	14	1.6	–	3.3	*8.1*	–	*14.3*	2.3	*10.9*	0.5	*1.9*	2.2	*11.3*
Claw 1 heights													
Branch	16	5.8	–	8.4	*30.3*	–	*42.0*	7.6	*37.5*	0.7	*2.8*	7.0	*35.9*
Spur	13	1.3	–	1.7	*5.4*	–	*8.2*	1.5	*7.2*	0.1	*0.8*	1.4	*7.2*
Spur/branch length ratio	13	16%	–	21%		–		18%	–	2%	–	0	–
Claw 2 heights													
Branch	16	6.7	–	8.6	*30.8*	–	*42.6*	7.6	*37.7*	0.5	*3.6*	7.5	*38.3*
Spur	14	1.0	–	1.5	*5.2*	–	*7.2*	1.3	*6.6*	0.1	*0.7*	1.4	*7.2*
Spur/branch length ratio	14	15%	–	20%		–		18%	–	1%	–	0	–
Claw 3 heights													
Branch	16	5.8	–	8.6	*30.6*	–	*45.1*	7.5	*37.3*	0.8	*4.0*	7.0	*35.6*
Spur	12	1.2	–	1.5	*5.4*	–	*7.5*	1.4	*6.6*	0.1	*0.8*	1.3	*6.7*
Spur/branch length ratio	12	16%	–	20%		–		18%	–	1%	–	0	–
Claw 4 lengths													
Branch	16	7.3	–	9.9	*38.3*	–	*48.2*	8.9	*43.8*	0.6	*3.7*	8.8	*45.2*
Spur	13	1.4	–	1.9	*6.3*	–	*8.9*	1.6	*7.8*	0.1	*0.7*	1.6	*8.0*
Spur/branch height ratio	13	15%	–	21%				18%	–	2%	–	0	–

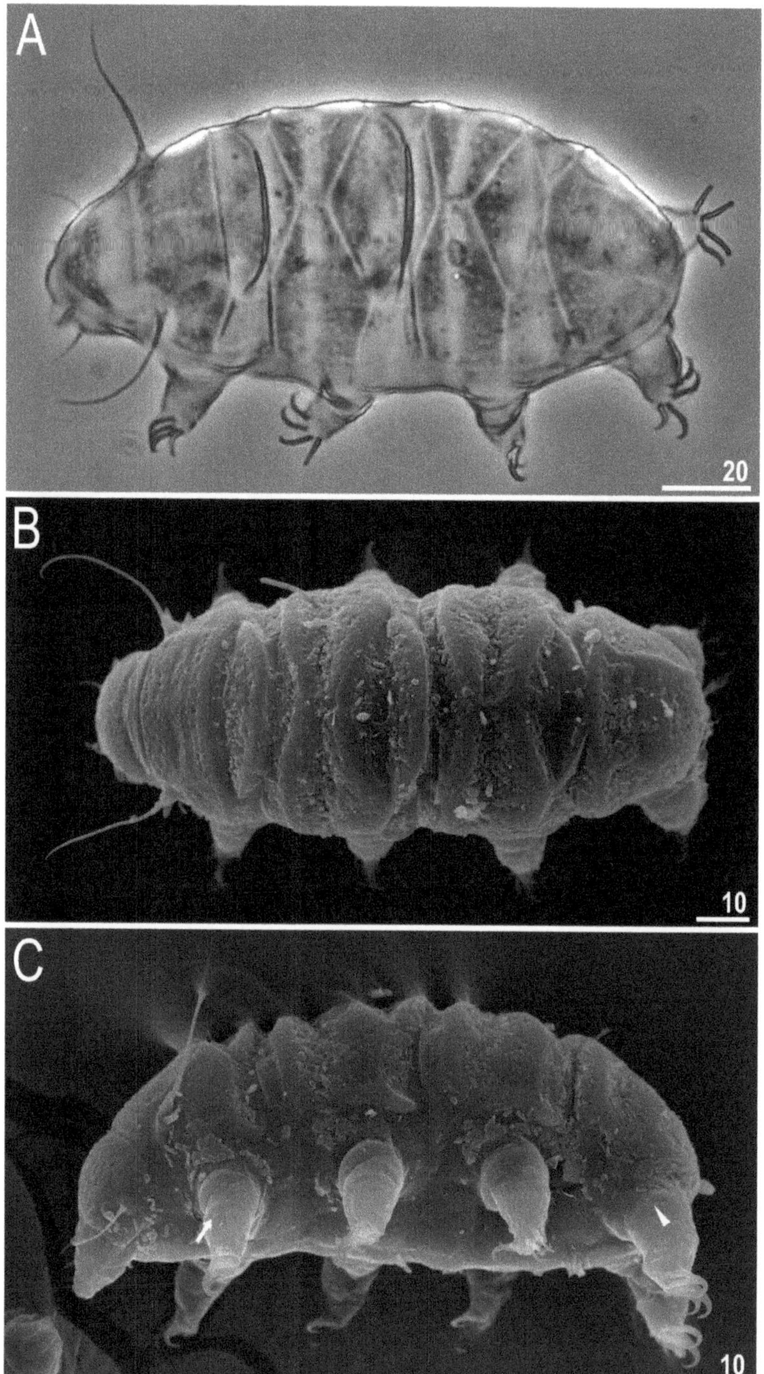

Figure 1. ***Bryodelphax mareki*** sp. nov.: habitus: (**A**) dorsal projection (holotype, PCM); (**B**) dorsal projection (SEM); (**C**) lateral projection; arrow indicates papilla-like structure on leg I, arrowhead indicates papilla on leg IV (SEM). Scale bars in μm.

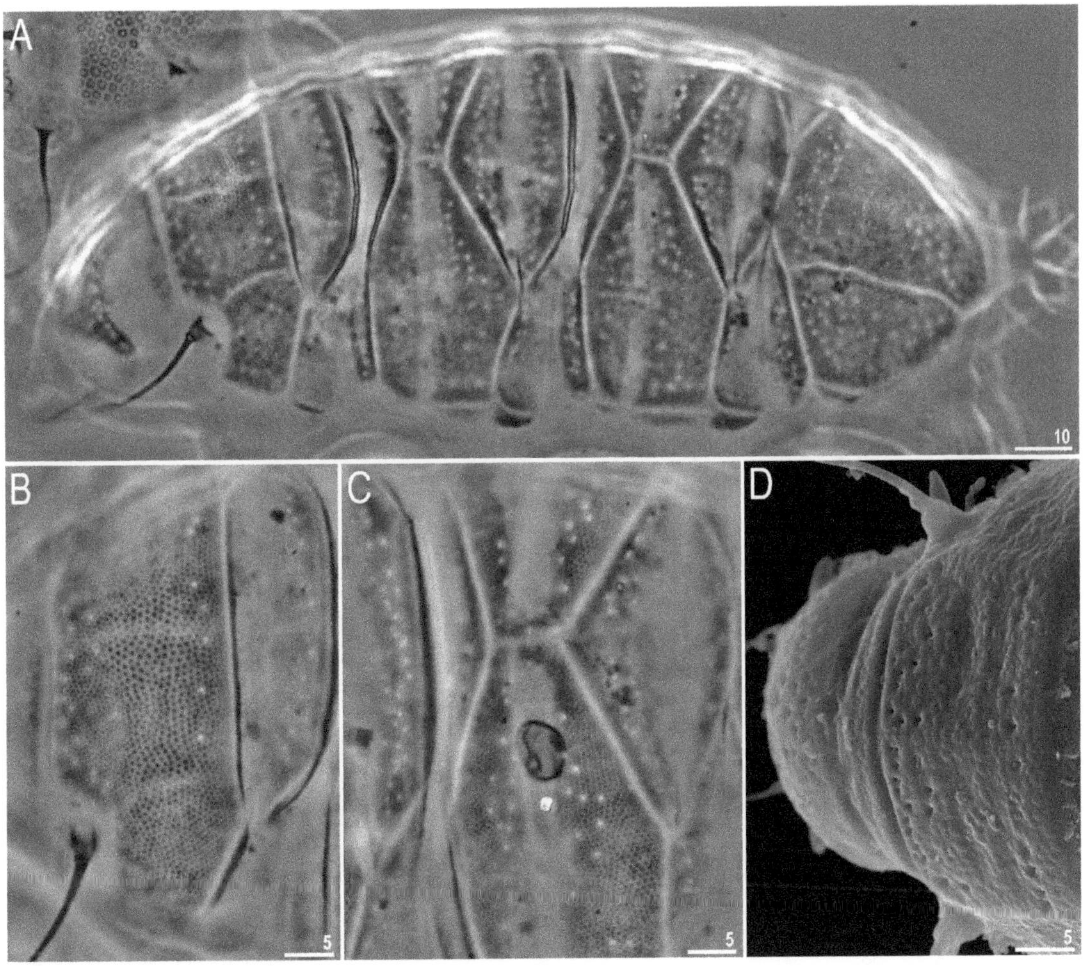

Figure 2. ***Bryodelphax mareki*** sp. nov.: (**A**) dorso-lateral view of the details of dorsal plates (paratype, PCM); (**B**) close up of the head plate (holotype, PCM); (**C**) close up of the pair plates II (holotype, PCM); (**D**) close up of the head and scapular plates (SEM). Scale bars in μm.

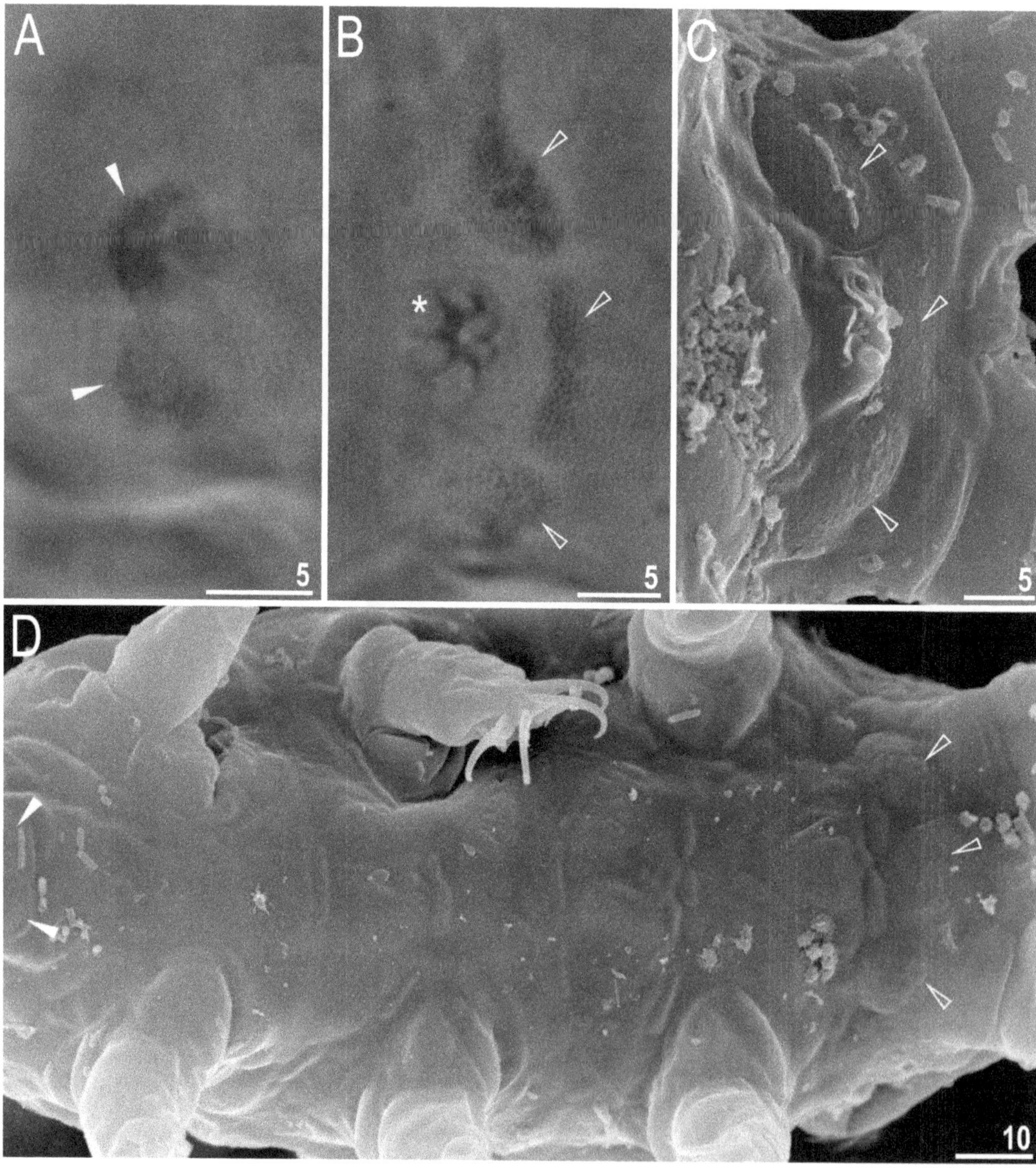

Figure 3. ***Bryodelphax mareki*** sp. nov.: ventral plates: (**A**) two ventral plates under head (filled arrowheads) (paratype, PCM); (**B**,**C**) three ventral plates around gonophore (empty arrowheads); asterisk indicates gonophore (paratypes, PCM and SEM, respectively); (**D**) ventral projection visible in SEM; filled arrowheads indicate two ventral plates under head, empty arrowheads indicate ventral plates around gonophore. Scale bars in μm.

Figure 4. ***Bryodelphax mareki*** **sp. nov.:** (**A**,**B**) claws of leg I with visible papilla-like structure (arrow) (paratypes, PCM and SEM, respectively); (**C**,**D**) claws of leg IV; arrowhead indicates papilla on leg IV (paratypes, PCM and SEM, respectively). Scale bars in μm.

Type Locality: 51°24′21″ N, 116°14′27″ W, 1900 m asl, Canada, Alberta, Banff National Park, near east end of the Louise Lake, moss on stone, May 2019, leg. Milena Roszkowska and Łukasz Kaczmarek.

Material examined: The 74 animals, i.e., holotype + 73 paratypes (females: 37; undefined sex: 34 and 2 exuviae) mounted on microscope slides in Hoyer's medium, 40 animals prepared for SEM and 20 animals prepared for molecular analyses (not included in the type series). However, DNA sequences were obtained from only five specimens (exoskeletons) which was later mounted on microscope slide in Hoyer's medium and included into type series.

Type depositories: Holotype (CN8.62) and 76 paratypes (slides: CN8.*, where the asterisk can be substituted by any of the following numbers: 46–55, 65–66, 17/S, 21/S, 22/S, 25/S, 28/S) are deposited at the Department of Animal Taxonomy and Ecology, Institute of Environmental Biology, Adam Mickiewicz University in Poznań, Uniwersytetu Poznańskiego 6, 61-614 Poznań, Poland. Two paratypes (two females; slides CN8.63 and CN8.64) are deposited in the Natural History Museum of Denmark, University of Copenhagen.

Etymology: The authors would like to dedicate this species to famous biochemist and last author's friend—Professor Marek Michalak, Faculty of Medicine and Dentistry, Biochemistry Department, University of Alberta, Edmonton, AB, Canada.

Description of the new species

Adult females (measurements and statistics in Table 3). Body light yellow in live specimens (transparent after mounting in Hoyer's medium) (Figure 1A–C), eyes absent or not visible after mounting on microscope slides. Small and conical primary and secondary clavae present. Cirri *internus* and *externus* with poorly developed cirrophores. Cirri *internus* always shorter than cirri *externus*. Cirri *A* of a typical length for *Bryodelphax*, i.e., up to 25% of the total body length. Only lateral appendages cirri *A* present apart from head appendages.

Dorsal sculpture, visible in PCM, composed of intra-cuticular pillars (visible as dark dots/granules) and pores (visible as white dots) (Figure 2A–C). The cuticular pillars (granules) on scapular plate 0.6–1.8 μm in diameter, on caudal plate 0.6–1.6 μm in diameter and on other plates 0.6–1.4 μm in diameter. Pores large and easily detectable (Figure 2A–D), distributed unevenly on scapular plate (0.3–0.6 μm in diameter; 0–14 pores/100 μm^2, $\overline{x}$ = 7.5, *N* = 10); on caudal plate (0.6–1.0 μm in diameter; 0–19 pores/100 μm^2, $\overline{x}$ = 4.44, *N* = 10) and on other plates (0.3–0.8 μm in diameter; 1–16 pores/100 μm^2, $\overline{x}$ = 6.3, *N* = 10). Median plates 1 and 2 divided by smooth transverse stripe, median plate 3 undivided. Median plate 2 largest among all median plates. Paired plates 1 and 2 also divided transversely into two parts by smooth stripes.

Ventral side with three rows of greyish plates (formula: III:2-2-1). First row with two plates just below the head (Figure 3A,D, filled arrowheads). Three genital plates surrounding the gonopore (two lateral, in line with the gonopore) and the third one situated posteriorly to the gonopore (Figure 3B–D, empty arrowheads)).

Papilla-like structure on leg I hardly visible under PCM but visible in SEM (Figures 1C and 4A,B, arrow). Papillae on leg IV present (Figures 1C and 4D, arrowhead). Dentate collar absent on leg IV (Figure 1A–C). All claws slender, claws IV always slightly longer than claws I–III. External claws smooth, internal ones with a small spur pointing downward and placed very close to the claw bases (Figure 4A–D). The female gonopore with the typical six-petal rosette.

Males and Juveniles. Not found.

DNA sequences

COI: three sequences; 584–672 bp;

18S rRNA: two sequences; 528 bp long;

28S rRNA: two sequences; 671 bp long.

Differential diagnosis. Presence of ventral plates attributes *Bryodelphax mareki* sp. nov. to the *weglarskae* group. Within this group, only *Bry. amphoterus* (Durante Pasa and Maucci [58]), *Bry. maculatus* Gąsiorek, Stec, Morek, Marnissi and Michalczyk [58] and *Bry. nigripunctatus* Degma, Gąsiorek, Vončina and Michalczyk [30] have a reduced number of ventral plate rows to two or three, as in the new species [30]. Adult females of *Bry. mareki* sp. nov. differ from:

Bry. amphoterus, known only from Croatia and Greece (McInnes [59]), by different formula of ventral plates (III:2-2-1 in the new species vs. II:2-2 in *Bry. amphoterus*), presence of papilla-like structure on leg I and papillae on leg IV and absence of dentate collar on leg IV.

Bry. maculatus, known only from Tunisia and Greece (Gąsiorek et al. [60]), by higher *sc* of clava (*20.4–33.0* in the new species vs. *11.5–19.1* in *Bry. maculaus*), longer cirrus *A* (35.2–46.7 μm in the new species vs. 27.3–34.9 μm in *Bry. maculatus*), higher *sc* of cirrus *A* (*174.1–212.9* in the new species vs. *114.8–152.5* in *Bry. maculatus*) and absence of dentate collar on leg IV.

Bry. nigripunctatus known only from Spain (Gąsiorek et al. [30]), by absence of epicuticular granules, longer clava (4.3–6.2 μm in the new species vs. 2.7–3.1 μm in *Bry. nigripunctatus*), higher *sc* of clava (*20.4–33.0* in the new species vs. *12.3–16.9* in *Bry. nigripunctatus*) and presence of papilla-like structure on leg I.

Genetic variability

Aligned sequences (obtained in present study and downloaded from GenBank) were trimmed to 591, 498 and 700 bp for COI (four sequences; two species), 18S rRNA (eight sequences; four species and two sequences of *Bryodelphax* sp.) and 28S rRNA (14 sequences; seven species and two sequences of *Bryodelphax* sp.) molecular markers, respectively. Only sequences of *Bryodelphax* downloaded from GenBank that coincided with our four aforementioned molecular markers were selected.

In the case of the COI molecular marker, only three sequences of *Bry. parvulus* [8] were available in GenBank. Analysis of the p-distances between our sequences (GenBank accession numbers: MW655785-87) and three sequences of *Bry. parvulus* was from 16% (GenBank accession numbers: JX683827, JX683826, unpublished) to 18% (GenBank accession number: HM193405 [61]). In the case of the 18S rRNA molecular marker (GenBank accession numbers of our sequences: MW680639-40), no genetic differences were observed when compared with *Bry. parvulus* (GenBank accession numbers: HM193371 [61]; JX676189, [62]) and p-distance between other sequences, i.e., *Bryodelphax* sp. (GenBank accession numbers: EU266963 and EF632433 [63]) and *Bry. tatrensis* [64] (GenBank accession numbers: JX676188 and JX676190 [62]) was 0.01%. The ranges of uncorrected genetic p-distances between our 28S rRNA sequences (GenBank accession numbers: MW680637-38) and the most similar *Bry.* cf. *parvulus* (GenBank accession number: MT333466 [30]) was 0.01% and the least similar *Bryodelphax* sp. (GenBank accession numbers: MH414964 [65]) was 0.16% (see Supplementary Materials—SM1).

Class: Eutardigrada Richters, 1926 [57]

Order: Parachela Schuster, Nelson, Grigarick and Christenberry, 1980 [66]

Superfamily: Macrobiotoidea Thulin, 1928 [8]

Family: Macrobiotidae Thulin, 1928 [8]

Genus: *Macrobiotus* C.A.S. Schultze, 1834 [11]

***Macrobiotus birendrai* sp. nov.**

(Tables 4 and 5, Figures 5–8)

LSID http://zoobank.org/urn:lsid:zoobank.org:act:5617E45B-865A-42DA-A70A-DB6FE1A52163

Table 4. Measurements (in μm) and *pt* values of selected morphological structures of individuals of ***Macrobiotus birendrai* sp. nov.** mounted in Hoyer's medium (N—number of specimens/structures measured; RANGE refers to the smallest and the largest structure among all measured specimens; SD—standard deviation, *pt*—ratio of the length of a given structure to the length of the buccal tube expressed as a percentage).

Character	N	Range						Mean		SD		Holotype	
		μm			*pt*			μm	*pt*	μm	*pt*	μm	*pt*
Body length	15	271	–	589	*815*	–	*1291*	449	*1055*	97	*133*	368	*978*
Buccopharyngeal tube													
Buccal tube length	16	33.2	–	52.6		–		42.7	–	5.9	–	37.7	–
Stylet support insertion point	16	26.2	–	43.0	*78.0*	–	*82.8*	34.3	*80.2*	5.2	*1.6*	29.6	*78.7*
Buccal tube external width	16	5.2	–	8.4	*13.5*	–	*17.0*	6.6	*15.3*	1.0	*1.1*	5.5	*14.7*
Buccal tube internal width	16	3.4	–	5.6	*8.7*	–	*11.9*	4.4	*10.2*	0.8	*1.0*	3.8	*10.0*
Ventral lamina length	13	19.9	–	32.9	*56.0*	–	*66.2*	25.7	*60.9*	4.2	*3.2*	22.9	*60.8*
Placoid lengths													
Macroplacoid 1	16	8.2	–	16.0	*23.0*	–	*31.6*	11.3	*26.2*	2.2	*2.6*	8.7	*23.0*
Macroplacoid 2	16	6.0	–	11.2	*17.2*	–	*22.0*	8.1	*18.9*	1.6	*1.5*	6.5	*16.3*
Microplacoid	16	2.4	–	4.9	*6.5*	–	*9.6*	3.5	*8.0*	0.8	*0.9*	2.8	*7.4*
Macroplacoid row	16	16.3	–	30.2	*49.1*	–	*57.9*	22.6	*52.6*	4.4	*3.4*	18.8	*49.9*
Placoid row	16	19.2	–	35.2	*57.9*	–	*67.9*	26.9	*62.7*	4.9	*3.4*	23.2	*61.5*
Claw 1 heights													
External primary branch	16	8.5	–	13.2	*20.6*	–	*29.8*	10.8	*25.3*	1.6	*2.3*	10.0	*26.5*
External secondary branch	16	6.1	–	10.3	*14.5*	–	*22.1*	8.1	*19.0*	1.4	*2.2*	7.8	*20.7*
Internal primary branch	16	8.2	–	12.7	*20.3*	–	*27.7*	10.3	*24.2*	1.5	*1.8*	9.3	*24.8*
Internal secondary branch	16	6.2	–	10.7	*16.7*	–	*24.3*	8.4	*19.6*	1.4	*2.3*	7.8	*20.8*

Table 4. *Cont.*

Character	N	Range µm	Range *pt*	Mean µm	Mean *pt*	SD µm	SD *pt*	Holotype µm	Holotype *pt*
Claw 2 heights									
External primary branch	16	8.1 – 14.0	*23.8 – 29.3*	11.5	*26.9*	1.7	*1.6*	10.3	*27.2*
External secondary branch	16	7.0 – 11.7	*17.5 – 23.9*	9.0	*21.1*	1.5	*1.9*	7.5	*19.8*
Internal primary branch	16	8.6 – 13.6	*23.0 – 30.7*	11.1	*26.0*	1.6	*2.1*	11.1	*29.5*
Internal secondary branch	16	6.8 – 11.0	*16.5 – 23.6*	8.5	*20.0*	1.3	*2.2*	8.8	*23.3*
Claw 3 heights									
External primary branch	16	8.7 – 14.8	*24.9 – 31.7*	11.7	*27.4*	1.7	*2.1*	10.4	*27.5*
External secondary branch	16	7.1 – 11.7	*17.5 – 25.2*	9.2	*21.5*	1.5	*1.7*	7.8	*20.8*
Internal primary branch	16	8.2 – 13.5	*22.7 – 30.3*	10.9	*25.5*	1.7	*1.9*	9.7	*25.8*
Internal secondary branch	16	6.5 – 11.1	*18.0 – 23.2*	8.8	*20.6*	1.5	*1.7*	8.3	*21.9*
Claw 4 lengths									
Anterior primary branch	16	9.5 – 15.8	*23.9 – 34.9*	12.5	*29.2*	1.9	*2.9*	11.7	*30.9*
Anterior secondary branch	16	6.7 – 13.0	*18.3 – 28.7*	9.5	*22.2*	1.7	*2.7*	8.9	*23.5*
Posterior primary branch	16	8.4 – 14.2	*23.7 – 32.2*	12.1	*28.3*	1.5	*2.4*	11.6	*30.8*
Posterior secondary branch	16	6.5 – 12.0	*18.3 – 24.3*	9.2	*21.5*	1.7	*1.9*	8.3	*22.1*

Table 5. Measurements (in µm) of selected morphological structures of eggs of ***Macrobiotus birendrai* sp. nov.** mounted in Hoyer's medium (N—number of specimens/structures measured; RANGE refers to the smallest and the largest structure among all measured eggs; SD—standard deviation).

Character	N	Range	Mean	SD
Egg bare diameter	5	74.9–86.9	87.2	7.6
Egg full diameter	5	87.8–103.9	102.4	9.3
Process height	33	5.9–9.4	7.9	1.1
Process base width	33	5.5–9.0	6.6	0.8
Process base/height ratio	33	65–100%	84%	10%
Terminal disc width	33	3.2–6.1	4.1	0.7
Inter-process distance	33	1.0–3.3	2.1	0.7
Number of processes on the egg circumference	5	22–26	23.6	2.0

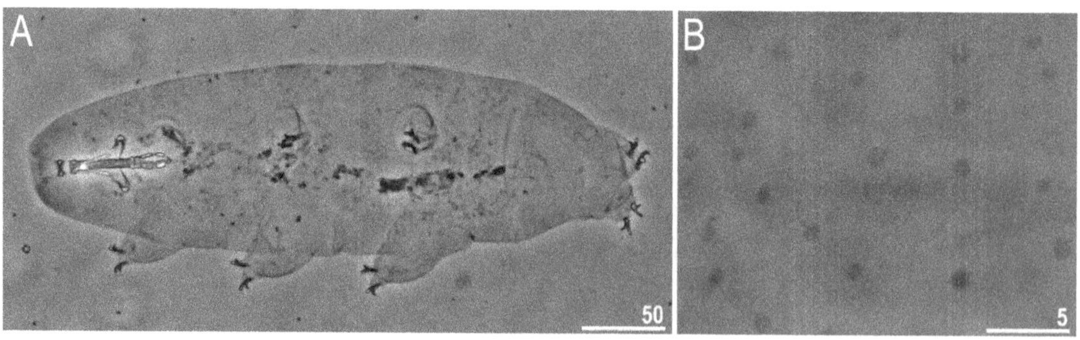

Figure 5. ***Macrobiotus birendrai*** sp. nov.: (**A**) dorso-ventral projection (holotype); (**B**) cuticular pores on dorsal side of the body (paratype). All PCM. Scale bars in µm.

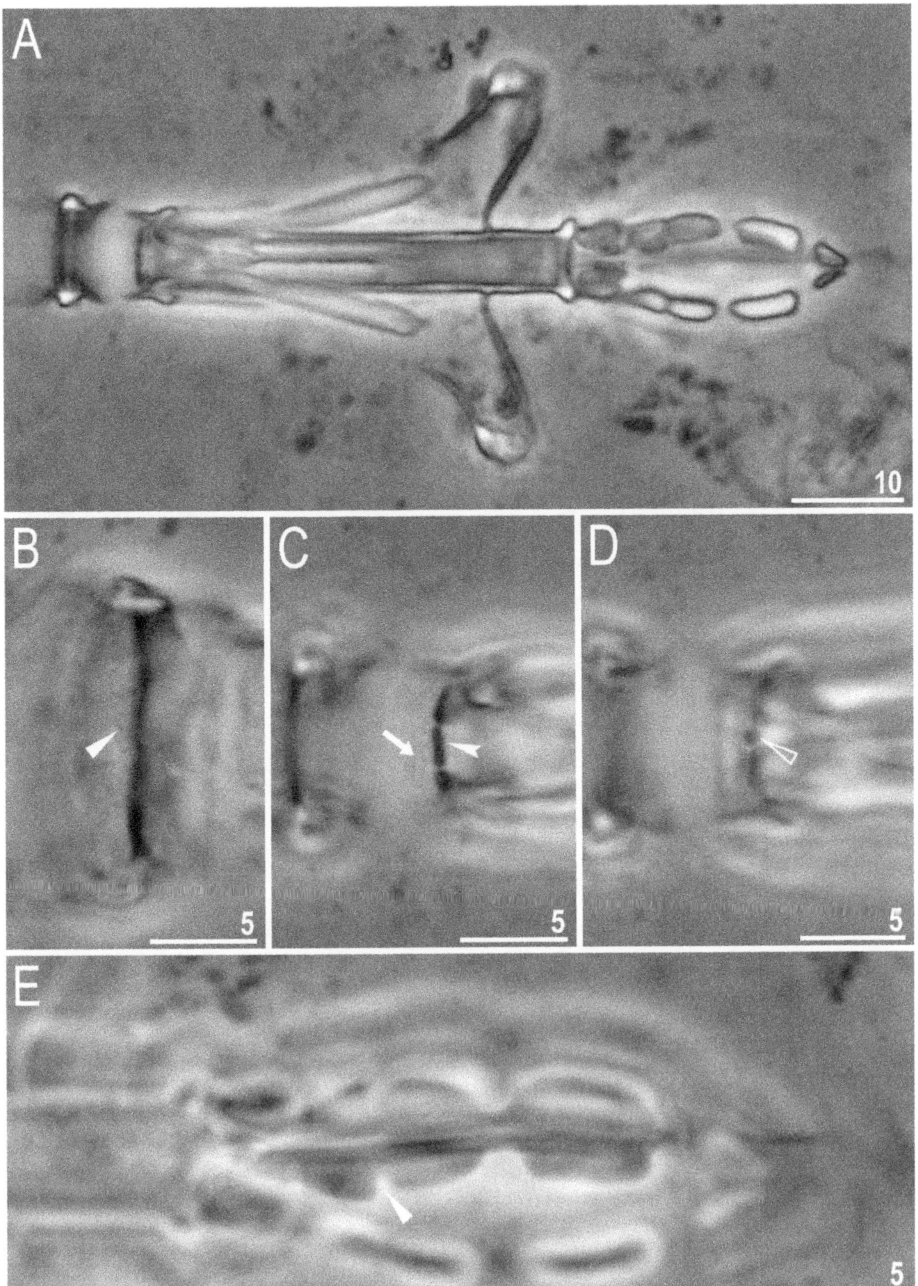

Figure 6. ***Macrobiotus birendrai*** sp. nov.: bucco-pharyngeal apparatus (dorso–ventral projection): (**A**) general view (paratype); (**B**) oral cavity armature with filled arrowhead indicating teeth of the first band (paratype); (**C**) oral cavity armature with arrow indicating teeth of the second band and indented filled arrowhead indicating dorsal teeth of the third band (paratype); (**D**) oral cavity armature with empty arrowhead indicating ventral teeth of the third band (paratype); (**E**) ventral placoids; the filled arrowhead indicates a first macroplacoid with central constriction (holotype). All PCM. Scale bars in μm.

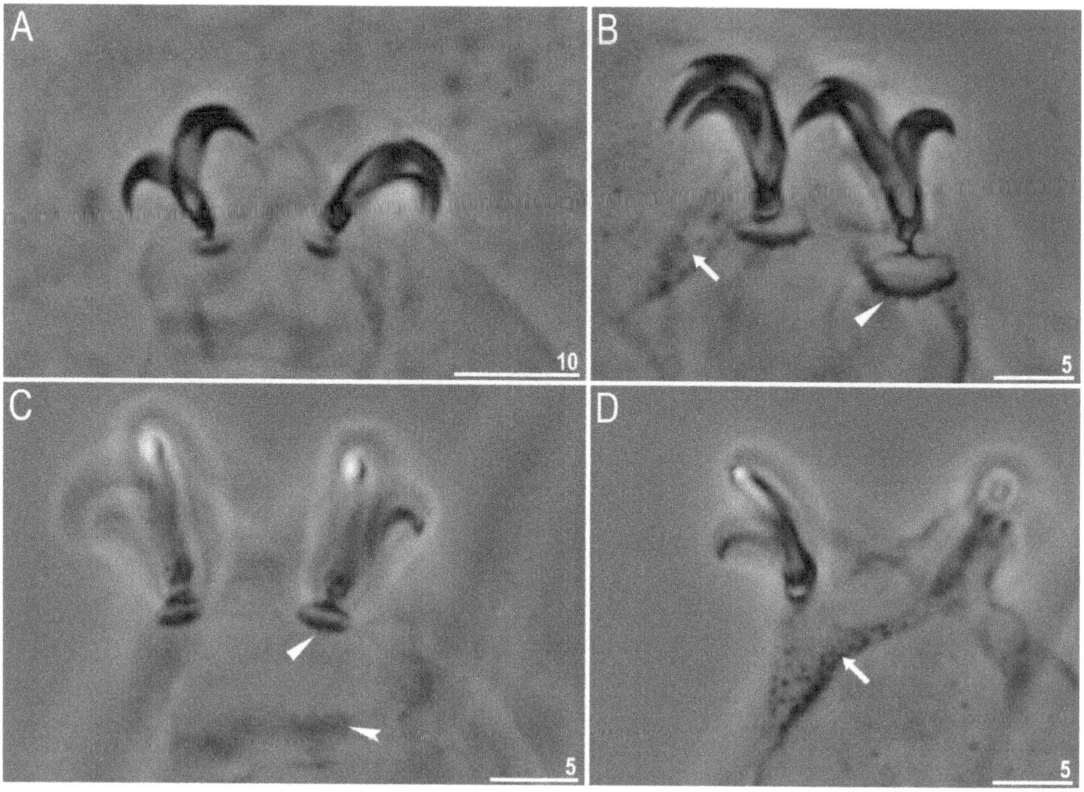

Figure 7. ***Macrobiotus birendrai*** sp. nov.: (**A**) claws III (paratype); (**B**) claws IV with dentate lunulas (arrowhead); arrow indicates granulation on legs IV (paratype); (**C**) lunulas under claws III with small teeth; indented arrowhead indicates cuticular bar under claws (paratype); (**D**) granulation on leg III (arrow) (holotype). All PCM. Scale bars in μm.

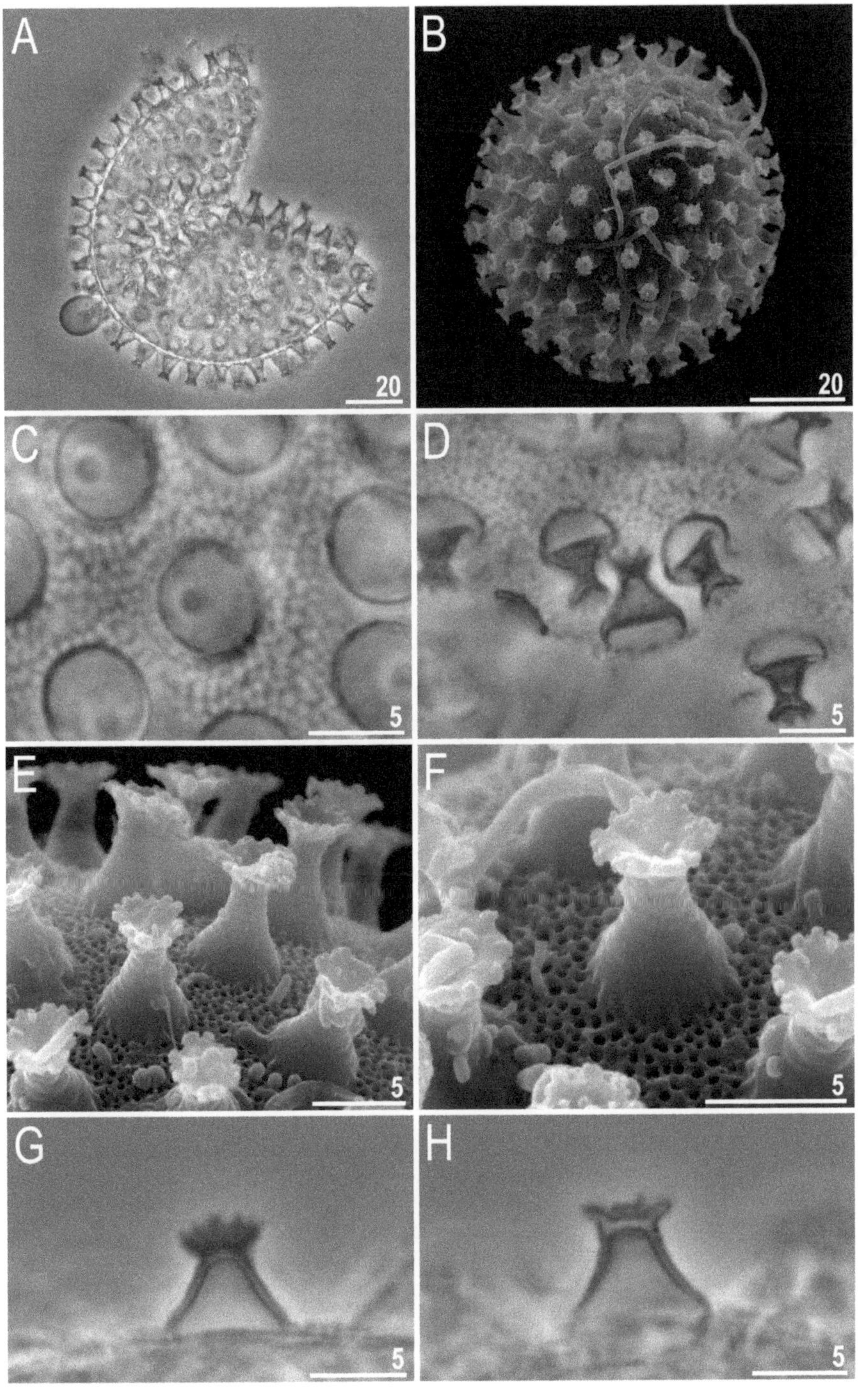

Figure 8. ***Macrobiotus birendrai*** sp. nov.: eggs: (**A**,**B**) egg chorion (PCM and SEM, respectively); (**C**) the surface between egg processes visible in PCM; (**D**) egg processes visible in PCM; (**E**,**F**) egg surface and processes visible in SEM; (**G**,**H**) egg processes visible in PCM. Scale bars in μm.

Type Locality: 51°24′21″ N, 116°14′27″ W, 1900 m asl, Canada, Alberta, Banff National Park, near east end of the Louise Lake, moss on stone, May 2019, leg. Milena Roszkowska and Łukasz Kaczmarek.

Material examined: The 57 specimens, i.e., holotype (slide: CN8.43) + 56 paratypes (adults: 44 and eggs: 12) were mounted on microscope slides in Hoyer's medium, four eggs prepared for SEM and five animals prepared for molecular analyses (not included in type series). However, DNA sequences were obtained from one female specimen (exoskeleton) which was later mounted on microscope slide in Hoyer's medium and included into type series.

Type depositories: Holotype (CN8.43) and 57 paratypes (slides: CN8.*, where the asterisk can be substituted by any of the following numbers: 3, 5–7, 15–17, 21, 29–33, 39, 40, 42–45, 101/S) are deposited at the Department of Animal Taxonomy and Ecology, Institute of Environmental Biology, Adam Mickiewicz University in Poznań, Uniwersytetu Poznańskiego 6, 61-614 Poznań, Poland. Six paratypes (five adults and one egg; slides CN8.41 and CN8.28) are deposited in the Natural History Museum of Denmark, University of Copenhagen.

Etymology: The first author would like to dedicate this species to her father—Birendra Prasad Lal Karna.

Description of the new species.

Adults (measurements and statistics in Table 4). Body transparent after fixation in Hoyer's medium, eyes present in all fixed specimens (Figure 5A). Entire cuticle covered with conspicuous round and lenticular pores (0.6–1.8 µm in diameter) distributed randomly (Figure 5B). However, larger pores present, on dorsal side, at the anterior and posterior part of the body. Bucco-pharyngeal apparatus of the *Macrobiotus* type, with ventral lamina and 10 peribuccal lamellae (Figure 6A). Mouth antero-ventral. Oral cavity armature of the *hufelandi* type, with first and the second band composed of numerous minute teeth (visible as granules in PCM) and third composed of three dorsal and three ventral transverse ridges (Figure 6B–D). Pharyngeal bulb spherical with triangular apophyses, two rod-shaped macroplacoids and a triangular microplacoid. Macroplacoid length sequence 2 < 1 (Figure 6A,E). The first macroplacoid with central constriction (Figure 6E, arrowhead), second with sub-terminal constriction. Claws of the *hufelandi* type (Figure 7A,B). Primary branches with distinct accessory points. Lunules under claws I–III with hardly visible teeth (visible only in bigger specimens) (Figure 7C, arrowhead) and dentate under claws IV (Figure 7B, arrowhead). Thin single continuous cuticular bars under claws I–III present (Figure 7C, indented arrowhead). Easily visible granulation present on legs I–IV (Figure 7B,D, arrows).

Eggs (measurements and statistics in Table 5). Eggs spherical, ornamented and laid freely with egg chorion of the *hufelandi* type (Figure 8A,B). Pores of egg surface mesh circular, similar in size and rather small, i.e., 0.2–0.8 µm in diameter (Figure 8C,E,F). Processes in the shape of inverted concave cups with terminal discs (Figure 8D–H). Terminal discs concave with serrated margins or with small irregular teeth (Figure 8D–H).

DNA sequences

We obtained good quality sequences for the applied molecular markers:

COI: single sequence; 609 bp long;

18S rRNA: single sequence; 553 bp long;

28S rRNA: single sequence; 721 bp long;

ITS-2: single sequence; 350 bp long.

Differential diagnosis. Based on egg processes morphology, the new species is most similar to *Mac. canaricus* Stec et al. [67], *Mac. hannae* Nowak and Stec [68], *Mac. crustulus* Stec et al. [12], *Mac. joannae* Pilato and Binda [69], *Mac. kamilae* Coughlan and Stec [70], *Mac. madegassus* Maucci [71], *Mac. noemiae* Roszkowska and Kaczmarek [72], *Mac. noongaris* Coughlan and Stec [70], *Mac. papei* Stec et al. [73], *Mac. paulinae* Stec et al. [16], *Mac. polypiformis* Roszkowska et al. [38] and *Mac. porifini* Kuzdrowska et al. [39], but differs specifically from:

Mac. canaricus **Stec, Krzywański and Michalczyk, 2018,** known only from the type locality on Canary Islands [67], by a different oral cavity armature (*hufelandi* type in the new species vs. *maculatus* type in *Mac. canaricus*), lunules I-III with hardly visible teeth, larger buccal tube internal width (3.4–5.6 μm in the new species vs. 2.1–3.3 μm in *Mac. canaricus*), higher *pt* of placoid row (*57.9–67.9* in the new species vs. *42.6–55.3* in *Mac. canaricus*), larger cuticular pores (0.7–1.8 μm in the new species vs. 0.4–0.7 μm in *Mac. canaricus*), different egg shell surface (porous shell in the new species vs. mesh shell in *Mac. canaricus*) and absence of granulation on terminal discs of the egg process.

Mac. crustulus **Stec, Dudziak and Michalczyk, 2020,** known only from the type locality in French Guiana [12], by a different oral cavity armature (*hufelandi* type in the new species vs. *lissostomus* type in *Mac. crustulus*), lunules I-III with hardly visible teeth, higher *pt* of stylet support insertion points (*78.0–82.8* in the new species vs. *69.3–72.7* in *Mac. crustulus*), larger buccal tube internal width (3.4–5.6 μm in the new species vs. 1.6–2.7 μm in *Mac. crustulus*), higher *pt* of buccal tube external and internal width (*13.5–17.0* and *8.7–11.9*, respectively, in the new species vs. *9.3–12.2* and *5.1–6.5*, respectively, in *Mac. crustulus*), higher *pt* of ventral lamina (*56.0–66.2* in the new species vs. *46.9–55.8* in *Mac. crustulus*), different egg shell surface (porous shell in the new species vs. mesh shell in *Mac. crustulus*), absence of granulation on convex central area of terminal discs and lower number of processes on the egg circumference (22–26 in the new species vs. 26–34 in *Mac. crustulus*).

Mac. hannae **Nowak and Stec, 2018,** known only from the type locality in Poland [68], by lunules I-III with hardly visible teeth, larger cuticular pores (0.6–1.8 μm in the new species vs. up to 0.55 μm in *Mac. hannae*), smaller distance between egg processes (1.0–3.3 μm in the new species vs. 4.0–8.1 μm in *Mac. hannae*), smaller egg bare diameter (74.9–86.9 μm in the new species vs. 88.6–109.2 μm in *Mac. hannae*) and absence of granules inside pores around egg processes.

Mac. joannae **Pilato and Binda, 1983,** known only from the type locality in Australia (Pilato and Binda [69]; see also comments in Nowak and Stec [68]), by lower average *pt* of stylet support insertion points (*80.2* in the new species vs. *81.9* in *Mac. joannae*), smaller size of macroplacoid 1 and microplacoid (8.2–16.0 and 2.4–4.9 μm, respectively, in the new species vs. 19.3 and 6.4 μm, respectively, in *Mac. joannae* in specimen of body length 400 μm), smaller size of claws I and II external and internal primary branch (I. 8.5–13.2 and 8.2–12.7 μm, respectively; II. 8.1–14.0 and 8.6–13.6 μm, respectively, in the new species vs. I. 25.0 and 23.5 μm, respectively; II. 27.0 and 24.5 μm, respectively in *Mac. joannae* in specimen of body length 400 μm) and smaller size of claw IV anterior and posterior primary branch (9.5–15.8 and 8.4–14.2 μm, respectively, in the new species vs. 28.0 and 27.0 μm, respectively, in *Mac. joannae* in specimen of body length 400 μm).

Mac. kamilae **Coughlan and Stec, 2019,** known only from the type locality in India [70], by a different oral cavity armature (*hufelandi* type in the new species vs. *patagonicus* type in *Mac. kamilae*), lunules I-III with hardly visible teeth, higher *pt* of stylet support insertion points (*78.0–82.8* in the new species vs. *71.6–75.9* in *Mac. kamilae*), shorter claws (for details see Table 4 in this paper and Table 4 in Coughlan and Stec [65]), absence of body granulation and absence of scattered granules on the terminal discs of egg process.

Mac. madegassus **Maucci, 1993,** known only from the type locality in Madagascar [71], by presence of eyes, different oral cavity armature (*hufelandi* type in the new species vs. *maculatus* type in *Mac. madegassus*), presence of cuticular pores, lunules I-III with hardly visible teeth, longer buccal tube (33.2–52.6 μm in the new species vs. up to 30.0 μm in *Mac. madegassus*), larger buccal tube external width (5.2–8.4 μm in the new species vs. up to 2.1 μm in *Mac. madegassus*), higher *pt* of buccal tube external width (*13.5–17.0* in the new species vs. 7.0 in *Mac. madegassus* in specimen of body length 316 μm), longer macroplacoids (I: 8.2–16.0 μm and II: 6.0–11.2 μm in the new species vs. I: up to 6.4 μm and II: up to 3.6 μm in *Mac. madegassus* in specimen of body length 316 μm), stylet supports inserted in more caudal position (26.2–43.0 μm in the new species vs. up to 20.4 μm in *Mac. madegassus* in specimen of body length 316 μm), higher *pt* of stylet support insertion points (*78.0–82.8* in the new species vs. up to *68.0* in *Mac. madegassus* in specimen of body length 316 μm) and lower number of processes on the egg circumference (20–26 in the new species vs. 30–34 in *Mac. madegassus*).

Mac. noemiae **Roszkowska and Kaczmarek, 2019**, known only from the type locality in Spain [72], by a different oral cavity armature (*hufelandi* type in the new species vs. *patagonicus* type in *Mac. noemiae*), lunules I-III with hardly visible teeth, higher *pt* of ventral lamina (*56.0–66.2* in the new species vs. *48.4–55.7* in *Mac. noemiae*), higher *pt* of macroplacoid row and placoid row (*49.1–57.9* and *57.9–67.9*, respectively, in the new species vs. *39.2–47.1* and *47.1–57.9*, respectively, in *Mac. noemiae*), smaller egg full and bare diameter (87.8–103.9 and 74.9–86.9 μm, respectively, in the new species vs. 118.5–123.5 100.6–105.7 μm, respectively, in *Mac. noemiae*), lower number of processes on the egg circumference (20–26 in the new species vs. 30–34 in *Mac. noemiae*), presence of discs on egg processes and absence of filaments on apisal part of egg processes.

Mac. noongaris **Coughlan and Stec, 2019**, known only from the type locality in Australia [70], by a different oral cavity armature (*hufelandi* type in the new species vs. *patagonicus* type in *Mac. noongaris*), lunules I-III with hardly visible teeth, larger maximum size of the cuticular pores (up to 1.8 μm in the new species vs. up to 0.8 μm in *Mac. noongaris*), absence of scattered granulation on the terminal discs of the egg processes, higher mean of egg bare diameter and egg full diameter (82.9 and 96.3 μm, respectively, in the new species vs. 70.7 and 82.1 μm, respectively, in *Mac. noongaris*), larger mean process height, process base width and terminal disc width (7.9, 6.6 and 4.1 μm, respectively, in the new species vs. 6.2, 5.0 and 3.3 μm, respectively, in *Mac. noongaris*), and smaller mean inter processes distance (2.1 μm in the new species vs. 3.4 μm in *Mac. noongaris*).

Mac. papei **Stec, Kristensen and Michalczyk, 2018** known only from type locality in Tanzania [73], by a different oral cavity armature (*hufelandi* type in the new species vs. *patagonicus* type in *Mac. papei*), lunules I-III with hardly visible teeth, absence of patches of cuticular granulation the internal surface of legs I–III, higher *pt* of macroplacoid II (*17.2–22.0* in the new species vs. *10.3–16.2* in *Mac. papei*), absence of flexible filaments on the terminal disc of the egg processes, lower mean of egg bare diameter and egg full diameter (82.9 and 96.3 μm, respectively, in the new species vs. 95.0 and 109.7 μm, respectively, in *Mac. papei*), smaller mean terminal disc width (4.1 μm in the new species vs. 4.3 μm in *Mac. papei*), and smaller mean inter processes distance (2.1 μm in the new species vs. 4.3 μm in *Mac. papei*).

Mac. paulinae **Stec, Smolak, Kaczmarek and Michalczyk, 2015** known only from type locality in Kenya [16], by lack of dorso-lateral patches of granulation, smaller maximum size of cuticular pores (up to 1.8 μm in the new species vs. up to 0.5 μm in *Mac. paulinae*), different oral cavity armature (*hufelandi* type in the new species vs. *maculatus* type in *Mac. paulinae*), different third band of teeth (three dorsal and three ventral teeth in the new species vs. single dorsal and single ventral tooth in *Mac. paulinae*), lunules I-III with hardly visible teeth, larger buccal tube external and internal width with higher *pt* (5.2–8.4 [*13.5–17.0*] and 3.4–5.6 μm [*8.7–11.9*], respectively, in the new species vs. 2.2–4.6 μm [*8.8–12.6*] and vs. 1.0–2.9 μm [*3.5–8.1*], respectively, in *Mac. paulinae*), longer ventral lamina (19.9–32.9 μm in the new species vs. 15.0–19.6 μm in *Mac. paulinae*), longer macroplacoid row with higher *pt* (16.3–30.2 μm [*49.1–57.9*] in the new species vs. 9.2–14.4 μm [*32.5–40.4*] in *Mac. paulinae*), longer placoid row with higher *pt* (19.2–35.2 μm [*57.9–67.9*] in new species vs. 10.8–17.4 μm [*37.8–48.9*] in *Mac. paulinae*), larger egg bare and full diameter (74.9–86.9 and 87.8–103.9 μm, respectively, in the new species vs. 57.0–70.5 and 66.3–85.6 μm, respectively, in *Mac. paulinae*) and absence of filaments on egg processes discs.

Mac. polypiformis **Roszkowska, Ostrowska, Stec, Janko and Kaczmarek, 2017** known only from type locality in Ecuador [38], by different oral cavity armature (*hufelandi* type in the new species vs. *maculatus* type in *Mac. polypiformis*), lunules I-III with hardly visible teeth, longer buccal tube (33.2–52.6 μm in the new species vs. 24.4–32.5 μm in *Mac. polypiformis*), stylet supports inserted in more caudal position with higher *pt* (26.2–43.0 μm

[*78.0–82.8*] in the new species vs. 17.1–23.5 μm [*70.1–72.9*] in *Mac. polypiformis*), larger buccal tube external and internal width and with higher *pt* (5.2–8.4 [*13.5–17.0*] and 3.4–5.6 μm [*8.7–11.9*], respectively, in the new species vs. 2.8–4.0 [*11.0–13.0*] and 1.6–2.4 μm [*6.1–8.6*], respectively, in *Mac. polypiformis*), longer ventral lamina with larger *pt* (19.9–32.9 μm [*56.0–66.2*] in the new species vs. 13.5–17.3 μm [*52.1–55.1*] in *Mac. polypiformis*), longer macroplacoid 1 (8.2–16.0 μm in the new species vs. 5.2–6.8 μm in *Mac. polypiformis*), longer macroplacoid 2 with higher *pt* (6.0–11.2 μm [*17.2–22.0*] in the new species vs. 2.8–4.1 μm [*11.4–14.5*] in *Mac. polypiformis*), longer microplacoid (2.4–4.9 μm in the new species vs. 1.5–2.3 μm in *Mac. polypiformis*), longer macroplacoid row with higher *pt* (16.3–30.2 μm [*49.1–57.9*] in the new species vs. 9.0–11.8 μm [*34.3–39.9*] in *Mac. polypiformis*) and longer placoid row with higher *pt* (19.2–35.2 μm [*57.9–67.9*] in the new species vs. 11.1–14.5 μm [*41.4–49.0*] in *Mac. polypiformis*), larger egg bare and full diameter (74.9–86.9 and 87.8–103.9 μm, respectively, in the new species vs. 61.9–70.5 and 70.4–81.2 μm, respectively, in *Mac. polypiformis*) and absence of filaments on egg processes discs.

Mac. porifini **Kuzdrowska, Mioduchowska, Gawalak, Bartylak, Kepel, Kepel and Kaczmarek 2021** known only from the type locality in Madagascar [39] by presence of eyes, different oral cavity armature (*hufelandi* type in the new species vs. *patagonicus* type in *Mac. porifini*), lunules I-III with hardly visible teeth, presence of dentate lunules on legs IV, higher *pt* of macroplacoid row (*49.1–57.9* in the new species vs. *36.2–47.7* in *Mac. porifini*), higher *pt* of placoid row (*57.9–67.9* in the new species vs. *43.5–57.7* in *Mac. porifini*), larger egg bare diameter (74.9–86.9 μm in the new species vs. 72.2–74.0 μm in *Mac. porifini*), absence of very small irregular granules on the surface of the discs and absence of micro granulation on the teeth of the terminal discs of egg process.

Genetic variability

Aligned sequences (obtained in our study and downloaded from GenBank) were trimmed to 609, 554, 682 and 200 bp for COI (21 sequences selected from GenBank—one sequence per species; (see Supplementary Materials—SM2)), 18S rRNA (20 sequences selected from GenBank—one sequence per species; see Supplementary Materials—SM3), 28S rRNA (16 sequences selected from GenBank—one sequence per species; see Supplementary Materials—SM4) and ITS2 (12 sequences selected from GenBank one sequence per species; see Supplementary Materials—SM5) molecular markers, respectively.

The analysis of the p-distances between our COI sequence (GenBank accession number: MW656266) and sequences of species from the genus *Macrobiotus* ranged from the most similar 17% for *Mac. canaricus* (GenBank accession number: MH063925 [67]), to least similar 25% for *Mac. kristenseni* Guidetti, Peluffo, Rocha, Cesari and Moly de Peluffo [74] (GenBank accession number: KC193575 [74]). In the conservative 18S rRNA gene fragment we observed no differences between our sequence (GenBank accession number: MW680641) and sequences of *Mac. hannae* (GenBank accession number: HQ604975 [75]). In turn, the uncorrected genetic p-distances between the least similar *Mac. polonicus* Pilato, Kaczmarek, Michalczyk and Lisi [37] (GenBank accession number: HM187580 [76]) was 4%. The analysis of the p-distances between our sequence of 28S rRNA (GenBank accession number: MW680644) and similar sequences of the genus *Macrobiotus* are as follows: the most similar was *Mac. hannae* (GenBank accession number: MH063924, Nowak and Stec [68]) with p-distance of 1% and the least similar was *Mac. polypiformis* (GenBank accession number: KX810009 [38]) and *Mac. scoticus* Stec, Morek, Gasiorek, Blagden and Michalczyk [77] (GenBank accession number: KY797266 [68]) with p-distance of 10%. In turn, the ranges of p-distances between our ITS-2 sequence (GenBank accession number: MW680418) and the most similar *Mac. hannae* (GenBank accession number: MH063923 [68]) was 11% and the least similar was *Mac. polypiformis* (GenBank accession number: KX810010 [38])—33%.

Genus: *Mesobiotus* Vecchi, Cesari, Bertolani, Jönsson, Rebecchi and Guidetti, 2016 [78]

Mesobiotus skorackii Kaczmarek, Zawierucha, Buda, Stec, Gawlak, Michalczyk and Roszkowska, 2018 [15]

(Tables 6 and 7)

Table 6. Measurements (in µm) and *pt* values of selected morphological structures of individuals of ***Mesobiotus skorackii*** mounted in Hoyer's medium (N—number of specimens/structures measured; RANGE refers to the smallest and the largest structure among all measured specimens; SD—standard deviation; *pt*—ratio of the length of a given structure to the length of the buccal tube expressed as a percentage).

Character	N	Range						Mean		SD	
		µm			*pt*			µm	*pt*	µm	*pt*
Body length	11	281	–	485	*818*	–	*1102*	418	*991*	62	*76*
Buccopharyngeal tube											
Buccal tube length	11	34.3	–	47.6		–		42.0	–	4.4	–
Stylet support insertion point	11	25.9	–	35.6	*74.6*	–	*77.0*	31.7	*75.5*	3.2	*0.8*
Buccal tube external width	11	5.9	–	8.0	*15.2*	–	*18.0*	7.0	*16.8*	0.7	*0.7*
Buccal tube internal width	11	3.8	–	5.6	*11.0*	–	*12.7*	4.9	*11.5*	0.6	*0.6*
Ventral lamina length	11	21.0	–	29.8	*54.4*	–	*62.9*	25.5	*60.6*	2.9	*2.7*
Placoid lengths											
Macroplacoid 1	11	4.3	–	6.7	*11.6*	–	*14.3*	5.5	*13.1*	0.7	*1.0*
Macroplacoid 2	11	3.7	–	5.4	*9.8*	–	*12.6*	4.7	*11.3*	0.5	*0.8*
Macroplacoid 3	11	4.1	–	6.4	*11.1*	–	*13.8*	5.3	*12.5*	0.7	*0.9*
Microplacoid	11	2.5	–	4.8	*6.7*	–	*10.3*	3.6	*8.5*	0.6	*1.2*
Macroplacoid row	11	15.3	–	22.9	*44.4*	–	*48.2*	19.4	*46.2*	2.4	*1.2*
Placoid row	11	19.7	–	27.6	*54.2*	–	*60.5*	24.4	*57.9*	2.9	*1.7*
Claw 1 heights											
External primary branch	11	6.9	–	11.2	*18.3*	–	*26.2*	9.5	*22.5*	1.4	*2.1*
External secondary branch	11	5.6	–	9.5	*14.5*	–	*20.7*	7.1	*16.9*	1.3	*2.1*
Internal primary branch	11	6.8	–	11.4	*19.8*	–	*25.9*	9.5	*22.6*	1.3	*1.7*
Internal secondary branch	11	5.3	–	8.9	*15.4*	–	*20.6*	7.7	*18.2*	1.0	*1.5*
Claw 2 heights											
External primary branch	11	6.9	–	11.4	*20.2*	–	*26.7*	9.9	*23.6*	1.4	*1.8*
External secondary branch	11	5.4	–	9.7	*15.8*	–	*22.7*	8.0	*19.0*	1.6	*2.6*
Internal primary branch	11	7.1	–	11.0	*20.1*	–	*25.6*	9.2	*21.9*	1.0	*1.6*
Internal secondary branch	10	5.5	–	10.3	*16.0*	–	*24.2*	7.6	*17.9*	1.3	*2.6*
Claw 3 heights											
External primary branch	11	7.8	–	11.6	*22.8*	–	*28.1*	10.4	*24.7*	1.2	*1.8*
External secondary branch	11	5.2	–	9.9	*15.2*	–	*23.1*	8.2	*19.4*	1.3	*2.7*
Internal primary branch	11	6.7	–	10.7	*19.6*	–	*27.2*	9.5	*22.6*	1.2	*1.9*
Internal secondary branch	11	5.3	–	9.6	*15.5*	–	*22.2*	8.0	*18.9*	1.1	*2.0*
Claw 4 lengths											
Anterior primary branch	11	7.2	–	13.0	*21.0*	–	*29.6*	11.4	*27.0*	1.7	*2.4*
Anterior secondary branch	11	5.9	–	10.3	*17.1*	–	*23.7*	9.0	*21.3*	1.3	*2.1*
Posterior primary branch	11	6.5	–	13.2	*18.9*	–	*30.7*	10.7	*25.3*	2.1	*3.6*
Posterior secondary branch	11	5.2	–	10.7	*15.1*	–	*24.4*	8.5	*20.1*	1.5	*2.6*

Table 7. Measurements (in µm) of selected morphological structures of eggs of *Mesobiotus skorackii* mounted in Hoyer's medium (N—number of specimens/structures measured; RANGE refers to the smallest and the largest structure among all measured eggs; SD—standard deviation).

Character	N	Range	Mean	SD
Egg bare diameter	16	62.8–89.5	74.1	8.1
Egg full diameter	16	87.6–113.6	102.2	7.6
Process height	48	13.5–20.4	17.1	1.7
Process base width	48	12.7–19.0	16.1	1.4
Process base/height ratio	48	72–127%	95%	12%
Inter-process distance	48	1.3–5.6	3.5	1.2
Number of processes on the egg circumference	16	9–13	11.1	1.2

Locality: 51°24′21″ N, 116°14′27″ W, 1900 m asl, Canada, Alberta, Banff National Park, near east end of the Louise Lake, moss on stone, May 2019, leg. Milena Roszkowska and Łukasz Kaczmarek.

Material examined: The 63 specimens; 44 animals and 19 eggs were mounted on microscope slides in Hoyer's medium, four eggs and four animals prepared for SEM and six animals prepared for molecular analyses. However, DNA sequences were obtained from one female specimen (exoskeleton) which was later mounted on microscope slide in Hoyer's medium and included into type series.

Depositories: All specimens (slides: CN8.*, where the asterisk can be substituted by any of the following numbers: 5–7, 31–40, 42–43, 45, 115/S) are deposited at the Department of Animal Taxonomy and Ecology, Institute of Environmental Biology, Adam Mickiewicz University in Poznań, Uniwersytetu Poznańskiego 6, 61-614 Poznań, Poland.

Short diagnosis:

Adults (measurements and statistics in Table 6). Body white in living animals and transparent after fixation in Hoyer's medium, eyes present, cuticle smooth. Bucco–pharyngeal apparatus of the *Macrobiotus* type, with ventral lamina and ten peribuccal lamellae. Mouth antero-ventral. Oral cavity armature of the *harmsworthi* type. Pharyngeal bulb spherical with triangular apophyses, three rod-shaped macroplacoids and a triangular microplacoid. Macroplacoid length sequence 2 < 3 < 1. The first macroplacoid narrower anteriorly, the second without constrictions and the third with a small, subterminal constriction. Claws of the *Mesobiotus* type. Lunules under claws I–III smooth and slightly dentated under claws IV. Thin cuticular bars under claws I–III present. Granulation hardly visible on legs I–III, whereas on legs IV always clearly marked.

Eggs (measurements and statistics in Table 7). Eggs laid freely, white and spherical. Egg processes in the shape of short and wide sharpened cones. Egg processes reticulated and surrounded by six areolae delimited by thin brims which are often discontinuous, thus areolae are not always fully formed (semi-areolation). Surface inside the areolae with clearly visible wrinkles.

DNA sequences

We obtained good quality sequences for the applied molecular markers:

COI: single sequence; 631 bp long;

18S rRNA: two sequences; 667–715 bp long;

28S rRNA: single sequence; 735 bp long.

Genetic variability

Aligned sequences (obtained in our study and downloaded from GenBank) were trimmed to 565, 474 and 713 bp for COI (14 sequences, selected from GenBank—one sequence per species; (see Supplementary Materials—SM6)), 18S rRNA (two sequences selected from GenBank—one sequence per species; see below) and 28S rRNA (13 sequences selected from GenBank—one sequence per species) molecular markers, respectively.

The ranges of uncorrected genetic p-distances between obtained COI sequence of *Meb. skorackii* (GenBank accession number: MW656257) and species of the genus *Mesobiotus*, for which sequences are available from GenBank, are as follows: 20–26%, with the most similar being *Meb.* cf. *barabanovi* (GenBank accession number: MN313170 [23]) and *Meb. occultatus* Kaczmarek, Zawierucha, Buda, Stec, Gawlak, Michalczyk and Roszkowska, [15] (GenBank accession number: MH195152 [15]) and the least similar being *Meb. dilimanensis* Itang, Stec, Mapalo, Mirano-Bascos and Michalczyk [79] (GenBank accession number: MN257047 [79]). In case 18S rRNA only two sequences were compared (because other 24 sequences deposited in GenBank were amplified using different sets of primers) and genetic p-distances between *Meb. harmsworthi* (at present undefined *Mesobiotus* species; GenBank accession number: MH079462 [66]) and *Meb. philippinicus* Mapalo, Stec, Mirano-Bascos and Michalczyk [80] (GenBank accession number: KX129793 [80]) was 0.01%. In turn, the ranges of p-distances between obtained 28S rRNA sequence (GenBank accession number: MW680636) and sequences downloaded from GenBank was: 5–13%, with the most similar being *Meb.* cf. *barabanovi* (GenBank accession number: MN310388 [23]) as well as *Meb. harmsworthi* (at present undefined *Mesobiotus* species; GenBank accession number: MH197264 [66]) and the least similar being *Meb. dilimanensis* (GenBank accession number: MN257049 [79]) (see Supplementary Materials—SM7).

4. Discussion

Out of 10 provinces and three territories of Canada, limno-terrestrial tardigrades have been reported in eight provinces and two territories. Up to now, no tardigrades have been reported from Northwest Territories, Nova Scotia nor Prince Edward Island. The highest number of tardigrade species were recorded from Nunavut (70) and the lowest Manitoba (only one). Moreover, 18 species were recorded from Alberta, 55 from British Columbia, 33 from New Brunswick, 29 from Newfoundland and Labrador, 12 from Ontario, 13 from Quebec, 3 from Saskatchewan and 5 from Yukon [13,14]. Only one species of the genus *Bryodelphax*, i.e., *Bry. parvulus* has been recorded from British Columbia and Nunavut. In case of the genus *Macrobiotus*, four species i.e., *Mac. echinogenitus* Richters [81], *Mac. hufelandi*, *Mac. occidentalis* Murray [82] and *Mac. virgatus* Murray [82] were recorded from Alberta, British Columbia, New Brunswick, Newfoundland and Labrador, Nunavut, Ontario and Quebec. Among these, only *Mac. hufelandi* belongs to the *hufelandi* group. Three species of the genus *Mesobiotus*, i.e., *Meb. harmsworthi* (Murray [83]), *Meb. montanus* (Murray [82]) and *Meb. pilatoi* (Binda and Rebecchi [84]) were recorded from Alberta, British Columbia, New Brunswick, Newfoundland and Labrador, Nunavut, Ontario and Quebec [14]. What is more intriguing, some species reported from Canada in the past are now considered as group of species or species with problematic taxonomical status [14].

Summarizing, from Canada only 121 tardigrade species and subspecies are known. Taking into consideration the area of the country (ca. 10 million km^2) and its diversity such as habitats and ecosystem, it is obvious that this number is highly underestimated. For comparison, USA, with the similar country area, has more than 220 species reported [14]. This contrast is even more spectacular while comparing Canadian tardigrade fauna with the number of tardigrade species from much smaller areas like e.g., Costa Rica (ca. 51,000 km^2 and 63 species known), Finland (ca. 340,000 km^2 and 68 species known), Italy (ca. 300,000 km^2 and 234 species known) or Poland (ca. 312,000 km^2 and 111 species known) [68,85–89]. The number of tardigrade species from Canada is expected to be much higher than reported up to date, especially that in the present study, in one analyzed sample, we found two species new for science and one new record for the country.

Supplementary Materials: The following are available online at https://www.mdpi.com/article/10.3390/d13080394/s1, SM1—Estimates of evolutionary divergence between 28S rRNA sequences of *Bryodelphax mareki* sp. nov based on p-distances, SM2—Estimates of evolutionary divergence between COI sequences of *Macrobiotus birendrai* sp. nov based on p-distances, SM3—Estimates of evolutionary divergence between 18S rRNA sequences of *Macrobiotus birendrai* sp. nov based on p-distances, SM4—Estimates of evolutionary divergence between 28S rRNA sequences of *Macrobiotus birendrai* sp. nov based on p-distances, SM5—Estimates of evolutionary divergence between ITS-2 sequences of *Macrobiotus birendrai* sp. nov based on p-distances, SM6—Estimates of evolutionary divergence between COI sequences of *Mesobiotus skorackii* based on p-distances, SM7—Estimates of evolutionary divergence between 28S rRNA sequences of *Mesobiotus skorackii* based on p-distances [13,15–17,23,30,38,39,41,60,67,70,73–80,90–103].

Author Contributions: Conceptualization, M.R. and Ł.K.; data curation, P.K.; formal analysis, P.K., M.R., M.M., M.G. and Ł.K.; investigation, P.K. and Ł.K.; methodology, M.R., M.M. and Ł.K.; supervision, Ł.K.; validation, P.K., M.R., M.M. and Ł.K.; visualization, M.R. and M.G.; writing—original draft, P.K., M.M. and Ł.K.; writing—review and editing, P.K., M.R., M.M., M.G. and Ł.K. All authors have read and agreed to the published version of the manuscript.

Funding: This research received no external funding.

Institutional Review Board Statement: Not applicable.

Informed Consent Statement: Not applicable.

Data Availability Statement: All data are presented in the article and in SM which are part of article.

Acknowledgments: Milena Roszkowska and Pushpalata Kayastha are scholarship holders of Passport to the future—Interdisciplinary doctoral studies at the Faculty of Biology, Adam Mickiewicz University, Poznań POWR.03.02.00-00-I006/17. The work of Monika Mioduchowska was supported by grant no. 2017/27/B/NZ8/01056 from the National Science Centre, Poland. Studies have been partially conducted in the framework of activities of BARg (Biodiversity and Astrobiology Research group).

Conflicts of Interest: The authors declare no conflict of interest.

References

1. Ramazzotti, G.; Maucci, W. Il Phylum Tardigrada. III edizione riveduta e aggiornata. *Mem. Ist. Ital. Idrobiol. Pallanza* **1983**, *41*, 1–1012.
2. Beasley, C.W. *The Phylum Tardigrada*, 3rd ed.; English Translation; Ramazzotti, G., Maucci, W., Eds.; Clark Beasley: Abilene, TX, USA, 1995; pp. 1–1014.
3. Nelson, D.R.; Guidetti, R.; Rebecchi, L.; Kaczmarek, Ł.; McInnes, S. Phylum Tardigrada. In *Thorp and Covich's Freshwater Invertebrates*; Elsevier: Amsterdam, The Netherlands, 2020; pp. 505–522.
4. Guidetti, R.; Bertolani, R.B. Tardigrade Taxonomy: An updated check list of the taxa and a list of characters for their identification. *Zootaxa* **2005**, *845*, 1–46. [CrossRef]
5. Degma, P.; Guidetti, R. Notes to the current checklist of Tardigrada. *Zootaxa* **2007**, *1579*, 41–53. [CrossRef]
6. Vicente, F.; Bertolani, R. Considerations on the taxonomy of the Phylum Tardigrada. *Zootaxa* **2013**, *3626*, 245–248. [CrossRef] [PubMed]
7. Degma, P.; Bertolani, R.; Guidetti, R. Actual Checklist of Tardigrada Species. Available online: https://iris.unimore.it/retrieve/handle/11380/1178608/357802/Actual%20checklist%20of%20Tardigrada%2040th%20Edition%2015-07-21.pdf (accessed on 15 July 2021).
8. Thulin, G. Über die phylogenie und das system der tardigraden. *Hereditas* **1928**, *11*, 207–266. [CrossRef]
9. Kristensen, R.M.; Michalczyk, Ł.; Kaczmarek, Ł. The first record of the genus Bryodelphax (Tardigrada: Heterotardigrada: Echiniscidae) from Easter Island, Rapa Nui (Pacific Ocean, Chile) with the description of a new species, *Bryodelphax aaseae*. *Zootaxa* **2010**, *2343*, 45–56. [CrossRef]
10. Gąsiorek, P.; Suzuki, A.C.; Kristensen, R.M.; Lachowska-Cierlik, D.; Michalczyk, Ł. Untangling the *Echiniscus* Gordian knot: *Stellariscus* gen. nov. (Heterotardigrada: Echiniscidae) from Far East Asia. *Invert. Syst.* **2018**, *32*, 1234–1247. [CrossRef]
11. Schultze, C.A.S. Macrobiotus Hufelandii, *animal e crustaceorum classe novum, reviviscendi post diuturnam asphyxiam et ariditatem potens*; Berolini Apud Carolum Curths: Berlin, Germany, 1834; pp. 1–8.
12. Stec, D.; Dudziak, M.; Michalczyk, Ł. Integrative descriptions of two new Macrobiotidae species (Tardigrada: Eutardigrada: Macrobiotoidea) from French Guiana and Malaysian Borneo. *Zool. Stud.* **2020**, *59*, e23. [CrossRef] [PubMed]
13. Collins, M.; Goudie, I. The tardigrade and associated micrometazoa of the textured lungwort lichen, *Lobaria scrobiculata*, in Eastern Newfoundland, Canada. *J. Agric. Life Sci.* **2020**, *7*. [CrossRef]
14. Kaczmarek, Ł.; Michalczyk, Ł.; Mcinnes, S.J. Annotated zoogeography of non-marine Tardigrada. Part III: North America and Greenland. *Zootaxa* **2016**, *4203*, 1–249. [CrossRef]
15. Kaczmarek, Ł.; Zawierucha, K.; Buda, J.; Stec, D.; Gawlak, M.; Michalczyk, Ł.; Roszkowska, M. An integrative redescription of the nominal taxon for the *Mesobiotus harmsworthi* group (Tardigrada: Macrobiotidae) leads to descriptions of two new *Mesobiotus* species from Arctic. *PLoS ONE* **2018**, *13*, e0204756. [CrossRef] [PubMed]
16. Stec, D.; Smolak, R.; Kaczmarek, Ł.; Michalczyk, Ł. An integrative description of *Macrobiotus paulinae* sp. nov. (Tardigrada: Eutardigrada: Macrobiotidae: *Hufelandi* Group) from Kenya. *Zootaxa* **2015**, *4052*, 501–526. [CrossRef] [PubMed]
17. Roszkowska, M.; Stec, D.; Gawlak, M.; Kaczmarek, Ł. An integrative description of a new tardigrade species *Mesobiotus romani* sp. nov. (Macrobiotidae: Harmsworthi group) from the Ecuadorian Pacific coast. *Zootaxa* **2018**, *4450*, 550–564. [CrossRef] [PubMed]
18. Dastych, H. A new species of the genus *Mopsechiniscus* Du Bois-Reymond Marcus, 1944 (Tardigrada) from the Venezuelan Andes. *Acta Biol. Ben.* **1999**, *10*, 91–101.
19. Fontoura, P.; Morais, P. Assessment of traditional and geometric morphometrics for discriminating cryptic species of the *Pseudechiniscus suillus* complex (Tardigrada, Echiniscidae). *J. Zool. Syst. Evol. Res.* **2011**, *49*, 26–33. [CrossRef]
20. Kaczmarek, Ł.; Zawierucha, K.; Smykla, J.; Michalczyk, Ł. Tardigrada of the Revdalen (Spitsbergen) with the descriptions of two new species: *Bryodelphax parvuspolaris* (Heterotardigrada) and *Isohypsibius coulsoni* (Eutardigrada). *Polar Biol.* **2012**, *35*, 1013–1026. [CrossRef]
21. Pilato, G.; Binda, M.G. Definition of families, subfamilies, genera and subgenera of the Eutardigrada, and keys to their identification. *Zootaxa* **2010**, *2404*, 1–54. [CrossRef]
22. Kaczmarek, Ł.; Michalczyk, Ł. The *Macrobiotus hufelandi* group (Tardigrada) revisited. *Zootaxa* **2017**, *4363*, 101–123. [CrossRef]
23. Kaczmarek, Ł.; Bartylak, T.; Stec, D.; Kulpa, A.; Kepel, M.; Kepel, A.; Roszkowska, M. Revisiting the genus *Mesobiotus* Vecchi et al., 2016 (Eutardigrada, Macrobiotidae)—Remarks, updated dichotomous key and an integrative description of new species from Madagascar. *Zool. Anz.* **2020**, *287*, 121–146. [CrossRef]
24. Michalczyk, Ł.; Kaczmarek, Ł. A description of the new tardigrade *Macrobiotus reinhardti* (Eutardigrada: Macrobiotidae, *harmsworthi* group) with some remarks on the oral cavity armature within the genus *Macrobiotus* Schultze. *Zootaxa* **2003**, *331*, 1–24. [CrossRef]
25. Kaczmarek, Ł.; Cytan, J.; Zawierucha, K.; Diduszko, D.; Michalczyk, Ł. Tardigrades from Peru (South America), with descriptions of three new species of Parachela. *Zootaxa* **2014**, *3790*, 357–379. [CrossRef] [PubMed]

26. Pilato, G. Analisi di nuovi caratteri nello studio degli eutardigradi. *Animalia* **1981**, *8*, 51–57.
27. Kiosya, Y.; Pogwizd, J.; Matsko, Y.; Vecchi, M.; Stec, D. Phylogenetic position of two Macrobiotus species with a revisional note on Macrobiotus sottilei Pilato, Kiosya, Lisi & Sabella, 2012 (Tardigrada: Eutardigrada: Macrobiotidae). *Zootaxa* **2021**, *4933*, 113–135. [CrossRef]
28. Perry, E.; Miller, W.R.; Kaczmarek, Ł. Recommended abbreviations for the names of genera of the phylum Tardigrada. *Zootaxa* **2019**, *4608*, 145–154. [CrossRef] [PubMed]
29. Michalczyk, Ł.; Kaczmarek, Ł. The Tardigrada Register: A comprehensive online data repository for tardigrade taxonomy. *J. Limnol.* **2013**, *72*, 175–181. [CrossRef]
30. Gąsiorek, P.; Vončina, K.; Degma, P.; Michalczyk, Ł. Small is beautiful: The first phylogenetic analysis of *Bryodelphax* Thulin, 1928 (Heterotardigrada, Echiniscidae). *Zoosystematics Evol.* **2020**, *96*, 217–236. [CrossRef]
31. Kaczmarek, Ł.; Michalczyk, Ł. A new species *Bryodelphax asiaticus* (Tardigrada: Heterotardigrada: Echiniscidae) from Mongolia (Central Asia). *Raffles Bull. Zool.* **2004**, 599–602.
32. Kaczmarek, Ł.; Michalczyk, Ł.; Degma, P. A new species of Tardigrada *Bryodelphax brevidentatus* sp. nov. (Heterotardigrada: Echiniscidae) from China (Asia). *Zootaxa* **2005**, *1080*, 33–38. [CrossRef]
33. Kaczmarek, Ł.; Parnikoza, I.; Gawlak, M.; Esefeld, J.; Peter, H.-U.; Kozeretska, I.; Roszkowska, M. Tardigrades from *Larus dominicanus* Lichtenstein, 1823 nests on the Argentine Islands (Maritime Antarctic). *Polar Biol.* **2018**, *41*, 283–301. [CrossRef]
34. Roszkowska, M.; Gawlak, M.; Draga, M.; Kaczmarek, Ł. Two new species of Tardigrada from Ecuador (South America). *Zootaxa* **2019**, *4545*, 511–530. [CrossRef]
35. Kaczmarek, Ł.; Michalczyk, Ł. Two new species of Macrobiotidae, *Macrobiotus szeptyckii* (*harmsworthi* group) and *Macrobiotus kazmierskii* (*hufelandi* group) from Argentina. *Acta Zool. Cracov.Ser. B Invert.* **2009**, *52*, 87–99. [CrossRef]
36. Kaczmarek, Ł.; Michalczyk, Ł. New records of Tardigrada from Cyprus with a description of the new species *Macrobiotus marlenae* (Eutardigrada: Macrobiotidae, *hufelandi* group). *Genus* **2004**, *15*, 141–152.
37. Pilato, G.; Kaczmarek, Ł.; Michalczyk, Ł.; Lisi, O. *Macrobiotus polonicus*, a new species of Tardigrada from Poland (Eutardigrada: Macrobiotidae, '*hufelandi* group'). *Zootaxa* **2003**, *258*, 1–8. [CrossRef]
38. Roszkowska, M.; Ostrowska, M.; Stec, D.; Janko, K.; Kaczmarek, Ł. *Macrobiotus polypiformis* sp. nov., a new tardigrade (Macrobiotidae; *hufelandi* group) from the Ecuadorian Pacific coast, with remarks on the claw abnormalities in eutardigrades. *Eur. J. Taxon.* **2017**, *327*. [CrossRef]
39. Kuzdrowska, K.A.; Mioduchowska, M.; Gawlak, M.; Bartylak, T.; Kepel, A.; Kepel, M.; Kaczmarek, Ł. Integrative description of *Macrobiotus porifini* sp. nov. (Macrobiotidae) from Madagascar and its phylogenetic position within the *hufelandi* group. *Eur. Zool. J* **2021**, *88*, 375–389. [CrossRef]
40. Pilato, G.; Kiosya, Y.; Lisi, O.; Sabella, G. New records of Eutardigrada from Belarus with the description of three new species. *Zootaxa* **2012**, *3179*, 39–60. [CrossRef]
41. Kayastha, P.; Berdi, D.; Mioduchowska, M.; Gawlak, M.; Łukasiewicz, A.; Gołdyn, B.; Kaczmarek, Ł. Some tardigrades from Nepal (Asia) with integrative description of *Macrobiotus wandae* sp. nov. (Macrobiotidae: Hufelandi Group). *Ann. Zool.* **2020**, *70*, 121–142. [CrossRef]
42. Casquet, J.; Thebaud, C.; Gillespie, R.G. Chelex without boiling, a rapid and easy technique to obtain stable amplifiable DNA from small amounts of ethanol-stored spiders. *Mol. Ecol. Resour.* **2012**, *12*, 136–141. [CrossRef]
43. Kaczmarek, Ł.; Grobys, D.; Kulpa, A.; Bartylak, T.; Kmita, H.; Kepel, M.; Kepel, A.; Roszkowska, M. Two new species of the genus *Milnesium* Doyère, 1840 (Tardigrada, Apochela, Milnesiidae) from Madagascar. *ZooKeys* **2019**, *884*, 1–22. [CrossRef]
44. Kaczmarek, Ł.; Roszkowska, M.; Poprawa, I.; Janelt, K.; Kmita, H.; Gawlak, M.; Fiałkowska, E.; Mioduchowska, M. Integrative description of bisexual *Paramacrobiotus experimentalis* sp. nov. (Macrobiotidae) from Republic of Madagascar (Africa) with microbiome analysis. *Mol. Phyl. Evol.* **2020**, *145*, 106730. [CrossRef]
45. Folmer, O.; Black, M.; Hoeh, W.; Lutz, R.; Vrijenhoek, R. DNA primers for amplification of mitochondrial cytochrome c oxidase subunit I from diverse metazoan invertebrates. *Mol. Mar. Biol. Biotechnol.* **1994**, *4*, 294–299.
46. Sands, C.J.; McInnes, S.J.; Marley, N.J.; Goodall-Copestake, W.P.; Convey, P.; Linse, K. Phylum Tardigrada: An "individual" approach. *Cladistics* **2008**, *24*, 861–871. [CrossRef]
47. Mironov, S.V.; Dabert, J.; Dabert, M. A new feather mite species of the genus *Proctophyllodes* Robin, 1877 (Astigmata: Proctophyllodidae) from the Long-tailed Tit *Aegithalos caudatus* (Passeriformes: Aegithalidae)—Morphological description with DNA barcode data. *Zootaxa* **2012**, *3253*, 54–61. [CrossRef]
48. White, T.J.; Bruns, T.; Lee, S.; Taylor, J. Amplification and direct sequencing of fungal ribosomal RNA genes for phylogenetics. In *PCR Protocols*; Academic Press: Cambridge, MA, USA, 1990; pp. 315–322. ISBN 9780123721808.
49. Hall, T. BioEdit: A user-friendly biological sequence alignment editor and analysis program for Windows 95/98/NT. *Nucleic Acids Symp.* **1999**, *41*, 95–98.
50. Rice, P.; Longden, I.; Bleasby, A. EMBOSS: The European molecular biology open software suite. *Trends Genet.* **2000**, *16*, 276–277. [CrossRef]
51. Goujon, M.; McWilliam, H.; Li, W.; Valentin, F.; Squizzato, S.; Paern, J.; Lopez, R. A new bioinformatics analysis tools framework at EMBL–EBI. *Nucleic Acids Res.* **2010**, *38*, 695–699. [CrossRef] [PubMed]
52. Altschul, S.F.; Gish, W.; Miller, W.; Myers, E.W.; Lipman, D.J. Basic Local Alignment Search Tool. *J. Mol. Biol.* **1990**, *215*, 403–410. [CrossRef]

53. Thompson, J.D.; Higgins, D.G.; Gibson, T.J. CLUSTAL W: Improving the sensitivity of progressive multiple sequence alignment through sequence weighting, position-specific gap penalties and weight matrix choice. *Nucleic Acids Res* **1994**, *22*, 4673–4680. [CrossRef]
54. Kumar, S.; Stecher, G.; Li, M.; Knyaz, C.; Tamura, K. MEGA X: Molecular evolutionary genetics analysis across computing platforms. *Mol. Biol. Evol.* **2018**, *35*, 1547–1549. [CrossRef]
55. Doyère, M. Mémoire sur les tardigrades. *Ann. Sci. Nat. Zool.* **1840**, *2*, 269–362.
56. Marcus, E. Zur Ökologie und Physiologie der Tardigraden. *Zool. Jahrb. Abt. Syst.* **1927**, *44*, 323–370.
57. Richters, F. Tardigrada. In *Handbuch der Zoologie*, 3rd ed.; Kükenthal, W., Krumbach, T., Eds.; Walter de Gruyter & Co.: Berlin, Germany; Leipzig, Germany, 1926; Volume 3, pp. 58–61.
58. Durante Pasa, M.V.; Maucci, W. Tardigradi muscicoli dell'Istria con descrizione di due specie nuove. *Mem. Ist. Ital. Idrobiol. Pallanza* **1975**, *32*, 69–91.
59. McInnes, S.J. Zoogeographic distribution of terrestrial/freshwater tardigrades from current literature. *J. Nat. Hist.* **1994**, *28*, 257–352. [CrossRef]
60. Gąsiorek, P.; Stec, D.; Morek, W.; Marnissi, J.; Michalczyk, Ł. The tardigrade fauna of Tunisia, with an integrative description of *Bryodelphax maculatus* sp. nov. (Heterotardigrada: Echiniscidae). *Afr. Zool.* **2017**, *52*, 77–89. [CrossRef]
61. Jørgensen, A.; Møbjerg, N.; Kristensen, R.M. Phylogeny and evolution of the Echiniscidae (Echiniscoidea, Tardigrada)–an investigation of the congruence between molecules and morphology. *J. Zool. Syst. Evol. Res.* **2011**, *49*, 6–16. [CrossRef]
62. Vicente, F.; Fontoura, P.; Cesari, M.; Rebecchi, L.; Guidetti, R.; Serrano, A.; Bertolani, R. Integrative taxonomy allows the identification of synonymous species and the erection of a new genus of Echiniscidae (Tardigrada, Heterotardigrada). *Zootaxa* **2013**, *3613*, 557–572. [CrossRef]
63. Węglarska, B. Die Tardigraden (Tardigrada) Spitzbergens. *Acta Zool. Cracov.* **1965**, *11*, 43–52.
64. Gąsiorek, P.; Stec, D.; Zawierucha, K.; Kristensen, R.M.; Michalczyk, Ł. Revision of *Testechiniscus* Kristensen, 1987 (Heterotardigrada: Echiniscidae) refutes the polar-temperate distribution of the genus. *Zootaxa* **2018**, *4472*, 261–297. [CrossRef]
65. Schuster, R.O.; Nelson, D.R.; Grigarick, A.A.; Christenberry, D. Systematic criteria of the Eutardigrada. *Trans. Am. Microsc. Soc.* **1980**, *99*, 284–303. [CrossRef]
66. Stec, D.; Krzywański, Ł.; Michalczyk, Ł. Integrative description of *Macrobiotus canaricus* sp. nov. with notes on *M. recens* (Eutardigrada: Macrobiotidae). *Eur. J. Taxon.* **2018**, 1–36. [CrossRef]
67. Nowak, B.; Stec, D. An integrative description of *Macrobiotus hannae* sp. nov.(Tardigrada: Eutardigrada: Macrobiotidae: Hufelandi group) from Poland. *Turk. J. Zool.* **2018**, *42*, 269–286. [CrossRef]
68. Pilato, G.; Binda, M. Descrizione du una nuova specie di eutardigrado d'Australia *Macrobiotus joannae* n. sp. *Animalia* **1983**, *10*, 262–272.
69. Coughlan, K.; Stec, D. Two new species of the *Macrobiotus hufelandi* complex (Tardigrada: Eutardigrada: Macrobiotidae) from Australia and India, with notes on their phylogenetic position. *Eur. J. Taxon..* **2019**, *573*. [CrossRef]
70. Maucci, W. Prime notizie su Tardigradi "terrestri" del Madagascar con descrizione di tre specie nuove. *Boll. Mus. Civ. Stor. Nat. Verona* **1993**, *17*, 381–392.
71. Roszkowska, M.; Kaczmarek, Ł. *Macrobiotus noemiae* sp. nov., a new tardigrade species (Macrobiotidae: *Hufelandi* group) from Spain. *Turk. J. Zool.* **2019**, *43*, 331–348. [CrossRef]
72. Stec, D.; Kristensen, R.M.; Michalczyk, Ł. Integrative taxonomy identifies *Macrobiotus papei*, a new tardigrade species of the *Macrobiotus hufelandi* complex (Eutardigrada: Macrobiotidae) from the Udzungwa Mountains National Park (Tanzania). *Zootaxa* **2018**, *4446*, 273–291. [CrossRef]
73. Guidetti, R.; Peluffo, J.R.; Rocha, A.M.; Cesari, M.; de Peluffo, M.C.M. The morphological and molecular analyses of a new South American urban tardigrade offer new insights on the biological meaning of the *Macrobiotus hufelandi* group of species (Tardigrada: Macrobiotidae). *J. Nat. Hist.* **2013**, *47*, 2409–2426. [CrossRef]
74. Bertolani, R.; Guidetti, R.; Marchioro, T.; Altiero, T.; Rebecchi, L.; Cesari, M. Phylogeny of Eutardigrada: New molecular data and their morphological support lead to the identification of new evolutionary lineages. *Mol. Phy. Evol.* **2014**, *76*, 110–126. [CrossRef]
75. Wełnicz, W.; Grohme, M.A.; Kaczmarek, Ł.; Schill, R.O.; Frohme, M. ITS-2 and 18S rRNA Data from *Macrobiotus polonicus* and *Milnesium tardigradum* (Eutardigrada, Tardigrada). *J. Zool. Syst. Evol. Res.* **2011**, *49*, 34–39. [CrossRef]
76. Stec, D.; Morek, W.; Gąsiorek, P.; Blagden, B.; Michalczyk, Ł. Description of Macrobiotus scoticus sp. nov. (Tardigrada: Macrobiotidae: Hufelandi Group) from Scotland by means of integrative taxonomy. *Ann. Zool.* **2017**, *67*, 181–197. [CrossRef]
77. Vecchi, M.; Cesari, M.; Bertolani, R.; Jönsson, K.I.; Rebecchi, L.; Guidetti, R. Integrative systematic studies on tardigrades from Antarctica identify new genera and new species within Macrobiotoidea and Echiniscoidea. *Invert. Syst.* **2016**, *30*, 303–322. [CrossRef]
78. Itang, L.A.M.; Stec, D.; Mapalo, M.A.; Mirano-Bascos, D.; Michalczyk, L. An integrative description of *Mesobiotus dilimanensis*, a new tardigrade species from the Philippines (Eutardigrada: Macrobiotidae: Furciger group). *Raffles Bull. Zool.* **2020**, *68*, 19–31.
79. Mapalo, M.A.; Stec, D.; Mirano-Bascos, D.; Michalczyk, Ł. *Mesobiotus philippinicus* sp. nov., the first limnoterrestrial tardigrade from the Philippines. *Zootaxa* **2016**, *4126*, 411–426. [CrossRef]
80. Richters, F. Nordische tardigraden. *Zool. Anz.* **1903**, 168–172.
81. Murray, J. Tardigrada. The British Antarctic Expedition 1907–1909. *Rep. Sci. Inv. 1 Biol.* **1910**, *Part V*, 83–187.

82. Murray, J. XXV.—Arctic Tardigrada, collected by Wm. S. Bruce. *Earth Environ. Sci. Trans. R. Soc. Edinb* **1907**, *45*, 669–681. [CrossRef]
83. Binda, M.G.; Rebecchi, L. Precisazioni su *Macrobiotus furciger* Murray, 1907, e descrizione di *Macrobiotus pilatoi* n. sp. (Eutardigrada, Macrobiotidae). *Animalia* **1992**, *19*, 101–109.
84. Kaczmarek, Ł.; Roszkowska, M. A new eutardigrade from Costa Rica with taxonomical and zoogeographical remarks on Costa Rican tardigrades. *N. Z. J. Zool.* **2016**, *43*, 234–245. [CrossRef]
85. Kaczmarek, Ł.; Kosicki, J.; Roszkowska, M. Tardigrada of Bory Tucholskie National Park, Zaborski Landscape Park, and their surroundings (Pomerania Province, Poland). *Turk. J. Zool.* **2018**, *42*, 6–17. [CrossRef]
86. Roszkowska, M.; Ostrowska, M.; Grobys, D.; Kmita, H.; Kaczmarek, Ł. Some tardigrades from Italy, with an updated checklist of limno-terrestrial species from the country. *Acta Zool. Bulg.* **2019**, *71*, 167–174.
87. Guidetti, R.; Gneuß, E.; Cesari, M.; Altiero, T.; Schill, R.O. Life-history traits and description of the new gonochoric amphimictic *Mesobiotus joenssoni* (Eutardigrada: Macrobiotidae) from the island of Elba, Italy. *Zool. J. Linnean Soc.* **2019**, *188*, 848–859. [CrossRef]
88. Vuori, T.; Massa, E.; Calhim, S.; Vecchi, M. Tardigrades of Finland: New records and an annotated checklist. *Zootaxa* **2020**, *4851*, 477–521. [CrossRef] [PubMed]
89. Bertolani, R.; Rebecchi, L.; Giovannini, I.; Cesari, M. DNA barcoding and integrative taxonomy of *Macrobiotus hufelandi* CAS Schultze 1834, the first tardigrade species to be described, and some related species. *Zootaxa* **2011**, *2997*, 19–36. [CrossRef]
90. Cesari, M.; Bertolani, R.; Rebecchi, L.; Guidetti, R. DNA barcoding in tardigrada: The first case study on *Macrobiotus macrocalix* Bertolani & Rebecchi 1993 (Eutardigrada, Macrobiotidae). *Mol. Ecol. Res.* **2009**, *9*, 699–706. [CrossRef]
91. Cesari, M.; Giovannini, I.; Bertolani, R.; Rebecchi, L. An example of problems associated with DNA barcoding in tardigrades: A novel method for obtaining voucher specimens. *Zootaxa* **2011**, *3104*, 42–51. [CrossRef]
92. Coughlan, K.; Michalczyk, Ł.; Stec, D. *Macrobiotus caelestis* sp. nov., a new tardigrade species (Macrobiotidae: Hufelandi group) from the Tien Shan Mountains (Kyrgyzstan). *Ann. Zool.* **2019**, *69*, 499–513. [CrossRef]
93. Stec, D.; Arakawa, K.; Michalczyk, Ł. An Integrative Description of *Macrobiotus shonaicus* sp. nov.(Tardigrada: Macrobiotidae) from Japan with notes on its phylogenetic position within the *hufelandi* group. *PLoS ONE* **2018**, *13*, e0192210. [CrossRef]
94. Stec, D.; Tumanov, D.T.; Kristensen, R.M. Integrative taxonomy identifies two new tardigrade species (Eutardigrada: Macrobiotidae) from Greenland. *Eur. J. Taxon.* **2020**, *614*. [CrossRef]
95. Stec, D.; Roszkowska, M.; Kaczmarek, Ł.; Michalczyk, Ł. An integrative description of a population of *Mesobiotus radiatus* (Pilato, Binda & Catanzaro, 1991) from Kenya. *Turk. Zool. Derg.* **2018**, *42*, 523–540. [CrossRef]
96. Mapalo, M.; Stec, D.; Mirano-Bascos, D.; Michalczyk, Ł. An integrative description of a limnoterrestrial tardigrade from the Philippines, *Mesobiotus insanis*, new species (Eutardigrada: Macrobiotidae: Harmsworthi group). *Raffles Bull. Zool.* **2017**, *65*, 440–454.
97. Stec, D.; Kristensen, R.M. An integrative description of *Mesobiotus ethiopicus* sp. nov. (Tardigrada: Eutardigrada: Parachela: Macrobiotidae: Harmsworthi group) from the northern Afrotropic region. *Turk. J. Zool.* **2017**, *41*, 800–811. [CrossRef]
98. Stec, D. *Mesobiotus datanlanicus* sp. nov., a new tardigrade species (Macrobiotidae: Mesobiotus harmsworthi group) from Lâm Đồng Province in Vietnam. *Zootaxa* **2019**, *4679*, 164–180. [CrossRef]
99. Jørgensen, A.; Faurby, S.; Hansen, J.G.; Møbjerg, N.; Kristensen, R.M. Molecular phylogeny of Arthrotardigrada (Tardigrada). *Mol. Phyl. Evol.* **2010**, *54*, 1006–1015. [CrossRef]
100. Schill, R.O.; Steinbrück, G. Identification and differentiation of Heterotardigrada and Eutardigrada species by riboprinting. *J. Zool. Syst. Evol. Res.* **2007**, *45*, 184–190. [CrossRef]
101. Guil, N.; Giribet, G. A comprehensive molecular phylogeny of tardigrades—Adding genes and taxa to a poorly resolved phylum-level phylogeny. *Cladistics* **2010**, *28*, 21–49. [CrossRef]
102. Tumanov, D.V. Integrative description of Mesobiotus anastasiae sp. nov.(Eutardigrada, Macrobiotoidea) and first record of Lobohalacarus (Chelicerata, Trombidiformes) from the Republic of South Africa. *Eur. J. Taxon.* **2020**, *726*, 102–131.
103. Czechowski, P.; Sands, C.J.; Adams, B.J.; D'Haese, C.A.; Gibson, J.A.; McInnes, S.J.; Stevens, M.I. Antarctic Tardigrada: A first step in understanding molecular operational taxonomic units (MOTUs) and biogeography of cryptic meiofauna. *Invertebr. Syst.* **2012**, *26*, 526–538. [CrossRef]

diversity

MDPI

Article

Historical Zooplankton Composition Indicates Eutrophication Stages in a Neotropical Aquatic System: The Case of Lake Amatitlán, Central America

Sarahi Jaime [1,*], Adrián Cervantes-Martínez [1], Martha A. Gutiérrez-Aguirre [1], Eduardo Suárez-Morales [2], Julio R. Juárez-Pernillo [3], Elena M. Reyes-Solares [3] and Victor H. Delgado-Blas [1]

1 Departamento de Ciencias y Humanidades, Campus Cozumel, Universidad de Quintana Roo (UQROO), Avenida Andrés Quintana Roo. Col. San Gervasio, Cozumel 77600, QRO, Mexico; adcervantes@uqroo.edu.mx (A.C.-M.); margutierrez@uqroo.edu.mx (M.A.G.-A.); blas@uqroo.edu.mx (V.H.D.-B.)

2 Departamento de Sistemática y Ecología Acuática, Campus Chetumal, El Colegio de la Frontera Sur (ECOSUR), Avenida Centenario Km 5.5, Chetumal 77014, QRO, Mexico; esuarez@ecosur.mx

3 Autoridad para el Manejo Sustentable de la Cuenca del Lago de Amatitlán (AMSA), Kilómetro 22 CA-9, Bárcenas, Villa Nueva 6624-1700, Guatemala; jjpernillo@gmail.com (J.R.J.-P.); elena11reyes@gmail.com (E.M.R.-S.)

* Correspondence: 1518305@uqroo.mx

Citation: Jaime, S.; Cervantes-Martínez, A.; Gutiérrez-Aguirre, M.A.; Suárez-Morales, E.; Juárez-Pernillo, J.R.; Reyes-Solares, E.M.; Delgado-Blas, V.H. Historical Zooplankton Composition Indicates Eutrophication Stages in a Neotropical Aquatic System: The Case of Lake Amatitlán, Central America. *Diversity* **2021**, *13*, 432. https://doi.org/10.3390/d13090432

Academic Editor: Michael Wink

Received: 8 July 2021
Accepted: 2 September 2021
Published: 8 September 2021

Abstract: This paper presents a study of freshwater zooplankton biodiversity, deemed as a reliable indicator of water quality. The Guatemalan Lake Amatitlán, currently used as a water source, has shown signs of progressive eutrophication, with perceptible variations of the local zooplankton diversity. Biotic and abiotic parameters were determined at four sites of Lake Amatitlán (Este Centro, Oeste Centro, Bahía Playa de Oro, and Michatoya) in 2016 and 2017. The local composition, the species richness and abundance of zooplankton, and the system environmental parameters were analyzed during both years surveyed. Biological data suggesting eutrophication of this tropical system were obtained, including a high rotifer abundance (11 species: the rotifers *Brachionus havanaensis* (109 ind L^{-1}) and *Keratella americana* (304 ind L^{-1}) were the most abundant species in this lake). The presumably endemic diaptomid copepod species, *Mastigodiaptomus amatitlanensis,* was absent in our samples, but we report the unprecedented occurrence of two Asian cyclopoid copepods (i.e., *Thermocyclops crassus* and *Mesocyclops thermocyclopoides*) for Lake Amatitlán and Guatemala. The presence of larger zooplankters like adults and immature copepods (i.e., *Arctodiaptomus dorsalis*) and cladocerans (*Ceriodaphnia* sp.) at site "Este Centro" indicates a relatively healthy zooplankton community and represents a focal point for managing the conservation of this lake.

Keywords: conservation; eutrophication; exotic species; tropical lakes; zooplankton

1. Introduction

The knowledge of zooplankton in the Neotropical region is growing with fragmented studies. Therefore, it is likely that the species richness of zooplanktonic taxa is underestimated because of the presumably high diversity and scarcity of zooplankton taxonomists [1–3]. In addition, the progressive destruction of aquatic habitat and the progressive spread of exotic species threaten native biodiversity, ecosystem health, and environmental services.

The zooplankton community and abundance are closely linked to the trophic state of the water system; for this reason, its diversity has been deemed as an indicator of water quality [4]. In eutrophicated systems (at tropical and temperate latitudes), the dominance of microzooplankton is common, compared with larger organisms, owing to the increased availability of food and water conditions [5,6].

For four decades, the Guatemalan Lake Amatitlán has shown signs of progressive eutrophication related to anthropic factors (i.e., peripheral population growth and urbanization, intensive use of water for agricultural irrigation), thus promoting the advancement towards eutrophication, related to the input of nearly 50% of the untreated residual urban and industrial waters from Guatemala City [7–10]. Because of this, some actions have been proposed to address this problem, either from the governmental level (i.e., Autoridad para el Manejo Sustentable de la cuenca del lago Amatitlán, AMSA 1996) or from descriptive studies of the lake involving the lake zooplankton biodiversity, like those by Basterrechea-Díaz (1997) [7] and Brandorff (2012) [11]; however, studies related with tropical epicontinental waterbodies have been more focused on environmental factors rather than biological community attributes or general limnology [12,13]; thus, the zooplankton biodiversity in Guatemala remains largely unknown [14], with only a few studies in Guatemalan lakes [15,16]. Most studies in Lake Amatitlán and Guatemala are more focused on current data instead of historical analysis.

Based on the analysis of both, historical and current data of zooplankton biodiversity and environmental conditions of Lake Amatitlán, we present information on the zooplankton distribution, species richness, abundance, and its relation with successive changes of its trophic state.

2. Materials and Methods

2.1. Study Sites and Sampling Methods

Lake Amatitlán is the fourth largest lake in Guatemala, Central America, and one of the most emblematic waterbodies of this country. This lake is a warm monomictic waterbody in the highland of Guatemala, located at an altitude of 1186 m above sea level (m.a.s.l.), with an area of 15.2 km^2 and 11 km length and a maximum depth of 23 m. Its formation originated from volcanic activity of Pacaya, Fuego, and Agua in the late Quaternary [10,14,17].

Four sampled sites were considered: Este Centro (EC), Oeste Centro (OC), Bahía Playa de Oro (BPO), and Michatoya (MICH) to analyze the zooplankton species that inhabit the eastern and western regions of Lake Amatitlán (Figure 1). The latter two sites (BPO and MICH) are in the runoff of Villalobos and Michatoya rivers, respectively [14]. Water samples for biotic and abiotic variables were collected for 2016 and 2017 in the rainy (May–October) and dry seasons (November–April).

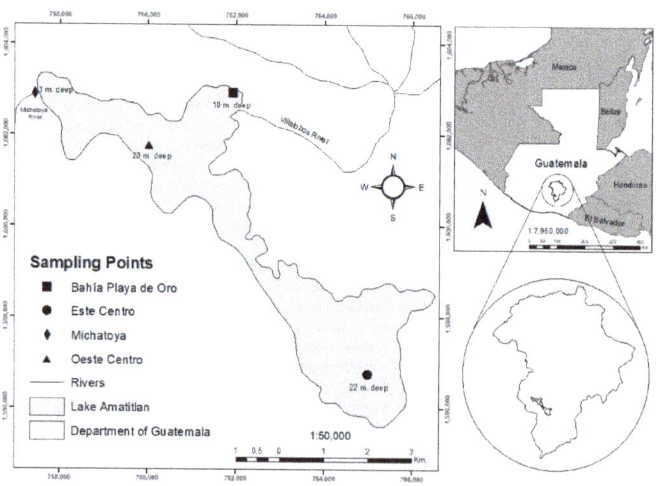

Figure 1. Location of Lake Amatitlán and sampling points for the biotic collection methodology. Water-filtered, vertical, and horizontal trawls as defined by Cervantes-Martínez & Gutiérrez-Aguirre (2015) [2].

2.1.1. Species Richness

Zooplankton samples ($n = 8$) were collected by vertical and horizontal trawls with a 45 µm plankton net between 1 and 22 m depth to ensure representative samples to evaluate the species richness in the lake, as it is well known that zooplankton tends to have vertical and horizontal migrations [2].

2.1.2. Species Abundance and Abiotic Variables

To estimate the zooplankton abundance, a known volume of water between 30 and 100 L was filtered through a 45 µm zooplankton net. The water was determined with a 2.1 L^{-1} capacity Van Dorn bottle [2,18]. Species abundance was determined by the account of two main groups: Rotifera and Copepoda, present in three aliquots of 1 mL each from the filtered samples, then the data were standardized as individuals per liter (ind L^{-1}) in each sampled site [19].

Abiotic variables were measured in situ monthly for both years of study and in all the water columns, with the multiparametric proves WTW Cond 197i, WTW Oxi 1970i, and HACH HQ for water temperature (°C), pH, oxygen concentration O_2 (mg L^{-1}), total dissolved solids (mg L^{-1}), and conductivity (µS cm^{-1}). With the actual environmental, richness, and zooplankton abundance data, a description of the trophic state of Lake Amatitlán was proposed.

2.2. Historical and Actual Records of Zooplankton and Environmental Parameters Analysis

Specific classification of Rotifera, Cladocera, and Copepoda of recently collected samples (collected in 2016 and 2017) was done according to Koste (1978) [20], Fontaneto & De Smet (2015) [21], Elías-Gutiérrez et al., (2008) [22], and Suárez-Morales et al., (2020) [23].

The presence/absence of the current zooplankton inventory was compared with previous surveys by Juday (1915) [24], Basterrechea-Díaz (1997) [7], and the record of copepods from the previous surveys of Wilson (1941) [25] and Brandorff (2012) [11], in order to analyze the historical composition of zooplankton of Amatitlán lake.

Historical environmental data recorded by Juday (1915) [24], Brezonik & Fox (1974) [26], Basterrechea-Díaz (1997) [7], and Ellenberg (2014) [27] were compared with the current data surveyed in this study.

3. Results

3.1. Species Richness

A total of 15 species of zooplankters including rotifers and crustaceans were found in the lake for 2016–2017 (Table 1); rotifers showed the highest species richness (80% of zooplankton species recorded), while copepods represented 20% of all zooplankton species in the lake.

We provide the first record of two cyclopoid exotic species (*Mesocyclops thermocyclopoides* and *Thermocyclops crassus*) for Lake Amatitlán and Guatemala. The endemic calanoid copepod, *Mastigodiaptomus amatitlanensis*, was absent in our current survey and the record of *Arctodiaptomus dorsalis* in Lake Amatitlán was confirmed here. Cladoceran crustaceans were very scarce in our samples; only a single specimen of *Ceriodaphnia* sp. was observed. The Brachionidae was the family with the highest species richness among rotifers in 2016 and 2017 (Table 1).

Nowadays, the east region (site EC) of Lake Amatitlán had the highest species richness in the lake (14 species), compared with the western region (9 species including the exotic *T. crassus* at OC). The largest zooplankters of the lake, including the cladoceran *Ceriodaphnia*, (~2 mm) [21], the calanoid copepod *A. dorsalis*, and the cyclopoid copepod *M. thermocyclopoides*, occurred in eastern region.

Table 1. Current and historical records of zooplankton species richness in Lake Amatitlán. Currently recorded species are shown in columns (1) EC, (2) OC, (3) BPO, and (4) MICH. Historical records are shown in columns 5–8, following data by Brandorff (2012) [11]; Basterrechea-Díaz (1997) [7]; Wilson (1941) [25]; and Juday (1915) [24], respectively. Presence (x), absence (-), new records (*).

Species	Current Data				Historical Data			
	1	2	3	4	5	6	7	8
Phylum: Rotifera Monogononta: Ploimida								
Family: Epiphanidae Harring, 1913								
Epiphanes macrourus Barrois & Daday, 1894 *	x	x	-	-	-	-	-	-
Family: Brachionidae Ehrenberg, 1838	-	-	-	-	-	-	-	-
Anuraeopsis fissa (Gosse, 1851) *	x	-	-	-	-	-	-	-
Brachionus angularis (Gosse, 1851) *	x	x	x	x	-	-	-	-
B. calyciflorus Pallas, 1766 *	x	x	x	x	-	-	-	-
B. plicatilis Müeller, 1786 *	x	-	-	x	-	-	-	-
B. havanaensis Rousselet, 1911 *	x	x	x	x	-	-	-	-
Keratella sp.	-	-	-	-	-	x	-	-
K. americana Carlin, 1943 *	x	x	x	x	-	-	-	-
K. cochleraris (Gosse, 1851)	-	-	-	-	-	-	-	x
Family: Trichocercidae Harring, 1913								
Trichocerca cf. *longiseta* (Schrank, 1802) *	-	x	x	x	-	-	-	-
T. pusilla (Lauterborn, 1898) *	-	x	x	x	-	-	-	-
Family: Asplanchnidae Eckstein, 1883								
Asplanchna sieboldi (Leydig, 1854) *	x	x	x	-	-	-	-	-
Flosculariaceae: Family: Trochosphaeridae Harring, 1913								
Filinia longiseta (Ehrenberg, 1834)	x	x	x	x	-	-	-	x
F. terminalis (Plate, 1886) *	x	x	x	-	-	-	-	-
Subclass: Bdelloidea *	x	-	-	-	-	-	-	-
Superclass: Crustacea Brachiopoda: Cladocera: Anomopoda								
Family: Daphniidae Straus, 1820								
Daphnia sp.	-	-	-	-	-	x	-	-
D. hyalina Leydig, 1860	-	-	-	-	-	-	-	x
Ceriodaphnia sp.	x	-	-	-	-	x	-	x
C. lacustris Birge, 1893	-	-	-	-	-	-	-	x
C. pulchella Sars, 1862	-	-	-	-	-	-	-	x
Family: Bosminidae Sars, 1865								
Bosmina sp.	-	-	-	-	-	x	-	-
Bosmina longirostris O. F. Müeller, 1776	-	-	-	-	-	-	-	x
Family: Chydoridae Stebbing, 1902								
Chydorus sphaericus (O.F. Müeller, 1785)	-	-	-	-	-	-	-	x
Copepoda: Calanoida Family: Diaptomidae G.O. Sars, 1932								
Subfamily: Diaptominae Kiefer, 1932								
Arctodiaptomus dorsalis (Marsh, 1907)	x	-	-	-	x	-	-	-
Mastigodiaptomus albuquerquensis (Herrick, 1895)	-	-	-	-	-	-	-	x
M. amatitlanensis (Wilson, 1941)	-	-	-	-	-	-	x	-
Copepoda: Cyclopoida Family: Cyclopidae Kiefer, 1927								
Subfamily: Cyclopinae Kiefer, 1927								
Thermocyclops crassus (Fischer, 1853) *	-	x	-	-	-	-	-	-
Mesocyclops thermocyclopoides Harada, 1931 *	x	-	-	-	-	-	-	-
Subfamily: Eucyclopinae Kiefer, 1927								
Eucyclops serrulatus (Fischer, 1851)	-	-	-	-	-	-	-	x
Nauplii	x	x	x	x	-	x	-	x
Juvenile Cyclopoid	x	x	x	x	-	-	-	-
Juvenile Calanoid	x	x	x	x	-	-	-	-

Our revision of the zooplankton community (Table 1) indicates that the historical data presented a great microcrustacean richness with the record of eight cladoceran species (*Dapnia* sp., *D. hyalina*, *Ceriodaphnia* sp. *C. lacustris*, *C. pulchella*, *Bosmina* sp., *B. longirostris*, and *Chydorus sphaericus*) and the three calanoid copepods: *A. dorsalis*, *Mastigodiaptomus albuquerquensis*, and the endemic *M. amatitlanensis*. The historical record of rotifers had the lowest species richness including three monogonont species. In our survey, the rotifer species richness increased significantly with 12 species not hitherto reported from the lake, including the record of organisms from the Subclass Bdelloidea.

3.2. Species Abundance

In this study, the total rotifer abundance was 522.7 ind L^{-1}. Rotifers represent the most abundant group in the lake; their numerical abundance is considerably higher than that recorded for copepods, including immature stages (7.1 ind L^{-1}). Cladocerans were almost absent from our samples.

Species with the highest abundance at all sites were the rotifers *B. havanaensis* (109 ind L^{-1}) and *K. americana* (304 ind L^{-1}), with a considerably lower abundance in the eastern area (9.3 and 121.8 ind L^{-1}, respectively). Species of the family Brachionidae were the most abundant mainly in the western region (sites OC, BPO, and MICH), whereas the lowest abundance of rotifers occurred in the eastern region (site EC) (see Table 2).

Table 2. Abundance (ind L^{-1}) calculated from zooplankton samples for all the studied points of Lake Amatitlán in 2017.

Species	Abundance (ind L^{-1})			
	EC	OC	BPO	MICH
Brachionusangularis	0.00	0.00	71.56	1.87
Brachionuscalyciflorus	0.70	0.70	85.56	4.67
Brachionusplicatilis	0.00	0.47	0.00	0.00
Trichocerca cf. *longiseta*	0.00	0.00	14.00	12.60
Trichocercapusilla	0.00	8.40	13.22	11.20
Asplanchnasieboldi	1.40	1.17	2.33	0.93
Filinialongiseta	0.00	1.40	28.00	13.07
Filiniaterminalis	8.17	49.23	35.00	60.20
Brachionushavanaensis	9.33	153.53	108.89	165.20
Keratellaamericana	121.80	432.60	265.22	408.33
Nauplii	3.50	2.57	3.11	1.40
Juvenile Cyclopoid	5.83	1.63	1.56	0.93
Juvenile Calanoid	2.80	0.47	3.89	0.47
M. thermocyclopoides	0.23	0.00	0.00	0.00

The local copepod abundance was represented mainly by nauplii and juvenile stages of Calanoida and Cyclopoida (average = 2.6, 2.5 and 1.9 ind L^{-1}, respectively), values resembling those recorded for the Rotifera like *B. plicatilis* (1.1 ind L^{-1}) and *A. sieboldi* (2.3 ind L^{-1}) in all the study sites, compared with adult copepods, where the abundance of the adult *M. thermocyclopoides* present only in EC was 0.23 ind L^{-1}.

3.3. Environmental Variables

Environmental variables values in both analyzed years, in general, presented basic pH values (>8 $\pm$ 0.33), dissolved oxygen showed an average of 4.76 $\pm$ 5.21 and 4.65 $\pm$ 4.92 mg L^{-1}, whereas temperature averaged 24 $\pm$ 1.31 °C, conductivity presented average values of 655.95 $\pm$ 59.52 and 678.23 $\pm$ 68.29 µS cm^{-1}, and finally TDS showed average values of 339.99 $\pm$ 47.35 and 341.43 $\pm$ 30.30 mg L^{-1}, respectively (Table 3).

Table 3. Historical and current environmental mean data of the water column recorded by previous surveys and this study. Juday (1915) [24], Brezonik & Fox (1974) [26], Basterrechea-Díaz (1997) [7], Ellenberg (2014) [27]. ND: no data available.

Environmental Variables	1910 [24]	1969 [26]	1985–1995 [7]	2008–2013 [27]	2016 ***	2017 ***
pH	ND	7.70	7.75	8.69	8.26	8.33
Water temperature (°C)	19.86	ND	22.75	25.00	24.46	24.23
Conductivity (μS cm^{-1})	ND	830	802	682	655.95	678.23
TDS (mg L^{-1})	ND	ND	610	ND	339.99	341.43
Dissolved oxygen O_2 (mg L^{-1})	4.74 *	8.40 **	4.20	8.90	4.76	4.65

* Data originally recorded in cubic centimeters per liter of water. ** Original data recorded at surface. *** Data recorded in this study.

The historical data presented in Table 3 show pH with slightly neutral values in 1969 to 1985–1995, whereas in the first two decades of the XXI century, the pH increased to reach clearly basic values, over 8. The water temperature changed along the time, 19.86 °C in 1910 to 24.23 °C in 2017. Conductivity and total dissolved solids decreased on average by 18.29 and 44.10%, respectively.

4. Discussion

The environmental parameters surveyed in this study can show the progressive eutrophication on Lake Amatitlán, according to the historical data recorded by authors like Juday (1915) [24], Brezonik & Fox (1974) [26] Basterrechea-Díaz (1997) [7], and Ellenberg (2014) [27]. The historical change in environmental and biological variables could reveal strong evidence of the current eutrophication of this lake. For instance, the observed changes of pH values, that is, an average of 8.26 and 8.33 in 2016–2017, differ in contrast from the values recorded in 1969 (7.70) [26], 1985–1995 (7.75) [7], and 2008 (9.3) [17].

The basic pH and the high concentration of dissolved oxygen at the surface promoted an increase of microzooplankters, like rotifers (especially *B. havanaensis* and *K. americana*), and a decrease of larger species like cladocerans and adult copepods, indicators of the system trophic state per se. Similar conditions have been recorded in American eutrophicated subtropical and tropical water bodies [4,28,29] as well as in other water bodies (i.e., temperate coastal water bodies) in which the replacement of larger copepod with smaller ones has been reported to the result from the eutrophication process [6].

Recently, phytoplankton blooming has been described as a consequence of this eutrophication progress in Lake Amatitlán, presenting a high concentration mainly in *Microcystis* sp. and *Dolichospermum* sp. cyanobacteria preceded by the diatom algae *Niszcha* sp. at the surface of the lake [9], which in turn allows herbivorous zooplankters like brachionid rotifers to become dominant organisms in eutrophicated epicontinental waterbodies [20].

In earlier studies on Lake Amatitlán, the zooplankton community was largely dominated by cladocerans and copepods. In 1915 [24], zooplankton had a widely different composition compared with our results: rotifers were then the less abundant zooplankton group in the lake (0.3 ind L^{-1}), preceded by copepods (11.6 ind L^{-1}) and cladocerans, the most abundant zooplankton group at that time (14.4 ind L^{-1}). The system trophic state is also related to the zooplankters body size; that is, a stronger level of eutrophication is frequently expressed by a greater abundance and species richness of microzooplankters like small rotifers [4,6,28,29]. A possible explanation of the local absence or scariness of larger zooplankters (i.e., *Ceriodaphnia* sp., adult cyclopoid and calanoid copepods, including *M. amatitlanensis*) could result from the competition for available food [5], eventually explaining the strong dominance of small brachionid herbivorous rotifers like *B. havanaensis* and *K. americana*.

The presence and high abundance of these latter species, together with another species of *Brachionus* and *Keratella* at the east region of Lake Amatitlán, suggest that eutrophic conditions that make food available for these microphagous species [30].

In the case of *A. dorsalis*, this species is widespread in America [31] and has been recorded as an invasive exotic copepod in Asiatic waterbodies [32,33]. The environmental conditions of Lake Amatitlán seem to be adequate for the development of this species because it shows a selective feeding on phytoplankton; thus, it frequently inhabits moderately to strongly eutrophicate environments [31,32], like Amatitlán lake.

It is well known that many diaptomid copepods tend to have restricted distributional patterns and endemic distributions in neotropical lakes [34]. Then, the local absence of the endemic copepod *M. amatitlanensis* in this study could be another indicator of the progressive eutrophication of Lake Amatitlán, because, since its description by Wilson (1941) [25], this species has not been recorded in other regional studies (i.e., Elías-Gutiérrez et al., 2008 [35]; Brandorff, 2012 [11]; and Gutiérrez-Aguirre, et al., 2020 [36]). It is probable that *M. amatitlanensis* occurs in other lakes of Guatemala (or Central America) and it is expected to be collected from adjacent systems. It is also probable that this species dwells at higher depths not easily reached by standard nets.

Our results showed a clear zonification; the eastern region (site EC) diverges from the other sites because of the absence of adjacent rivers (see Figure 1), its distance from the other sampling points (the closest site is OC, 7.04 km away), and its separation from other sites owing to a train riel that divides the lake in two [14]. Therefore, the EC area has the best conservation status of the lake, precisely where we found the greatest species richness and the larger zooplankters, with the copepods *T. crassus* (average body length of 0.56–0.93 mm) [37], *M. thermocyclopoides* (0.78–0.89 mm) [38], and *A. dorsalis* (0.77–1.13 mm) [31] among them. Thus, it is convenient to consider EC as a potential conservation site as it has better environmental conditions for the conservation and preservation of zooplankton biodiversity.

On the other hand, we report the presence of two exotic cyclopoid copepod species for the Central American Lake Amatitlán and Guatemala country, *M. thermocyclopoides* and *T. crassus*. *M. thermocyclopoides* is a native species from Taiwan and is well spread in Asia and Africa, and commonly widespread at tropical latitudes. This species has been recorded in lakes from South Mexico in epicontinental waterbodies from Chiapas state, Mexico, considering that their introduction may be related to anthropic factors (i.e., agriculture and aquaculture) [37,38]. This is the second record of the invasion of this species in Central American countries, as it has been recorded before in Costa Rican water bodies by Collado et al. (1894) [39], and the ecological potential of *Mesocyclops* use as biocontrol of vector mosquitoes like *Aedes aegypti* is well known [40–42]. Therefore, its finding in Guatemalan lakes represents a source for mass culture of this copepod to be used as biocontrol.

Thermocyclops crassus is commonly spread at tropical latitudes in Africa, Australia, and Asia; it was also recorded in Laurentian great lakes in the United States of America [43]; recorded for the first time in tropical lakes from Tabasco state, Mexico [37]; as well as in small ponds of San José Province in Costa Rica [39]. Being a thermophilic species, *T. crassus* has a narrow temperature tolerance [44], so it may be a local indicator of the temperature changes in the lake along time.

Finally, the physical, chemical, and biological conditions of the lake have clearly changed over time, from being a lake with oligotrophic characteristics to one with hypertrophic conditions in a relatively short period of time (100 years, approximately), allowing us to follow and describe the stages and speed of the eutrophication process of a large neotropical lake.

5. Conclusions

The historical analysis of zooplankton composition in the lake presented in this study reinforces the knowledge of its eutrophic state, suggesting a useful role of the zooplankton as a bioindicator and making possible the visualization of the changes in its composition over time, showing the progressive trophic state towards eutrophic or hypereutrophic conditions.

It is likely that the absence of the endemic species *M. amatitlanensis* is a warning sign regarding the accelerated loss of biodiversity and reinforces the idea that zooplankton is a great tool as a bioindicator of the health status for continental aquatic ecosystems, in both tropical and temperate latitudes.

Further studies analyzing bottom sediments to search resting eggs of zooplankton in Lake Amatitlán and around it can answer the question of the absence of *M. amatitlanensis*, where this type of knowledge is also scarce in inland aquatic systems of the region.

Finally, is convenient to consider the isolated site EC as a focal point for conservation as it presents better environmental conditions for the conservation and preservation of zooplankton biodiversity, owing to the record of the largest zooplankters found in this site.

Author Contributions: Conceptualization, S.J. and A.C.-M.; methodology, S.J., A.C.-M., M.A.G.-A., E.S.-M., J.R.J.-P., E.M.R.-S. and V.H.D.-B.; software, S.J.; validation, S.J., A.C.-M., M.A.G.-A., E.S.-M., J.R.J.-P., E.M.R.-S. and V.H.D.-B.; formal analysis, S.J., A.C.-M., M.A.G.-A., E.S.-M., J.R.J.-P., E.M.R.-S. and V.H.D.-B.; investigation, S.J., A.C.-M., M.A.G.-A., E.S.-M., J.R.J.-P., E.M.R.-S. and V.H.D.-B.; resources, J.R.J.-P. and E.M.R.-S.; data curation, S.J.; writing—original draft preparation, S.J.; writing—review and editing, S.J., A.C.-M., M.A.G.-A., E.S.-M., J.R.J.-P., E.M.R.-S. and V.H.D.-B.; visualization, S.J. and A.C.-M.; supervision, S.J. and A.C.-M.; project administration, S.J., A.C.-M., M.A.G.-A., E.S.-M., J.R.J.-P., E.M.R.-S. and V.H.D.-B. All authors have read and agreed to the published version of the manuscript.

Funding: This research was funded by the National Council of Science and Technology (CONACYT) with the Investigator SNI-III Assistant Program, Grant Number 19812.

Institutional Review Board Statement: Zooplankton collected in this study was provided by AMSA in Guatemala. Zooplankton is not under any protection by Guatemalan laws; therefore, no specific permits are required for this kind of field.

Data Availability Statement: The data presented in this study are available on request from the corresponding author due to the need of further use to complete the first author's postgraduate requirements.

Acknowledgments: We gratefully acknowledge the invitation of Manuel Elías-Gutiérrez to contribute to this special issue of the Selected Papers from 1st International Electronic Conference on Biological Diversity, Ecology, and Evolution, and his valuable suggestions that made possible this research. We acknowledge AMSA for all the biological material and data bases that made possible the present study. Anonymous reviewers made valuable comments and suggestions. University of Quintana Roo, Cozumel with the investigation group vulnerability of continental and coastal water systems and the National Council of Science and Technology (CONACYT) with the Investigator SNI-III Assistant Program (Number 19812) provided financial support.

Conflicts of Interest: The authors declare no conflict of interest.

References

1. José de Paggi, S.B.; Wallace, R.; Fontaneto, D.; Marinone, M.C. Phylum Rotifera. In *Thorp and Covich's Freshwater Invertebrates*; Damborenea, C., Rogers, C.D., James, T., Eds.; Elsevier Science Publishing Co Inc.: San Diego, CA, USA, 2020; pp. 145–200.
2. Cervantes-Martínez, A.; Gutiérrez-Aguirre, M.A. Physicochemistry and Zooplankton of Two Karstic Sinkholes in the Yucatan Peninsula, Mexico. *J. Limnol.* **2015**, *74*, 382–393. [CrossRef]
3. Suárez-Morales, E.; Reid, J.W.; Ilige, T.; Fiers, F. *Catálogo de Los Copépodos (Crustacea) Continentales de La Península de Yucatán, México*; Chetumal, Quintana Roo, México; Comisión Nacional para el Conocimiento y Uso de la Biodiversidad: Ciudad de México, Mexico, 1996; 296p.
4. Gómez-Márquez, J.L.; Peña-Mendoza, B.; Guzmán-Santiago, J.L.; Gallardo-Pineda, V. Composición, Abundancia Del Zooplancton y Calidad de Agua En Un Microreservorio En El Estado de Morelos. *Hidrobiologica* **2013**, *23*, 227–240.
5. Gama Flores, J.L.; Sarma, S.S.S.; López Rocha, A.N.; Nandini, S. Effects of Cladoceran-Conditioned Medium on the Demography of Brachionid Rotifers (Rotifera: Brachionidae). *Hydrobiologia* **2018**, *844*, 21–30. [CrossRef]
6. Uye, S.I. Replacement of Large Copepods by Small Ones with Eutrophication of Embayments: Cause and Consequence. *Hydrobiologia* **1994**, *292–293*, 513–519. [CrossRef]
7. Basterrechea-Díaz, M. *El Lago de Amatitlán: Década de Estudios Limnológicos 1985–1995*; Academia de Ciencias Médicas, Físicas y Naturales de Guatemala: Guatemala City, Guatemala, 1997; 45p.
8. Sigui, N. ¿Por Qué Continúa La Contaminación de Aguas En Guatemala? *Cienc. Tecnol. Salud* **2016**, *3*, 167–176. [CrossRef]
9. Rodas-Pernillo, E.; Vasquez-Moscoso, C.A.; García, O.F. Dinámica Del Consumo y Aporte de Nutrientes de Fitoplancton, Dominado Por *Microcystis* sp. (Cyanophyceae) Del Lago de Amatitlán. *Cienc. Tecnol. Salud* **2020**, *7*, 2409–3459.
10. Richardson-Varas, R.; Muñoz-Luza, M.; Landeros-Cáceres, F.; Contreras-Celis, V.; Carranza-González, J.; Nuñez, M.; Hernández, E.; Alfonso-Álvarez, R.; Ramos, E.; Mazariegos, J.; et al. Índice de Fragilidad Ambiental En Las Cuencas Hidro-Geomorfológicas Del Lago Peñuelas, Chile y Del Lago Amatitlán, Guatemala. *Rev. Geográfica* **2015**, *156*, 98–109.

11. Brandorff, G.O. Distribution of Some Calanoida (Crustacea: Copepoda) from the Yucatán Peninsula, Belize and Guatemala. *Rev. Biol. Trop.* **2012**, *60*, 187–202. [CrossRef] [PubMed]
12. Cervantes-Martínez, A.; Elías-Gutiérrez, M.; Suárez-Morales, E. Limnological and Morphometrical Data of Eight Karstic Systems "cenotes" of the Yucatan Peninsula, Mexico, during the Dry Season (February–May, 2001). *Hydrobiologia* **2002**, *482*, 167–177. [CrossRef]
13. Schmitter-Soto, J.J.; Comín, F.A.; Escobar-Briones, E.; Herrera-Silveira, J.; Alcocer, J.; Suárez-Morales, E.; Elías-Gutiérrez, M.; Díaz-Arce, V.; Marín, L.E.; Steinich, B. Hydrogeochemical and Biological Characteristics of Cenotes in the Yucatan Peninsula (SE Mexico). *Hydrobiologia* **2002**, *467*, 215–228. [CrossRef]
14. Jiménez, S.; Juárez, J.; Trujillo, L.; Dubón, S.; Valenzuela, O.; Castro, A.M. *Calidad de Agua de La Cuenca y Lago de Amatitlán*; División de Control, Calidad Ambiental y Manejo de Lagos: Guatemala City, Guatemala, 2015; 32p.
15. Elías-Gutiérrez, M.; Kotov, A.A.; Garfias-Espejo, T. Cladocera (Crustacea: Ctenopoda, Anomopoda) from Southern Mexico, Belize and Northern Guatemala, with Some Biogeographical Notes. *Zootaxa* **2006**, *1119*, 1–27. [CrossRef]
16. García-Morales, A.E.; Elías-Gutiérrez, M. The Rotifer Fauna of Guatemala and Belize: Survey and Biogeographical Affinities. *Rev. Biol. Trop.* **2007**, *55*, 569–584.
17. Pérez, L.; Bugja, R.; Lorenschat, J.; Brenner, M.; Curtis, J.; Hoelzmann, P.; Islebe, G.; Scharf, B.; Schwalb, A. Aquatic Ecosystems of the Yucatán Peninsula (Mexico), Belize, and Guatemala. *Hydrobiologia* **2011**, *661*, 407–433. [CrossRef]
18. Cervantes-Martínez, A.; Elías-Gutiérrez, M.; Gutiérrez-Aguirre, M.A.; Kotov, A.A. Ecological Remarks on *Mastigodiaptomus Nesus* Bowman, 1986 (Copepoda: Calanoida) in a Mexican Karstic Sinkhole. *Hydrobiologia* **2005**, *542*, 95–102. [CrossRef]
19. Ramírez García, P.; Nandini, S.; Sarma, S.S.S.; Robles Valderrama, E.; Cuesta, I.; Hurtado, M.D. Seasonal Varations of Zooplankton Abundance in the Freshwater Reservoir Valle de Bravo (Mexico). *Hydrobiologia* **2002**, *467*, 99–108. [CrossRef]
20. Koste, W. *Rotatoria: Die Rädertiere Mitteleuropas: Ein Bestimmunswerk Bengründet von Max Voigt Überordnung Monogononta*; Gebrüder Borntraeger: Berlin, Germany; Stuttgart, Germany, 1978; 234p.
21. Fontaneto, D.; De Smet, W.H. Rotifera. In *Handbook of Zoology Gastrotricha, Cycloneuralia and Gnathifera*; Rhaera, A.S., Ed.; De Gruyter-GmbH: Berlin, Germany, 2015; Volume 3, pp. 196–217.
22. Elías-Gutiérrez, M.; Suárez-Morales, E.; Gutiérrez-Aguirre, M.A.; Silva-Briano, M.; Granados-Ramírez, J.G.; Garfias-Espejo, T. *Cladocera y Copepoda de Las Aguas Continentales de México. Guía Ilustrada*; Universidad Autónoma de México: Ciudad de México, Mexico, 2008; 322p.
23. Suarez-Morales, E.; Gutiérrez-Aguirre, M.A.; Gómez, S.; Perbiche-Neves, G.; Previattelli, D.; Dos Santos-Silva, N.; da Rocha, C.E.F.; Mercado-Salas, N.F.; Manriquez, T.M.; Cruz-Quintana, Y.; et al. Class Copepoda. In *Thorp and Covich's Freshwater Invertebrates: Volume 5: Keys to Neotropical and Antarctic Fauna*; Damborenea, C., Rogers, C.D., James, T., Eds.; Elsevier Science Publishing Co Inc.: San Diego, CA, USA, 2020; pp. 663–796.
24. Juday, C. Limnological Studies on Some Lakes in Central America. *Wisconsin Acad. Sci. Arts Lett.* **1915**, *18*, 214–250.
25. Wilson, M.S. New Species and Distribution Records of Diaptomid Copepods from the Marsh Collection in the United States National Museum. *J. Washingt. Acad. Sci.* **1941**, *31*, 509–515.
26. Brezonik, P.L.; Fox, J.L. The Limnology of Selected Guatemalan Lakes. *Hydrobiologia* **1974**, *45*, 467–487. [CrossRef]
27. Ellenberg, R.L. Limnology of Lake Amatitlán in Guatemala and Its Eutrophication Process. Ph.D. Thesis, Technical University of Berlin, Berlin, Germany, 2014; 128p.
28. Frutos, S.M.; Poi, A.S.G.; Neiff, J.J. Zooplankton Abundance and Species Diversity in Two Lakes with Different Trophic States (Corrientes, Argentina) Abundância e Diversidade Específica Do Zooplâncton Em Dois Lagos. *Acta Limnol. Bras.* **2009**, *21*, 367–375.
29. Moreno-Gutiérrez, R.M.; Sarma, S.S.S.; Sobrino-Figueroa, A.S.; Nandini, S. Population Growth Potential of Rotifers from a High Altitude Eutrophic Waterbody, Madín Reseivoir (State of Mexico, Mexico): The Importance of Seasonal Sampling. *J. Limnol.* **2020**, *77*, 441–451. [CrossRef]
30. Obertegger, U.; Smith, H.A.; Flaim, G.; Wallace, R.L. Using the Guild Ratio to Characterize Pelagic Rotifer Communities. *Hydrobiologia* **2011**, *662*, 157–162. [CrossRef]
31. Reid, J.W. *Arctodiaptomus dorsalis* (Marsh): A Case History of Copepod Dispersal. *Banisteria* **2007**, *1860*, 3–18.
32. Papa, R.D.S.; Li, H.; Tordesillas, D.T.; Han, B.; Dumont, H.J. Massive Invasion of *Arctodiaptomus dorsalis* (Copepoda, Calanoida, Diaptomidae) in Philippine Lakes: A Threat to Asian Zooplankton Biodiversity? *Biol. Invasions* **2012**, *14*, 2471–2478. [CrossRef]
33. Metillo, E.B.; Masorong, A.M.; Macabangkit, S.A.N.; Licayan, J.R.U.; Tordesillas, D.T.; Papa, R.D.S. First Record of the Invasive *Arctodiaptomus dorsalis* (Marsh, 1907) (Copepoda: Calanoida: Diaptomidae) in Lake Lanao (Mindanao Is., Philippines). *Acta Manila. Ser. A* **2014**, *62*, 19–23.
34. Perbiche-Neves, G.; Previattelli, D.; Pie, M.R.; Duran, A.; Suárez-Morales, E.; Boxshall, G.A.; Nogueira, M.G.; da Rocha, C.E.F. Historical Biogeography of the Neotropical Diaptomidae (Crustacea: Copepoda). *Front. Zool.* **2014**, *11*, 36. [CrossRef]
35. Elías-Gutiérrez, M.; Jerónimo, F.M.; Ivanova, N.V.; Valdez-Moreno, M.; Hebert, P.D.N. DNA Barcodes for Cladocera and Copepoda from Mexico and Guatemala, Highlights and New Discoveries. *Zootaxa* **2008**, *1839*, 1–42. [CrossRef]
36. Gutiérrez-Aguirre, M.A.; Cervantes-Martínez, A.; Elías-Gutiérrez, M.; Lugo-Vázquez, A. Remarks on *Mastigodiaptomus* (Calanoida: Diaptomidae) from Mexico Using Integrative Taxonomy, with a Key of Identification and Three New Species. *PeerJ* **2020**, *8*, e8416. [CrossRef]

37. Gutiérrez-Aguirre, M.A.; Suárez-Morales, E. The Eurasian *Thermocyclops crassus* (Fischer, 1853) (Copepoda, Cyclopoida) Found in Southeastern Mexico. *Crustaceana* **2000**, *73*, 705–713. [CrossRef]
38. Gutiérrez-Aguirre, M.A.; Reid, J.W.; Suárez-Morales, E. An Afro-Asian Species of *Mesocyclops* (Copepoda: Cyclopoida) in Central America and Mexico. *J. Crustacean Biol.* **2003**, *23*, 352–363. [CrossRef]
39. Collado, C.; Defaye, D.; Dussart, B.H.; Fernando, C.H. The Freshwater Copepoda (Crustacea) of Costa Rica with Notes on Some Species. *Hydrobiologia* **1984**, *119*, 89–99. [CrossRef]
40. Gutiérrez-Aguirre, M.A.; Suárez-Morales, E.; Cervantes-Martínez, A.; Elías-Gutiérrez, M.; Previattelli, D. The Neotropical Species of *Mesocyclops* (Copepoda, Cyclopoida): An Upgraded Identification Key and Comments on Selected Taxa. *J. Nat. Hist.* **2006**, *40*, 549–570. [CrossRef]
41. Suárez-Morales, E.; Gutiérrez-Aguirre, M.A.; Elías-Gutiérrez, M. Observations on the Structure of Mandibular Gnathobase in Some American *Mesocyclops* (Copepoda: Cyclopidae). *Proc. Biol. Soc. Wash.* **2003**, *116*, 742–753.
42. Cervantes-Martínez, A.; Gutiérrez-Aguirre, M.A.; Delgado-Blas, V.H.; Ruíz-Ramírez, J.-D. *Especies de Zooplancton Dulceacuícola de Cozumel;* Quintana Roo, Universidad de Quintana Roo (UQROO): Cozumel, Mexico; Chetumal, Mexico, 2018; 86p.
43. Connolly, J.K.; Watkins, J.M.; Hinchey, E.K.; Rudstam, L.G.; Reid, J.W. New Cyclopoid Copepod (*Thermocyclops crassus*) Reported in the Laurentian Great Lakes. *J. Great Lakes Res.* **2017**, *43*, 198–203. [CrossRef]
44. Verbitsky, V.B.; Lazareva, V.I.; Medyantseva, E.N.; Malysheva, O.A.; Zhdanova, S.M.; Verbitskaya, T.I.; Grishanin, A.K. The Preferred and Avoidance Temperatures of *Thermocyclops crassus* (Fischer, 1853) and Their Relation to the Temperature of Optimal, Pessimal and Normal Performance of the Species. *J. Therm. Biol.* **2018**, *78*, 106–113. [CrossRef] [PubMed]

diversity

MDPI

Article

Biodiversity Evaluation: From Endorsed Indexes to Inclusion of a Pollinator Indicator

Sergio Albertazzi [1], Elisa Monterastelli [2], Manuela Giovanetti [1,*], Emanuele Luigi Zenga [1], Simone Flaminio [1], Marta Galloni [3], Marino Quaranta [1] and Laura Bortolotti [1]

1 CREA Research Centre for Agriculture and Environment, Via di Corticella 133, 40128 Bologna, Italy; sergio.albertazzi@crea.gov.it (S.A.); emanuele.zenga@crea.gov.it (E.L.Z.); simone.flaminio@crea.gov.it (S.F.); marino.quaranta@crea.gov.it (M.Q.); laura.bortolotti@crea.gov.it (L.B.)

2 Independent Researcher, Via di Corticella 85, 40128 Bologna, Italy; elisa.monterastelli@gmail.com

3 Deptartment of Biological, Geological and Environmental Sciences, University of Bologna, Via Irnerio 42, 40126 Bologna, Italy; marta.galloni@unibo.it

* Correspondence: manuela.giovanetti@crea.gov.it

Abstract: There is increasing interest in evaluating biodiversity to preserve ecosystem services. Researchers can sustain policymakers by providing tools, such as indexes and indicators, that need constant implementation to become accepted standards. Implementation may vary from re-evaluation of existing indicators to introduction of new ones based on emerging threats to biodiversity. With the aim of contributing to the compelling need to estimate and counterbalance pollinator loss, we screened existing bioindicators. We first selected indexes/indicators applied to agricultural contexts and concurrently endorsed by a regulatory agency. We then extended our analysis to indexes/indicators based on arthropod taxa and formally recognized at least by national bodies. Our procedure identified a combination of surveys of various animal taxa and remote landscape analyses (e.g., using a GIS and other cartographic tools). When the animals are arthropods, most indexes/indicators can only address confined environments (e.g., grasslands, riversides). Indicator strength was improved by the simultaneous inclusion of biotic and abiotic components. Pollinator sensitivity to changes at micro-habitat level is widely appreciated and may help distinguish agricultural practices. A biodiversity index based on pollinators, including a wide monitoring scheme supplemented by citizen science, is currently fostered at the European level. The results obtained using such an index may finally enable focusing of strategic funding. Our analysis will help to reach this goal.

Keywords: biodiversity; agroecosystems; arthropods; environment; pollinators; indicators; RDP measures

Citation: Albertazzi, S.; Monterastelli, E.; Giovanetti, M.; Zenga, E.L.; Flaminio, S.; Galloni, M.; Quaranta, M.; Bortolotti, L. Biodiversity Evaluation: From Endorsed Indexes to Inclusion of a Pollinator Indicator. *Diversity* **2021**, *13*, 477. https://doi.org/10.3390/d13100477

Academic Editor: Michael Wink

Received: 9 August 2021
Accepted: 25 September 2021
Published: 29 September 2021

Publisher's Note: MDPI stays neutral with regard to jurisdictional claims in published maps and institutional affiliations.

1. Introduction

The biodiversity of the agroecosystems is becoming crucial in European legislation as a key to tackle food security, human and environmental health and climate change. Biodiversity conservation has been introduced among the specific objectives of the European Common Agriculture Policy (CAP). The full title, maintained in the current version of the CAP, is "Contribute to the protection of biodiversity, enhance ecosystem services and preserve habitats and landscape" [1,2]. A complex system of indicators (of context, impact, result and output) has been implemented in subsequent CAP schemes. These indicators are basic tools for addressing EU achievements, especially those linked to rural development programs (RDPs). When biodiversity is no longer just a matter of scientific enquiry, it is necessary to discriminate between existing tools and/or promote development of new ones tailored to policy. These tools need to have a scientific background, but they also need to adjust to widely accepted economic and political criteria. They therefore need to be endorsed (i.e., tested by a scientific regulatory agency) and/or officially recognized (i.e., included in an existing regulation). Endorsement and/or official recognition can vary from one country to another, and information in the grey literature and national languages may be difficult to collect. At the European level, a superimposed list of tools may need to be included in national regulations.

For example, measuring biodiversity in agricultural systems has been tackled by previous versions of CAP through the Farmland Bird Index (FBI—impact indicator I.08 for CAP 2014–2020) and High Nature Value farming (HNV—impact indicator I.09 for CAP 2014–2020) [3]. FBI has been a pioneer among indexes, addressing farmland birds as elements of the quality of farmed environments. Birds are considered optimal subjects because they are near the top of the food chain. They have therefore been monitored and trends identified over the course of time through on-site bird surveys. The concept of HNV farming was established in the early 1990s [4–6] and refers to farming systems that favor biodiversity by traditional agricultural practices applied to wide areas. Rural development measures to preserve and develop HNV farming were fostered, and an impact indicator, mostly based on land cover and farming criteria, was applied. Depending on its employment, the index components may vary [7]. These indexes did not prove to be perfectly suited to the aims of CAP when it came to evaluating individual farm performance and RDPs. HNV farming will not be included in CAP post-2020 [8]. Furthermore, although FBI proved satisfactory at regional/national level, it was less relevant to local rural development measures [9]. Bird biology, based on areas usually larger than target farms, could be unsuitable for understanding the biodiversity of small–medium enterprise (SME) properties. With the specific objective mentioned above, CAP post-2020 establishes three indexes for biodiversity protection. FBI (now impact indicator I.18) will be retrieved as a proxy with a view to increasing farmland bird populations. Another indicator that will be employed is I.19, "Percentage of species and habitats of Community interest related to agriculture with stable or increasing trends", which assesses the trends of habitats and species listed in the relevant Habitats Directive annexes, on the basis of a strong connection between the presence and persistence of such species and habitats, and the sustainability and good environmental quality of a given agricultural ecosystem. The third indicator is I.20, "Agricultural land with landscape features", that estimates the area of landscape features of farmland (as a percentage of shared Utilized Agricultural Area—UAA). Landscape features that support biodiversity and ecosystem services, such as hedgerows, patches of trees, woodland, ponds, water bodies, streams and moderately managed areas like field margins, may be included. The unit of measurement, based almost entirely on cartographic information, has yet to be defined, but will hopefully remedy the lack of acknowledgement of farm-level RDPs for biodiversity improvement. I.19 and I.20 are not fully defined but will certainly be implemented through national/transnational debate. Despite intensive research, not all the indicators for evaluating the resulting impact of RDPs have yet been identified.

In recent decades, the relevance of other groups of animals, grouped under the term *pollinators*, acquired eminence when discussing biodiversity loss. In temperate areas, pollinators are mainly insects belonging to the orders Hymenoptera (especially bees), Diptera (especially hoverflies) and Lepidoptera (especially butterflies). Other insect groups (moths, beetles) and birds are also included. Merging with the above, an index on pollinators may soon be included in the list of indicators related to biodiversity. The Intergovernmental Science-Policy Platform on Biodiversity and Ecosystem Services (IPBES) released an assessment [10] intended to underline how closely pollinators are linked to agricultural productivity and food production. The assessment also fosters national pollinator strategies and action plans, since despite their importance, pollinators constantly bear the brunt of human activity. Since pollinators are a fundamental component of ecosystems, pressure on them needs to be monitored and quantified. The EU Pollinators Initiative began in 2018 and focused on the loss of wild pollinators and the possible consequences of losing wildflower species linked to them. It also focuses on the fact that a large part of annual agricultural production depends directly on insect pollinators. Some of the actions in the report [11] are particularly consistent with the context of our work: Action 1—Support monitoring and assessment; Action 5—Improve pollinator habitats on and around farmland. Accordingly, the EU Biodiversity Strategy for 2030 focuses on reversing the decline of pollinators [12], and the European Parliament has asked for a pollinator index to be included in CAP post-2020 [13]. The term *pollinators* includes many animal species and groups, all of which are insects at temperate latitudes. For specific insect groups that are also pollinators, e.g., butterflies [14] and syrphid flies [15], indexes/indicators already exist. However, there is not yet an index/indicator comprehensive for all pollinator species. The "Proposal for an EU Pollinator Monitoring Scheme" (EU-PoMS; [16]) intends to implement a cost-effective core scheme to foster EU national monitoring of all essential pollinators (wild bees, butterflies, hoverflies, moths, rare and threatened pollinator species) by standardized methods, to create a sound scientific base for "a general EU indicator of the status and trends of pollinators and a CAP-specific indicator to evaluate the impacts of Common Agricultural Policy on pollinators and pollination" [16]. This proposal will soon be tested in the field in a pilot study.

At the Italian level, the 2019 Directive on the conservation of biodiversity of the Ministry for Environment, Land and Sea Protection [17] provided funding and enhanced research on pollinator populations in Italian National Parks, with special acknowledgement of threats driven by agricultural practices. ISPRA (Istituto Superiore per la Protezione e la Ricerca Ambientale—the Italian institute for environmental protection and research) highlights the complexity of using pollinators as indicators of environmental health and proposes an approach involving in situ sampling followed by simplified identification of samples, without reaching the rank of species [18]. Other projects focusing on pollinator diversity have been funded in Italy, and our research group is involved in two of them: the European LIFE 4 POLLINATORS (LIFE18 GIE/IT000755), led by the University of Bologna, and the national BeeNet, led by CREA (Research Centre for Agriculture and Environment). One of the objectives of the former is to define the protocol of a new biodiversity indicator for assessing rural development plans [19] through pollinator monitoring and direct involvement of farmers in the Emilia-Romagna region. The main objective of the latter is to extend a large regional monitoring scheme for honeybees and wild bees to the whole of Italy. These projects will help to fill gaps in our understanding of pollinator ecology, especially that of bees (Apoidea, [16]), since we are lacking much information on species-specific requirements. This is in line with the European situation described in the European Red List, where about 56% of pollinator species are "data deficient" [20].

Since our long-term goal is to define an indicator based on pollinators and since we are directly involved in national and European projects, we decided to investigate existing indicators and their characteristics. We considered their scientific status and current political–legal acceptance. We started with indicators already used to investigate biodiversity and continued with those focused on arthropods, assuming a similarity of these animals with pollinators active in temperate areas.

2. Materials and Methods

We conducted a search of the literature on official websites of the regulatory agencies: scientific agencies responsible for the development and endorsement of indicators and political agencies responsible for the recognition of biodiversity indicators in regional, national and European legislation. We surveyed webpages and reports of the European Environment Agency (EEA), the ETC on Biological Diversity (ETC/BD), the Directorate General for Environment and Directorate General for Agriculture and Rural Development (European Commission), identifying the main indicators related to our goal. At the national level, we included the indexes and/or indicators mentioned in the Italian "Testo Unico Ambientale" [21], a text adopting numerous European directives on environmental issues; we surveyed protocols drawn up by ISPRA. We also perused any scientific literature directly linked to the indexes/indicators.

2.1. *Definitions and Criteria*

During our search, it emerged that the terms index/indicator are employed differently depending on the context (ecological, political, etc.), especially the term indicator, usage of which can vary widely [22]. For our purposes, we define the terms as follows:

- "Index" is an instrument that sets a value to describe a measurable phenomenon (e.g., the value of a sampled population in relation to the expected value).
- "Indicator" is a more complex and often composite tool for evaluating a phenomenon that cannot be measured directly.

Since our goal was to weigh information included in indexes/indicators used to evaluate the efficacy of biodiversity and national funding in agricultural contexts, and with a view to developing a pollinator indicator, we selected indexes/indicators based on two criteria:

- Their present or past potential to reflect biodiversity in agroecosystems;
- Their proximity to arthropod lifestyles.

To select indexes/indicators reflecting the first point, we surveyed official documents at the European and national level because whatever a new pollinator indicator might include, it must be suitable for future applications in fields that are not purely scientific. Since Italy is a European member state, European legislation and its indicators had to be considered. In some cases, however, they are applied flexibly, and national documents may explain how. Regarding the second criterion, we selected indexes/indicators already included in Italian national/regional regulation on arthropods or already considered to be bio-indicator organisms, at least by Italian government agencies. This second criterion refers to the fact that in temperate latitudes, pollination is mainly carried out by insects [16]. Bees, hoverflies and butterflies are constant visitors to flowers, where they forage for resources such as pollen and nectar. Like other insects, pollinator ranges of activity may be limited spatially and strongly influenced by vegetation/landscape characteristics. Some indicators referring to spatial context were therefore also considered here.

2.2. *Parameters*

We analyzed the indexes/indicators on the basis of:

1. *Taxonomic groups*: We indicated the taxa of the study species and their ecological/biological similarity with pollinator lifestyles.
2. *Monitoring type*: Monitoring can be ongoing at regular intervals so that data for the indicator flows from the monitoring dataset itself, or it can be spot data collection for specific needs and comparison with existing information on the subject.
3. *Spatial and habitat context*: We defined the spatial scale (regional, local, codified habitats, portions of habitats) and the parameters used to define it (arbitrary or ecological, i.e., application of a rigid sampling scheme, adaptation of the sampling scheme to land characteristics, individual case studies).
4. *Background*: We evaluated the amount of ecological/biological knowledge on the study species, (i.e., is there an expected population/list of species typical of a given habitat in the absence of disturbance?).
5. *Sampling effort and level of taxonomic identification*: We identified the type of sampling protocol and subsequent taxonomic effort, the taxonomic level of identification and the skills required for these activities.
6. *Final output*: We underlined how outputs are reported (i.e., descriptive, ratio or class/category).

3. Results

3.1. *Selected Indicators*

Here we describe the nine indexes/indicators we selected, providing with information regarding their position in the European/national context (Figure 1):

- **Farmland Bird Index (FBI)** originates from widespread European monitoring of birds and targets common species more closely connected with agricultural environments. It has been widely used at the European level [23] and applied differently in various CAP proposals, until the recent CAP post-2020 [1,2]. In CAP 2014–2020 [3], it was incorporated as an impact indicator (I.08) and as a context indicator (C.35). In proposed CAP post-2020 [1,2] it is retained as an impact indicator (I.18). At the Italian level, data begins from the period 2000–2005 with a report linked to national rural development plans 2007–2013 [23].
- **High Nature Value Farming (HNV)** is related to the concept that European low intensity farming systems contribute more to the preservation of biodiversity. However, development of an indicator proved difficult, since HNV applies to totally different landscapes in different countries, and the results were contradictory. HNV was applied until CAP 2014–2020 [3]. Like FBI, it was incorporated as an impact indicator (I.09) and as a context indicator (C.37) in CAP 2014–2020 but will be discontinued.
- **CAP post-2020 I.19 (I.19)** is a newly introduced indicator [1,2], expected to enhance biodiversity protection. It is based on species and habitats and their trends. No technical information is yet available.

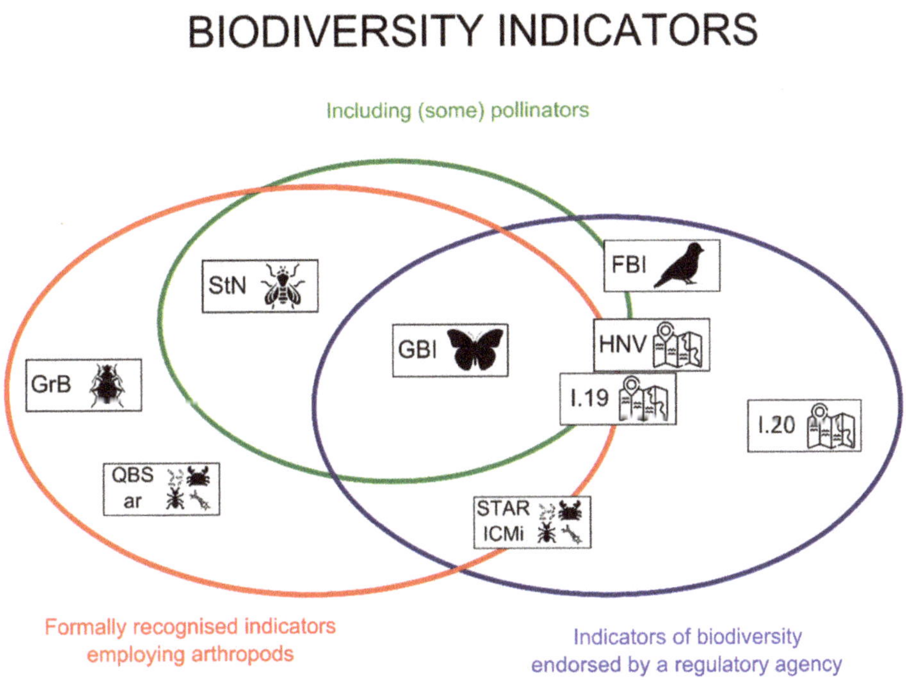

Figure 1. Relative position of selected indicators when considering their recognition and/or endorsement in current legislations (national and/or international). In the green circle fall those indicators including (at least some) groups of pollinators. This image has been designed using resources from Flaticon.com (authors of icons: Freepik, Eucalyp, Darius Dan, surang, geotatah).

- **CAP post-2020 I.20 (I.20)** is a newly introduced indicator [1,2], expected to enhance provision of ecosystem services. It refers to the share of UAA with given landscape features. No technical information is yet available.
- **Grassland Butterfly Index (GBI)** is based on European species trends and, when applied at the European level, on supranational species trends [24]. Different European countries started monitoring in a random way, some in just a few regions/areas, the

oldest data being collected in the UK in 1976. Italy started monitoring recently as part of the pilot project ABLE (Assessing Butterflies in Europe; [25]) and the data collected are still preliminary. GBI is mentioned in the Strategic Plan 2020–2024 of the DG Environment as an indicator of result (2.2; [26]).

- **Freshwater macrobenthos index (STAR ICMi)** has a complex name because it is the product of the transnational project, STAR (Standardisation of river classifications: Framework method for calibrating different biological survey results against ecological quality classifications to be developed for the Water Framework Directive; [27]). It includes multimetric indexes (normalized and weighted metrics) concerning the fresh water macrobenthos. In Italy, this indicator has been studied by ISPRA that prepared a technical handbook [28]. Italian legislation started by incorporating some water monitoring parameters [21] but now also includes this indicator [29].
- **Soil Macrobenthos Index (QBS-ar)** is based on the functional traits of all organisms that have developed morphofunctional structures to live in soil. Convergence of characters is therefore expected and is the point of focus. The environment is soil, and the organisms are various arthropods in their adult and larval stages. This indicator is applied at the regional level in Italy and has been described in detail by ARPA Piedmont (Agenzia Regionale per la Protezione Ambientale; [30]). To our knowledge, it has not yet been included in European monitoring or indicators.
- **"Syrph the Net" (StN)** is a large database proposed to evaluate ecosystems on the basis of the presence of syrphid flies in relation to what is to be expected in a given environment (based on CORINE and EUNIS systems). Martin Speight has been the promoter of the site and the system. Syrphids are potentially very good bioindicators due to their many species and adaptations, relatively easy identification and stable taxonomy. In Italy, ISPRA prepared a manual [15] with indications on how to collect and evaluate data.
- **"Ground Beetle Index" (GrB)**: Carabids are frequently used to indicate habitat alterations, since the species are expected to react to disturbance according to whether they have generalist or specialist lifestyles. Carabid beetles are well known taxonomically and ecologically and have been used widely in different studies. Their use has also been criticized (reviewed in [31]). In Italy, they are employed for regional biodiversity assessment, especially by national environment protection agencies who have also prepared a technical handbook [32].

3.2. Selected Indicators: Details

Details of our analysis of the indicator parameters follow and are summarized in Table 1.

1. *Taxonomic groups*. Among the nine indexes/indicators, we found almost all taxa of pollinators active at temperate latitudes. We considered the taxa of pollinators included in the recent EU guidelines on pollinator monitoring [16]: e.g., bees, butterflies, flies and beetles. GBI, HNV and I.19 focus on butterflies; StN and I.19 include hoverflies (the full list has not yet been completed; [33]) and I.19 also includes bees. Only a few species of beetles are recognized as pollinators, and they are currently not considered in any indicator (GrB focuses on ground-living species). Pollinators are rarely included, and not all groups are considered in the same indicator or at the same taxonomic level. The only exception could be I.19, though it only includes threatened species. The indicators incorporating butterflies and hoverflies include all species at the species level since knowledge of these groups is good.

Table 1. Main characteristics of the nine indicators. For STAR ICMi, *various* indicates Anellida, Arthropoda, Bryozoa, Cnidaria, Mollusca, Nematoda, Nemertea, Porifera, Rotatoria, Platyhelminthes; for QBS-ar, *various* indicates Arachnida, Chilopoda, Diplopoda, Insecta, Malacostraca, Pauropoda, Symphyla.

Index/Indicator	Taxonomic Groups		Monitoring Type		Spatial/Habitat Context	Final Output	
	Taxa	Pollinators Included?	Frequency	Ongoing Monitoring Plans		Type of Data	Type of Output
FBI	Aves	not in temperate areas	regular (annual schedule)	yes, European	regional	species abundance	ratio
HNV	Aves, Insecta	yes, butterflies	regular (but depending on cartographic system updates)	depending on cartography employed	regional/local	% area	ratio
I.19	any endangered living species	yes, butterflies, syrphid flies, Apoidea	regular (but depending on cartographic system updates)	depending on cartography employed	regional/local	% area	ratio
I.20		no	regular (but depending on cartographic system updates)	depending on cartography employed	regional/local	% area	ratio
STAR ICMi	various	no	regular (every 5 years, with 3–4 replicates during the year of monitoring)	yes, European	portion of habitats	abundance or pres-ence/absence	classes
GBI	Insecta (Lepidoptera)	yes, butterflies	regular (annual schedule)	yes, European	codified habitats	abundance	ratio
QBS-ar	various	no	spot (local sampling)	no	local	presence/absence	ratio/classes
StN	Syrphidae	yes, syrphid flies	spot (local sampling)	no	local	presence/absence	ratio
GrB	Carabidae	no	spot (local sampling)	no	local	abundance	descriptive

Birds are the only vertebrates that emerged among existing indicators. We have already mentioned their role in agroecosystems. At temperate latitudes, few cases of ornithophilous pollination have been documented [34]. There have been occasional reports of birds feeding on flower nectar [35]. The ecological importance of bird–flower visitation in Europe is still uncertain, especially for plant reproductive output; however, effective pollination has been confirmed for several native and exotic plant species [36,37].

2. *Monitoring type.* Monitoring can be: (a) based on international monitoring (FBI, STAR-ICMi and GBI), (b) based on cartographic analysis, especially the Geographical Information System—GIS (HNV, I.19, I.20), or (c) the output of local sampling (QBS-ar, StN, GrB).

FBI, GBI and STAR-ICMi are based on monitoring programs defined by the European Bird Census Council (EBCC), the European Butterfly Monitoring Scheme (eBMS) and Directive 2000/60/EC, respectively. These schemes are supervised by European agencies but carried out directly and independently by each European country. They occur at constant intervals, yearly for FBI and GBI and every 5 years for STAR ICMi (which includes repetition over the course of the year). FBI and GBI collect information on any available species, but few are used to discuss trends. Instead, STAR ICMi considers the macroinvertebrate population of given stretches of river. FBI monitoring uses specific 50 km^2 grids, where trained observers report all records of visual recognition or birdsong; in the absence of major impediments, monitoring will be replicated at the same site in future years. The Italian regulation for the application of STAR ICMi includes all water stands, while for GBI there is no predefined grid to follow and not all grasslands are monitored. Other indicators are based on local sampling in relation to similar situations from the literature (QBS-ar, StN, GrB). In some cases, such as StN, the data of all studies using the same methods are constantly updated, thus increasing the efficacy of comparison with an expected population. The database is not linked to any national agency and does not require external funds. Contributions to the database are voluntary, and no study is funded by StN. However, even without funding, the abundance of data contributed by volunteers helps to complete the overall geographic or habitat information. QBS-ar values are interpreted on the basis of available literature concerning a given pollutant, but no dataset including the entire literature is available as in the case of StN. In Italy, CREA, ISPRA, University of Parma and the private agency Timesis s.r.l. set up a permanent working group on soil science [32]. An aim is to improve application of the index through definition of local values. Finally, GrB can only count on the technical reports of ISPRA that standardize monitoring methods but allow free interpretation of results by external experts.

Table 2 contains examples of various past and current monitoring schemes in Italy.

Table 2. Past and ongoing monitoring in Italy; distribution maps obtained from different sources as reported.

		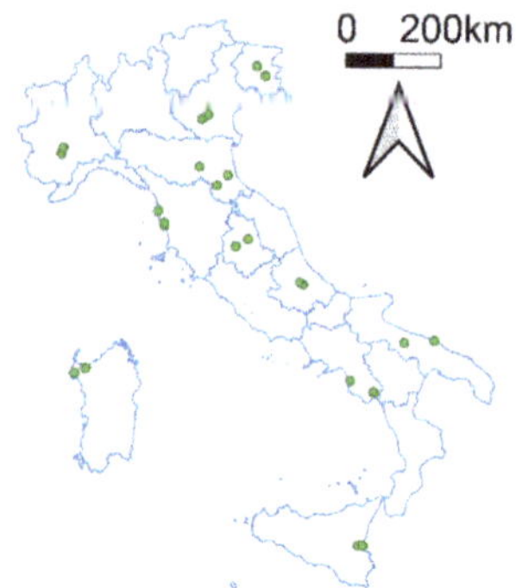
Example of distribution of monitoring sites for birds in Italy. Each dot indicates an agroecosystem site (10 × 10 km) where records have been taken for more (red) or less (yellow dots) than 11 years. The image (slightly modified) can be found in a national report under various authors (professional and volunteer) who collaborated with Lega Italiana per la Protezione degli Uccelli (LIPU) on the project MITO2000. The report is freely available at https://www.reterurale.it/flex/cm/pages/ServeBLOB.php/L/IT/IDPagina/22311 (accessed on 29 July 2021). Reference: Rete Rurale Nazionale & LIPU (2021) Farmland Bird Index nazionale e andamenti di popolazione delle specie in Italia nel periodo 2000–2020. p. 11	Example of distribution of monitoring sites for butterflies in Italy. The image can be found at https://butterfly-monitoring.net/it/mydata (accessed on 29 July 2021) (base map and data from OpenStreetMap and OpenStreetMap Foundation) and is regularly updated. The version shown was downloaded on 29 July 2021. Each dot represents a transect set up by professional or volunteer workers; data are transferred to the site and corresponding database by Butterfly Conservation Europe and the Centre for Ecology & Hydrology.	Example of distribution of monitoring sites for wild bees in Italy. Site selection was by the ongoing project BeeNet, in which the authors are directly involved. The sites are situated in agroecosystems, intensive and seminatural, monitored once a month by experts in 11 Italian regions.

3. *Spatial and habitat context*. The spatial context is not usually part of the indicator itself but is included in the process of site selection. The most common cartographic system employed is that of the CORINE Land Cover Project (CLC). GBI and GrB use CLC to identify the areas to monitor, and likewise StN applies the CORINE European Habitat Classification System (for macrohabitats) and the EU Habitats Directive, further describing the microhabitats where syrphid larvae develop. FBI prefers a regular spatial distribution of the sampling sites, geolocated and later classified on the basis of land use and bird classes. STAR ICMi and QBS-ar use different cartographic systems, including CLC and regional ones. STAR ICMi identifies regions characterized by water/climate/rocky features. QBS-ar is especially linked to soil and pedoclimatic features: a level of detail (fourth level of CLC) would ideally be necessary, but it is not available for all areas. In a few cases or for certain studies (HNV and I.20), other cartographic systems are used, such as those of protected areas (Natura2000) or areas of special interest for certain butterfly or bird species or the threatened species cartography (I.19).

4. *Background*. Knowing the ecology and biology of target species is very important. I.19 will consider extinction risk, while STAR-ICMi and QBS-ar evaluate morphological adaptations to individual microhabitats. QBS-ar will add adaptation to soil characteristics

and sensitivity to pollutants, scored from 0 to 20 (eco-morphological index). StN associates species with their ecological status, assigning a code (blank-3, depending on non-preferred habitat); code 1 allows species association with a habitat when that habitat is linked to another. Syrphid flies, like butterflies, are linked to certain habitats, especially during their larval stage (not mobile), which makes prediction of species assemblages easier. Ground beetles are defined as generalist or specialist (GrB): Detailed knowledge of their diet and mobility can help define habitat alterations. Extensive knowledge of the ecological needs of different species allows selection of flag species that may provide information about the characteristics of an environment and the content of other more common species linked to it. An example is GBI, based on 10 generalist and seven specialist species that provide clues to the potential presence of about another 100 butterflies.

5. *Sampling effort and taxonomic identification*. Sampling effort is established by monitoring protocols, while taxonomic identification can be carried out in the field or in the laboratory. EBCC monitoring plans include a different pool of bird species in each country (233 nesting species in Italy, although we only have enough information for the indicator in the case of 99 species). eBMS investigates 435 European butterfly species, identified at the species level directly in the field. Both are coordinated and supervised by regulatory agencies through the work of thousands of trained professional and volunteer workers. An opposite situation is that of samplings that require microscope identification in the laboratory: STAR-ICMi at family level, QBS-ar at order level and StN and GrB at the species level. QBS-ar not only identifies the order but creates 29 morpho-functional groups that couple different orders or distinguish adult and larval stages. The implementation of the citizen science (the involvement of the public/volunteers—citizen scientists—in scientific surveys) is emerging as a key to successful monitoring programs [38], which develop support materials often translated into different languages and adapted to local/regional fauna. However, this may influence the level of identification that can be reached, depending on the complexity provided by the different target organisms.

Finally, another parameter that may vary is the type of data collected: abundance (FBI, GBI, STAR-ICMi), or presence/absence (occupancy) (StN, HNV, I.19, I.20, GrB, QBS-ar).

6. *Final output*. Indexes usually compare a value with a reference. GBI and FBI apply a population trend (the latter since 1980 in some European countries, since 2000 in Italy). For HNV, I.19 and I.20, the reference is the entire area covered by the administration grid or the farm. It may also be a given local population (StN, STAR ICMi, GrB, FBI). Ideally, the value of the index indicates the disturbance suffered by the environment and recorded by the sampled population. FBI and GBI consider few species, as already mentioned, while StN and GrB consider all species sampled. In some cases, expert opinion is needed to interpret rough data and estimate disturbance (GrB). In other cases, indexes transform the data into a well-defined qualitative scale (STAR-ICMi), or a set of user-friendly values, so that even non-experts can compare results on a national/European basis (QBS-ar and StN).

Some indicators are better employed in association with others that describe the habitat/environment. For example, STAR ICMi helps describe the environment when considered in association with other indicators based on algae, plants and fish; chemical and physical parameters (water, pollutants) or geomorphological features. Physical/chemical parameters are also employed for QBS-ar. FBI can be coupled with the Woodland Bird Index (WBI, evaluating 18 species), All Common Species Index (CSI, 99 species) or birds especially sensitive to phytochemicals (15 widespread species and another six in specific environments). eBMS is working to define indicators for agro-environments and forests, to be coupled with GBI currently available only for grasslands.

4. Discussion

With a view to a future pollinator indicator that could integrate existing ones to survey biodiversity and to direct RDP actions, we assessed past and recent indexes/indicators used for biodiversity assessments. Indicators have become a common tool to evaluate goals, especially at government level [22]. The choice and targeting of indicators are constantly revised. Good examples are the past and current indicators used to monitor biodiversity at the European level (CAP 2014–2020 and CAP post-2020), where FBI has been retained and HNV discarded. While overall international pressure can drive the selection of some indicators, others may be employed at the national level, according to national laws or national mitigation measures to be evaluated. We therefore decided to target nine indexes/indicators applied (developed and endorsed) at the European level, at the national level (Italy) and involving other invertebrates.

Our analysis considered some index/indicator characteristics related to biological and environmental contexts, as well as their practical use. The taxonomic groups considered by the indicator, for example, may span in the entire animal kingdom. However, the indicator must consider the dimension of the employed variable: e.g., insects can be expected to interact with the environment very differently from birds. When considering pollinators, we found that some of them are included in past (HNV) or current (GBI, StN) indicators, or predicted in indicators yet to be defined (I.19). However, not all pollinators' groups are considered at once. Our goal was restricted to verifying how biodiversity is tackled and to what extent pollinator groups are currently included. We intentionally did not discuss here the complexity of pollinator, the ecosystem service they provide (pollination) and the interplay among these variables and the environmental characteristics (in natural or agricultural systems). For an interested reader, how published literature addresses the topic is clearly reviewed in a very recent paper [39]. The authors underlined the importance of clarifying definitions, the pollination studies' context and the focus element of the pollination system and concluded by highlighting the need for developing comparable indicators and standardized methods.

Some efforts in the direction of standardization have been made especially concerning cartography. Martin and colleagues [40] synthesized results from 49 studies to investigate how the spatial arrangement of crop fields and associated landscape features (e.g., field margins) impacts arthropods and their functions. Advances in landscape analysis make it possible to optimize descriptions of land/soil use, albeit at different levels of detail, depending on the database used. Proper identification of landscape can no longer be excluded from monitoring plans [41], and this is also currently supported by the number of indicators that include cartographic information (Tables 1 and 2). Similarly, standardization is important to monitoring. Concerning pollinators, the EU Pollinator Monitoring Scheme [16] is addressing the issue with the contribution of experts in different pollinator groups and from different countries, placing special emphasis on the monitoring plan and procedure through the pilot project SPRING. In Italy, the national BeeNet project replicates a fixed protocol that includes concurrent monitoring of wild bees and plants in 24 agro-environments. Sites were selected in intensive or seminatural ecosystems by landscape analysis using first the standard CLC and then checking on-site actual conditions.

Our analysis highlights two critical points: the background knowledge on the target and the efforts related to sampling and taxonomic identification. For pollinators, the situation is evolving fast. Public interest has increased sharply in recent decades: society is alarmed by pollinator decrease and interested in initiatives to understand the current situation and to sustain pollinator conservation [42]. Our research team participates in the LIFE 4 POLLINATORS project, which applies some specific actions aimed at data collection on pollinators (bees, wasps, hoverflies, beeflies, beetles, butterflies and moths) in various environments. Direct involvement of citizens, students and farmers includes using a web-tool platform for uploading photos of pollinators visiting flowers, participation at "mini-Bioblitzes" in natural parks and application of specific observation protocols in schools, botanical gardens and farms.

Possibly the most interesting result of our survey is the inclusion of citizen science in data collection. Citizen science brings important added value that makes it possible to implement datasets for establishing trends and baselines useful for indexes/indicators of species. Some successful monitoring programs and indicators rely largely on volunteer citizen-science activity, which consists of involving the public (citizen scientists) in scientific surveys [38]. Citizen science is used widely in various fields of natural science, where the data collected are used to monitor species, trace populations, design distribution maps and define conservation and management plans [43]. As mentioned, Hymenoptera are a very complex taxonomic group and their identification may be even more difficult for citizens, due to their small size, many similar species and the lack of easy-to-use identification tools. To overcome this problem, some bee citizen-science projects focus on a single species in relation to the plant pollinated by it. An example is squash bee (*Eucera* (*Peponapis*) *pruinosa* (Say, 1867)) monitoring on cucurbit flowers for impacts of farm management on bee nesting [44]. Other projects rely on a single plant, which is the selected "site" for observations quantifying the pollinator service [45]. Likewise, projects may focus on a single kind of nesting site, such as nest boxes, to limit the number of insects observed to cavity-nesting species [46]. Low taxonomic data quality is generally considered a main limitation to volunteer biodiversity monitoring. However, such data can be highly informative too, if methods and protocols are developed to allow for the inaccuracy of data provided by volunteers in place of experts [47]. Kremen et al. [48] compared the data collected on pollinators at the rank of order and superfamily by citizen scientists and bee specialists, respectively, in 17 sites; a positive correlation was found between the two datasets with regard to the difference in abundance and richness of pollinator groups between sites. The result was consistent, although citizen scientists observed only half the bee groups detected by professional scientists. The few existing citizen-science projects that monitor pollinator biodiversity in a given natural, urban or agronomic ecosystem train citizens to identify bees at a higher taxonomic rank than species, and often pool pollinator species into few easily identifiable groups [48,49]. A way to reduce errors due to misidentification is to employ a "verified method", in which all observations collected or sent by citizens are verified by experts to increase the accuracy of the data collected [50].

Difficulties in species identification are not the same for all pollinators. Identification of syrphid flies and butterflies is very advanced, as demonstrated by the success of the indicators StN and GBI. Bees are in fact the largest group of pollinators and are also the most difficult to identify [51], due to the large number of species with very different characteristics [51,52] and difficult identification with wide variations in different countries: e.g., in Europe more species live in the Mediterranean area. Regarding identification, some help may soon come from metabarcoding techniques [53]. DNA analyses may enable us to avoid training experts in all taxa, although the support of morphological taxonomy will undoubtedly still be prominent for many years [54].

To conclude, a future pollinator indicator should include: (1) elements that deal with the reduced mobility of pollinators and thus read the landscape; (2) implementation of data collection through citizen science, thus supporting data verification and (3) spot distribution of RDP funding, considering national levels. Actual tools that incorporate information on land use into indicators need to be sharpened to include greater detail. We should also pay attention to the relationship between environmental parameters and the target taxa of pollinators. To overcome the many large gaps in our knowledge of the pollinator biology and ecology of certain species, we suggest broadening the environmental parameters, possibly by building a complex indicator based on several indexes. Among them, those more strictly linked to pollinators should be included (e.g., vegetation type, crops, agricultural practices, climatic context, etc.). In some cases, a reduced number of species could be selected. For example, species sensitive to pesticides can be the main target, or those reacting differently to certain agricultural practices. The ideal approach could be to incorporate information on the abundance and occupancy of sampled species, widening the range of endorsed methodologies.

Author Contributions: Conceptualization, S.A., E.M., M.G. (Manuela Giovanetti); methodology, S.A., E.M., M.G. (Manuela Giovanetti); literature analysis, S.A., E.M., S.F., E.L.Z., M.G. (Marta Galloni); resources M.G. (Marta Galloni), L.B., M.Q.; writing—original draft preparation, S.A., E.M., L.B., M.G. (Marta Galloni), M.G. (Manuela Giovanetti), M.Q.; writing—review and editing, M.G. (Manuela Giovanetti), L.B.; visualization, S.A., E.M., M.G. (Manuela Giovanetti); supervision, M.G. (Manuela

Giovanetti), L.B. funding acquisition, L.B., M.Q. All authors have read and agreed to the published version of the manuscript.

Funding: This research was funded by MIPAAF (Ministry for Environment, Land and Sea Protection grants to CREA "Research Centre for Agriculture and Environment) through the project BeeNet 2019–2023 (Italian National Funds under FEASR 2014–2020).

Institutional Review Board Statement: Not applicable.

Data Availability Statement: Not applicable.

Acknowledgments: We are indebted to Antonella Trisorio (CREA "Agricultural Policies and Bioeconomy") who provided updated information and important material. We also thank the project "LIFE 4 POLLINATORS" (LIFE18 GIE/IT/000755) for support. We finally thanks Helen Ampt, who kindly revised the English spelling and grammar, and four anonymous reviewers for the interesting points they raised and discussed with us.

Conflicts of Interest: The authors have no conflict of interest, and the funders had no role in the design of the study; in the collection, analysis or interpretation of data; in the writing of the manuscript or in the decision to publish the results.

References

1. European Commission. COM/2018/392 Final. In *Proposal for a Regulation of the European Parliament and of the Council Establishing Rules on Support for Strategic Plans to Be Drawn up by Member States under the Common Agricultural Policy (CAP Strategic Plans)*; European Commission: Brussels, Belgium, 2018. Available online: https://eur-lex.europa.eu/legal-content/EN/TXT/?uri=COM%3A2018%3A392%3AFIN (accessed on 29 July 2021).
2. European Commission. *COM/2018/392 Final—2018/0216 Annexes to the Proposal for a Regulation of the European Parliament and of the Council*; European Commission: Brussels, Belgium, 2018.
3. DG Agriculture and Rural Development. *Technical Handbook on the Monitoring and Evaluation Framework of CAP 2014–2020*; European Commission: Brussels, Belgium, 2017.
4. Keenleyside, C.; Beaufoy, G.; Tucker, G.; Jones, G. *High Nature Value Farming throughout EU-27 and Its Financial Support under the CAP*; Institute for European Environmental Policy: London, UK, 2014; ISBN 978-92-79-37958-1.
5. Beaufoy, G.; Baldock, D.; Dark, J. *The Nature of Farming: Low Intensity Farming Systems in Nine European Countries*; Institute for European Environmental Policy: London, UK, 1994; ISBN 1-873906-01-3.
6. Bignal, E.M.; McCracken, D.I. The nature conservation value of European traditional farming systems. *Environ. Rev.* **2000**, *8*, 149–171. [CrossRef]
7. Paracchini, M.L.; Petersen, J.E.; Hoogeveen, Y.; Bamps, C.; Burfield, I.; van Swaay, C. *High Nature Value Farmland in Europe: An Estimate of the Distribution Patterns on the Basis of Land Cover and Biodiversity Data*; EC Joint Research Centre, Institute for Environment and Sustainability: Luxembourg, 2008; ISBN 978-92-79-09568-9.
8. DG Agriculture and Rural Development. *The Post-2020 Common Agricultural Policy: Environmental Benefits and Simplification*; European Commission: Brussels, Belgium, 2019.
9. Calvi, G.; Campedelli, T.; Tellini Florenzano, G.; Rossi, P. Evaluating the benefits of agri-environment schemes on farmland bird communities through a common species monitoring programme. A case study in northern Italy. *Agric. Syst.* **2018**, *160*, 60–69. [CrossRef]
10. Intergovernmental Science-Policy Platform on Biodiversity and Ecosystem Services (IPBES). *The Assessment Report of the Intergovernmental Science-Policy Platform on Biodiversity and Ecosystem Services on Pollinators, Pollination and Food Production*; Potts, S.G., Imperatriz-Fonseca, V.L., Ngo, H.T., Eds.; Secretariat of the IPBES: Bonn, Germany, 2016; ISBN 978-92-807-3567-3.
11. European Commission. COM/2021/261 final. In *Progress in the Implementation of the EU Pollinators Initiative*; European Commission: Brussels, Belgium, 2021.
12. European Commission. COM/2020/380 final. In *EU Biodiversity Strategy for 2030 Bringing Nature back into Our Lives*; European Commission: Brussels, Belgium, 2020.
13. European Parliament. *European Parliament Resolution on the EU Pollinators Initiative 2019/2803(RSP)*; European Parliament: Strasbourg, France, 2019; Available online: https://eur-lex.europa.eu/legal-content/EN/TXT/?uri=CELEX%3A52019IP0104 (accessed on 29 July 2021).
14. van Swaay, C.A.M.; Dennis, E.B.; Schmucki, R.; Sevilleja, C.G.; Balalaikins, M.; Botham, M.; Bourn, N.; Brereton, T.; Cancela, J.P.; Carlisle, B.; et al. *The EU Butterfly Indicator for Grassland Species: 1990–2017: Technical Report. Butterfly Conservation Europe*. 2019; ABLE/eBMS. Available online: www.butterfly-monitoring.net (accessed on 29 July 2021).
15. Burgio, G.; Sommaggio, D.; Birtele, D. *I Sirfidi (Ditteri): Biodiversità e Conservazione*; ISPRA, Manuali e Linee Guida 128/2015; ISPRA: Roma, Italy, 2015; ISBN 978-88-448-0743-6.

16. Potts, S.; Dauber, J.; Hochkirch, A.; Oteman, B.; Roy, D.; Ahnre, K.; Biesmeijer, K.; Breeze, T.; Carvell, C.; Ferreira, C.; et al. *Proposal for an EU Pollinator Monitoring Scheme*; EUR 30416 EN; Publications Office of the European Union: Luxembourg, 2020; ISBN 978-92-76-23859-1.
17. Ministero dell'Ambiente. Decreto Ministeriale n. 43 del 26 febbraio 2019. In *Direttiva Generale Contenente le Priorità Politiche e l'indirizzo per lo Svolgimento Dell'azione Amministrativa e per la Gestione del Ministero dell'Ambiente per l'anno 2019*; Ministero dell'Ambiente: Roma, Italy, 2019.
18. ISPRA. *Metodi di Campionamento Proposti per L'attuazione dei Progetti per il Monitoraggio e la Tutela Degli Impollinatori nei Parchi Nazionali. Direttiva del Ministero dell'Ambiente 2019*; ISPRA, Ministero dell'Ambiente: Roma, Italy, 2019.
19. LIFE 4 Pollinators—Involving People to Protect Wild Bees and Other Pollinators in the Mediterranean. Available online: https://www.life4pollinators.eu (accessed on 19 September 2021).
20. Nieto, A.; Roberts, S.P.M.; Kemp, J.; Rasmont, P.; Kuhlmann, M.; García Criado, M.; Biesmeijer, J.C.; Bogusch, P.; Dathe, H.H.; De la Rúa, P.; et al. *European Red List of Bees*; Publication Office of the European Union: Luxembourg, 2014; ISBN 978-92-79-44512-5.
21. Ministero dell'Ambiente. *Decreto Legislativo n. 152 del 3 Aprile 2006; Norme in Materia Ambientale. GU Serie Generale n.88 del 14-04-2006—Suppl. Ordinario n. 96*; Ministero Dell'Ambiente: Roma, Italy, 2006.
22. Wieck, C.; Hausmann, I. Indicators everywhere: The new accountability of agricultural policy? In Proceedings of the 172nd EAAE Seminar "Agricultural Policy for the Environment or Environmental Policy for Agriculture?" Brussel, Belgium, 28–29 May 2019. [CrossRef]
23. de Carli, E.; Gustin, M.; Rossi, P.; Trisorio, A. *Analisi del Farmland Bird Index nel Periodo 2000–2005 e Individuazione dei Fattori Positivi e Negativi per la Specie Agricole Contenuti nei Piani di Sviluppo Rurale 2007–2013*; INEA: Roma, Italy, 2009. Available online: http://dspace.inea.it/handle/inea/707 (accessed on 29 July 2021).
24. European Environment Agency. *The European Grassland Butterfly Indicator: 1990–2011*; Publications Office of the European Union: Luxembourg, 2013; ISBN 978-92-9213-402-0.
25. Roy, D.B.; Bourn, N.; Collins, S.; Dennis, E.B.; Schmucki, R.; Settele, J.; Sevilleja, C.G.; Van Swaay, C.A.M.; Wynhoff, I. *Assessing Butterflies in Europe: Executive Summary*; Butterfly Conservation Europe: Wageningen, The Netherlands, 2020.
26. DG Environment. *Strategic Plan 2020–2024*; European Commission: Brussels, Belgium, 2020. Available online: https://ec.europa.eu/info/publications/strategic-plan-2020-2024-environment_en (accessed on 29 July 2021).
27. Szoszkiewicz, K.; Zbierska, J.; Staniszewski, R.; Jusik, S.; Zgoła, T.; Kupiec, J. Standardisation of River Classifications: Framework Method for Calibrating Different Biological Survey Results against Ecological Quality Classifications to be Developed for the Water Framework Directive; Deliverable under the STAR project of the 5th Framework Programme Energy, Environment and Sustainable Development, Contract No: EVK1-CT 2001-00089. 2005. Available online: http://www.eu-star.at/ (accessed on 29 July 2021). [CrossRef]
28. Buffagni, A.; Erba, S.; Archi, F.; Bussettini, M.; Piva, F. *Linee Guida per la Valutazione Della Componente Macrobentonica Fluviale ai Sensi del DM 260/2010*; ISPRA: Roma, Italy, 2014; ISBN 978-88-448-0645-3.
29. Ministero dell'Ambiente. *Decreto n. 260, dell' 8 Novembre 2010,; Regolamento Recante i criteri Tecnici per la Classificazione Dello Stato dei Corpi Idrici Superficiali, per la Modifica Delle Norme Tecniche del Decreto Legislativo 3 Aprile 2006, n. 152, Recante Norme in Materia Ambientale, Predisposto ai Sensi Dell'Articolo 75, Comma 3, del Medesimo Decreto Legislativo. Gazzetta Ufficiale Serie Generale n.30 del 07-02-2011—Suppl. Ordinario n. 31*; Ministero Dell'Ambiente: Roma, Italy, 2010.
30. Angelini, P.; Fenoglio, S.; Isaia, M.; Jacomini, C.; Migliorini, M.; Morisi, A. *Tecniche di Biomonitoraggio Della Qualità del Suolo*; ARPA Piemonte: Torino, Italy, 2002; ISBN 88-7479-003-1.
31. Rainio, J.; Niemelä, J. Ground beetles (Coleoptera: Carabidae) as bioindicators. *Biodivers. Conserv.* **2003**, *12*, 487–506. [CrossRef]
32. Brandmayr, P.; Zetto, T.; Pizzolotto, R. *I Coleotteri Carabidi per la Valutazione Ambientale e la Conservazione Della Biodiversità*; APAT: Roma, Italy, 2005; ISBN 88-448-0152-3.
33. European Red List of Hoverflies. Available online: https://www.iucn.org/regions/europe/our-work/biodiversity-conservation/european-red-list-threatened-species/european-red-list-hoverflies (accessed on 19 September 2021).
34. Ortega-Olivencia, A.; Rodríguez-Riaño, T.; Valtueña, F.J.; López, J.; Devesa, J.A. First confirmation of a native bird-pollinated plant in Europe. *Oikos* **2005**, *110*, 578–590. [CrossRef]
35. Schwilch, R.; Mantovani, R.; Spina, F.; Jenni, L. Nectar consumption of warblers after long-distance flights during spring migration. *Ibis* **2008**, *143*, 24–32. [CrossRef]
36. da Silva, L.P.; Ramos, J.A.; Olesen, J.M.; Traveset, A.; Heleno, R.H. Flower visitation by birds in Europe. *Oikos* **2014**, *123*, 1377–1383. [CrossRef]
37. Jaca, J.; Nogales, M.; Traveset, A. Reproductive success of the Canarian Echium simplex (Boraginaceae) mediated by vertebrates and insects. *Plant Biol.* **2019**, *21*, 216–226. [CrossRef]
38. Kullenberg, C.; Kasperowski, D. What Is Citizen Science?—A Scientometric Meta-Analysis. *PLoS ONE* **2016**, *11*, e0147152. [CrossRef]
39. Bartholomée, O.; Lavorel, S. Disentangling the diversity of definitions for the pollination ecosystem service and associated estimation methods. *Ecol. Indic.* **2019**, *107*, 105576. [CrossRef]
40. Martin, E.A.; Dainese, M.; Clough, Y.; Báldi, A.; Bommarco, R.; Gagic, V.; Garratt, M.P.D.; Holzschuh, A.; Kleijn, D.; Kovács-Hostyánszki, A.; et al. The interplay of landscape composition and configuration: New pathways to manage functional biodiversity and agroecosystem services across Europe. *Ecol. Lett.* **2019**, *22*, 1083–1084. [CrossRef]

41. Giovanetti, M.; Albertazzi, S.; Flaminio, S.; Ranalli, R.; Bortolotti, L.; Quaranta, M. Pollination in Agroecosystems: A Review of the Conceptual Framework with a View to Sound Monitoring. *Land* **2021**, *10*, 540. [CrossRef]
42. Schatz, B.; Maxime, D.; Mickael, H.; Benoît, G.; Fabrice, A.; Colette, S.; Maxence, G.; Michez, D. Pollinator conservation in the context of global changes with a focus on France and Belgium. *Acta Oecol.* **2021**, *112*, 103765. [CrossRef]
43. Frigerio, D.; Richter, A.; Per, E.; Pruse, B.; Vohland, K. Citizen Science in the Natural Sciences. In *The Science of Citizen Science*; Vohland, K., Land-zandstra, A., Ceccaroni, L., Lemmens, R., Perelló, J., Ponti, M., Samson, R., Wagenknecht, K., Eds.; Springer: Cham, Switzerland, 2021; ISBN 978-3-030-58277-7. [CrossRef]
44. Appenfeller, L.R.; Lloyd, S.; Szendrei, Z. Citizen science improves our understanding of the impact of soil management on wild pollinator abundance in agroecosystems. *PLoS ONE* **2020**, *15*, e0230007. [CrossRef] [PubMed]
45. Birkin, L.; Goulson, D. Using citizen science to monitor pollination services. *Ecol. Entomol.* **2015**, *40*, 3–11. [CrossRef]
46. Bloom, E.H.; Crowder, D.W. Promoting data collection in pollinator citizen science projects. *CSTP* **2020**, *5*, 1–12. [CrossRef]
47. Schmeller, D.S.; Henry, P.-Y.; Julliard, R.; Gruber, B.; Clobert, J.; Dziock, F.; Lengyel, S.; Nowicki, P.; Déri, E.; Budrys, E.; et al. Advantages of volunteer-based biodiversity monitoring in Europe. *Conserv. Biol.* **2009**, *23*, 307–316. [CrossRef]
48. Kremen, C.; Ullman, K.S.; Thorp, R.W. Evaluating the quality of citizen-scientist data on pollinator communities. *Conserv. Biol.* **2011**, *25*, 607–617. [CrossRef]
49. Ullman, K.M.; Vaughan, M.; Kremen, C.; Shih, T.; Shepherd, M. *California Pollinator Project: Citizen Scientist Pollinator Monitoring Guide*; The Xerces Society for Invertebrate Conservation: Portland, OR, USA, 2010.
50. Gardiner, M.M.; Allee, L.L.; Brown, P.M.; Losey, J.E.; Roy, H.E.; Smyth, R.R. Lessons from lady beetles: Accuracy of monitoring data from US and UK citizen-science programs. *Front. Ecol. Environ.* **2012**, *10*, 471–476. [CrossRef]
51. Zurbuchen, A.; Landert, L.; Klaiber, J.; Müller, A.; Hein, S.; Dorn, S. Maximum foraging ranges in solitary bees: Only few individuals have the capability to cover long foraging distances. *Biol. Conserv.* **2010**, *143*, 669–676. [CrossRef]
52. De Palma, A.; Kuhlmann, M.; Roberts, S.P.M.; Potts, S.G.; Börger, L.; Hudson, L.N.; Lysenko, I.; Newbold, T.; Purvis, A. Ecological traits affect the sensitivity of bees to land-use pressures in European agricultural landscapes. *J. Appl. Ecol.* **2015**, *52*, 1567–1577. [CrossRef]
53. Ratnasingham, S.; Hebert, P.D. BOLD: The Barcode of Life Data System (http://www.barcodinglife.org). *Mol. Ecol. Notes* **2007**, *7*. [CrossRef]
54. Wägele, H.; Klussmann-Kolb, A.; Kuhlmann, M.; Haszprunar, G.; Lindberg, D.; Koch, A.; Wägele, J.W. The taxonomist—An endangered race. A practical proposal for its survival. *Front. Zool.* **2011**, *8*, 25. [CrossRef] [PubMed]

diversity

MDPI

Article

Microbiome Changes of Endemic Lake Baikal Sponges during Bleaching Syndrome Development

Valeria Itskovich [1,*], Oxana Kaluzhnaya [1], Olga Glyzina [1], Ragothaman Prathiviraj [2], George Seghal Kiran [3] and Joseph Selvin [2]

1 Limnological Institute, Siberian Branch of the Russian Academy of Sciences, 3 Ulan-Batorskaya, 664033 Irkutsk, Russia; x-sun77@rambler.ru (O.K.); glyzina@lin.irk.ru (O.G.)
2 Department of Microbiology, School of Life Sciences, Pondicherry University, Puducherry 605014, India; raj.prathivi@gmail.com (R.P.); josephselvinss@gmail.com (J.S.)
3 Department of Food Science and Technology, School of Life Sciences, Pondicherry University, Puducherry 605014, India; seghalkiran@gmail.com
* Correspondence: itskovich@mail.ru

Abstract: The sponge (Porifera) microbiome is an indicator of both natural and anthropogenic stressors. Studying Baikal sponge microbial communities could help reveal if there is a connection between bacterial symbionts and a mass sponge bleaching event that was recently detected; 16S rRNA sequencing was performed among healthy and diseased freshwater sponges of *Lubomirskia baikalensis* and *Baikalospongia intermedia*, which were collected from Lake Baikal, Russia. A phylum-based taxonomic classification showed that Chlorophyta, Acidobacteria, Bacteroidetes, Actinobacteria and Cyanobacteria were most abundant across samples. When comparing healthy and diseased *L. baikalensis* samples, large variations in microbial composition were found at the phylum level. Comparative analyses, which were performed for the first time for *B. intermedia*, showed a decrease in Chlorophyta (unicellular green algae) and an increase in Bacteroidetes and Cyanobacteria in diseased specimens. At the genus level, the *Opitutus* (Verrucomicrobia), *Planctomyces*, and *Nitrospira* content increased in all diseased sponges, which reflected a general tendency toward an increase in Cyanobacteria in diseased sponges. Comparative analysis of the diseased and healthy sponge metagenomes showed that diseased sponges underwent various nonspecific changes in bacterial composition. The bacterial community composition is probably influenced by sponge type and degree of disease affection.

Keywords: Porifera; microbiome; sponge disease; 16S rRNA sequencing; Baikal

Citation: Itskovich, V.; Kaluzhnaya, O.; Glyzina, O.; Prathiviraj, R.; Seghal Kiran, G.; Selvin, J. Microbiome Changes of Endemic Lake Baikal Sponges during Bleaching Syndrome Development. *Diversity* **2021**, *13*, 653. https://doi.org/10.3390/d13120653

Academic Editor: Michael Wink

Received: 16 November 2021
Accepted: 2 December 2021
Published: 8 December 2021

1. Introduction

Sponges (Porifera) constitute an important component of marine and freshwater ecosystems because of their species' richness, abundance, and key functional roles [1–4]. Sponges are inhabited by a wide variety of microorganisms, including archaea, heterotrophic bacteria, cyanobacteria, microscopic algae (green, red, cryptophytic, diatoms), dinoflagellates, and fungi; these microorganisms account for up to 50% of their biomass [5–7]. Sponge bacterial communities tend to be dominated by Gamma-, Alphaproteo-bacteria, Actinobacteria, Cyanobacteria, Chloroflexi and Poribacteria [8–10].

Sponge symbiotic communities are based on complex functional relationships that were formed during the adaptation of the entire community to environmental conditions [11,12]. Microorganisms can be alternative sources of energy and carbon for the sponge, take part in the nitrogen cycle, protect against oxidative stress, and produce various bioactive metabolites [9,12,13]. In aquatic ecology, metagenomic approaches make it possible to investigate complex microbial communities and their interactions with the host and with the environment.

Studies on the effects of increased temperature and ocean acidification on marine Porifera revealed that microbial symbiotic communities play an important role in maintaining sponge health and survival [11,14]. Under stress, sponge species undergo compositional and functional shifts in the microbiome [10,15–19]. Most research has been conducted on the microbial associations of marine sponges. Symbiotic associations of freshwater sponges have been insufficiently studied, even though these sponges also have diverse microbiomes including Actinobacteria, alpha-, beta-proteobacteria, Verrucomicrobia, and Flavobacterium [20–26].

Lake Baikal is the world's oldest and deepest lake, estimated to be 35 million years old with a maximum depth of 1647 m. The Lake Baikal endemic sponge family Lubomirskiidae constitutes the bulk of the benthic biomass and includes 14 described species and two subspecies [27–29]. Sponges form the main part of the benthos biomass and, as sedentary biofiltrators, play an important role in the lake's ecology. *Lubomirskia baikalensis Pallas, 1776* is a benthic littoral and sublittoral species that inhabits depths from 3–120 m. *Lubomirskia baikalensis* is a branched sponge that can grow to over 1 m and forms underwater forests at depths of 8–15 m. *Baikalospongia intermedia* Dybowsky,1880 is an overgrowing sponge that occurs at all depths, including the deep-water zone. *B. intermedia* and *L. baikalensis* are the most widespread species [29,30].

An ecological crisis has been observed in the littoral area of the lake since 2011 [31]. An important sign of the ecological crisis is the mass death of the sponges, and in some regions of Lake Baikal, 100% of the individuals are affected [31–34]. Studying Baikal sponge microbial communities could help reveal if there is a connection between bacterial symbionts and a mass sponge bleaching event that was recently detected.

Climate warming, coastal eutrophication, accumulation of toxic industrial contaminants, or infection with pathogens are indicated as possible reasons for the mass death of Baikal sponges [31–36]. The revealed narrow temperature optimum of *L. baikalensis* indicates the probable effect of increased temperature in the lake on disease development [33]. These findings correlate with the findings of numerous studies of sea sponges and corals, which also showed the influence of climate warming on the development of several diseases of these organisms [11,37–41]. However, metagenomic studies of the effect of disease on the composition of the symbiotic community of Baikal sponges are rare.

The diversity of 16S rRNA genes in the microbial community of diseased *L. baikalensis* revealed the predominance of Cyanobacteria and low abundance of Bacteroidetes and Betaproteobacteria [32]. It was also found that the mucous films on the surface of diseased sponges were formed by cyanobacteria of the order Oscillatoriales, which included representatives of the genera *Tychonema*, *Phormidium*, and *Leptolyngbya* [42]. Other results of 16S rRNA gene sequencing revealed that, in diseased sponges, the most represented OTUs belong to the families Oscillatoriaceae, Cytophagaceae, Flavobacteriaceae, Chitinophagaceae, Sphingobacteriaceae, Burkholderiaceae, Rhodobacteraceae, Comamonadaceae, Oxalobacteraceae and Xanthomonadaceae [35]. A comparison of healthy and diseased sponge microbiomes showed an increase in the number of Bacteroidetes and Proteobacteria, and the absence of a specific pathogen in diseased samples [36]. Bacteroidetes and Proteobacteria (families Flavobacteriaceae, Burkholderiaceae and Moraxellaceae) are abundant in diseased sponges [43]. It was confirmed that mat-forming cyanobacteria *Tychonema* plays a special role in the disease and death of Baikal sponges [44]. Moreover, studies of diseased sponge microbiomes were mostly carried out for *L. baikalensis*, even though other species are also susceptible to disease. Thus, the currently available data on the Baikal sponge microbial communities are limited and contradictory.

In this work, we carried out a comparative metagenomic analysis of the diseased and healthy sponges of two Lubomirskiidae species, *L. baikalensis* and *B. intermedia*, which were collected at two different time points.

2. Materials and Methods

Samples of healthy and diseased *L. baikalensis* and *B. intermedia* were collected by SCUBA in Southern (Bolshie Koty 51°53′54.4″ N, 105°04′15.3″ E and Chertov most 51°54′41.9″ N, 105°13′29.2″ E) and Northern (Severobykalsk 55°60′93.4″ N, 109°34′80.3″ E) basins of Lake Baikal in 2015 and 2006 at 10m depth. The description of the samplesare shown in Table 1. Sponge samples were frozen in liquid nitrogen for molecular analysis and were fixed in 70% ethanol for morphological examination. Spicule and skeleton preparation were performed as previously described [27] and were examined using an Olympus CX22 microscope. DNA extraction was performed using the RIBO-sorb RNA/DNAextraction kit (InterLabService, Moscow, Russia). DNA quality was assessed by running a subsample on a 1% agarose gel and the quantity of DNA was measured using a NanoVue (GE Healthcare).

Table 1. A detailed experimental determination of the effect of healthy and bleached freshwater sponges *Lubomirskiabaikalensis* and *Baikalospongia intermedia* collected from Baikal Lake, Russia.

S. No.	BioProject ID	BioSample ID	Sample Name	Species	Nature of the Sample	Base Pair Count	Seq. Count	Location
1	PRJNA665339	SAMN16252099	LBH1	*L. baikalensis*	Healthy	64,423,962	254,909	Bolshie Koty, Baikal, Russia
2		SAMN16252100	LBB1	*L. baikalensis*	Bleached	92,784,188	367,304	Bolshie Koty, Baikal, Russia
3		SAMN16252101	BIH1	*B. intermedia*	Healthy	72,441,375	286,657	North Baikal, Russia
4		SAMN16252102	BIB1	*B. intermedia*	Bleached	53,745,195	212,772	North Baikal, Russia
5		SAMN16252103	LBH2	*L. baikalensis*	Healthy	42,470,245	168,170	Bolshie Koty, Baikal, Russia
6		SAMN16252104	LBH3	*L. baikalensis*	Healthy	51,347,284	203,100	Chertov most, South Baikal, Russia
7		SAMN16252105	LBB2	*L. baikalensis*	Bleached	63,233,204	250,215	Chertov most, South Baikal, Russia

The V3-V4 region of the 16S rRNA gene was amplified with primers 341F-806R [45] using Illumina MiSeq 250 bp chemistry. The PCR conditions were as follows: initial denaturation at 95 °C for 2min; followed by 25 cycles: denaturation at 95C for 30 s, anneal at 55C for 30 s, extension at 72C for 30 s; the final extension at 72C for 8 min. The complete dataset of seven paired-end healthy (LBH1, LBH2, LBH3, BIH1) and bleached (LBB1, LBB2, BIB1) freshwater sponge samples were generated by a fastQ file format. The primary analysis of NGS sequencing data, the removal of short and chimeric sequences, clustering in OTUs (operational taxonomic units), an assessment of biodiversity by calculating ACE, Chao1, and Shannon indices were carried out using the Mothur v.1.22.0 program (http://www.mothur.org, accessed on 1 November 2021). The Pyrosequencing pipeline program (http://pyro.cme.msu.edu, accessed on 1 November 2021) was used to determine species diversity and taxonomic composition and to compare communities. Data cluster analysis was performed using the Complete Linkage Clustering program, which is part of the Pyrosequencing pipeline.

The metagenomic analysis was performed using the MG-RASTserverV4.0.3 software package [46]. Sequence similarity search of the 16s rRNA gene was performed against the available biological database SILVA. The best hits were classified based on the percentage of identity and sequence query coverage. The rarefaction curve plot was used to annotate the species richness among the samples based on Rarefaction Curve Analysis [47–50]. A genus-level relative abundance of an operational taxonomic unit (OTUs) was computed and compared for a better understanding of the composition of a microbial community among the selected samples.

The relative abundance of OTUs at genus level comparison was performed using the Morpheus online tool (https://software.broadinstitute.org/morpheus/, accessed on 1 November 2021). We carried out the clustering heat map with a complete linkage method using the Euclidian distance metric [51], and One minus Pearson correlation metric [52] implemented in the Morpheus tool. The genus identified in OTUs from seven selected samples was further compared with five-set and two-set interactive Venn diagrams using the InteractiVenn tool (http://www.interactivenn.net/, accessed on 1 November 2021).

3. Results

3.1. Analysis of Essential Parameters and Taxonomic Hit Distribution

Out of seven analyzed samples, five belong to *L. baikalensis* (three healthy and two bleached samples) and *B. intermedia* (one healthy and one bleached sample). Sequences were deposited in the NCBI-SRA portal (PRJNA665339). The basic nucleotide features (Base pair and total sequence count) of the selected samples are shown in Table 1.

We performed a taxonomic hits distribution of taxa using a contigLCA algorithm finding a single consensus taxonomic entity for all features on each individual sequence. The bacterial community was dominated by Chlorophyta, Acidobacteria, Bacteroidetes, Actinobacteria and Cyanobacteria as shown in Figure 1a. It shows that the overall 21 phyla play a major role in the growth physiology and maintenance of their cell survivability in freshwater environments. Samples LBH3 and BIH1 have a high relative abundance of Chlorophyta (80%) while this is 50% or less in other samples. Acidobacteria, Proteobacteria, Bacteroidetes, Cyanobacteria and Actinobacteria are found to be abundant and distributed nonspecifically between healthy and bleached samples (Figure 1b).

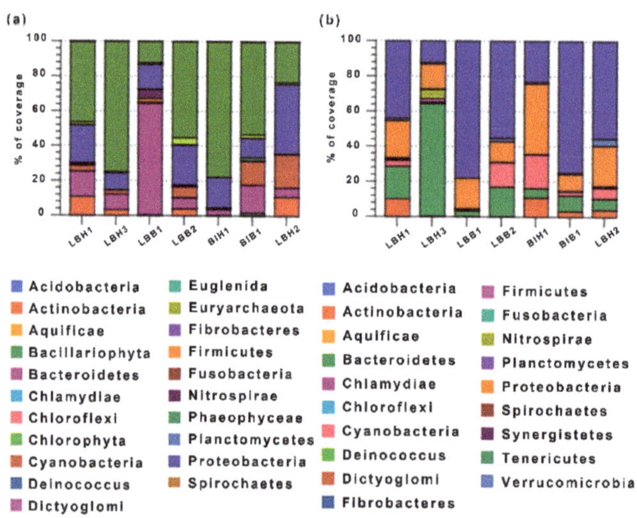

Figure 1. A phylum based bar diagram represents overall reads (**a**) and separated bacterial reads (**b**) classification was predicted among healthy and bleached freshwater sponges *Lubomirskia baikalensis* and *Baikalospongia intermedia*.

3.2. Analysis of Rarefaction Curve

The rarefaction curve plot was used to annotate the species richness among the samples. From our analysis, we identified that the species count is increased in bleached samples of *L. baikalensis* (LBB1) upon 360,000 sequence reads. Whereas LBH3 has increased its species count to 1000 within a short number of sequence reads (250,000). It clearly indicates that a larger amount of bacterial diversity was found in the LBH3 sample. Whereas other samples have 600–800 species count within a short read. The samples LBH2 are found to be a low species diversity sample (Figure 2). Bacterial diversity increases in bleached samples of *B. intermedia*.

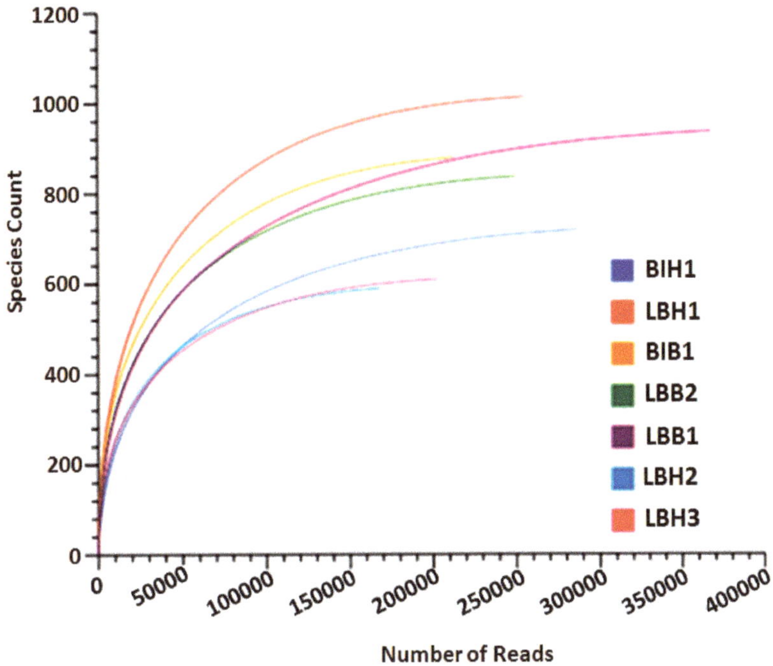

Figure 2. A phylum based rare-fraction curve was predicted among healthy and bleached fresh-water sponges *Lubomirskia baikalensis* and *Baikalospongia intermedia*.

3.3. Analysis of Relative Abundances of OTUs at Genus Level

The genus based rank abundance and operational taxonomic units were computed for top hits from the selected samples. It shows that a predominance of bacterial genera, such as *Flavobacterium, Eubacterium, Candidadus* and *Tetrasphaera,* are found to be abundant in a healthy sample of *L. Baikalensis* (LBH1), while *Synechococcus* was highly abundant in *L. baikalensis* (LBH2). While compared to the bleached samples, few genera, such as *Nitrospira, Planctomyces, Prolixibacter* and *Clostridium,* are found to be abundant (Figure 3a). We further compared the top 50 relative abundance genera; it shows that the overall 16 genera are commonly found in both healthy and bleached *L. baikalensis* samples (Figure 3b). Whereas 30 common and 20 unique genera are found in two healthy *L. baikalensis* samples (LBH1 and LBH2) (Figure 3c).

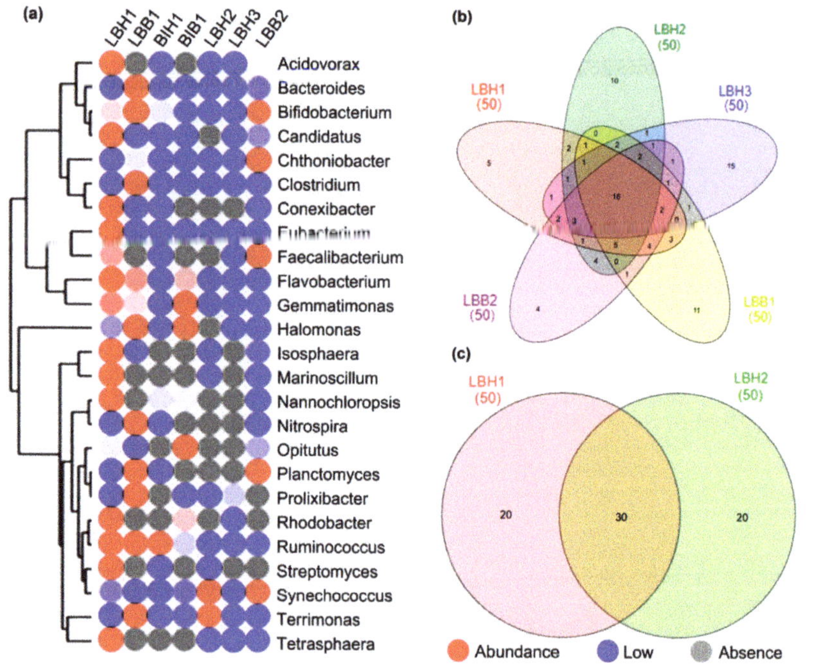

Figure 3. Top hits rank abundance plot comparison based on genus-level for selected healthy and bleached sponges. (**a**) The heat map represents the abundances of annotations of taxonomic richness and evenness on a log scale. (**b**,**c**) The Venn diagram represents the common and unique genus found in the samples.

3.4. Analysis of OTUs within Genus Level

We further classified and compared the top best enriched operational taxonomic units found between the samples. The heat map represents the distribution of taxa using a contig LCA algorithm, finding a single consensus taxonomic entity for all features on each individual sequence. This shows that the genera *Candidatus, Eubacterium, Flavobacterium, Nitrospira, Plactomyces, Prolixibacter, Terrimonas, Bifidobacterium* are abundant in healthy and bleached *L. baikalensis* samples. Whereas in *B. intermedia* (bleached) samples, the genera *Coptotermes, Gemmatimonas, Halomonas, Opitutus* are found to be abundant (Figure 4a). More than 15 genera from healthy and nine genera from bleached samples are uniquely found in *L. baikalensis* (Figure 4b) whereas in *B. intermedia,* 23 common and 27 unique genera are found in both healthy and bleached samples (Figure 4c).

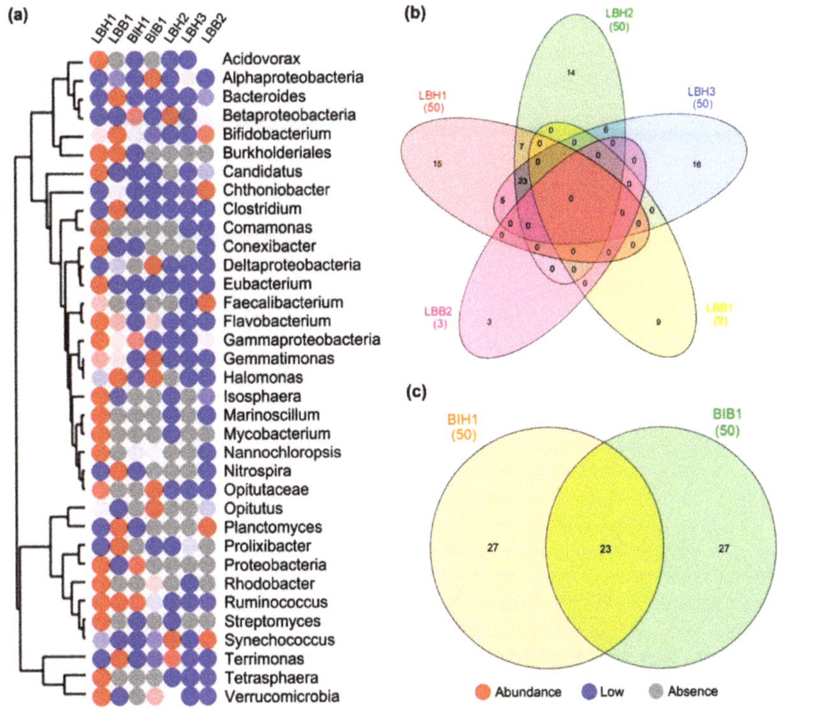

Figure 4. Top hits operational taxonomic units comparison based on genus-level for selected healthy and bleached sponges. (**a**) The heat map represents the abundances of annotations of taxonomic richness and evenness on a log scale. (**b**,**c**) The Venn diagram represents the common and unique genus found in the samples.

4. Discussion

A phylum-based taxonomic classification showed that the microbiome genera that were most abundant across all samples belonged to Acidobacteria, Bacteroidetes, Actinobacteria and Cyanobacteria. Chlorophyta is also abundant in microbiomes. Moreover, the composition of the microbiome is nonspecifically altered in diseased *L. baikalensis*. These phyla were found in moderate abundance in the samples of both healthy and diseased individuals. Our samples of healthy and diseased *L. baikalensis* were collected at two different collection points and at different times (2006 and 2015). Therefore, fluctuations in the microbial community composition among healthy sponges can be caused by different environmental conditions in Lake Baikal during a given time period. In marine sponges, the microbial composition has substantial intraspecific and interspecific variability, and varies by depth and season [15]. Thus, when comparing healthy and diseased *L. baikalensis* samples, we did not reveal specific changes in diseased sponge bacterial composition. We have shown that, among healthy *L. baikalensis*, there are significant differences in the content of Chlorophyta, Bacteroidetes, Cyanobacteria and other phyla. These results are consistent with some previously obtained results. When studying the 16SrRNA gene diversity of *L. baikalensis* microbial communities in discolored tissue areas, Cyanobacteria were dominant and there was a low abundance of Bacteroidetes and Betaproteobacteria compared to healthy samples [32]. Other 16S rRNA gene sequencing results revealed that, in diseased sponges, Cyanobacteria, Bacteroidetes, Alphaproteobacteria, Betaproteobacteria, and Gammaproteobacteria showed more than five-fold increases in abundance compared to healthy sponges [35]. Simultaneously, a comparison of the microbiomes of healthy and diseased sponges collected in 2011 and 2015 showed an increase in the number of Bacteroidetes and Proteobacteria [36]. There was no clear difference between the microbiomes of diseased and healthy sponges; for example, a high content of Verrucomicrobia (class Methylacidiphilae) was characteristic of both diseased and healthy sponges [36].

Cyanobacteria probably play an important role in the development of the disease of the Baikal sponges [42,44]. Proliferation of benthic cyanobacteria on Lake Baikal can be observed during all seasons [31,44]. The influence of the season on the microbiome and diseases of the Baikal sponges has not been studied previously. Although other studies analyzed samples collected in different seasons, there was no comparative analysis between them. With our limited number of samples, we demonstrate a difference in the microbiome of healthy *L. baikalensis* collected in different seasons, but more samples need to be analyzed.

Most previous studies of diseased sponge microbiomes were only carried out for *L. baikalensis*, even though other species are also susceptible to disease. For the first time, we performed a comparative analysis of diseased and healthy *B. intermedia*; we identified a decrease in Chlorophyta and an increase in Bacteroidetes and Cyanobacteria in diseased specimen. Until now, very few studies have been conducted on the microbiome in *B. intermedia* [21,24,26]. Proteobacteria, Actinobacteria, Planctomycetes, Cloroflexi, Verrucomicrobia, Acidobacteria, Chlorobi and Nitrospirae were identified within the *B. intermedia* microbiome [21]. It was noted that deep-sea habitat conditions affected the taxonomic diversity of microbiome bacteria and the presence of functionally significant microorganisms in communities. In microbial associations of *Baikalospongia* sp., the bacterial phyla Bacteroidetes, Proteobacteria, and Actinobacteria were predominantly identified [24]. Seo et al. [26] found differences in the bacterial species composition and diversity among *B. intermedia, L. baikalensis*, and *S. papyracea*. The bacterial communities in *B. intermedia* and *L. baikalensis* were highly similar because both species were collected from shallow zones, and Cyanobacteria and Proteobacteria accounted for the highest proportion [26]. Our data showed that Chlorophyta, Acidobacteria, Bacteroidetes, Proteobacteria and Cyanobacteria were predominant in healthy *B. intermedia*. Thus, different Baikal sponge species and different samples of the same species have differences in microbial composition; however, at present, we cannot say whether they are host- or habitat-specific.

At the genus level, the Opitutus (Verrucomicrobia) content increased or appeared in all diseased sponges. Verrucomicrobia, which are predominantly heterotrophic microorganisms that decompose hydrocarbon substrates, are widespread in marine, freshwater, soil, and hot spring ecosystems. Verrucomicrobia are characteristic of Baikal sponge communities [32]. It has been shown that Verrucomicrobia abundance is positively correlated with an increased nutrient content, phosphate availability, and seasonal algal blooms, and can change depending on the season [24,32,42]. The number of Planctomyces also increased in diseased *L. baikalensis* samples. Planctomycetes in freshwater ecosystems are generally considered minor phyla. Representatives of this phylum participate in the an-aerobic oxidation of ammonium and have the ability to degrade hydrocarbons produced by phytoplankton [53]. The genus *Nitrospira* also appeared or increased in abundance in diseased *L. baikalensis*, which reflected a general trend toward an increase in cyanobacteria in diseased sponges. Additionally, in diseased sponges, tendencies toward a decrease in the number or disappearance of some proteobacteria and planctomycetes were noted. *Streptomyces* and *Acidovorax* disappeared from diseased sponges. The number of Isosphaera, which are acidophilic planktomycetes capable of degrading numerous heteropolysaccharides, decreased in two diseased *L. baikalensis* samples.

In marine sponges, elevated temperature can disrupt the functionally important microbial symbionts [39,54,55]. Microbial community changes upon exposure to elevated temperature can manifest as a loss of specific bacterial and archaeal taxa, and increases in opportunistic microorganisms [11,39,55]. Global warming has led to several sponge mortality events [37,40,41,56,57]; in Lake Baikal, a probable cause of sponge disease is also increased water temperature. The occurrence of diseased sponges throughout Lake Baikal, including in ecologically less disturbed areas [34], also supports this idea. Previously, a decrease in the heat shock protein (HSP70) content of diseased sponges with various types of lesions was shown, which indicates a suppression of their physiological and energetic states [33]. This suppression may occur even before the development of visible signs such as lesions, which are the result of the weakening of the spongy immunity. For example, in corals, the microbiome can shift prior to bleaching [58,59]. The composition of the sponge bacterial community is probably also influenced by the degree of disease involvement. Because the extent of disease involvement is difficult to visually determine when collecting sponges, samples identified as healthy may already be at the stage of early destruction of the microbiome. This may explain the lack of a clear picture of microbial composition changes in diseased sponges. In addition, it is possible that studies of changes in the eukaryotic community composition will also elucidate the reasons for the destruction of diseased sponge microbiomes.

5. Conclusions

These results help to clarify the nature of changes in symbiotic relationships during discoloration syndrome development in freshwater sponges. We showed large fluctuations in microbial composition in diseased and healthy sponges based on our previously published data, which may be due to different collection points, depths and collection seasons. Given the limited number of samples, we analyzed experiments under the same conditions that are needed to identify the change in microbiome composition between healthy and diseased individuals and its causes. We also showed interspecific differences in discoloration syndrome development in Baikal sponges. Thus, more data are also needed from samples at different depths and seasons, and from different Lubomirskiidae species.

Author Contributions: Conceptualization, V.I. and J.S.; methodology, V.I., O.G. and O.K.; software, R.P., G.S.K. and J.S.; validation, V.I. and J.S.; formal analysis, R.P. and G.S.K.; investigation, V.I. and O.G.; data curation, V.I., R.P. and G.S.K.; writing—original draft preparation, V.I., O.K., R.P., G.S.K. and J.S.; writing—review and editing, V.I., O.K., R.P., G.S.K. and J.S.; visualization, R.P., G.S.K. and J.S.; supervision, V.I. and J.S.; project administration, V.I. and J.S.; funding acquisition, V.I. and J.S. All authors have read and agreed to the published version of the manuscript.

Funding: The reported study was funded by RFBR and DST according to the research project № 19-54-45034, by RFBR according to the research project № 20-04-00868 and basic funding, project number 0279-2021-0011.

Institutional Review Board Statement: Not applicable.

Informed Consent Statement: Not applicable.

Data Availability Statement: The raw sequencing reads were deposited in the NCBI-SRA portal (PRJNA665339).

Acknowledgments: We thank Igor Khanaev and Alexandr Kupchinsky for the help with samples collection. We thank Mallory Eckstut, from Edanz (https://www.edanz.com/ac, accessed on 1 November 2021) for editing a draft of this manuscript. We thank the three anonymous reviewers for their valuable comments. This study was carried out in the Large-Scale Research Facilities "Experimental Freshwater Aquarium Complex for Baikal Hydrobionts" (http://www.lin.irk.ru/aqua, accessed on 1 November 2021).

Conflicts of Interest: The authors declare no conflict of interest.

References

1. Diaz, M.C.; Rützler, K. Sponges: An essential component of Caribbean coral reefs. *Bull. Mar. Sci.* **2001**, *69*, 535–546.
2. Bell, J.J.; Biggerstaff, A.; Bates, T.; Bennett, H.; Marlow, J.; McGrath, E.; Shaffer, M. Sponge monitoring: Moving beyond diversityand abundance measures. *Ecol. Indic.* **2017**, *78*, 470–488. [CrossRef]
3. Bell, J.J. The functional roles of marine sponges. *Estuar. Coast. Shelf Sci.* **2008**, *79*, 341–352. [CrossRef]
4. Van Soest, R.W.M.; Nicole Boury-Esnault, J.V.; Dohrmann, M.; Erpenbeck, D.; De Voogd, N.J.; Santodomingo, N.; Vanhoorne, B.; Kelly, M.; Hooper, J.N.; Hooper, A. Global diversity of sponges (Porifera). *PLoS ONE* **2012**, *7*, e35105. [CrossRef]
5. Hentschel, U.; Usher, K.M.; Taylor, M.W. Marine sponges as microbial fermenters. *FEMS Microbiol. Ecol.* **2006**, *55*, 167–177. [CrossRef] [PubMed]

6. Webster, N.S.; Blackall, L.L. What do we really know about sponge-microbial symbioses. *ISME J.* **2009**, *3*, 1–3. [CrossRef] [PubMed]
7. Cervino, J.M.; Winiarski-Cervino, K.; Poison, S.W.; Goreau, T.; Smith, G.W. Identification of bacteria associated with a disease affecting the marine sponge Ianthella basta in New Britain, Papua New Guinea. *Mar. Ecol. Prog. Ser.* **2006**, *324*, 139–150. [CrossRef]
8. Choudhury, J.D.; Pramanik, A.; Webster, N.; Llewellyn, L.E.; Gachhui, R.; Mukherjee, J. Assessing the complex sponge microbiota: Core, variable and species-specific bacterial communities in marine sponges. *ISME J.* **2012**, *6*, 564–576. [CrossRef]
9. Hentschel, U.; Piel, J.; Degnan, S.M.; Taylor, M.W. Genomic insights into the marine sponge microbiome. *Nat. Rev. Microbiol.* **2012**, *10*, 641–654. [CrossRef]
10. Webster, N.S.; Thomas, T. The sponge hologenome. *MBio* **2016**, *7*, e00135-16. [CrossRef] [PubMed]
11. Fan, L.; Liu, M.; Simister, R.; Webster, N.S.; Thomas, T. Marine microbial symbiosis heats up: The phylogenetic and functional response of a sponge holobiont to thermal stress. *ISME J.* **2013**, *7*, 991–1002. [CrossRef]
12. Webster, N.S.; Luter, H.M.; Soo, R.M.; Botte, E.S.; Simister, R.L.; Abdo, D.; Whalan, S. Same, same but different: Symbiotic bacterial associations in GBR sponges. *Front. Microbiol.* **2012**, *3*, 444. [CrossRef] [PubMed]
13. Erwin, P.M.; Pita, L.; Lopez-Legentil, S.; Turon, X. Stability of sponge-associated bacteria over large seasonal shifts in temperature and irradiance. *Appl. Environ. Microbiol.* **2012**, *78*, 7358–7368. [CrossRef]
14. Morrow, K.M.; Fiore, C.L.; Lesser, M.P. Environmental drivers of microbial community shifts in the giant barrel sponge, Xestospongia muta, over a shallow to mesophotic depth gradient. *Environ. Microbiol.* **2016**, *18*, 2025–2038. [CrossRef]
15. Selvin, J.; Priya, S.; Seghal Kiran, G.; Thangavelu, T.; Sapna Bai, N. Sponge associated marine bacteria as indicators of heavy metal pollution. *Microbiol. Res.* **2009**, *164*, 352–363. [CrossRef] [PubMed]
16. Bourne, D.G.; Dennis, P.G.; Uthicke, S.; Soo, R.M.; Tyson, G.W.; Webster, N. Coral reef invertebrate microbiomes correlate with the presence of photosymbionts. *NPG* **2013**, *7*, 1452–1458. [CrossRef]
17. Bourne, D.G.; Morrow, K.M.; Webster, N.S. Insights into the coral microbiome: Underpinning the health and resilience of reef ecosystems. *Annu. Rev. Microbiol.* **2016**, *70*, 317–340. [CrossRef] [PubMed]
18. Botté, E.S.; Nielsen, S.; Abdul Wahab, M.A.; Webster, J.; Robbins, S.; Thomas, T.; Webster, N.S. Changes in the metabolic potential of the sponge microbiome under ocean acidification. *Nat. Commun.* **2019**, *10*, 4134. [CrossRef]
19. Kiran, G.S.; Sekar, S.; Ramasamy, P.; Thinesh, T.; Hassan, S.; Lipton, A.N.; Ninawe, A.S.; Selvin, J. Marine sponge microbial association: Towards disclosing unique symbiotic interactions. *Mar. Environ. Res.* **2018**, *140*, 169–179. [CrossRef] [PubMed]
20. Kaluzhnaya, O.V.; Krivich, A.A.; Itskovich, V.B. Diversity of 16S rRNA Genes in Metagenomic Community of the Freshwater Sponge Lubomirskia baicalensis. *Russ. J. Genet.* **2012**, *48*, 851–854. [CrossRef]
21. Kaluzhnaya, O.V.; Itskovich, V.B. Phylogenetic Diversity of Microorganisms Associated with the Deep-Water Sponge Baikalospongia intermedia. *Russ. J. Genet.* **2014**, *50*, 667–676. [CrossRef]
22. Kaluzhnaya, O.V.; Itskovich, V.B. Distinctive features of the microbial diversity and the polyketide synthase genes spectrum in the community of the endemic Baikal sponge Swartschewskia papyracea. *Russ. J. Genet.* **2016**, *52*, 38–48. [CrossRef]
23. Costa, R.; Keller-Costa, T.; Gomes, N.C.; da Rocha, U.N.; van Overbeek, L.; van Elsas, J.D. Evidence for selective bacterial Communitystructuring in the freshwater sponge Ephydatia fluviatilis. *Microb. Ecol.* **2013**, *65*, 232–244. [CrossRef]
24. Gladkikh, A.S.; Kaluyzhnaya, O.V.; Belykh, O.I.; Ahn, T.S.; Parfenova, V.V. Analysis of bacterial communities of two Lake Baikal endemic sponge species. *Mikrobiologiia* **2014**, *83*, 682–693. [CrossRef] [PubMed]
25. Gaikwad, S.; Shouche, Y.S.; Gade, W.N. Microbial community structure of two freshwater sponges using Illumina MiSeq sequencing revealed high microbial diversity. *AMB Expr.* **2016**, *6*, 40. [CrossRef] [PubMed]
26. Seo, E.Y.; Jung, D.; Belykh, O.I.; Bukshuk, N.A.; Parfenova, V.V.; Joung, Y.; Kim, I.C.; Yim, J.H.; Ahn, T.-S. Comparison of bacterial diversity and species composition in three endemic Baikalian sponges. *Ann. Limnol.* **2016**, *52*, 27–32. [CrossRef]
27. Efremova, S.M. New genus and new species of sponges from family Lubomirskiidae Rezvoj, 1936. In *Index of Animal Species Inhabiting Lake Baikal and Its Catchment Area*; Timoshkin, O.A., Ed.; Nauka: Lake Baikal, Russia, 2004; Volume 1, pp. 1261–1278.
28. Bukshuk, N.A.; Maikova, O.O. A new species of Baikal endemic sponges (Porifera, Demospongiae, Spongillida, Lubomirskiidae). *ZooKeys* **2020**, *906*, 113–130. [CrossRef] [PubMed]
29. Itskovich, V.; Kaluzhnaya, O.; Veynberg, Y.; Erpenbeck, D. Endemic Lake Baikal sponges from deep water. 2: Taxonomy and bathymetric distribution. *Zootaxa* **2017**, *4236*, 335–342. [CrossRef]
30. Masuda, Y. Studies on the taxonomy and distribution of freshwater sponges in Lake Baikal. In *Biosilica in Evolution, Morphogenesis and Nanobiotechnology*; Müller, W.E.G., Grachev, M.A., Eds.; Springer: Berlin/Heidelberg, Germany, 2009; pp. 81–110. [CrossRef]
31. Timoshkin, O.A.; Samsonov, D.P.; Yamamuro, M.; Moore, M.V.; Belykh, O.I.; Malnik, V.V.; Sakirko, M.V.; Shirokaya, A.A.; Bondarenko, N.A.; Domysheva, V.M.; et al. Rapid ecological change in the coastal zone of Lake Baikal (East Siberia): Is the site of the world's greatest freshwater biodiversity in danger? *J. Great Lakes Res.* **2016**, *42*, 487–497. [CrossRef]
32. Kaluzhnaya, O.V.; Itskovich, V.B. Influence of Baikal sponge bleaching to taxonomic composition of symbiotic microorganisms. Russ. *J. Genet.* **2015**, *51*, 1335–1340.
33. Itskovich, V.B.; Shigarova, A.M.; Glyzina, O.Y.; Kaluzhnaya, O.V.; Borovskii, G.B. Heat shock protein 70 (Hsp70) response to elevated temperatures in the endemic Baikal sponge *Lubomirskia baicalensis*. *Ecol. Indic.* **2018**, *88*, 1–7. [CrossRef]

34. Khanaev, I.V.; Kravtsova, L.S.; Maikova, O.; Bukshuk, N.; Sakirko, M.V.; Kulakova, N.V.; Butina, T.V.; Nebesnykh, I.A.; Belikov, S.I. Current state of the sponge fauna (Porifera: Lubomirskiidae) of Lake Baikal: Sponge disease and the problem of conservation of diversity. *J. Great Lakes Res.* **2018**, *44*, 77–85. [CrossRef]
35. Kulakova, N.V.; Sakirko, M.V.; Adelshin, R.V.; Khanaev, I.V.; Nebesnykh, I.A.; Perez, T. Brown Rot Syndrome and Changes in the Bacterial community of the Baikal Sponge Lubomirskia baicalensis. *Microb. Ecol.* **2018**, *75*, 1024–1034. [CrossRef] [PubMed]
36. Belikov, S.; Belkova, N.; Butina, T.; Chernogor, L.; Martynova-Van Kley, A.; Nalian, A.; Rorex, C.; Khanaev, I.I.; Maikova, O.; Feranchuk, S. Diversity and shifts of the bacterial community associated with Baikal sponge mass mortalities. *PLoS ONE* **2019**, *14*, e0213926. [CrossRef]
37. Cebrian, E.; Uriz, M.J.; Garrabou, E.; Ballesteros, E. Sponge mass mortalities in a warming Mediterranean. Are cyanobacteria harboring species worse off? *PLoS ONE* **2011**, *6*, e20211. [CrossRef] [PubMed]
38. Sokolow, S. Effects of a changing climate on the dynamics of coral infectious disease: A review of the evidence. *Dis. Aquat. Org.* **2009**, *87*, 5–18. [CrossRef]
39. Ramsby, B.D.; Hoogenboom, M.O.; Whalan, S.; Webster, N.S. Elevated seawater temperature disrupts the microbiome of an ecologically important bioeroding sponge. *Mol. Ecol.* **2018**, *27*, 2124–2137. [CrossRef] [PubMed]
40. Vicente, V.P. Regional commercial sponge extinction in the West Indies: Are recent climatic changes responsible? *Mar. Ecol. Prog. Ser.* **1989**, *10*, 179–191. [CrossRef]
41. Cerrano, C.; Bavestrello, G.; Bianchi, C.N.; Cattaneovietti, R.; Bava, S.; Morganti, C.; Morri, C.; Picco, P.; Sara, G.; Schiaparelli, S.; et al. A catastrophic mass-mortality episode of gorgonians and other organisms in the Ligurian Sea (North-Western Mediterranean), summer 1999. *Ecol. Lett.* **2000**, *3*, 284–293. [CrossRef]
42. Kaluzhnaya, O.; Itskovich, V. Molecular identification of filamentous cyanobacteria overgrowing the endemic sponge Lubomirskia baicalensis. *Inland Waters* **2017**, *7*, 267–271. [CrossRef]
43. Chernogor, L.; Klimenko, E.; Khanaev, I.; Belikov, S. Microbiome analysis of healthy and diseased sponges Lubomirskia Baicalensis by using cell cultures of primmorphs. *PeerJ* **2019**, *7*, e27851v1. [CrossRef] [PubMed]
44. Sorokovikova, E.; Belykh, O.; Krasnopeev, A.; Potapov, S.; Tikhonova, I.; Khanaev, I.; Kabilov, M.; Baturina, O.; Podlesnaya, G.; Timoshkin, O. First data on cyanobacterial biodiversity in benthic biofilms during mass mortality of endemic sponges in Lake Baikal. *J. Great Lakes Res.* **2020**, *46*, 75–84. [CrossRef]
45. Takahashi, S.; Tomita, J.; Nishioka, K.; Hisada, T.; Nishijima, M. Development of a prokaryotic universal primer for simultaneous analysis of Bacteria and Archaea using next-generation sequencing. *PLoS ONE* **2014**, *9*, e105592. [CrossRef]
46. Meyer, F.; Paarmann, D.; D'Souza, M.; Olson, R.; Glass, E.M.; Kubal, M.; Paczian, T.; Rodriguez, A.; Stevens, R.; Wilke, A.; et al. The metagenomics RAST server—A public resource for the automatic phylogenetic and functional analysis of metagenomes. *BMC Bioinform.* **2008**, *9*, 386. [CrossRef]
47. Hughes, J.B.; Hellmann, J.J.; Ricketts, T.H.; Bohannan, B.J.M. Counting the uncountable: Statistical approaches to estimating microbial diversity. *Appl. Environ. Microbiol.* **2001**, *67*, 4399–4406. [CrossRef] [PubMed]
48. Schloss, P.D.; Handelsman, J. Introducing DOTUR, a computer program for defining operational taxonomic units and estimating species richness. *Appl. Environ. Microbiol.* **2005**, *71*, 1501–1506. [CrossRef] [PubMed]
49. Prathivraj, R.; Rajeev, R.; Fernandes, H.; Rathna, K.; Lipton, A.N.; Selvin, J.; Kiran, G.S. A gelatinized lipopeptide diet effectively modulates immune response, disease resistance and gut microbiome in Penaeus vannamei challenged with Vibrio parahaemolyticus. *Fish. Shellfish Immunol.* **2021**, *112*, 92–107. [CrossRef]
50. Schloss, P.D.; Handelsman, J. Toward a census of bacteria in soil. *PLoS Comput. Biol.* **2006**, *2*, e92. [CrossRef] [PubMed]
51. Dattorro, J. *Convex Optimization, Euclidean Distance Geometry 2ε*; Mεβoo Publishing: Palo Alto, CA, USA, 2015.
52. Wagner, J.G.; Aghajanian, G.K.; Bing, O.H. Correlation of performance test scores with tissue concentration of lysergic aciddiethylamide in human subjects. *CPT* **1968**, *9*, 635–638.
53. Woebken, D.; Teeling, H.; Wecker, P.; Dumitriu, A.; Kostadinov, I.; Delong, E.F.; Amann, R.; Glцckner, F.O. Fosmids of novel marine Planctomycetes from the Namibian and Oregon coast upwelling systems and their cross_comparison with planctomycete genomes. *ISME J.* **2007**, *1*, 419–435. [CrossRef]
54. Lemoine, N.; Buell, N.; Hill, A.L.; Hill, M.S. Assessing the utility of sponge microbial symbiont communities as models to study global climate change: A case study with Halichondria bowerbanki. In *Porifera Research Biodiveristy, Innovation and Sustainability*; Custodio, M.R., Lobo-Hajdu, G., Hajdu, E., Muricy, G., Eds.; Museu Nacional: Rio De Janeiro, Brazil, 2007; pp. 419–425.
55. Webster, N.S.; Soo, R.; Cobb, R.; Negri, A.P. Elevated seawater temperature causes a microbial shift on crustose coralline algae with implications for the recruitment of coral larvae. *ISME J.* **2011**, *5*, 759–770. [CrossRef]
56. Sweet, M.J.; Bulling, M.T. On the importance of the microbiome and pathobiome in coral health and disease. *Front. Mar. Sci.* **2017**, *4*, 9. [CrossRef]
57. Lopez-Legentil, S.; Song, B.; McMurray, S.E.; Pawlik, J.R. Bleaching and stress in coral reef ecosystems: Hsp70 expression by the giant barrel sponge Xestospongia muta. *Mol. Ecol.* **2008**, *17*, 1840–1849. [CrossRef] [PubMed]
58. Bourne, D.; Iida, Y.; Uthicke, S.; Smith-Keune, C. Changes in coral-associated microbial communities during a bleaching event. *ISME J.* **2008**, *2*, 350–363. [CrossRef]
59. Lee, S.T.; Davy, S.K.; Tang, S.L.; Fan, T.-Y.; Kench, P.S. Successive shifts in the microbial community of the surface mucus layer and tissues of the coral Acropora muricata under thermal stress. *FEMS Microbiol. Ecol.* **2015**, *91*, fiv142. [CrossRef] [PubMed]

diversity

MDPI

Article

Cydalima perspectalis in Poland—8 Years of Invasion against the Background of Three Other Invasive Species

Paweł K. Bereś [1], Patrycja Ziętara [2], Mirosław Nakonieczny [2,*], Łukasz Kontowski [3], Michał Grzbiela [4] and Maria Augustyniak [2]

1 Regional Experimental Station in Rzeszów, Institute of Plant Protection-NRI, ul. Langiewicza 28, 35-101 Rzeszów, Poland; p.beres@iorpib.poznan.pl
2 Institute of Biology, Biotechnology and Environmental Protection, Faculty of Natural Sciences, University of Silesia in Katowice, ul. Bankowa 9, 40-007 Katowice, Poland; patrycja.zietara@us.edu.pl (P.Z.); maria.augustyniak@us.edu.pl (M.A.)
3 Agricultural Holding in Szałkowo, ul. Szałkowo 36A, 14-200 Iława, Poland; lukaszkont@wp.pl
4 United Phosphorus Limited, ul. Stawki 40, 01-040 Warszawa, Poland; michal.grzbiela@onet.eu
* Correspondence: miroslaw.nakonieczny@us.edu.pl; Tel.: +48-600-377-027

Abstract: The box tree moth (*Cydalima perspectalis*) origins from East Asia. In Europe, it was recorded for the first time in 2007, and in Poland in 2012. By the end of 2020, it was found all over Poland. There are no published data on the range of *C. perspectalis* occurrence in Poland because it is not a quarantine pest in the European Union and is not subject to official monitoring. Data collected in 2018–2020 via a website dedicated to monitoring, for the first time, illustrate the current range and its largest concentrations in southern and central Poland. The monitoring confirmed that the main directions of the invasion are related to the main communication routes of Poland (south-north) and are of a long-distance character. The dispersal pattern corresponds to the model developed for *Cameraria ohridella*: a stratified dispersal model that considers long-distance road/rail transport. The second important factor contributing to the invasion of *C. perspectalis* are large human communities enabling rapid local dispersion (a diffusion model). Comparing its invasion with the monitoring data from 2007–2013 of two other invasive pests of Poland: *Ostrinia nubilalis* and *Diabrotica virgifera*, shows that a diffusion model best describes the spatial spread of these pests only to uninhabited neighboring areas.

Keywords: Poland; Europe; *Cydalima perspectalis*; box tree moth; invasive species

Citation: Bereś, P.K.; Ziętara, P.; Nakonieczny, M.; Kontowski, Ł.; Grzbiela, M.; Augustyniak, M. *Cydalima perspectalis* in Poland—8 Years of Invasion against the Background of Three Other Invasive Species. *Diversity* **2022**, *14*, 22. https://doi.org/10.3390/d14010022

Academic Editor: Luc Legal

Received: 24 November 2021
Accepted: 23 December 2021
Published: 31 December 2021

1. Introduction

The natural range of the box tree moth (*Cydalima perspectalis* Walker, 1859) is in Southeast Asia's humid subtropical regions. Described in the mid-nineteenth century from China, it was found in India half a century later. At the turn of the 20th and 21st centuries, its presence was confirmed in Korea, Japan, and Russia's the Far East [1–3]. *C. perspectalis* was recorded in Europe for the first time in 2007 in Germany (Baden-Württemberg, the city of Weil am Rhein) and in the Netherlands. Considering the size of the damage done at the place of its first finding should be assumed that it was brought to Europe at least two years earlier. The places of dispersion to Europe are most probably the distribution centers of ornamental plants imported from China in Germany and the Netherlands [1]. Eggs and caterpillars, especially of the earlier stages, move easily along with the boxwood bushes (including cuttings) transported for commercial purposes. Molecular studies of the mitochondrial cytochrome oxidase I and II genes indicate that the source of European populations is multiple introductions of insects from eastern China due to the rapid, long-distance transport of boxwood shrubs as part of the ornamental plant trade from this country to Europe. The lack of precise legal regulations in the trade of ornamental plants, the liberalization of the existing law, and the general trade globalization facilitated such a rapid spread of the species to and in Europe [4]. By 2020, *C. perspectalis* had infested almost all of Europe (Figure 1). It is recorded from Great Britain through southern Scandinavia, Lithuania, Western Ukraine, and Russia in the east, to the Balkans and Portugal in the south [4–13]. It has also been reported in Turkey, Georgia, and Dagestan [14–16]. The current range of occurrence of this species in Europe is consistent with the bioclimatic model (CLIMEX®) prepared in 2014 for *C. perspectalis*, in which its potential range was established based on the lower development threshold temperature, which depends on latitude and altitude (up to 2000 m above sea level in Georgia). The range of *C. perspectalis* in the north is limited by the low temperature, which prevents the closure of one life cycle per year. In the south, by the requirements related to obligatory diapause [5].

Figure 1. "Twist of fate"—a boxwood hedge in front of the European Museum in Schengen, damaged by an East Asian invasive box tree moth (September 2019). Photo: M. Nakonieczny.

The biology of the species in Poland has not been fully described yet. In Europe, *C. perspectalis* can develop up to five generations a year depending on the latitude. The main factor limiting its development is temperature, as the lower developmental threshold for the immature stages is 8–12 °C. However, it is known that it has a variable number of generations during the year, depending on weather (mean temperature) conditions too. The female lays gelatinous, transparent eggs with up to 20 in the batches. A maximum of a female can lay up to 500 eggs [3,17]. Depending on the temperature, 2–3 mm long larvae hatch from the eggs after a few to several days. Then, 3–4 weeks later, they reach their maturity size of about 4 cm. At a temperature of 20 °C, the development from hatching to the adult takes about 40 days [5,17]. At 25 °C, the average duration of each larval stage is three days, with an extended photoperiod shortening the development time of the larvae [18]. The pre-pupa stage lasts approximately eight days. At 25 °C, the imago appears after about ten days. The adult insect lives up to two weeks. In autumn, when the temperature drops, the caterpillars build cocoons from the leaves and hibernate [5,17,19]. In Central Europe, where there are usually two generations, the insect needs 518 degree days for the over wintering generation, and 430 for the second generation to complete the cycle. This difference translates into a different mean growth rate, higher in the first generation of insects feeding in spring than in larvae feeding in autumn [17].

Hatching larvae feed as a group first and then disperse. The foraging during the boxwood growing season varies spatially. In spring, most of the larvae feed in the lower parts of the shrubs. In summer, they feed on the middle parts, and in autumn, they feed on leaves in the top part, which causes of plant death [20]. Last instar larvae, in the leaves absence, may feed on the shoots by gnawing out phloem. Foraging is accompanied by the formation of filaments that form dense, hydrophobic clusters of leaves, which facilitates the identification of attacked plants, but makes it difficult to control the larvae, both by chemical and biological methods. The pupa is tied in a cocoon of leaves entwined with yarn, protecting it against low temperatures and pesticides [17,19]. In Central Europe, the insect undergoes obligatory diapause in winter, lasting at least eight weeks. The larva of the

3rd stage of the second generation of insects usually hibernates [5,18]. It forms leaf cocoons, usually inside the bush, making it difficult to assess the degree of plant infestation and making it easier to withstand frosts. Over wintering caterpillars can survive a temperature drop-down to −30 °C, which is not uncommon in north-eastern Poland. Shortening of the photoperiod and the growing food shortage caused by the intensive feeding of the larvae of both generations, exceeding the regenerative capacity of the host plant, may cause diapause as well [3,18].

Recently, Suppo and his team [21] introduced the first phenology model for the box tree moth based on a temperature and photoperiod to simulate the life cycle of this insect and the developmental rate for each life stage. Minimum temperature appeared the most critical parameter for the model, while diapause was negligible. The model shows that higher temperatures associated with global warming can significantly increase the number of pest generations during the year and may lead to the overlapping of successive generations and the over wintering of caterpillars at various instars [21]. Results obtained by Poitou et al. [22] confirmed the positive effect of increasing temperature on the rate and dynamics of diapause termination and lack of stimulatory effect of photoperiod. The recent models of *C. perspectalis* spreading in Catalonia showed the importance of its host plant presence, dispersal capacity, and climate suitability [23].

On the map of the range of the box tree moth in Europe from 2012, prepared by a team of researchers from CABI–Switzerland [24], Poland was noted as a country where this species does not appear yet. The first information (published in Polish) about *C. perspectalis* in Poland comes from the village of Michałkowa in the Sowie Mountains (southwestern Poland, Lower Silesia), where the insect was detected in 2012 [25]. The following data on the places where the pest appeared come from 2015. This year, Blaik et al. [25] detected *C. perspectalis* in Suchy Bór near Opole (Opole voivodeship) and in the downtown Kraków (Lesser Poland), which indicates that the pest infested the southern part of the country. Further confirmed information on the range of *C. perspectalis* in Poland comes from south-eastern Poland (Subcarpathian voivodeship). In the years 2016–2017, insects' presence was detected in the following towns: Grabiny, Rzeszów, Umieszcz, and Zgłobień [26]. This expansion clearly indicated the latitudinal direction of the insect spread in Poland, parallel to the Carpathian arc until the 2017 year.

An example of a quick and destructive boxwood moth invasion is the Botanical Garden in Łódź (central Poland). The first signs of foraging were found in the autumn of 2018, while in June 2021, the last, utterly dead boxwood hedges were removed. The garden does not currently have any boxwood collection (Figure 2a–c).

An important document that authorized us to undertake monitoring activities is the "Proposal for a resolution of the European Parliament on the boxwood moth (*Cydalima perspectalis*)", which encourages the Commission to:

- Recognize the box tree moth as a harmful organism under Directive 2000/29/E.C.;
- Support research into biological controls for the box tree moth through existing funding programs;
- Promote joint monitoring of the box tree moth by the competent European authorities [27].

The recognition by the EU of the need to monitor the boxwood moth in Europe indicates that the importance of this species is very high and noticeable in individual countries where *C. perspectalis* can pose a severe threat to boxwood hedges and topiaries as well as wild plants.

Figure 2. About a 50-year-old boxwood hedge damaged by *C. perspectalis* in the Botanical Garden in Łódź (Central Poland): (**a**) traces of feeding found one year after the first box tree moth was noticed in the garden despite the use of an insecticide (Łódź, October 2019); (**b**) wholly defoliated hedge in the Botanical Garden in Łódź in June 2021; (**c**) dead boxwood shoots without bark (June 2021). Photo (**a**–**c**): M. Nakonieczny.

In the last 30 years, the invasion of many insect species, alien to the Polish entomofauna, have been found in Poland [28]. However, some of them were particularly spectacular and of real economic importance. These include the invasions of *Ostrinia nubilalis* (European corn borer), *Diabrotica virgifera virgifera* (Western corn rootworm), and *Cameraria ohridella* (Horse-chestnut leaf miner). There are many differences and similarities to that found for *C. perspectalis* in their pace and spread directions.

The study aims were to determine the current range of *C. perspectalis* occurrence in Poland, and the pace and degree of its invasiveness. Obtained data will allow a more precise definition of the threats to the cultivated boxwood, especially those of historical and cultural importance. A comparison of the invasions of the box tree moth with other invasive species in Poland from the last 20 years should enable the use of models of spatial spreading and occurrence developed for other species. Although in 2020, *C. perspectalis* was recorded almost all over Poland, there are no officially published data on this subject.

2. Materials and Methods

2.1. Cydalima Perspectalis

In Poland, the official monitoring of pest's occurrence conducted by the Main Inspectorate of Plant Health and Seed Inspection does not cover *C. perspectalis*. In the European Union, it is not a quarantine organism. For this reason, since its first finding in several locations in the southwest and southern Poland in 2012 [25,26], there is no reliable and systematic information about the directions of the spread of this species. Because box trees are cultivated all over Poland, especially *Buxus sempervirens*, many scenarios for this insect's spread were possible.

Because there is no system for monitoring the spread of the box tree moth in Poland, in 2018–2020, it was decided to establish its range. Since 2018, there has been a dramatic increase in the number of reports from gardeners and plant breeders in southern and central Poland about damaged box trees as a result of foraging a new, unknown boxwood pest. However, the problem was that in the national database, which is responsible for tracking the spread of various species in Poland, operating as part of the Biodiversity Map conducted by The Polish Biodiversity Information Network (PolBIN, KSIB) in 2018, only one town was listed (Warsaw) to be infested by *C. perspectalis* [29].

To identify the current range of C. *perspectalis* occurrence in Poland, the database of the internet website "Allotment and Garden Our Passion" [30], which brings together over 15,000 gardeners from Poland, was used. For this purpose, hobby gardeners gathered around this website provided information via social media (Facebook, Instagram) about the towns where they found the box tree moth on their plants from April 2018 to November 2020. Using the e-mail address provided on the main page of the website using the sub-website "Contact", they provided the name of the place in their e-mail correspondence, including additional information, e.g., the date of the pest's appearance, photographs, or information (photographs) about the condition of the boxwoods. Each year, these places were verified by analyzing the photos of the damage and/or insects sent and field trips. Inspection visits were carried out annually from April to September, especially to locations from which the submitted documentation raised any doubts (Figure 3a,b). The presence of the box tree moth was verified in all voivodeships (the highest-level administrative division of Poland) from which the reports came. Particular attention was paid to the places farthest from where the box tree moth had already been found in previous reports. The resulting maps of the range of the box tree moth do not cover all the places where C. *perspectalis* occurs but show critical areas for subsequent reports of this insect's presence.

(a)

(b)

Figure 3. *Buxus sempervirens* bushes damaged by C. *perspectalis*: (**a**) an example of a positively verified photograph uploaded to the website "Allotment and Garden Our Passion"; (**b**) an example of verification of a report on the presence of C. *perspectalis* in the field: Trzebownisko near Rzeszów, September 2020, Photo: P. Bereś.

2.2. *Ostrinia nubilalis, Diabrotica virgifera virgifera, Cameraria ohridella* (Selected Methodological Issues for Comparative Purposes)

- *Ostrinia nubilalis*. In Poland, monitoring of this species was carried out by inspectors of the State Plant Health and Seed Inspection Service (SPHSIS) and the Institute of Plant Protection-National Research Institute, Regional Experimental Station in Rzeszów (IPP-NRI). The monitoring started in 2004, and was completed in 2013. As it would not be a quarantine pest, the observations made by inspectors were only used to signal the presence of the pest in the second-degree administrative division units in Poland (the first-degree are voivodships), such as counties and cities with the rights of counties (there are 380 counties in Poland). Therefore, 50 to 150 plants at the ripening dough stage (BBCH 85) were inspected depending on the cultivation area. On small maize crops on up to 1 ha, 50 plants were observed, while on fields of more than 1 ha, the number of plants increased to 150. Inspections were carried out at the end of August and the beginning of September. Observations were carried out based on the methodology provided in an annex to a manual for plant inspection services concerning forecasting, alerting, and data recording drawn up by SPHSIS. In all the years of the study, observations were terminated when any signs of caterpillars feeding were recorded. The county taken by *O. nubilalis* was considered any in which symptoms of caterpillars feed on at least one maize plantation. Hence, the monitoring was qualitative rather than quantitative [31,32].

- *Diabrotica virgifera virgifera.* Since it was a quarantine species subject to European monitoring, the State Plant Health and Seed Inspection Service (SPHSIS) was responsible for its monitoring in Poland. The species was first detected in Poland in 2005, and official state monitoring started in 2006 and was completed in 2013 (end of quarantine in EU). Changes in the number of *D. virgifera* adults were monitored with two types of trap: PTs–CSALOMON® PAL transparent sticky "cloak" trap, baited with synthetic sex pheromone (lures only males), and FTs–CSALOMON® PALs yellow sticky "cloak" trap (lures females and males). In 2006–2013, the monitoring was qualitative (finding the presence of the pest in a given crop area) and quantitative to take appropriate measures to protect maize crops and control the pest under the European Union Directives [33,34].
- *Cameraria ohridella.* There were no monitoring studies conducted. Information about the occurrence was often obtained from the press and popular science literature. There is no possibility of recreating the migration routes for this species in Poland, which would be determined based on monitoring or scientific research. Apart from scientific articles, the spread map of *C. ohridella* was prepared using the Polish Biodiversity Map database "BioMap", where scientists officially confirm the observed appearance of the pest [35,36].

3. Results

3.1. *Cydalima Perspectalis*

In total, in 2018–2020, thanks to gardeners and plant breeders, users of the website "The Allotment and the Garden Our Passion" were collecting information from 166 places (towns) in Poland in the form of 674 documented reports on the presence of *C. perspectalis* in all voivodeships (Figures 4–6 and Tables S1 and S2).

Table 1. List of voivodeships and numbers assigned to them in the publication.

Number	Voivodeship (PL)	Voivodeship (ENG)	Number	Voivodeship (PL)	Voivodeship (ENG)
1	zachodniopomorskie	West Pomeranian	9	dolnośląskie	Lower Silesian
2	pomorskie	Pomeranian	10	łódzkie	Łódź
3	warmińsko-mazurskie	Warmian-Masurian	11	świętokrzyskie	Holy Cross
4	lubuskie	Lubusz	12	lubelskie	Lublin
5	wielkopolskie	Greater Poland	13	opolskie	Opole
6	kujawsko-pomorskie	Kuyavian-Pomeranian	14	śląskie	Silesian
7	mazowieckie	Masovian	15	małopolskie	Lesser Poland
8	podlaskie	Podlaskie	16	podkarpackie	Subcarpathian

(a) (b) (c)

Figure 4. (**a**) The hypsometric and administrative map of Poland (source: https://pl.wikipedia.org/wiki/) (accessed on 17 November 2021); (**b**) main road routes in Poland (source: https://www.google.pl/maps/---modified) (accessed on 17 November 2021); (**c**) administrative division of Poland into 16 voivodeships. (names of voivodships—see Table 1; 1–16: numbers assigned to individual voivodships).

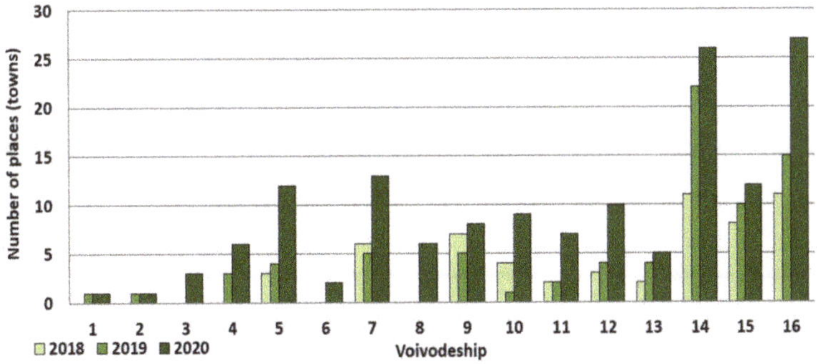

Figure 5. The number of places (towns) where *Cydalima perspectalis* was recorded in individual voivodships in 2018–2020. Axis X: 1–16—numbers assigned to individual voivodships; see Table 1.

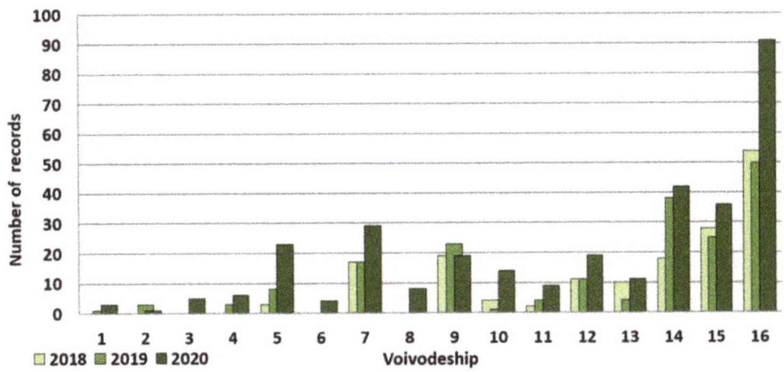

Figure 6. The number of confirmed records of *Cydalima perspectalis* occurrence in individual voivodeships in 2018–2020. Axis X: 1–16—numbers assigned to individual voivodships; see Table 1.

Three years of social monitoring of *C. perspectalis* in Poland allow us to conclude that there was an increase in the population size in all voivodeships each year, both in moth-controlled cities (new locations) and in the reported records. In the last 2020 year of research, we revealed the presence of box tree moths in Warmian-Masurian, Kuyavian-Pomeranian, and Podlaskie voivodeships (No.: 3, 6, and 8). The most significant increase in the number of new locations and reported records in 2020 compared to the previous year was found for voivodeships: Greater Poland, Masovia, Silesian, and Subcarpathian (No.: 5, 7, 14, and 16) (Figures 5 and 6).

In 2018, based on confirmed data from the gardening website, *C. perspectalis* was reported from 57 places. They were located within ten voivodeships (out of 16): Greater Poland, Masovian, Lower Silesian, Łódź, Holy Cross, Lublin, Opole, Silesian, Lesser Poland, and Subcarpatian (No.: 5, 7, 9, 10, 11, 12, 13, 14, 15, and 16). Most information about damaged box trees came from the Silesian and Subcarpathian voivodeships (No.: 14 and 16) (Figure 7). The box tree moth spread in Poland in 2012–2017 from West to East (along with the arch of the Sudetes and the Outer Western Carpathians)—green arrows. It is the main road axis of southern Poland (A4 motorway) from West to East. However, the large number of discovered sites of *C. perspectalis* in 2016–2017 in the Subcarpathian Region (green oval) may be the result of an invasion of the boxwood moth from South Slovakia where its presence was recorded the first time in Zvolen, Košice, and Presov in 2013 [20,37].

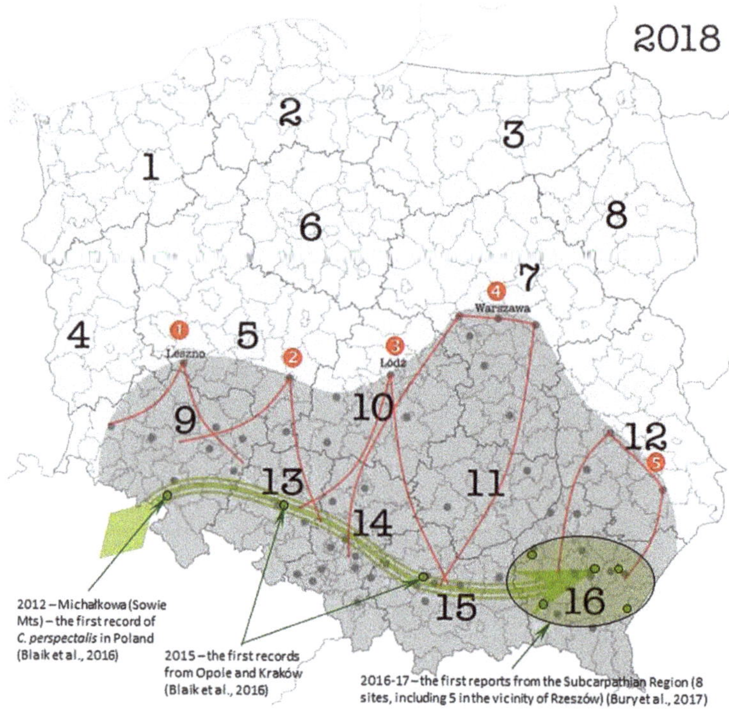

Figure 7. The range of *Cydalima perspectalis* occurrence in Poland in 2018 and its possible routes of spread. Each grey point indicates the presence of a pest found in social monitoring in 2018; the green arrow marks the route for insects to spread until 2017 [25,26]; red lines and numbers indicate the main directions of the species expansion to the north in 2018; 1–16: numbers assigned to individual voivodships; see Table 1; ❶–❺: extreme, key locations for the expansion of *C. perspectalis* in Poland, recorded in 2018—see the text below.

The northernmost points of the *C. perspectalis* range recorded in 2018 correspond to the main national and international south-north communication routes. These routes indicate their essential role in the spread of this species to the north of Poland. The most important of them are: the S5 clearway Wrocław-Poznań (Leszno: ❶); S11-Opole-Poznań (Kalisz: ❷); motorways: latitudinally running A4 from Opole via Katowice to Kraków along with the A1 motorway running north from Katowice to Łódź (❸) and further to Warsaw (❹) and with the S7 clearway from Kraków to Warsaw (❹); S19 from Rzeszów to Lublin with a branch to Zamość (❺) (Figure 7).

In 2019, the presence of *C. perspectalis* was found in 77 localities located in 13 voivodeships. These were the same voivodeships as in the previous years. Additionally, the pest appeared in the country's west and the north in the following voivodeships: West Pomeranian, Pomeranian, and Lubusz (No.: 1, 2, and 4). The pest's appearance in Szczecin and Gdańsk, on the coast of the Baltic Sea, was a big surprise. This year, the most information about the species' appearance came from the south of Poland in the following voivodeships: Silesian, Lesser Poland, and Subcarpathian (No.: 14, 15, and 16) (Figure 8). The confirmed presence of *C. perspectalis* in Polish port cities on the Baltic Sea (Szczecin and Gdańsk) may result both from the movement along communication routes along the Odra and Warta river valleys (Szczecin) and/or from sea transport from Western Europe (or even from Asia)—the presence of moths in Gdańsk and Szczecin.

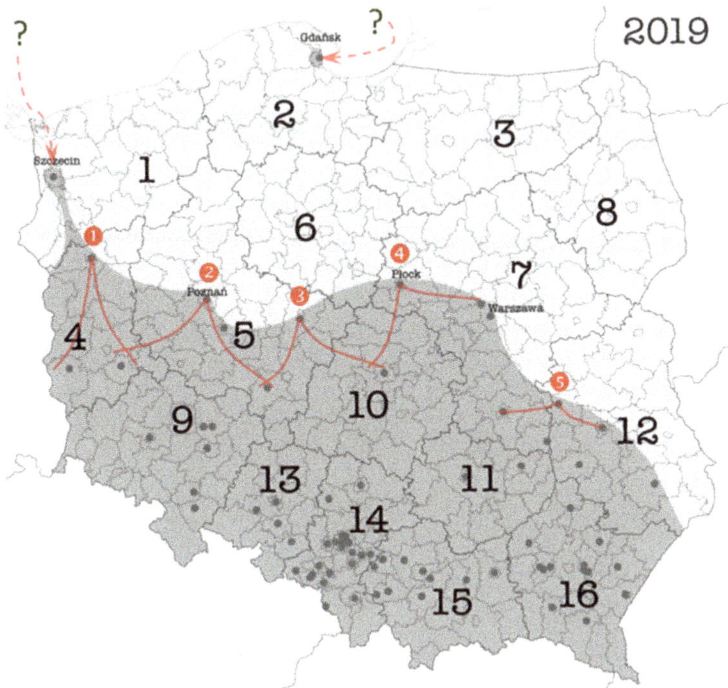

Figure 8. *Cydalima perspectalis* occurrence range in Poland in 2019 and its other possible routes of spread. Each grey point indicates the presence of a pest found in the course of social monitoring in 2019; red lines and numbers indicate the main directions of the species expansion to the north in 2019; 1–16: numbers assigned to individual voivodships; see Table 1; ❶–❺: extreme, key locations for the expansion of *C. perspectalis* in Poland, recorded in 2019—see the text below.

Once again, major communication routes reflect the further spread of *C. perspectalis* to the north. These include: S3 Wrocław/Legnica-Szczecin clearway (report from Gorzów Wielkopolski ❶); clearway S5 to Poznań (❷); latitudinally running A2 Poznań-Warsaw motorway (report from Konin: ❸). On the other hand, road routes of local importance, such as Warsaw-Bydgoszcz or Lublin-Puławy allowed insects to reach Płock (❹) and Puławy (❺)—places only about 100 km away from Warsaw or Łódź or 50 km from Lublin (Figure 8).

In 2020, there was a further large-scale expansion of *C. perspectalis* in Poland. Information about the pest's outbreak came from as many as 148 localities located in all 16 voivodeships that are part of Poland's administrative division. The last three voivodeships are Warmian-Masurian, Kuyavian-Pomeranian, and Podlaskie (No.: 3, 6, and 8). In addition to Gdańsk and Szczecin, the pest was recorded in such towns in the north of the country as Elbląg, Olsztyn, Ełk, Suwałki, and Bielsk Podlaski. The year 2020, however, confirmed earlier observations that currently *C. perspectalis* is most abundant in southern Poland, where it is warmer than in the north and where the vegetation period is slightly more extended (Figure 9).

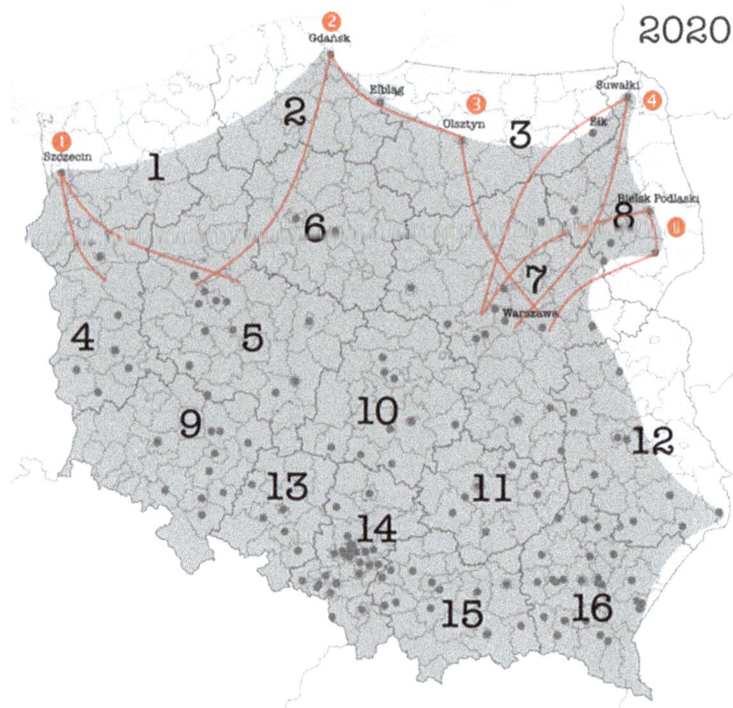

Figure 9. *Cydalima perspectalis* occurrence range in Poland in 2020, and its final possible routes of spread. Each grey point indicates the presence of a pest found in the course of social monitoring in 2020; red lines and numbers indicate the main directions of the species expansion to the north in 2020; 1–16: numbers assigned to individual voivodships; see Table 1; ❶–❺: extreme, key locations for the expansion of *C. perspectalis* in Poland, recorded in 2020—see the text below.

Data from 2020 confirms the role of the main national and international communication routes. The presence of *C. perspectalis* in Szczecin (❶) reflects the route of clearway S3 (from Gorzów Wielkopolski) and S11 (from Poznań). The presence of insects in Gdańsk (❷) indicates a significant role of the A1 motorway along the main south-north communication axis and the S7 clearway from Warsaw to Gdańsk via Olsztyn and Elbląg (❸). The main communication axes from Poland to the northeast to the border with Lithuania (clearway S62) made it easier for insects to reach Suwałki and Ełk (❹) and to the border with Belarus (S8) to Bielsk Podlaski (❺) (Figure 9).

The number of *C. perspectalis* records reported in 2020 is extremely interesting if the number of inhabitants is considered. Comparing only two voivodeships: Silesian and Subcarpathian (No.: 14 and 16)), where the number of reported records is similar (Figure 7), after such recalculation, the number of records from Silesia is comparable to other voivodeships in southern and western Poland. Despite its small area, the Silesian voivodeship (after the Masovian voivodeship with the capital of Warsaw) has the second largest number of inhabitants. A large number of records from the Subcarpathian voivodeship (No.: 16), with a relatively small number of inhabitants, probably results from the fact that the authors of the monitoring and website come from Rzeszów, where its excellent promotion took place (Figure 10).

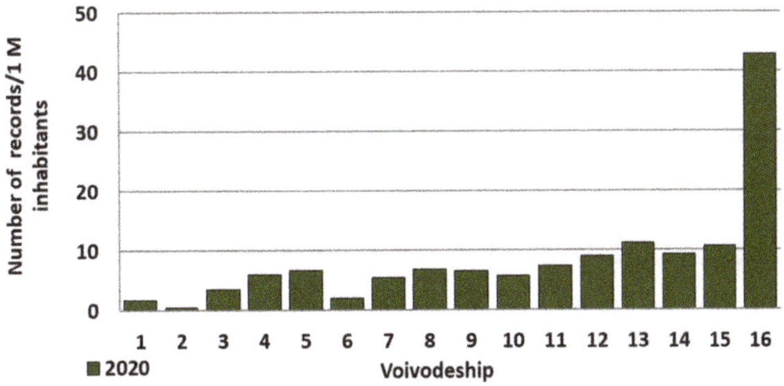

Figure 10. The number of confirmed records of *Cydalima perspectalis* occurrence in individual voivodeships in 2020 per 1 M inhabitants. Axis X: 1–16—numbers assigned to individual voivodships; see Table 1.

3.2. Ostrinia nubilalis, Diabrotica virgifera virgifera, Cameraria ohridella (Selected Result Issues for Comparative Purposes)

Ostrinia nubilalis. Although the species was recorded for the first time in Poland on millet (*Panicum miliaceum*) in the 1930s, and then in the 1950s on *Zea mays*, it became a severe pest of maize cultivation in Poland beginning of the 21st century. Published data show that from the 1950s to the 1990s, this pest was recorded in a limited number only in south-eastern Poland (No.: 15 and 16) and was closely related to the cultivation of maize in areas with optimal conditions for its cultivation. However, in the last two decades, the total area of maize cultivation in Poland has been systematically growing. In 2008, its cultivation area was over 730 thousand hectares, but in 2012 this area increased very quickly to over 1 million hectares [38,39]. Such a rapid increase in the area of maize sown was caused by the freezing of winter cereals and winter rape at the turn of 2011/2012, which prompted farmers to cultivate maize [40]. Since 2012, maize has been sown every year in Poland on more than one million hectares [41]. Data from the Central Statistical Office (CSO) show that maize is sown in all 16 voivodeships. There is, however, regional differentiation of maize sowing in terms of the use of the crop. In the south, grain is dominated, and in the north, silage is dominated [42]. The great interest in maize cultivation resulted in the densification of crops and the reduction of the distance between the cultivation of this plant in all provinces. The above-presented trend in the change of maize cultivation in Poland was favorable for the spread of pests of this crop.

Monitoring carried out in 2004–2008 stated that in the first decade of the 21st century, *O. nubilalis* began to quickly infect maize cultivation towards the northwest (No.: 1, 4, and 5)—Figure 11. Up to the end of 2008, damaged plants caused by caterpillars were recorded in 185 counties located in 14 voivodeships [31]. From 2009, it could be found in almost the entire country, except for central Poland and the eastern part of Pomerania (No.: 2 and 7). Notably, the expansion of this pest to the north of Poland reflects the increase in maize acreage in the first decade of the 21st century in central and northern Poland. The monitoring carried out in 2010–2012 showed the permanent presence of this species in almost all of Poland [32]—Figure 11. Apart from the increase in the area and the spread of maize cultivation, it is difficult to indicate for *O. nubilalis* a dispersion factor other than a slow, local expansion into subsequent adjacent areas of cultivation with often limited crop protection against pests.

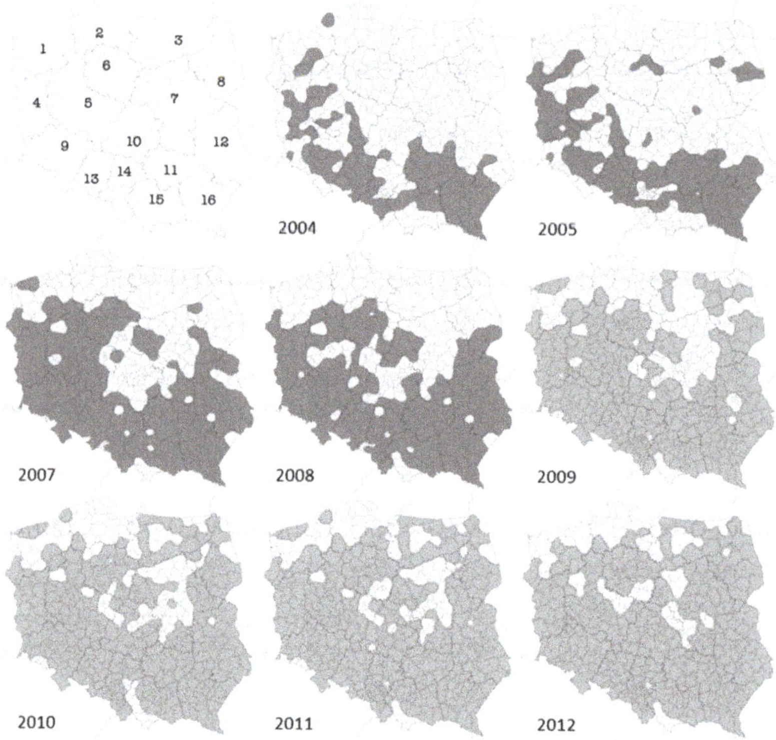

Figure 11. Stages of expansion of *Ostrinia nubilalis* in Poland (grayed out areas) identified as part of the national monitoring of this species carried out by the State Plant Health and Seed Inspection Service (SPHSIS) and the Institute of Plant Protection-National Research Institute, Regional Experimental Station in Rzeszów (IPP-NRI) in 2004–2012 [31,32]-modified; 1–16: numbers assigned to individual voivodships; see Table 1.

Diabrotica virgifera virgifera in Poland was first recorded in August 2005 in the south-eastern part of the country in Subcarpathian voivodeship (No.: 16) in three outbreaks, near Dukla, Łąka, and Jasionka, where six adults were collected [33]—Figure 12. The first recording of this species in Poland is closely related to the transit of freight from Hungary via Slovakia to Poland, the site at the Polish-Slovak border crossing in Dukla, and maize cultivation in the areas adjacent to the airport in Rzeszów (No.: 16). By 2007, *D. virgifera* had infested entire south-eastern Poland and the border areas with the Czech Republic, including almost the entire Opole Voivodeship (No.: 13, 15, and 16, and partially 9, 11, 12, and 14) [34]. Unpublished data from the SPHSIS archive, in which data from the national monitoring of this species in 2008–2013 are stored, clearly indicate a relatively slow but successive infestation of subsequent maize crops in neighboring administrative units (Figure 12).

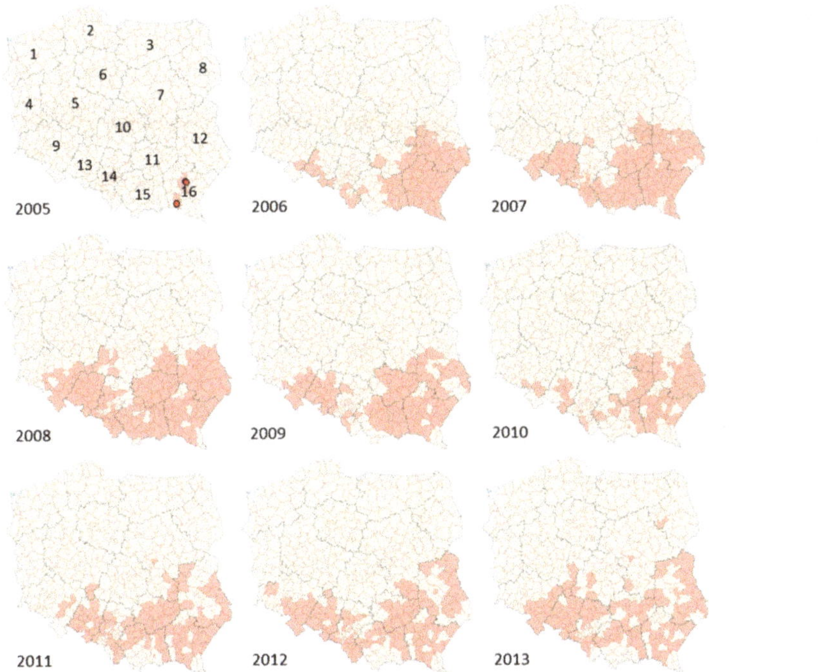

Figure 12. Stages of expansion of *Diabrotica virgifera virgifera* in Poland identified as part of the national monitoring of this species by the State Plant Health and Seed Inspection Service (SPHSIS) in 2005–2013 [34]-modified and SPHSIS archives. The red points indicate the first described localities of *D. virgifera* in Poland in 2005 [33]; 1–16: numbers assigned to individual voivodships; see Table 1.

Cameraria ohridella. Horse chestnut-*Aesculus hippocastanum* is a tree initially found only in park and roadside plantings as an ornamental plant. Currently, it is also found in the wild almost all over Poland (Figure 13a) [13]. The horse chestnut pest, Horse-chestnut leaf miner-*C. ohridella*, was first recorded in Poland in 1998 in the Botanical Garden in Wojsławice, 50 km south of Wrocław (No.: 9) near the S8 clearway from the Czech border to Wrocław and on to Warsaw [35]. A year later, a mass appearance of Horse-chestnut leaf miner was found, and trees were heavily damaged in the country's southern regions, in the vicinity of Cieszyn, Pszczyna, and Racibórz (No.: 14). These are towns located in the border zone with the Czech Republic. Therefore, another unintentional introduction from the Czech Republic using the natural south-north migration route, the Moravian Gate and the communication routes running through it from the Czech Republic (A1 motorway and S52 clearway) cannot be ruled out. In 2001, insects of this species were noted in the provinces of southern Poland (No.: 9, 15, and 16) and, also, in central Poland (No.: 5, 7, 10, and 11), reaching Masuria in the north (No. 3). In 2002, the first damage to the chestnut tree was recorded on the German border in Słubice (No.: 4), on the Baltic coast (Kołobrzeg and Sopot—No.: 1 and 2) and in eastern and north-eastern Poland (Lublin and Białystok—No.: 8 and 12). These data indicate that from the site near Wrocław, where it was found for the first time (and perhaps also from the sites described in 1999 in Silesia—No.: 14), the expansion of insects was two-way-eastwards towards Ukraine and through Central Poland to the north-east towards Belarus, and the west towards the Polish-German border (Słubice) and Pomerania (Figure 13b). The explanation for such a rapid appearance of the Horse-chestnut leaf miner in Warsaw and Masuria is possible, only taking into account the use of road transport and the main south-north communication axes, such as the A1 motorway (Cieszyn-Katowice-Warsaw and further north) and the A4 motorway running through Krakow to Rzeszów and the Polish-Ukrainian border. Fifteen years later, the same communication routes were used in their expansion by *Cydalima perspectalis* (Figures 7–9). In 2002 *C. ohridella* was already recorded nationwide in many scattered locations, confirmed by the anthropogenic dispersion pattern of this species in Poland, which is estimated on average at least 200 km per year. Currently, this species inhabits all of Poland, and its range coincides with the distribution of the common horse chestnut in Poland (Figure 13a).

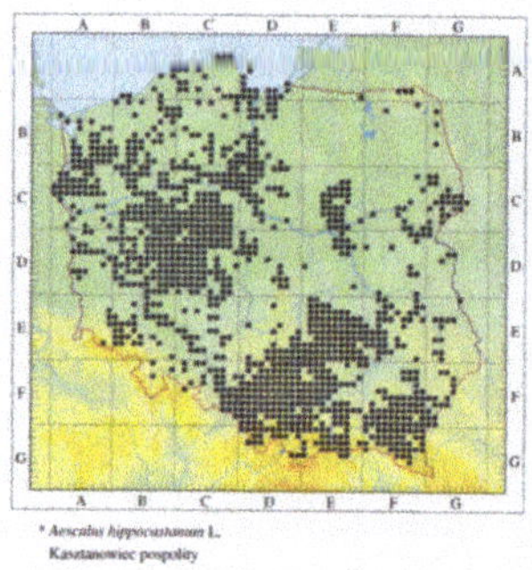

(a)

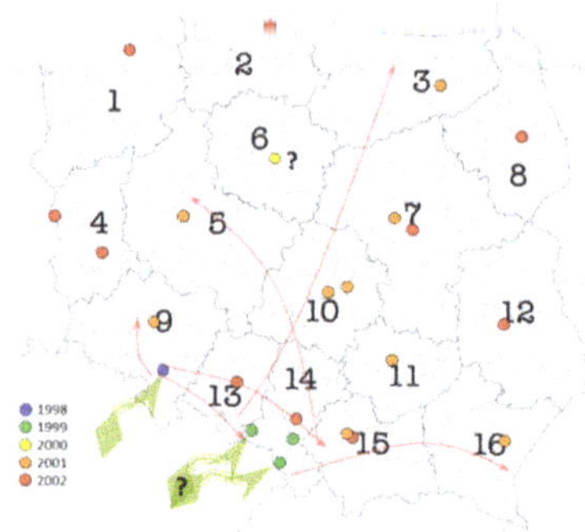

(b)

Figure 13. (**a**) A map of the distribution of the horse chestnut (*Aesculus hippocastanum*) in Poland [43]; *: each black point represents confirmed sites of *A. hippocastanum* in Poland; (**b**) the first recorded sites of the occurrence of *Cameraria ohridella* in Poland in the years 1998–2002, the arrows indicate the areas of the main directions of insect spread [35,36]-modified. 1–16: numbers assigned to individual voivodships; see Table 1.

4. Discussion

Such a sudden increase of knowledge in the *Cydalima perspectalis* range, which covered entire Poland for a few years, might be related to the fact that various media started talking about the appearance of the box tree moth due to the information campaign. For this reason, gardeners were more aware of the presence of specimens of a new pest species and were more likely to observe the boxwood plants, including detecting them more often.

Assuming that it usually takes two years from the first appearance of insects to boxwood until they are entirely defoliated, we can assume that the data sent by users of the gardening website are just such a consequence of a two- or three-year delay in detecting the pest [19,44]. Information from observers indicates that the natural expansion of this insect in Poland was a secondary factor, an example of which is the 8-year long settlement route in southern Poland in the latitudinal direction along the Carpathian arc. It should be assumed that the main factor was accidental, untargeted transfers of insects with infected plants through the use of road transport and resale of infected plants in subsequent parts of the country. An example is that in 2019 the presence of *C. perspectalis* was noted in Gdańsk, at a distance of over 300 km from the previous year, the closest place of the outbreak in Płock (Figures 8 and 9). It cannot be ruled out that *C. perspectalis* reached the Baltic coast independently by sea transport. It can also be stated that most of the new insect occurrence sites are associated with large cities. The rapid expansion in the last three years was also favored by warmer, above the long-term norms, average daily and monthly temperatures. The hot and long autumn of 2019 was the reason for the third generation of insects' mass appearance.

In Poland, the box tree moth easily survives during the winter period. In spring, as the temperature rises, the over wintering caterpillars start foraging, so the first adults appear at the beginning of April [45]. In Central Europe, the insect develops 2–3 generations a year, depending on weather conditions [17]. In 2018 in south-eastern Poland, due to the dry and hot summer, the 4th generation of this pest was likely to be developed because active moths were found in November [45]. Adults *C. perspectalis* can fly up to 10 km per year, then a long-range invasion of the pest (especially in Europe) is favored by bulk freight or boxwood cargo transport. Local dispersion is facilitated by, e.g., hedges, horticulture, nursery gardens, or internet sales [46].

In the last 30 years, over 30 species of insects, alien to the Polish entomofauna, has been found in Poland [28]. However, some of them were particularly spectacular and of real economic importance. These include the invasions of *Ostrinia nubilalis* (European corn borer), *Diabrotica virgifera virgifera* (Western corn rootworm), and *Cameraria ohridella* (Horse-chestnut leaf miner). There are many differences and similarities to that found for *C. perspectalis* in their pace and spread directions.

For example, in the monitoring studies carried out in the United States on *D. virgifera*, it was found that adult insects, under favorable conditions, move an average of several dozen kilometers per year (under the most favorable conditions, a natural dispersion of insects over a distance of over 100 km was recorded). The effect of such a pattern of spreading the Western corn rootworm was the appearance of insects only in the areas adjacent to the previously infected [47,48]. In Poland, for *D. virgifera*, spreading over much smaller distances was observed, up to several dozen kilometers per year. However, spreading was favored by lowland terrain and the fragmentation of field crops of maize, which, with the doubling of maize crops in 2005–2013, caused a progressive invasion of *D. virgifera* to the north and north-east. The only exception was the presence of *D. virgifera* in 2013 in Podlaskie voivodeship (No.: 8), which was undoubtedly brought to this area using transport or plant material. Studies on the spread of *D. virgifera* in Europe in countries such as Italy, France, Germany, Austria, and Switzerland have confirmed that the rapid spread of the species is related to the total acreage of maize cultivation in a given area. Areas that cover more than 50% of the acreage are classified as "high-risk areas" for this species' invasions [49]. These data fully corresponded to the *D. virgifera* dispersion pattern in Poland, which confirm the rapid invasion of the species in traditional maize cultivation areas (south-eastern Poland) and the increase in the range of occurrence with the increase in the acreage after 2008. A similar pattern of insects spread from pest-infested areas to nearby pest-free areas was observed for *O. nubilalis*. An additional factor contributing to the spread of this species was the abrupt increase in the acreage of maize cultivation in 2012 [41].

A completely different spreading strategy was observed for the third mentioned above invasive pest. The high rate of spread of *C. ohridella* and the rapid increase in its occurrence range is mainly attributed to road transport. In many cases, the favorable factors were tree stands located near the main communication routes, from which people, animals, or wind further transmitted insects to other trees. It is estimated that the distribution range of Horse-chestnut leaf miners increased in Europe at a rate of about 60 km to 114 km per year [50].

With large-scale spatial data of the occurrence of a given invasive species, several spatial models of the spread of insects to new, uninfected areas can be distinguished. These include, among other things, a diffusion model, a leptokurtic dispersal model and a stratified dispersal model [50,51]. Comparing the pace and directions of the insect's habitat mentioned above species in new areas of Poland, it should be stated that two basic ways of infesting new areas can be distinguished. The first is related to pests of arable crops, good examples of which are *O. nubilalis* and *D. virgifera*. These insects, closely related to crops, in this case, maize, take over successive areas gradually, usually over short distances (about 50 km in the direction of neighboring crops). An important factor contributing to the dispersion is the high density of the host plant crops and the short distance between them. On the other hand, the slowing down of the dispersion rate is associated with plant protection treatments and the applicable phytosanitary regulations. A diffusion model can best describe the spread of *O. nubilalis* and *D. virgifera* [50,51].

On the other hand, species of no economic importance, such as, for example, *C. ohridella*, take over new areas in a jump-like manner, sometimes over long distances, often hundreds of kilometers, using land transport means (including water and air transport). Five years was enough for this species to be recorded throughout Poland, both on trees growing in areas not cared for by humans and those of a decorative and recreational nature. In the analyzes of the models of the spread of this species in Germany and France, it was found that this process is best described by a leptokurtic dispersal and a stratified dispersal model. It was also found that a stratified dispersal model incorporating the effect of human population density provides the best description of the spread of *C. ohridella* in many countries of Europe [50,51].

The use of social networks and dedicated websites dedicated to societies and interest groups' activities is not the first time this type of approach has been used to research the distribution of insects. *C. perspectalis* meets most of the insect criteria suitable for this type of social monitoring. It is a species that feeds close to humans, causes specific and massive boxwood dieback symptoms, visible to everyone, even to people who are not interested in entomology. It is only necessary to consider the time insects need from the first colonization of plants to their death-about two years [19]. The lack of natural enemies enables a more precise determination of the year of insects' appearance in a given area. Previously, this type of approach was used in the British Isles where, in addition to the official operating The British and Irish network of County Moth Recorders (CMRs), which was the primary source of fully reliable records of the species, the website of the European Boxwood and Topiary Society (EBTS) provides a facility for users to report occurrences of this species and we have accessed all such data for 2018 (www.ebts.org/bmctracker) (accessed on 22 February 2021) [11]. A similar approach that reflects Citizen Science's idea has been successfully used in recent years in Toronto (Canada), where the first appearance of *C. perspectalis* was recorded in August 2018 [52].

Observations made by Blaik et al. [25] and Bury et al. [26] were used by EPPO to map the distribution of *C. perspectalis* in Europe [53]. In turn, the map of *C. perspectalis* distribution in Europe conducted by CABI lacks detailed information on the occurrence of the species in Poland, including its first appearance [54].

Genetic studies based on mtDNA for cytochrome oxidase genes revealed two haplotypes—HTA and HTB—in Europe out of twelve identified in Chinese populations of *C. perspectalis*. These results support the hypothesis of multiple introductions of this pest from eastern China to Europe. Lack of precise trade regulations for ornamental plants and trade globalization facilitated the rapid spread of the pest [4,7]. The lack of genetic research data on the population in Poland does not allow for an unambiguous statement whether the spread of *C. perspectalis* is the result of a single introduction to the southwest of Poland from the Czech Republic or it is multiple introductions (from the Czech Republic, Slovakia, and Germany, or through the Baltic ports from Denmark or Sweden). Research of this type could provide an answer about the origin of this species in the cities of northern voivodeships (Szczecin, Gdańsk—Figure 8). It cannot be ruled out that these cities may have been infested with insects from Denmark or Sweden, where this species was first recorded in 2013 and 2016 [54].

Due to the lack of nationwide monitoring of the box tree moth's occurrence in Poland, data on the presence of the pest came only from random observations. Without the involvement of state services dealing with the monitoring of alien origin species, it was impossible to create an accurate map of the range of this species in Poland. Our observations in 2018–2020 clearly show the growing range of *C. perspectalis* in Poland. The obtained data indicate that the main directions of the species spread in Poland were the main communication axes of the country. A good example to justify such a thesis is the sudden appearance of several confirmed positions in the Subcarpathian Voivodeship (No.: 16) in 2016–2017. Data on the spread of this insect in Hungary and Slovakia confirm its presence in large Hungarian cities as early as 2011, and the following year it will appear in Slovakia. In 2013 and 2014, the insect was recorded in Prešov (a large communication junction in eastern Slovakia) and Košice. These localities lie on the main communication axis from Slovakia to Poland (Košice-Prešov-Rzeszów) [37,55]. On the other hand, the detected presence of *C. perspectalis* in Lithuania in 2018 in Vilnius is unrelated to the spread of insects along the axis of transit transport from southern Europe to the Baltic countries running through Poland [56]. *C. perspectalis* was recorded in the north-eastern voivodships bordering Lithuania only in 2020 in towns lying on the extension of the Warsaw-Ełk/Suwałki-Vilnius clearway. This assumption strengthens the thesis that the box three moth reaches new areas of Europe by sea transport, which may explain the earlier appearance of this species in large port towns (Szczecin, Gdańsk, Poland) in Poland as early as 2019 (Figure 8).

For *C. perspectalis*, an invasive pest of boxwood, a plant that grows in Poland only due to artificial plantings, a similar spreading pattern can be observed as in *C. ohridella*. It has been proven that the spread of this species is related to the use of land transport and the density of the human population—a stratified dispersal model. This model best explains the unintentional long-distance movement of the pest within one growing season, which may have resulted from the long-distance trade of infected plants. At the same time, an essential factor influencing the rate of spread of a species is the density of the human population, favoring local spread over short distances, locally matching the diffusion model [50,51]. Such a dispersion model is well reflected in the published maps for Switzerland, Hungary, and Slovakia, where the first recorded sites of this species begin with large cities and then along the main communication routes between them [7,11,37,55,57]. The time it took for *C. perspectalis* to infect all of Poland was only twice as long as that stated for *C. ohridella*. The years 2018–2020, in which citizen monitoring of the occurrence of the box tree moth was carried out, were crucial to the invasion of the whole of Poland.

Proposed by Nacombo et al. [5], a bioclimatic (CLIMEX®) model for *C. perspectalis* distribution in Europe, based on climate, ecological, and developmental parameters described for this species (e.g., diapause termination, thermal requirements and phenology) well reflects the directions of expansion of this pest in Poland. The places of occurrence of the box tree moth confirmed in 2020, as well as the number of raised records, correctly reflect the predictions obtained using the bioclimatic model (compare with Figure 9), pointing to the regions of south-eastern (No: 12, 14, and 16), central (No.: 5 and 10), and western (No.: 4) Poland. Similar analyses performed for the Slovak population confirm the model's usefulness [5,55]. Perhaps the model used should have been enriched, as was conducted for *C. ohridella*, with the human population density parameter, as it was conducted in the model to predict this invasive species for horse chestnut in a stratified dispersal model [5,50,51,55].

However, it should be noted that most of the obtained data were provided by gardeners and plant breeders, who often did not know about the appearance in Poland in the area where a new, alien species of pest lived. Some gardeners, boxwood growers, and institutions dealing with urban greenery and parks lost their boxwood bushes, topiaries, and hedges, which caterpillars utterly destroyed. Such a rapid and spectacular invasion of *C. perspectalis* in Poland makes it necessary to research to understand the biology of this species under Polish conditions. In this, it is essential to develop comprehensive methods of its control using biological and chemical methods, which will take into account Poland's climatic and weather conditions and the methods of growing boxwood [40,58,59].

5. Conclusions

The collected data indicate that the box tree moth (*Cydalima perspectalis*) took over its entire area from its first discovery in 2012 in Poland in 2020. Insects have spread throughout the country, often over long distances, using major communication routes. Currently, it is the greatest threat to boxwood in southern and central Poland. The lack of nationwide monitoring of *C. perspectalis* makes it challenging to control its spread and combat it, especially in regions where it appears the first time. The developed coverage maps, together with the data on recording the presence of *C. perspectalis*, allow gardeners and plant breeders to analyze the situation on an ongoing basis and undertake adequate control and eradication methods.

Supplementary Materials: The following supporting information can be downloaded at: https://www.mdpi.com/article/10.3390/d14010022/s1: Table S1: The number of places where the appearance of *Cydalima perspectalis* was recorded and the number of reports of insects in individual voivodeships, in 2018–2020; Table S2: Detailed list of places (towns) in which the occurrence of *Cydalima perspectalis* was recorded and confirmed in 2018–2020.

Author Contributions: P.K.B. and M.N. conceived and designed the experiments; Ł.K., M.G. and P.K.B. performed the experiments; M.A., M.N. and P.K.B. analyzed the data; M.N., P.K.B. and P.Z. wrote the paper. All authors have read and agreed to the published version of the manuscript.

Funding: This research received no external funding.

Institutional Review Board Statement: Not applicable.

Informed Consent Statement: Not applicable.

Data Availability Statement: All data used in the publication are available at the WWW addresses of the links quoted. If there are no citation, the data used are not in any public database. Raw data on the presence of *C. perspectalis*, *O. nubilalis* and *D. virgifera* in Poland are available only in internal reports of the Institute of Plant Protection.

Acknowledgments: The authors would like to thank all the gardeners cooperating with the gardening site "Allotment and Garden Our Passion (DIONP)" for the data provided on the range of *C. perspectalis* occurrence in Poland in 2018–2020. We want to express special thanks to the owner of this gardening website, Beata Bereś. We also thank Tomasz Konefał from the Main Inspectorate of Plant Health and Seed Inspection in Toruń for his help in the graphic development of maps of *C. perspectalis* occurrence range in Poland.

Conflicts of Interest: The authors declare no conflict of interest.

References

1. EPPO Global Database: Reporting Service No. 11—2007. Num. Article: 2007/215 Incursion of *Diaphania perspectalis* in Germany and Addition to the EPPO Alert List. Available online: https://gd.eppo.int/reporting/article-1295 (accessed on 3 December 2020).
2. Mally, R.; Nuss, M. Phylogeny and nomenclature of the box tree moth, *Cydalima perspectalis* (Walker, 1859) comb. n., which was recently introduced into Europe (Lepidoptera: Pyraloidea: Crambidae: Spilomelinae). *Eur. J. Entomol.* **2010**, *107*, 393–400. [CrossRef]
3. Wan, H.; Haye, T.; Kenis, M.; Nacambo, S.; Xu, H.; Zhang, F.; Li, H. Biology and natural enemies of *Cydalima perspectalis* in Asia: Is there biological control potential in Europe? *J. Appl. Entomol.* **2014**, *138*, 715–722. [CrossRef]
4. Bras, A.; Avtzi, D.N.; Kenis, M.; Li, H.; Vétek, G.; Bernard, A.; Courtin, C.; Rousselet, J.; Roques, A.; Auger-Rozenberg, M.-A. A complex invasion story underlies the fast spread of the invasive box tree moth (*Cydalima perspectalis*) across Europe. *J. Pest Sci.* **2019**, *92*, 1187–1202. [CrossRef]
5. Nacambo, S.; Leuthardt, F.L.G.; Wan, H.; Li, H.; Haye, T.; Baur, B.; Weiss, R.M.; Kenis, M. Development characteristics of the box-tree moth *Cydalima perspectalis* and its potential distribution in Europe. *J. Appl. Entomol.* **2014**, *138*, 14–26. [CrossRef]
6. Strachinis, I.; Kazilas, C.; Karamaouna, F.; Papanikolaou, N.E.; Partsinevelos, G.K.; Milonas, P.G. First record of *Cydalima perspectalis* (Walker, 1859) (Lepidoptera: Crambidae) in Greece. *Hell. Plant Prot. J.* **2015**, *8*, 66–72. [CrossRef]
7. Matošević, D.; Lukić, I.; Bras, A.; Lacković, N.; Pernek, M. Spatial distribution, genetic diversity and food choice of box tree moth (*Cydalima perspectalis*) in Croatia. *South-East Eur. Forest.* **2017**, *8*, 41–46. [CrossRef]
8. Nagy, A.; Szarukán, I.; Csabai, J.; Molnár, A.; Molnár, B.P.; Kárpáti, Z.; Szanyi, S.; Tóth, M. Distribution of the box tree moth (*Cydalima perspectalis* Walker 1859) in the north-eastern part of the Carpathian Basin with a new Ukrainian record and Hungarian data. *EPPO Bull.* **2017**, *47*, 279–282. [CrossRef]
9. Gómez-Undiano, I.; Martínez-Ovejero, P.; Villegas, S.; Prieto, N.; Herrero, A.; Moreno, A.V. First record of *Cydalima perspectalis* (Walker, 1859) for Madrid, Spain (Lepidoptera: Crambidae, Spilomelinae). *SHILAP-Rev. Lepidopt.* **2018**, *46*, 585–591.
10. EPPO Global Database: Reporting Service No. 06—2019. Num. Article: 2019/118, Update on the Situation of *Cydalima perspectalis* in the EPPO Region. Available online: https://gd.eppo.int/reporting/article-65482019 (accessed on 3 December 2020).
11. Plant, C.W.; Poole, C.; Salisbury, A.; Bird, S. The box-tree moth *Cydalima perpectalis* (Walker, 1859) in Britain: An overview of its spread and current status. *Entomol. Rec. J. Var.* **2019**, *131*, 122–147.
12. CABI–Invasive Species Compendium, *Cydalima perspectalis* (Box Tree Moth). Available online: https://www.cabi.org/ISC/datasheet/118433 (accessed on 3 December 2020).
13. Vieira, V. First record of *Cydalima perspectalis* (Walker, 1859) from São Miguel Island, Azores (Portugal) (Lepidoptera: Crambidae). *SHILAP-Rev. Lepidopt.* **2020**, *48*, 141–146.

14. Hizal, E. Two invasive alien insect species, *Leptoglossus occidentals* (Heteroptera: Coreidae) and *Cydalima perspectalis* (Lepidoptera: Crambidae), and their distribution and host plants in Istanbul province, Turkey. *Fla. Entomol.* **2012**, *95*, 344–349. [CrossRef]
15. Hizal, E.; Kose, M.; Yesil, C.; Kaynar, D. The new pest *Cydalima perspectalis* (Walker, 1859) (Lepidoptera: Crambidae) in Turkey. *J. Anim. Vet. Adv.* **2012**, *11*, 400–403. [CrossRef]
16. Poltavsky, A.N.; Ilyina, E.V. New finds of alien Lepidoptera species in Dagestan. *Russ. J. Biol. Invasions* **2017**, *8*, 347–350. [CrossRef]
17. Leuthardt, F.L.G.; Baur, B. Oviposition preference and larval development of the invasive moth *Cydalima perspectalis* on five European box-tree varieties. *J. Appl. Entomol.* **2013**, *137*, 437–444. [CrossRef]
18. López, C.; Eizaguirre, M. Diapause and biological cycle of *Cydalima perspectalis* (Walker) in the eastern Pyrenees. *J. Appl. Entomol.* **2019**, *143*, 1096–1104. [CrossRef]
19. Korycinska, A.; Eyre, D. *Box Tree Caterpillar Cydalima perspectalis*; External Factsheets; The Food and Environment Research Agency (FERA): York, UK, 2009; p. 4, (Updated 2011). Available online: https://webarchive.nationalarchives.gov.uk/20100713154951/http://fera.defra.gov.uk/plants/plantHealth/pestsDiseases/documents/boxTreeCaterpillar.pdf (accessed on 22 February 2021).
20. Kulfan, J.; Dzurenko, M.; Parák, M.; Sarvašová, L.; Saniga, M.; Brown, P.; Zach, P. Larval feeding of *Cydalima perspectalis* on box trees with a focus on the spatial and temporal distribution. *Plant. Prot. Sci.* **2020**, *56*, 197–205. [CrossRef]
21. Suppo, C.; Bras, A.; Robinet, C.A. Temperature- and photoperiod-driven model reveals complex temporal population dynamics of the invasive box tree moth in Europe. *Ecol. Modelling* **2020**, *432*, 109229. [CrossRef]
22. Poitou, L.; Bras, A.; Pineau, P.; Lorme, P.; Roques, A.; Rousselet, J.; Auger-Rozenberg, M.-A.; Laparie, M. Diapause regulation in newly invaded environments: Termination timing allows matching novel climatic constraints in the box tree moth, *Cydalima perspectalis* (Lep.: Crambidae). *Insects* **2020**, *11*, 629. [CrossRef]
23. Canelles, T.Q.; Bassols, E.; Vayreda, J.; Brotons, L. Predicting the potential distribution and forest impact of the invasive species *Cydalima perspectalis* in Europe. *Ecol. Evol.* **2021**, *11*, 5713–5727. [CrossRef]
24. Kenis, M.; Nacambo, S.; Leuthardt, L.G.F.; Di Domenico, F.; Haye, T. The box tree moth, *Cydalima perspectalis*, in Europe: Horticultural pest or environmental disaster? *Aliens: Invasive Species Bull.* **2013**, *33*, 38–41.
25. Blaik, T.; Hebda, G.; Masłowski, J. *Cydalima perspectalis* (Walker, 1859)—Inwazyjny gatunek motyla w faunie Polski (Lepidoptera: Crambidae). (*Cydalima perspectalis* (Walker, 1859)—An invasive butterfly species in the fauna of Poland (Lepidoptera: Crambidae). *Przyr. Sudetów* **2016**, *19*, 121–124.
26. Bury, J.; Olbrycht, T.; Mazur, K.; Babula, P.; Czudec, P. First records of the invasive box tree moth *Cydalima perspectalis* (Walker, 1859) (Lepidoptera: Crambidae) in south-eastern Poland. *Fragm. Faun.* **2017**, *60*, 101–106. [CrossRef]
27. D'Ornano, M.; Goddyn, S.; Jalkh, J.-F.; Ferrand, E. Motion for a European Parliament Resolution on the Box Tree Moth (*Cydalima perspectalis*). European Parliament B8-1209/2016. Available online: https://www.europarl.europa.eu/doceo/document/B-8-2016-1209_EN.pdf (accessed on 22 February 2021).
28. Soika, G.; Łabanowski, G. *Organizmy Inwazyjne Wykrywane w Polskich Szkółkach Instrukcja Rozpoznawania Roztoczy i Owadów Inwazyjnych na Podstawie Wyglądu i Cech Diagnostycznych (Invasive Organisms Detected in Polish Nurseries. Instructions for Recognizing Mites and Invasive Insects on the Basis of Appearance and Diagnostic Features)*; Instytut Ogrodnictwa (The National Institute of Horticultural Research): Skierniewice, Poland, 2014.
29. BioMap. Biodiversity Map Taxa: *Cydalima perspectalis* (F. Walker, 1859). Available online: https://baza.biomap.pl/en/taxon/species-cydalima_perspectalis/default/tr/y/cf/y (accessed on 22 February 2021).
30. DIONP—Działka i Ogród Naszą Pasją (DIONP—Allotment and Garden Our Passion). Available online: https://www.dionp.pl (accessed on 22 February 2021).
31. Bereś, P.K.; Konefał, T. Distribution range of the european corn borer (*Ostrinia nubilalis hbn.)* on maize in 2004–2008 in Poland. *J. Plant Prot. Res.* **2010**, *50*, 326–334. [CrossRef]
32. Bereś, P.K.; Konefał, T. Występowanie *Ostrinia nubilalis* (Hbn.) na kukurydzy (*Zea mays* L.) w Polsce w latach 2009–2012. (Occurrence of *Ostrinia nubilalis* (Hbn.) on maize (*Zea mays* L.) in Poland in 2009–2012). *Prog. Plant Prot.* **2015**, 3, 296–304.
33. Sahajdak, A.; Bereś, P.K.; Konefał, T. *Diabrotica virgifera* Le Conte—A new threat to maize crops in Poland and measures taken against the pest. *J. Plant Prot. Res.* **2006**, *46*, 157–161.
34. Konefał, T.; Bereś, P.K. *Diabrotica virgifera* Le Conte in Poland in 2005–2007 and regulations in the control of the pest in 2008. *J. Plant Prot. Res.* **2008**, *49*, 129–134. [CrossRef]
35. Łabanowski, G.; Soika, G. *Cameraria ohridella* damages horse chestnut trees in Poland. *Ochr. Roślin* **1998**, *42*, 12–16.
36. BioMap. Biodiversity Map Taxa: *Cameraria ohridella* Deshka et Dimić. 1986. Available online: https://baza.biomap.pl/pl/taxon/species-cameraria_ohridella/default/tlpl/y (accessed on 9 November 2021).
37. Vétek, G.; Zach, P.; Matošević, D.; Tuba, K.; Lakatos, F.; Kulfan, J.; Csóka, G.; Gomboc, S.; Nagy, S.; Glavendekić, M.; et al. Invasion by the box tree moth, *Cydalima perspectalis* (Lepidoptera: Crambidae), in southeastern Europe. In *Invasive Dendrophilous Organisms: Challenges and Protection Operations*; All-Russian Research Institute of Silviculture and Mechanization of Forestry East Palearctic Regional Section International Organization for the Biological Control of Pests and Animals: Pushkino, Russia, 2019; pp. 17–28.
38. *Użytkowanie Gruntów, Powierzchnia Zasiewów i Pogłowie Zwierząt Gospodarskich w 2008 r. (Land Use, Sown Area and Livestock Population in 2008)*; Central Statistical Office (CSO): Warsaw, Poland, 2008. Available online: www.stat.gov.pl (accessed on 4 November 2021).

39. *Crop Production in 2012*; Central Statistical Office (CSO): Warsaw, Poland, 2012. Available online: www.stat.gov.pl (accessed on 4 November 2021).
40. Syp, A. Projekcja zmian wielkości plonów kukurydzy w Polsce i w Unii Europejskiej w 2030 roku (Projected changes in grain maize yields in Poland and European Union in 2030). *Stow. Ekon. Rol. I Agrobiz. (Pol. Assoc. Agric. Agribus. Econ.)-Rocz. Nauk.* **2015**, *17*, 373–378.
41. *Crop Production in 2017*; Central Statistical Office (CSO): Warsaw, Poland, 2018. Available online: www.stat.gov.pl (accessed on 4 November 2021).
42. Księżak, J. Regionalne zrónicowanie uprawy kukurydzy w Polsce w latach 2000–2006 (Regional diversification of maize cultivation in Poland in 2000–2006). *Acta Sci. Pol. Agric.* **2008**, *7*, 47–60.
43. Zając, A.; Zając, M. (Eds.) *Atlas Rozmieszczenia Roślin Naczyniowych w Polsce. Dodatek (Distribution Atlas of Vascular Plants in Poland: Appendix)*; Instytut Botaniki Uniwersytetu Jagiellońskiego (Institute of Botany of the Jagiellonian University): Kraków, Poland, 2019; pp. 25, 182–183.
44. Raineri, V.; Bonechi, F.; Caracciolo, D.; Cresta, P.; Mariotti, M. *Cydalima perspectalis* (Walker, 1859) (Lepidoptera, Crambidae) and the threats for the Nature 2000 habitat 5110 in Liguria (NW-Italy). *Boll. Mus. Ist. Biol. Univ. Genova.* **2017**, *79*, 215–236.
45. Bereś, P. *Ćma Bukszpanowa. Azjatycki Najeźdzca Niszczy Bukszpany w Polsce. (Box Tree Moth. An Asian Invader Destroys Box Trees in Poland)*; Wiedza i Praktyka Sp. z o.o.: Warszawa, Poland, 2019; pp. 1–11.
46. Leuthardt, F.L.G.; Glauser, G.; Baur, B. Composition of alkaloids in different box tree varieties and their uptake by the box tree moth *Cydalima perspectalis*. *Chemoecology* **2013**, *23*, 203–212. [CrossRef]
47. Spencer, J.L.; Hibbard, B.E.; Moeser, J.; Onstad, D.W. Recent research on the western corn rootworm. Behaviour and ecology of the western corn rootworm (*Diabrotica virgifera virgifera* LeConte). *Agric. For. Entomol.* **2009**, *11*, 9–27. [CrossRef]
48. Meinke, L.J.; Sappington, T.W.; Onstad, D.W.; Guillemaud, T.; Miller, N.J.; Komáromi, J.; Levay, N.; Furlan, L.; Kiss, J.; Toth, F. Recent research on the western corn rootworm. Western corn rootworm (*Diabrotica virgifera virgifera* LeConte) population dynamics. *Agric. For. Entomol.* **2009**, *11*, 29–46. [CrossRef]
49. Baufeld, P.; Enzian, S. Maize growing, maize high-risk areas and potential yield losses due to Western corn rootworm (*Diabrotica virgifera virgifera*) damage in selected European countries. In *Western Corn Rotworm–Ecology and Management*; Vidal, S., Kuhlmann, U., Edwards, C.R., Eds.; CABI Publishing: Walingford, CT, USA; Cambridge, UK, 2005; pp. 285–302.
50. Gilbert, M.; Grégoire, J.C.; Freise, J.F.; Heitland, W. Long-distance dispersal and human population density allow the prediction of invasive patterns in the horse chestnut leafminer *Cameraria ohridella*. *J. Anim. Ecol.* **2004**, *73*, 459–468. [CrossRef]
51. Gilbert, M.; Guichard, S.; Freise, J.; Grégoire, J.C.; Heitland, W.; Straw, N.; Tilbury, C.; Augustin, S. Forecasting *Cameraria ohridella* invasion dynamics in recently invaded countries: From validation to prediction. *J. Appl. Ecol.* **2005**, *42*, 805–813. [CrossRef]
52. Guarrasi, M.A. Potential Impacts and Control of the Non Native Box Tree Moth in Canada. Frans Eggermont. 2018. Available online: https://www.cif-ifc.org/wp-content/uploads/2019/10/2019ConfPres_S05P2_Potential-Impacts-and-Management-of-the-Non-Native-Box-Tree-Moth-in-Canada-_Mariaelana-A-Guarrasi.pdf (accessed on 3 December 2020).
53. EPPO Global Database: *Cydalima perspectalis* (DPHNPE). Available online: https://gd.eppo.int/taxon/DPHNPE (accessed on 22 February 2021).
54. Bengtsson, B.Å. Anmärkningsvärda fynd av småfjärilar (Microlepidoptera) i Sverige 2016. (Remarkable records of Microlepidoptera in Sweden during 2016). *Entomol. Tidskr.* **2017**, *138*, 1–24.
55. Kulfan, J.; Zach, P.; Holec, J.; Brown, P.M.J.; Sarvašová, L.; Skuhrovec, J.; Martinková, Z.; Honěk, A.; Vál'ka, J.; Holecová, M.; et al. The invasive Box tree moth five years after introduction in Slovakia: Damage risk to Box trees in urban habitats. *Forests* **2020**, *11*, 999. [CrossRef]
56. Paulavičiūtė, B.; Mikalauskas, D. First record of the box tree moth *Cydalima perspectalis* (Walker, 1859) (Lepidoptera: Crambidae) in Lithuania. *Liet. Entomol. Draugijos Darb.* **2018**, *2*, 55–59.
57. Leuthardt, F.L.G.; Billen, W.; Baur, B. Ausbreitung des Buchsbaumzünslers *Diaphania perspectalis* (Lepidoptera, Pyralidae) in der Region Basel—EIne für die Schweiz neue Schädlingsart (Spread of the Box-tree pyralid *Diaphania perspectalis* (Lepidoptera: Pyralidae) in the region of Basel—A pest species new for Switzerland. *Entomo Helv.* **2010**, *3*, 51–57.
58. Bereś, P.K.; Siekaniec, Ł.; Kontowski, Ł.; Kucharska-Świerszcz, M. Przydatność *Bacillus thuringiensis*, *Beauveria bassiana* oraz spinosadu w biologicznej ochronie bukszpanu pospolitego przed ćmą bukszpanową w południowo-wschodniej Polsce (Usefulness of *Bacillus thuringiensis*, *Beauveria bassiana* and spinosad in the biological control of *Cydalimia perspectalis* on boxwood in southeastern Poland). In Proceedings of the 60th Scientific Session, Institute of Plant Protection—National Research Institute, Poznań, Poland, 11–13 February 2020; pp. 178–179.
59. Bereś, P.K.; Zawada, D.; Siekaniec, Ł. Efekty chemicznego zwalczania gąsienic ćmy bukszpanowej na bukszpanie pospolitym z wykorzystaniem acetamiprydu oraz mieszaniny acetamiprydu z lambda-cyhalotryną (Effects of chemical control of *Cydalima perspectalis* on *Buxus sempervirens* using acetamiprid and mixture of acetamiprid with lambda-cyhalothrin). In Proceedings of the 60th Scientific Session, Institute of Plant Protection—National Research Institute, Poznań, Poland, 11–13 February 2020; p. 169.

MDPI
St. Alban-Anlage 66
4052 Basel
Switzerland
Tel. +41 61 683 77 34
Fax +41 61 302 89 18
www.mdpi.com

Diversity Editorial Office
E-mail: diversity@mdpi.com
www.mdpi.com/journal/diversity

www.ingramcontent.com/pod-product-compliance
Lightning Source LLC
Chambersburg PA
CBHW061837210926
51850CB00006B/914
9783036573625